Engineering with Polymers

Engineering with Polymers

Peter C. Powell

and

A. Jan Ingen Housz

University of Twente
The Netherlands

Second edition

Stanley Thornes (Publishers) Ltd

First edition 1983

Second edition published 1998 by:
Stanley Thornes (Publishers) Ltd
Ellenborough House
Wellington Street
CHELTENHAM
GL50 1YW
United Kingdom

98 99 00 01 02 / 10 9 8 7 6 5 4 3 2 1

A catalogue record for this book is available from the British Library.

ISBN 0-7487-3987-4

The views in this book are expressed in good faith to the best of our knowledge and in accordance with the current state of the art, but necessarily directed to the task of making the fundamentals of the discipline of engineering with polymers understood and presenting them as a basis for an introductory course. This implies that aspects that could be of importance for specific applications of this discipline are not treated. Specific data are only provided in order to support the process of learning and understanding and need not necessarily be exact. Moreover, any use of the knowledge and information expressed falls outside the control of the authors. They cannot accept any legal responsibility or liability for any use that is made of the contents of this book.

The cover picture shows an impression of a one-piece universal joint coupling between two axles based on twelve integral hinges as functional elements, compare problem 4 in section 10.2 (courtesy Philips C.C.P.).

Typeset by Pure Tech India Limited, Pondicherry
Printed and bound in Great Britain by St Edmundsbury Press, Bury St Edmunds, Suffolk

Contents

Preface to second edition

In the 15 years since the appearance of the first edition of *Engineering with Polymers* there have been many changes relevant to the objectives stated in that preface. These changes include:

1. A substantial increase in the amount of teaching of polymers within the university engineering curriculum up to about graduate level, particularly in mechanical engineering courses, with reference to the teaching of materials behaviour, mechanics of polymers and polymer–fibre composites, design of polymer-based products, and (perhaps to a lesser extent) teaching of polymer melt fluid mechanics and heat transfer and the design of polymer melt processing machinery.
2. An increasing recognition of the contribution which solid-body mechanics can make to the design of cost-effective products within specialist courses in polymer technology up to about graduate level.
3. The number of books on polymers has perhaps doubled over the past 15 years, running now to some 3000 or so, many of them being extremely specialist in narrow areas and properly so. The number of titles aiming to provide an overview relevant to the general needs of engineering students remains much smaller, just a few.

It follows then that there has been emerging an increasing number of students with an improved elementary understanding of polymer properties and processing and the relevance of this knowledge to the design of products to be made from polymers. Many of these graduates have already begun to make a substantial contribution to the more successful use of polymers as their careers develop. This will doubtless continue. The meaningful questions these graduates raise, and the value of their mechanical engineering background skills, contribute directly to the expansion of the use of polymers within the polymer and polymer-using industries. It is hoped that the first edition of this book, and similar titles, have made some modest contribution towards these positive developments.

With such expansion of activity and publications, the question naturally arises: why a second edition of *Engineering with Polymers*?

We recognize that the academic scene continues to change and develop. Within the mechanical engineering curriculum we recognize that polymers were being introduced in a usable way much earlier than 15 years ago. It is encouraging that there is now scope for the incorporation within the engineering syllabus of more material and in different ways compared with 15 years ago. In particular the increasing emphasis on design activities and learning-by-doing provide new challenges. The then rather new feature in the first edition of introducing additional learning material through the solving of problems has become much more widely accepted.

Our new edition naturally builds on the old edition. Large parts of the first edition remain, with some amendments to correct mistakes, to bring up-to-date, and to include a few more problems which experience has suggested will enhance the building of understanding. In retaining the overview of polymer engineering, we recognize that much of the basics was rightly stated.

The bringing up-to-date is particularly noticeable in Chapter 9, where the influence of processing on properties and product behaviour has been substantially revised in the light of recent research. Chapter 7 on fibre–polymer composites has been considerably extended, in recognition that the mechanics appeals strongly to mechanical engineers as a natural extension to the well-known conventional solid-body mechanics for isotropic materials, and the concepts pervade much of the behaviour of manufactured polymer products, whether composites or just polymers with a measure of molecular orientation.

We see the first nine chapters as providing a minimum literacy in polymer background for a mechanical engineer. Chapter 10 is a completely new addition to the book. In this chapter we introduce various tools for the designer. Some of these are substantially practical. Some of them introduce more advanced material relevant to specific issues, on the basis that if readers can follow them, they can develop a similar level of reasoning in other topics: to cover all possible developments in this depth is impossible in one modest volume. We have had to make a selection. Some of the chapter provides material which suggests how it may be used in the design context, starting with incomplete design requirements and using the information in the rest of the book to suggest what sorts of questions need to be asked and what answers are needed before going on to any design calculations and reaching conclusions of practical interest.

We are well aware that there has been an explosion in the use of computers, both in academia and in industry; this development makes a substantial contribution to product design in polymers as well as in most other technologies. We have, however, resisted a temptation to cast this book in the form of numerical modelling of engineering with polymers. We are reluctant to put the emphasis on the use of numerical methods before the basic physical principles are understood. What we have tried to do is to present a coherent account of the physics and engineering of polymers: we hope that the theoretical analysis is clear and provides a framework which can be used to solve simple problems within the scope of the book. We are convinced that computers can then be used to make more detailed investigations of the phenomena described in each chapter of this book and to solve engineering problems with polymers. We are also convinced by our colleagues who do so, that others can successfully translate the principles into numerical techniques far better than we can.

The first edition was written by Peter Powell as sole author, although he enjoyed support from a large number of colleagues and constructive critics during that time and afterwards. He has since gained further experience in teaching the subject material of the book within the UK (especially at Imperial College, London, at the Manchester Metropolitan University and as Director of the UK Integrated Graduate Development Scheme in Polymer Engineering at Manchester) and also outside the UK in continental Europe, North Africa, and North and South America.

During the past 10 years Peter Powell has enjoyed increasing interaction with the University of Twente, where Jan Ingen Housz had developed courses on the same subjects from his position on the mainland of Europe; at first by giving invited lectures on short courses at Twente. In 1991 Peter Powell moved to Twente to develop full-time a new Chair in Engineering Design in Plastics within the Mechanical Engineering department at Twente, and from September 1991 they collaborated not least in the preparation of this second edition. This collaboration, together with exposure to and direct involvement with the Dutch approach to design teaching, project-centred learning and polymer engineering education, has indeed been most stimulating and fruitful.

We both acknowledge our indebtedness to many colleagues (in particular Remko Akkerman, Durk van Dijk, Kaspar Jansen and Peter Reed) who have suggested analysis and examples and to many students who have worked through early versions and who have always courteously suggested revisions and corrections which have made for greater clarity. From them we have learned much.

Peter C. Powell
A. Jan Ingen Housz

Enschede, August 1997

Preface to first edition

Products made from plastics and rubber materials are based on polymers, and contribute strongly to the national economy not least in terms of performance, reliability, cost-effectiveness and high added value. Among the many reasons why polymers are widely used, two stand out. First, polymers operating in a variety of environments have useful ranges of deformability and durability which can be exploited by careful design. Secondly, polymers can often readily, rapidly and at an acceptable (low) cost be transformed into usable products having complicated shapes and reproducible dimensions. Moreover the volume of polymers used in the Western economy already exceeds that of metals. It therefore follows that engineers come into contact with polymers in terms of designing, making or using products.

Unfortunately most engineers who graduated in the early 1980s or before still know little about polymers and are not therefore well equipped (from their education alone) to exploit them within the broad range of available materials. Universities, polytechnics and colleges are aware of the need for some polymer content within undergraduate engineering courses. Most of this teaching should be broadly based and devolve on existing portfolios of standard engineering skills rather than provide an advanced course for polymer specialists. A modest number already meet this need, the remainder seek to do so.

There appears to be no consensus of what might constitute a minimum useful knowledge of engineering with polymers within a mechanical engineering degree course. And, although there are some 200 titles on polymers, no one book seems to cover the field at the right level, although one or two provide useful guidance. Neither has there as yet emerged a preferred way of teaching this topic – not that this is at all surprising.

In the long term it would be very satisfactory if the main concepts of engineering with polymers pervaded undergraduate courses, taking their natural place alongside other established materials in illustrating the themes of engineering science, technology, design and practice. Achievement of this on a national basis will take considerable time and effort. However, in many institutions it seems more practicable to offer a self-contained lecture course in polymer engineering to begin with. The success of such an approach can then be judged not least by the rate of pervasion of polymers into the degree course, probably via design and project work. Such an introductory lecture course has the advantage of requiring only modest resources.

The material in this book has been developed over many years and is a slightly extended version of a well supported 20 lecture course offered to final-year undergraduate students in the Mechanical Engineering Department at Imperial College, London. The book outlines and illustrates the main principles of engineering with polymers whilst recognizing the superficiality of a broad but shallow and largely descriptive approach and the intellectual rigour of advanced theoretical analysis. Furthermore, where practicable the book introduces skills which permit a simplified quantitative analysis of the behaviour of polymers during processing and in use.

Within the constraints of time and space, the coverage must inevitably be selective rather than comprehensive. It has been assumed that the student or engineer picking up this book is familiar with the basic principles of engineering. On mechanics of isotropic solids, this includes Hooke's law in two to three dimensions, plane stress, plane strain and constraint, transformation of axes for stress and strain, yield criteria and the basic concept of mode I linear elastic fracture mechanics. The simple micromechanics of unidirectionally aligned strong fibres in a weak flexible matrix are revised but briefly. On the transport side, the reader will be familiar with the isothermal developed one-dimensional laminar flow of incompressible Newtonian fluids within pipes and between parallel plates, and with steady and non-steady heat conduction within isotropic continua. This book recognizes that many students of engineering currently have little or no polymer background and respond without enthusiasm to a chemistry-based materials approach.

The contents of this book embrace four main themes. The first four chapters introduce the language, terminology and technology of polymers. After an introduction to the relevant polymer physics there is an account of the general properties of polymers and the need for compounds or recipes on which commercial plastics and rubber materials are based. The first section of the book ends with a description in engineering terms of the main methods used to make articles from polymers.

The second section, Chapters 5 and 6, gives an account of the stiffness, strength and fracture of nominally isotropic polymers, and describes how these properties can depend markedly on the duration or rate of loading, on temperature and on the nature of the environment. The design procedures for avoiding load-bearing failure are standard; the values of properties may seem strange – moduli in the range 10^6 to $10^9\,N/m^2$, design stresses of $10\,MN/m^2$ and fracture toughnesses of the order of $1\,MN/m^{3/2}$. Acceptable strains up to 1% are commonplace in plastics, and can frequently be more than an order of magnitude higher in rubber. For reasons of space and with great regret, these two chapters do not discuss procedures for designing rubber components which operate under cyclic loading or at very large strain (where large strain non-linear elasticity is required).

The third section recognizes the growing importance of materials such as fibre-reinforced plastics and rubbers that have properties which can vary dramatically according to the direction in which they are measured. Chapter 7 outlines the intriguing properties of panels of these materials and introduces elementary design methods for avoiding failure. The analysis here inevitably looks more complicated than that used for isotropic polymers but, beneath the disguise of some algebraic manipulations, the principles are but a modest extension of well known theory. Beyond fibre-reinforced systems, this chapter provides a basis for the exploitation of molecular orientation of unreinforced polymers in plastics packaging, and helps to develop an understanding of the causes of some faults in plastic products arising during manufacture.

The final section, Chapters 8 and 9, describes the thermal and flow properties of polymer melts relevant to the main melt processing methods. This leads to the development of simple fluid mechanics and heat transfer, to predict relationships between pressure and output rate, and energy requirements, both important in sizing equipment to achieve cost-effective production. An account is also given of the common effects of processing on the properties of polymeric articles.

Students welcome opportunities to test their understanding of the principles of a subject, not least where these relate to examination practice or professional practice. Problems are an integral part of this book, not just tacked on to the end of each chapter. Some of the early problems in each group are simple algebraic ones included to give a 'feel' for the orders of magnitude of those quantities such as forces, pressures, deformations, deflections, flow rates and cooling times commonly encountered in designing with polymers. Others involve modest development of the material in the text to provide additional insight into problems which arise in polymer engineering and technology. There is a rather larger number of problems here than could reasonably be associated with most 20 lecture courses.

A further selection of problems could have been devised for solution using computing or numerical techniques, for example in designing with composites where the level of algebra is fairly trivial but where repeated manual calculations are extremely tedious and error-prone. This approach relies on the basic principles, however, and the line has been drawn there. Those who wish to explore the application of modern methods of calculation would find ample opportunities when engineering with polymers.

There are of course many ways of approaching the teaching of the topics mentioned in this book. The self-contained single lecture course represents one proven approach within which a reasonably self-consistent body of material can be covered; this presents challenges recognized by both the teacher and the student. There is of course the attendant risk in this tidy approach that the subject of engineering with polymers is seen to be compartmentalized and set aside from other topics on which it depends or to which it can contribute. Design studies and project work can show this rather too clearly – and do provide opportunities for integrating the main themes of engineering. Another proven approach has the objective of integrating polymers within the engineering curriculum, throughout the degree course. Provided the polymer aspects of the syllabus are actually covered, this has much to commend it. Some members of staff teaching topics such as mechanics of materials or fluid mechanics have a well developed background in such established materials as steel, concrete, water and air, and may feel, understandably, less confident about teaching aspects of engineering with polymers, even though they realize that the threshold of essential polymer knowledge in their field is often quite modest.

However, whatever the style of teaching, I hope that my colleagues in academia, who have an important role in polymer engineering, will find here a useful source of material which they can adapt or develop to meet their perceptions of local needs.

The scope of engineering with polymers is vast. This book is small, and it is apparent that its contents reflect a wish to integrate polymer topics within several standard engineering techniques, especially mechanics of solids and fluid mechanics. It would have been possible to provide other slants embracing such important aspects as machinery operation, manufacturing systems and control, but these have not been pursued. The emphasis has been on the principles which should be of lasting value. Current practice and technology change rapidly – practical details including costings can be readily grasped and developed for specific industries when the reader needs them, and for these reasons are excluded. But, even where topics have been included as central to the principles of engineering with polymers, limitations of space and teaching time have dictated worthy omissions. These include the behaviour of rubber and fibre–polymer composites under cyclic loading, designing rubber products which operate at large strains, joints and load transfer into polymer structures, thermal stresses and interactions between heat and momentum transfer.

I would like to thank Professor Arthur Birley of Loughborough University and Professor Graham Ellison of Bristol University for their helpful advice and comments on drafts of this book and for encouraging me to complete the project. Dr Roger Fenner kindly commented on Chapter 8. I am grateful to Professor Gordon Williams of Imperial College for stimulating and fruitful discussions on the teaching of engineering with polymers to mechanical engineering students in both undergraduate and postgraduate courses, and for the opportunity to develop these ideas into their present form. I would like to thank successive years of students at Imperial College and elsewhere who have responded generously with helpful feedback. I am indebted to Mrs Liz Hall who so efficiently transformed my manuscripts into a magnificent typescript; her courtesy, sense of humour and impeccable professionalism provided a source of inspiration and encouragement which I greatly value.

Peter C. Powell
Bristol 1983

Introduction 1

1.1 ENGINEERING WITH POLYMERS

Plastics and rubber materials are based on polymers. Unfortunately it is all too easy to associate them with flimsy articles which seem to become unserviceable at the slightest knock. After all, what use could an engineer possibly make of a material with a modulus in the range 10^9 down to $10^6\,N/m^2$ or less, and a breaking stress of $10^7\,N/m^2$ or so?

Yet, as representative examples in the following paragraphs show, properly designed and manufactured polymer-based articles play an invaluable role in traditional engineering areas such as fluid containment, springs and suspension units, power transmission, electrical and thermal insulation, and in load-bearing structures up to the size of ship hulls. In addition, a large proportion of the applications of plastics and rubber materials are in areas which are not obviously the province of traditional engineering. However, to penetrate the market, the design, development and manufacture of these products has called for a substantial application of engineering principles: many polymer products in service have to be designed with confidence to lower safety factors than are commonly encountered in general engineering.

1.1.1 Examples of engineered products

A basically toroidal pressure vessel contains air at about $+0.2\,MN/m^2$ for some years, subject to a wide range of temperatures, say -20 to $+80°C$, and has to withstand a superposed radial load of at least 2 kN applied at a frequency up to 25 Hz. The unit must be sufficiently deformable to seal well against a curved metal surface to contain the pressure, deformable enough to cope with repeated impact, yet stiff enough not to deform uncontrollably under tangential shear forces of 5 kN or more. It must have sufficient abrasion resistance to survive at least 30 000 km of rolling contact with rough and sharp surfaces, a high coefficient of friction to prevent tangential slip under abusive loading, and chemical resistance to water, oil, petrol, antifreeze, salt and battery acid. This is a description of part of the duty of a pneumatic car tyre, and signifies a well-engineered polymer product which is reliable, cost effective and well regarded.

Other examples where polymer products have to withstand pressures include pipes, tanks and pressure vessels. They may be made from plastics having moduli of about $1\,\mathrm{GN/m^2}$, a design stress seldom exceeding $10\,\mathrm{MN/m^2}$ or so, and a design fracture toughness substantially below $1\,\mathrm{MN/m^{3/2}}$. Plastics systems reinforced with long, stiff, strong fibres may have design stresses and fracture toughnesses up to an order of magnitude higher. Resistance to an exacting range of chemicals is often a key advantage of polymers, for example in parts of chemical plants. Flexible hoses made from fabric-reinforced rubber for coupling car engine and radiator have to withstand antifreeze circulating at temperatures up to about 100°C and, depending on the climate, down to −25°C or less. Liquid-tight joints in pressurized systems rely extensively on rubber sealing strips and rings.

Polymers are used to make springs. This is mainly the province of rubber, where its low modulus (about $1\,\mathrm{MN/m^2}$), and the ability to undergo large strains of 100% or so reversibly, provide scope for efficient and compact designs. The rubber tyre is part of a car's suspension system. Bridge bearing units can each carry 10 kN or more in vertical compression whilst readily permitting horizontal shear to accommodate thermal deformations in the span. Dock fender systems are designed to absorb very large amounts of energy when ships of large displacement come into slow-moving contact with them. Antivibration mountings are used to isolate the motion of moving machinery from its surroundings: examples include engine mounting blocks for cars and lorries (where the inherent damping capacity of rubber minimizes the transmission of vibration at resonance during start-up), and blocks to isolate buildings from traffic vibration near roads and (underground) railways. Railway suspension systems illustrate a common need of springs to operate with different stiffnesses in different directions, sometimes by an order of magnitude, for example vertical stiffness under the weight of the carriage or wagon, horizontal stiffness caused by acceleration or deceleration during travel along the track, and sideways stiffness caused by cornering forces. Springs are made in plastics too: for example, light-duty springs in mechanisms are moulded from thermoplastics having a modulus of the order of $1\,\mathrm{GN/m^2}$, and leaf springs for lorry chassis are made from fibre-reinforced plastics having a modulus of about $40\,\mathrm{GN/m^2}$.

Polymers have a role to play in power transmission, although not at the high-power end of the spectrum. Engineers have used durable plastics gear wheels for over 20 years for transmitting up to 40 kW. Road vehicle transmission shafts made from fibre-reinforced plastics have successfully undergone trials to explore their performance envelope under searching conditions, and now play an important role in motor car development. Helicopter rotor blades based on reinforced plastics are lighter in weight, more robust and attractive in cost because they last 10 times longer under fatigue than the metal ones they have replaced. Rubber plays a major role in power transmission too, not least in compact but mechanically simple flexible coupling units, in belt drives in machinery and in conveyor belting for mechanical handling plant. In a slightly different context, distributor head assemblies which form a vital part of the electric spark generation system in the internal combustion engine are a well-established use for thermoset plastics.

On the structural side there are some spectacular examples of engineering interest. Complete ship hulls some 50 m long are made in one piece from glass fibre-reinforced plastics having a modulus of the order of $10\,GN/m^2$: the hull is non-magnetic, an essential attribute in a minesweeper. The low density of polymers is often exploited; one example includes a lattice girder 'kit' which can be assembled to make a guyed radio mast 36 m high which carries an axial compressive load of 4 tonnes – it weighs only 140 kg (compared with 800 kg for the equivalent steel mast) and is made from a composite having a modulus of up to $40\,GN/m^2$.

Polymers are good insulators, among the best available to the engineer. A major application is in electrical insulation of wire and cable, ranging from run-of-the-mill domestic wire coverings and sheathings having volume resistivities of the order of $10^{16}\,\Omega\,m$ and operating at about ambient temperature, to dielectrics for long-distance communication cables where dielectric losses have to be kept down to a few microradians; plastics films a few micrometres thick are used as dielectrics for capacitors. The thermal conductivity of polymers is of the order of 0.1 W/m K, some four orders lower than that of copper, and even lower for cellular polymers. Thermal insulation is widely exploited in building panels, often having a sandwich structure with a foam core – with the added advantage of good flexural stiffness per unit weight. Pipes and tanks for containing hot or cold fluids do not lose as much heat to ambient when made from polymers compared with metal ones, because of the substantial temperature drop through the wall: less lagging, or none at all, is required.

Returning to the theme of structural performance, the low stiffness and strength of polymers have dictated a disciplined approach to the design of many articles which would not at first sight be regarded as 'load-bearing' by most engineers. Examples outside the scope of previous paragraphs include stacking containers such as bottle crates and 200 l drums, bottles for carbonated drinks, chair shells and other items of furniture. To achieve the stiffness or strength required in service, a careful analysis of loads and stresses has been made. In particular, the ease with which products of complicated shape can be made has led to a full use of geometrical stiffness in flexure based on second moment of area. Many polymer items bristle with ribs, e.g. bottle crates and extruded panels to replace corrugated board, and stepped features or curvature, e.g. in cladding panels for buildings. Sandwich structures involving foamed polymer cores are widely used in boxes, business machine housings, furniture and even some land drainage pipes. This theme is also well explored in packaging where chocolate box liners and egg boxes for the retail market achieve their stiffness with admirable economy within the constraints imposed by manufacture.

1.1.2 Engineering the making of polymeric products

Whether or not the product has an engineering appeal and whether or not the design relies on the principles of engineering, it still has to be made from the raw plastics or rubber material. This processing almost always involves a substantial amount of standard mechanical or production engineering technology.

At some stage in their history, most plastics or rubber products have been shaped in the fluid state and then caused to solidify, so the processing equipment consists of pipes or flow channels through which material is pumped or delivered in a controlled way to a shaping stage involving a mould or die.

The fluid may have the viscosity of water (about 1×10^{-3} N s/m^2), but often is much more viscous, with a consistency more like that of treacle (100 to 10^4 N s/m^2). Substantial quantities of energy are often needed to heat the polymer to make it readily shapable – and the heat then has to be removed: this involves carefully designed heaters or heat exchangers. The process has to be controlled to achieve products of closely toleranced dimensions, which is not made easier by the low thermal diffusivities of polymers, of the order of 10^{-7} m^2/s. The features governing the rate of production of articles of acceptable quality depend on the nature of the product and the method of manufacture.

Some products are fabricated from stock shapes – sheet plastics for large structures such as ductings, liquid storage tanks and advertising display panels; and rubber sheet for building up reinforced hose, and for the preliminary stages of constructing pneumatic tyres for vehicles. But the stock shapes still have to be made, usually by the routes mentioned above.

1.2 PREDICTING PERFORMANCE

A central theme in design is to ensure that a product will perform as required with the maximum of economy and efficiency. The designer is accustomed to checking the likely stiffness and strength of a proposed article under load or deformation, using the concepts of equilibrium and compatibility, and a relationship between stress and strain or a failure criterion. On the transport side energy supplies and output rates are checked using the equations of energy, momentum and continuity, together with a relationship between stress and strain rate, or a failure criterion.

In the first 2 years of an engineering degree or course of similar level, the main emphasis is on the behaviour of isotropic, homogeneous, linear elastic solids, and there is discussion of the isothermal laminar flow of Newtonian time-independent fluids. In both situations the general theory is exemplified by a number of simple examples of direct practical importance.

The question now is, 'Do these principles apply in the design of polymer products and of machinery for making them?' To explore this question, it is helpful to take some simple examples, and see how successful we can be in solving polymer design problems with the knowledge we already have.

1.2.1 Extruded plastics pipes

About 10% of all thermoplastics are extruded as a viscous melt through a die to make pipes. Typical applications range from hypodermic needles through drinking straws, rainwater downpipes and land drainage to pipe services for cold water, gas, sewage and industrial effluent.

Suppose a pipe of outside diameter 168 mm is to be made from unplasticized PVC such that it can safely withstand an internal pressure of $1.2\,MN/m^2$. How thick should the wall be? The tensile strength of the unplasticized PVC is $50\,MN/m^2$ according to BS 2782. Using a safety factor of 2.1 (which is accepted as reasonable in uPVC pipe design and will be discussed later in the book), together with a simple hoop stress analysis for a thin-walled pipe, leads to a minimum wall thickness of about 4 mm. But a check on BS 3505:1968, *Unplasticized PVC for cold water services*, reveals that the minimum acceptable wall thickness is about 9.2 mm, in spite of the same value of safety factor. Why was the original calculation wrong by more than a factor of 2?

The main point here is that the yield strength (and indeed almost all the mechanical properties) of plastics depends on temperature and on the timescale of loading. It happens that both BS 2782 and BS 3505 describe behaviour measured at 20°C; but whereas BS 2782 involves a tensile test at constant crosshead speed lasting about 1 min, BS 3505 requires the pipe to withstand a lower stress for 50 years. It is the difference of more than seven orders of magnitude in the timescales of loading which has caused the drop in tensile strength, and hence the increase in wall thickness, in this example. Had the pipe been made from a high-density polyethylene, the difference between the short-term and the long-term strengths would be about threefold – this is bigger than the difference for uPVC (Fig. 1.1), largely because there is a change in mode of failure in the polyethylene from ductile to brittle. This change is obviously important, and the reasons for it are discussed in some detail later.

1.2.2 Extrusion of pipe through a die

Suppose a pipe of outside diameter 168 mm and 10 mm thick is to be extruded from a polypropylene melt at 230°C. At this temperature the

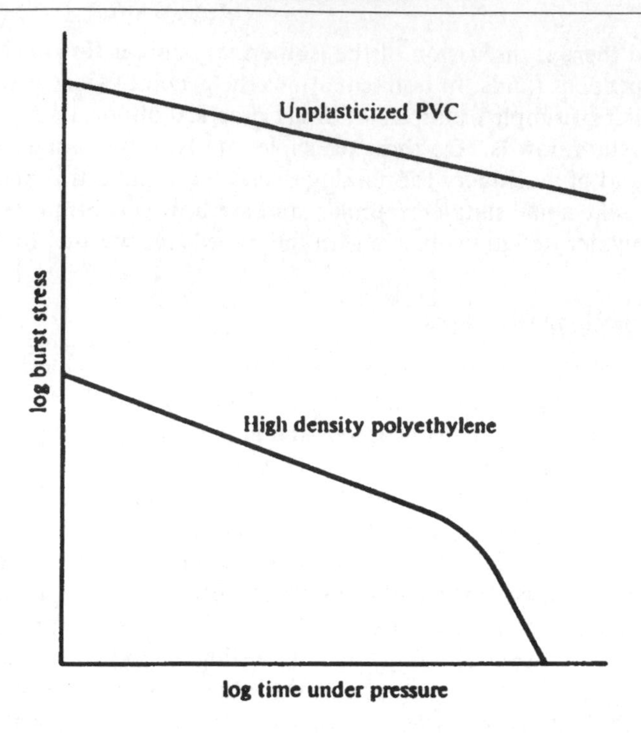

Fig. 1.1 Creep rupture data for pipe at 20°C made from two thermoplastics

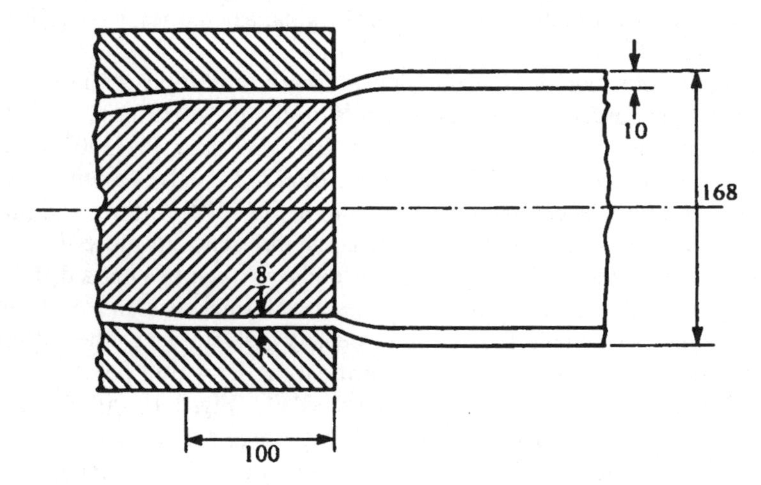

Fig. 1.2 Molten plastics tube swells on leaving the die

shear viscosity η of the melt is about $10^4\,\mathrm{N\,s/m^2}$ and the density ρ about $800\,\mathrm{kg/m^3}$.

Inspection of the die reveals that the die gap H_d in the land is only 8 mm (Fig. 1.2). The reason for this is that as the molten pipe emerges from the die, it swells in diameter and wall thickness (and shortens if the swelling occurs without change in volume). There is no comparable die swell in the solid-phase extrusion of metals.

If the average velocity in the die land is $\bar{v}_d = 10\,\text{mm/s}$, then the Reynolds number Re is given by

$$Re = \bar{v}_d H_d \rho / \eta \simeq 6.4 \times 10^{-6}$$

Thus even though the flow rate is about $0.12\,\text{m}^3/\text{h}$ or 100 kg/h, the high viscosity ensures that the flow is very definitely laminar. Assuming (not too unreasonably) that the flow may be modelled as incompressible, laminar, isothermal and Newtonian, then for pressure flow within closely spaced stationary plates, the shear rate at the wall $\dot{\gamma}_w$ is given by

$$\dot{\gamma}_w = 6Q/WH_d^2 \simeq 6\bar{v}_d/H_d = 7.5/\text{s}$$

where Q is the volume flow rate and W is the width of the flow channel. The concept of a Newtonian shear viscosity leads to the corresponding value of shear stress at the wall, τ_w:

$$\tau_w = \eta\dot{\gamma}_w = 7.5 \times 10^4\,\text{N/m}^2$$

If the length of the die land L_d is taken as 0.1 m, the pressure drop in the land ΔP_d is given by

$$\Delta P_d = 2L_d\tau_w/H_d \simeq 2\,\text{MN/m}^2$$

This order of pressure drop is typical and confirms for this extrusion system that the choice of flow model was a reasonable one.

1.2.3 Melt flow in an injection mould

Injection moulding is the plastics analogue of pressure die casting of metals. The polymer melt (at a temperature usually in the range 150–300°C) is injected from a reservoir through a runner into a cavity in a closed split mould. The cavity defines the shape and dimensions of the product, which could be a gear wheel, a ball-bearing cage, a housing for an electric drill, a shoe sole, or even a one-piece dinghy hull. The mould is cool or cold if a thermoplastics article is to be made.

Suppose the runner for a mould to make a polypropylene pipe fitting is 3 mm in radius (R) and 100 mm long (L) and that the same polypropylene is used for the pipe and the fitting. If the melt is injected at a typical constant volumetric flow rate Q of $10^{-4}\,\text{m}^3/\text{s}$, the shear rate at the wall of the runner is

$$\dot{\gamma}_w = 4Q/\pi R^3 \simeq 5000/\text{s}$$

The shear viscosity of the melt at 230°C has already been quoted as $10^4\,\text{N s/m}^2$. Even though the average velocity in the runner is about 4 m/s, the Reynolds number is still only about 2×10^{-3} and therefore comfortably laminar.

Modelling the flow as before, the shear stress at the wall is about 50 MN/m^2, giving a pressure drop along the runner of some 3 GN/m^2. This is unreasonably high, and certainly not needed in practice, both in terms of pumping duties and the implied thickness of the reservoir (which is a tube having an inside diameter in the range 40–60 mm). Now it would be possible to reduce the viscosity by a factor of about 3 by increasing the melt temperature to say 280°C, but this would still give estimates of pressure drop which are too high by much more than an order of magnitude. (It would also be possible to use an inherently more fluid version of the same polymer, but this idea will not be pursued here.)

The main error in the above calculation was the incorrect assumption that the shear viscosity of the melt at 230°C was constant over a range of shear rates. In fact the shear viscosity of polymer melts is a function of shear rate or shear stress (Fig. 1.3). At low shear rates the viscosity of this material is about 10^4 N s/m^2, and thus the extrusion estimate was reasonable. But at the high shear rate in the runner wall, 5000 /s, the shear viscosity drops to only 100 N s/m^2 and hence the pressure drop along the runner is now two orders less than that previously calculated; this now agrees with the values measured in instrumented runner systems in injection moulds. The overall pressure required to fill the mould cavity is greater than 30 MN/m^2 because this value does not include losses incurred in flow through the gate and within the cavity itself.

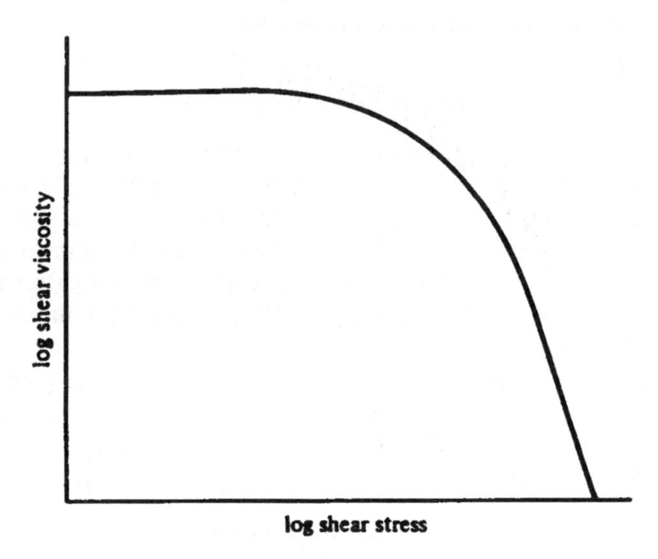

Fig. 1.3 Dependence of shear viscosity on shear stress for a typical polymer melt

1.2.4 Fibre–plastics composites

For some applications, plastics having moduli of about 1 GN/m^2 and design strengths of about 10 MN/m^2 are inadequate. One way of achieving higher stiffness and perhaps strength is to incorporate stiff strong fibres. Glass fibres are most widely used because of their cheapness and availability, with a modulus of about 70 GN/m^2 and a usable strength of the order of 1.5 GN/m^2. Carbon fibres have a modulus of at least 200 GN/m^2 but, being more expensive, they are used mainly where high stiffness per unit weight or low inertia of moving parts is at a premium.

The fibres work best when loads are applied along their axis. Loads are usually applied to the polymer forming the surface of the composite product. Loads are then transferred from the polymer to the fibres across the fibre–plastic interface where there must be a strong bond. The performance of the composite depends on the degree of alignment of the fibres. If the fibres are all aligned in one direction, it is quite easy to achieve a modulus of 50 GN/m^2 and a tensile strength of 600 MN/m^2, although the shear modulus is at best only a few giganewtons per square metre. However, this very same composite has properties which can vary by an order of magnitude or more depending on the direction in which the loads are applied in the plane relative to the fibres. The modulus and strength along the fibre direction have just been quoted; in the transverse direction the modulus is little better than that of the plastic – typically 3 GN/m^2 – and the value of Poisson's ratio can vary in a known manner from 0.005 to well over unity (legitimately!).

It comes as no surprise that calculations based on the familiar elasticity and stress analysis for isotropic materials can give totally wrong predictions of the performance of fibre–plastics composites.

Essentially there are two extreme approaches to designing with composites. The first approach, widely used in practice, is to try to reduce the strong directionality of properties by spreading the fibres more or less randomly in the plane of the sheet. This suggests that the composite is then sensibly isotropic in the plane and that isotropic analysis can be used in design. The in-plane tensile modulus of such a glass-reinforced plastic would be about 5–8 GN/m^2 with a tensile strength of about 100 MN/m^2. There are no fibres running through the thickness of the sheet, so the shear strength is only that of the plastics matrix, and a check must *always* be made that interlaminar shear will not occur. This seems straightforward but the simplicity of design calculations has only been achieved at the expense of poor utilization of the fibres. The volume fraction must be low, and most of the fibres are loaded at a considerable angle to their axis.

The second approach is used where exacting specifications of performance have to be met. This approach tries to make the very best use of the fibres, at the expense of more complicated analysis and the need for careful control of production techniques. Fibres are deliberately lined up to match the combinations of applied loads and the desired responses, often in the form of bonded stacks of unidirectional plies (with a high volume fraction of fibres) or in the less efficient but easier-to-handle form of woven cloth. Neither form overcomes the poor interply shear properties. The analysis, usually described as composite mechanics, or anisotropic elasticity and strength, explains features of the behaviour of laminates which seem 'odd' by the yardsticks used for isotropic materials. These 'odd' features include the following:

- the strong directionality of properties of most laminates;
- the shear which accompanies direct stress applied in the plane at angles to the fibre direction in a ply or laminate;
- the possibility that a non-symmetrical angle-ply laminate will not only shear but even twist or bend under an in-plane tensile stress;
- the stiffest and strongest laminate does not necessarily result from aligning the fibres in the direction(s) of maximum stress;
- non-symmetrical laminates may warp when heated or cooled.

1.2.5 Rubber springs

For many years rubber has been a familiar engineering material. Applications exploit one or more of the following main characteristics: low modulus, the ability to sustain and completely recover from high strain, abrasion resistance, toughness and efficient vibration absorption.

There are a few special rubber products which do operate at strains of several hundred per cent, e.g. toy balloons and rubber bands for domestic use, catapult elastic for arresting aircraft landing on ship decks, and foam rubber for upholstery. Designers must take fully into account the non-linear shape of the stress–strain curve. However, many engineering applications of rubber involve what rubber engineers call 'low' strains: typically up to 20% in compression and 35% or so in shear. Under these circumstances engineers assume (not too unreasonably) that the stress–strain curve for rubber is effectively linear and that the stiffness of the rubber can be described by a modulus. Standard elasticity theory then works quite well (within 15%), even though the strains are 'large' by traditional standards, where the modes of deformation are tensile or shear.

Suppose a coupling consists of a rubber disc 50 mm in diameter and 25 mm thick, to the circular faces of which are bonded rigid steel plates. What is the torsional stiffness of the unit if the rubber has a shear modulus G of 0.5 MN/m^2?

$$T/\theta = \pi G R^4/2H$$

where R is the radius and H is the rubber thickness. This leads to a stiffness of about 200 N m/rad, which confirms well what is measured in practice up to a twist of about 0.7 rad.

A similar type of problem would be to calculate the compressive stiffness of a rectangular block 10 mm thick and 80 mm square made from a rubber having a tensile modulus E of 2 MN/m^2. The unit has steel plates bonded to the square rubber surfaces to which the load is applied. What is the stiffness?

On the basis of load per unit of cross-sectional area, the stiffness K_c would seem to be

$$K_c = EA/H \simeq 1.3\,MN/m$$

but measurements indicate a stiffness of nearly 9 MN/m. Admittedly the calculation assumes that the modulus in tension is the same as the modulus in compression, which becomes less reasonable at high strains, but this is not a major source of error. The real problem is that the simple analysis ignores the restraint imposed by the bonded steel plates. These plates prevent the lateral expansion of the rubber at the loaded faces, and this stiffens the block by a factor of about 7. A thin block would obviously be restrained more than a thick block.

The shear and compression units are made by sandwiching the required quantity of rubber compound between pretreated steel plates and compressing the rubber within a heated mould to the required shape. The mould is hot, so that the rubber can be vulcanized or crosslinked (typically at about 130–150°C) to make it elastic and form-stable. This process takes some time, especially in a thick block, because the thermal diffusivity of the rubber is only about $10^{-7}\,m^2/s$. The time taken to heat up the rubber from, say, 20 up to 130°C when in perfect contact with a mould running at 160°C may be calculated from non-steady-state heat conduction theory. For only one-dimensional heat flow, the dimensionless temperature of 0.15 corresponds to a Fourier number of about 0.8 at the midplane. For the compression unit the half-thickness is 5 mm, giving a heating time of some 200 s. The circular coupling has a half-thickness of 12.5 mm and the midplane would take over 20 min to reach 130°C.

Rubber is often reinforced with fibres. Examples include steel and textile fibres in tyres, hose, conveyor belting for materials or people, and power transmission belts. Most rubber composites are vulcanized hot. If the fibres are laid in a non-symmetrical fashion and have a low thermal expansion coefficient compared with that of the rubber, then if it is stress-free at the moulding temperature the product will show a tendency to warp as it cools.

1.3 FUNCTIONALITY OF PRODUCTS

In order to design products that will fulfil all functional requirements the designer has to deal with the full range of materials, process and product engineering. The authors of this book necessarily had to choose coherent fields from the disciplines of polymer engineering and to guide readers through them such that they would feel at home as much as possible by the end and would be able to deal with other problems in these disciplines on their own, based on their engineering skills in general. The examples mentioned in the previous section give an idea of our choice.

The design examples and problems in this book deal only superficially with a number of aspects of value engineering, even though these aspects should not be neglected in practice. They include

- materials choice and compounding;
- process choice in relation to cost and tolerances, for example;
- material and product properties such as:
  - friction and wear;
  - electrical, optical and acoustical performance,
  - transfer or insulation of heat and matter;
- aesthetic (colour, surface) and ergonomic aspects;
- recyclability.

These aspects are mentioned here just to make the reader aware that design is an integrated process, where all aspects of engineering knowledge should contribute to creating pleasing and functional products.

1.4 CONCLUDING REMARKS

Section 1.2 examined the possibility of using standard engineering approaches to assign leading dimensions to articles made from polymers or to estimate sizes and energy requirements for machinery used to make them. Sometimes the approach worked quite well – examples include the estimate of pressure drop in the pipe extrusion die, and the stiffness in torsion of the rubber spring unit. Sometimes the analysis was appropriate – for example, the pressure containment in the pipe – but the values of properties were inappropriate. On another occasion the properties selected were appropriate within engineering limits, but the boundary conditions were overlooked, as in the restraint offered by the steel plates in the rubber spring under compression.

There is another category where the characteristics of polymers dictate new constitutive relationships between stress and strain or strain rate. This is exemplified by the need for a power-law relationship between stress and strain rate in a fluid, leading to algebraically more complicated equations for pressure drops and energy requirements compared with those of a Newtonian fluid. There are also those linear solids having properties which vary markedly according to the directions in which they are measured: these require a considerably more elaborate form of analysis, of which conventional isotropic elasticity is a special case.

It is quite unsatisfactory to have engineering methods which sometimes work well and sometimes do not – unless it is known in advance which to expect. This book introduces the main themes of engineering with polymers so that the reader will have sufficient background to discern the main properties of polymers, with emphasis on load-bearing, and to appreciate the appropriate types of analysis where calculations of performance need to be made.

In a short book on a vast subject, it is obvious that topics can only be introduced rather than developed. The reader who needs to go further into particular aspects is encouraged to consult the books listed in the further reading lists at the end of each chapter and then to explore the rich, diverse and fascinating research literature.

2 Aspects of Polymer Physics

2.1 INTRODUCTION

There are many metals. Some are well known and widely used, e.g. iron, aluminium, zinc and copper, while others have special qualities, e.g. mercury, titanium and gold. Each metal has its own portfolio of properties. Most metals are not used in the pure state but contain deliberate additions to confer desirable properties; thus there is a range of steels, and of aluminium alloys. Properties such as strength, toughness and hardness can depend markedly not only on composition but also on microstructure induced by thermal or other treatments. Structure and properties may change with time in the unstressed condition (e.g. crystallization of tin or aging of some aluminium alloys) or in the stressed condition (e.g. creep of lead at 20°C, of aluminium at >100°C and of steels at high temperatures).

Similarly with polymers there is a wide range of natural and synthetic polymers. Natural polymers exist in plants and animals, and include starch, proteins, lignin, cellulose, collagen, silk and natural rubber. Synthetic polymers derive mainly from oil-based products and include polyethylene, nylon, epoxies, phenolics, synthetic 'natural rubber' and styrene–butadiene rubber. Polymers form the basis of plastics, rubbers, fibres, adhesives and paints; this book is concerned with plastics and rubbers and mainly with synthetic polymers. The range of properties is wide with moduli ranging from below 1 MN/m^2 to 50 GN/m^2 over the range of solid polymers. Even within a given type of polymer, e.g. polyethylene, the shear viscosities of the melt can be varied over six orders of magnitude. Properties depend on composition (e.g. mixtures and alloys) and on microstructure.

The aim of this chapter is to introduce the language and terminology of polymers without delving into the detailed underlying chemistry. This leads to a broad approach. The examples given are illustrative and representative, but not exhaustive; further details are given in several standard texts.

The main terms involved are linear and network polymers; thermoplastic, thermosetting and crosslinked polymers; amorphous and partially crystalline states; molecular mobility in the rubbery and glassy states; and orientation. These terms help to explain the general behaviour of plastics and rubbery materials, and to distinguish the different properties of different polymers and the different properties of variants of a given polymer type. These terms also provide a basis for identifying interactions between properties and processing. The emphasis here will be on mechanical properties.

2.2 LINEAR AND NETWORK POLYMERS

The basic building block of polymers is the mer or monomer. These can be joined chemically end-on-end to produce a long **linear** chain-like polymer with many thousands of repeat units or mers. An example is polyethylene which consists of a long string of covalently bonded carbon atoms, each sprouting two small hydrogen atoms (Fig. 2.1). The aspect ratio could be 10^4:1 or higher. The chains are not joined together chemically. Different types of mer lead to different characters of polymer having different mechanical, thermal, melt and corrosion-resistant properties.

These linear systems are thermoplastic: heat them and they will melt (and can be shaped under pressure); cool them and they will solidify and harden. This process can be repeated, with care, so that process scrap can be recycled. Over 80% of plastics are thermoplastic in nature, the main ones being polyethylene (PE), polypropylene (PP), polyvinyl chloride (PVC) and polystyrene (PS); all the other thermoplastics only account for about 7% of the thermoplastics market.

If all the mers are the same, then the long-chain molecule is called a homopolymer. If two types of mer are used, a copolymer is produced, e.g. styrene–butadiene copolymer. It is even possible to use three types of mer, e.g. acrylonitrile–butadiene styrene terpolymer (ABS). The properties of these materials depend on the proportion of each mer and how they are joined together.

Most rubber polymers are also based on linear chains, the important ones (in terms of tonnage produced) being *cis*-polyisoprene (natural rubber, NR), styrene–butadiene copolymer rubber (SBR) and polybutadiene rubber (BR). These are thermoplastic in character as polymers but they are never used as such, being chemically modified to make them much more useful.

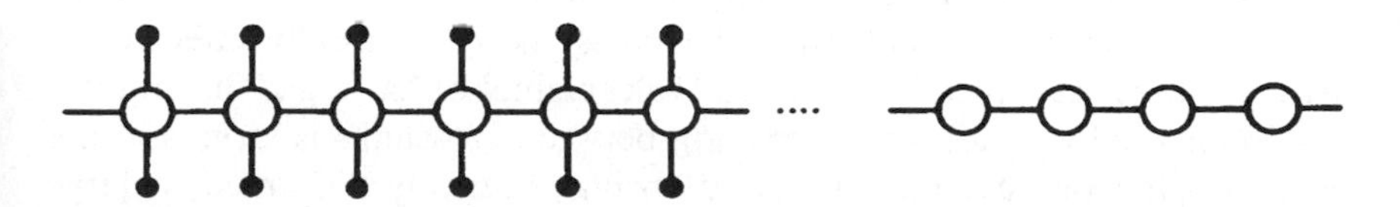

Fig. 2.1 The basic linear chain

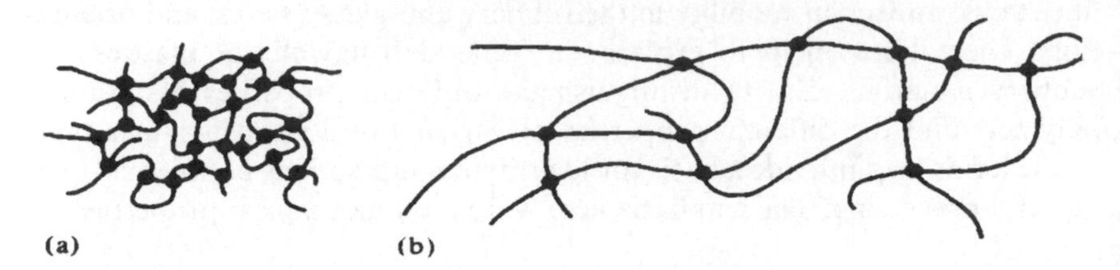

Fig. 2.2 The network polymer: (a) short chain and (b) long chain

Network polymers are based on a different type of mer which can be chemically joined up not only at each end but elsewhere too. These mers can therefore form covalently bonded three-dimensional networks rather than just linear chains. If the nodal points or links are close together, a stiff material results; if widely separated, then the network may be flexible (Fig. 2.2).

One stiff system is based on phenol and formaldehyde, usually supplied for convenience as a flake or powder in which only a few monomer units have been joined (called a prepolymer). On heating, the prepolymer melts and can be shaped in a mould, when the prepolymer units join up with each other randomly to produce one large, densely crosslinked molecule. The mass has been thermoset and is irreversibly crosslinked: phenol–formaldehyde (polymer) is rigid, rather brittle, will not melt on reheating, and has good chemical resistance. This 'phenolic' is used in the distributor head of a car's ignition system. Polymers based on formaldehyde and various comonomers account for more than half the crosslinked plastics.

Another example is based on a short-chain unsaturated polyester (UP) having seven or eight repeat units per chain: this is a syrupy liquid. There are active sites along the chain, each of which can be linked to an active site in an adjacent chain using styrene as a link or bridge. The result is a densely crosslinked rigid solid used with glass fibres to make ship hulls, and pressure vessels and pipes for chemical plants. The properties of the crosslinked system depend on the type of polyester used and on the amount of styrene in the mix: the user can choose what best suits the purpose. The systems can be formulated to crosslink hot (i.e. thermoset) or at room temperature (chemically crosslinked or 'chemiset').

Rubber polymers form the third example of a network system. Active sites in adjacent styrene–butadiene copolymer rubber (SBR) molecules can be permanently bridged by sulphur. The crosslinking is usually effected by heating, the process then being called 'vulcanizing' or 'curing'. The amount of sulphur used is small, so the spacing between crosslinks is large and the result is a flexible material capable of being reversibly deformed to large strains up to several hundred per cent. Thus, cured SBR is widely used to make tyre treads for cars.

There are a few systems which admit to so many variations that different polymers in the same generic class can be either linear, or network, or even ambivalent.

The two basic constituents of polyurethane can be chosen to contain long chains or short chains, each with many or few active sites. The family of polyurethanes contains linear thermoplastics, lightly crosslinked flexible long-chain rubbers and densely crosslinked rigid systems. Applications range from shoe soles, foam for rigid or flexible upholstery and foam for insulation panels in building and refrigeration, to structural frames for seats.

Thermoplastic rubbers demonstrate ambivalent behaviour. Thermoplastics can be reprocessed, while crosslinked rubbers cannot. Thermoplastic rubbers effect a suitable compromise. One system consists of a butadiene rubber with the ends of the long rubber molecules capped with polystyrene: many rubber chains meet in one small polystyrene blob. The blob acts as a crosslink so the solid material is elastic – like rubber. On heating, the polystyrene melts and can be shaped under pressure. On cooling, the polystyrene solidifies. This process is repeatable. The main use of thermoplastic rubbers is in shoe soling.

The hydrocarbon nature of polymers is responsible for their low density (about 1000 kg/m^3), low thermal conductivity (of the order of 0.1 W/m^2 K) and high expansion coefficients ($50–200 \times 10^{-6}/K$). The high melt viscosities (often more than 100 N s/m^2) are caused by the long-chain structure.

2.3 NAMES OF POLYMERS

Polymers are chemicals which have useful mechanical and other properties. Polymers therefore have chemical names which are precise but often long and unattractive. There is considerable incentive to seek more convenient forms and abbreviations for everyday use and for marketing.

It is convenient and often done to indicate polymers by their ISO-standardized initials, e.g. PVC for polyvinyl chloride, PTFE for polytetrafluoroethylene, SBR for styrene–butadiene copolymer rubber. Some polymers are almost always described by their colloquial name, usually a contraction of the formal chemical one: polythene for polyethylene (PE), acrylic for polymethylmethacrylate, or a common name: nylon-6,6 (for polyamide-6,6 (six-six), PA-6,6 or polyhexamethylene adipamide) or nylon-6 (for polycaprolactam); there are other nylons, such as nylon-11 (eleven) and nylon-12, with rather different properties.

Sometimes there is the muddling practice of calling the polymer by the name of its monomer or only one of its monomers, e.g. styrene, acetal, urea, epoxy and urethane (PUR); the context usually makes it plain whether the monomer or polymer is intended. Rubber is always used crosslinked, so it is common shorthand to omit the adjectives 'cured', 'vulcanized' or

Table 2.1 Some basic polymer names

Abbreviation	*Chemical name*	*Common name, if any*	*Maximum crystallinity (%)*	T_g (°C)	T_m (°C)
(a) *Thermoplastics*					
PVC	Polyvinyl chloride	(Vinyl)	10	+80	
LDPE	Low-density polyethylene	Polythene	65	−90	115
PS	Polystyrene	(Styrene)	0	+100	
PP	Polypropylene		75	−27	176
HDPE	High-density polyethylene	Polythene	90	−110	135
PC	Polycarbonate		0	+149	
PA	Polyhexamethylene adipamide	Polyamide-6,6, nylon-6,6	65	[a]	265
PMMA	Polymethylmethacrylate	Acrylic	0	+100	
PTFE	Polytetrafluoroethylene		75	(+20)	327
ABS	Acrylonitrile–butadiene–styrene terpolymer		0	≃ 100	
POM	Polyoxymethylene or polyacetal	Acetal or Acetal copolymer	85	−75	175
PETP	Polyethylene terephthalate	Thermoplastic polyester[b]	65	+67	
	Polysulphone[b]		0		+180–220
(b) *Crosslinked plastics*					
PF	Phenol–formaldehyde	Phenolic	0	Decomposes first	–
UF	Urea–formaldehyde	Urea or amino	0	Decomposes first	–
MF	Melamine formaldehyde	Melamine or amino	0	Decomposes first	–
UP	Unsaturated polyester[b]	Polyester	0		–
EP	Epoxide[b] or epoxy[b]	Epoxy	0	+120–190[c]	–
PU	Polyurethane[b]		0		–
(c) *Crosslinked rubber polymers*					
NR	*cis*-Polyisoprene	Natural rubber	0[d]	−70	–
SBR	Styrene–butadiene copolymer (25/75)		0	−55	–
BR	*cis*-Polybutadiene		0	−85	–
CR	Polychloroprene		0[d]	−50	–
IIR	Isobutylene–isoprene copolymer	Butyl	0	−79	–
EPDM	Ethylene–propylene–diene monomer rubber		0	−60	–

[a] Strongly dependent on moisture content
[b] A family of polymers
[c] Common typical values, not complete range
[d] See section 2.8

'crosslinked' to describe the state of the rubber in the finished product. This seldom causes difficulty once the convention is recognized.

Trade names are widely used to market and sell polymers. They roll off the tongue easily, and indicate the seller's identity, but are not usually informative. A few companies adopt the policy of not disclosing the identity of the polymer, and rely exclusively on their brand name; this is extremely unhelpful from a technical viewpoint.

A list of the names and ISO abbreviations of some common and representative polymer types is given in Table 2.1.

2.4 THERMOPLASTICS

This section identifies in more detail the reasons for making different types of linear polymer.

The basic polymer chain is the simple linear polyethylene: a long string of covalently bonded carbon atoms with small pendant hydrogen atoms. These covalent chemical bonds, which hold the chain together, are called primary bonds. They are very strong, as apparent from the high strength of highly oriented high-density polyethylene fibres.

It is tempting (but incorrect) to model these linear chains as a set of aligned rigid rods (Fig. 2.3). Such an orthotropic array would be stiff and strong along the axis of alignment because of the covalent bonds, and weak and flexible in the transverse direction because of the much weaker secondary bonds between chains. However, the 'rods' are short – a micrometre or two at most – so axial loads would only be transferred down the line from rod to rod by secondary forces, giving a low strength. In fact, the chains do not have a zig-zag configuration, as they would have had when fully oriented, but have an untidily coiled configuration, with adjacent molecules severely entangled and linked, where they are close, by secondary forces. The entanglement adds extra stiffness and strength to those links. Such a mass is in principle isotropic and homogeneous.

The external shape of this configuration looks like a ball, the diameter of which can be estimated. For this purpose we need the restriction that the chain elements are fully free to move, e.g. in a very dilute solution. This enables the links to be randomly oriented in space (within the restraints of the basic chain unit). Thermally induced movement will help to form such an

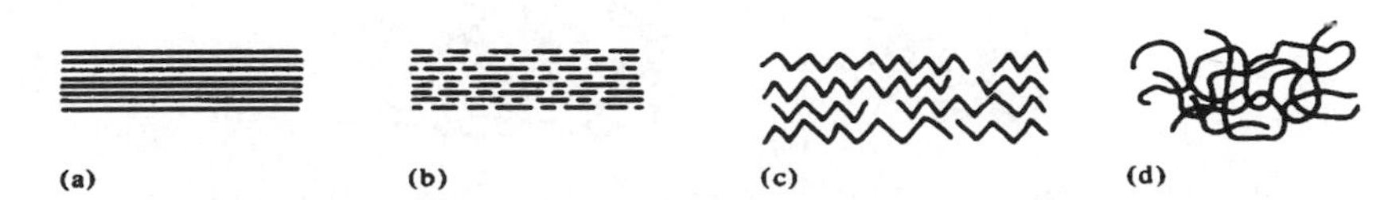

Fig. 2.3 Diagrammatic models of the polymer mass

ever-changing configuration, which might look like an irregular coil if it were enlarged enough. The estimation is carried out by statistical means. For a linear polyethylene molecule with 10 000 CH_2 chain elements (with a zig-zag stretched length of 1.26×10^{-6} m) the diameter of that coil could be 44×10^{-9} m. On the basis of its diameter, the coil, if alone, would have a density of only 1/160 of that of the amorphous fraction in polyethylene, so that at the same place lots of other (parts of) coils would coincide. This indicates a rather strong entanglement of a chain by other chains, which also is quite probable for undissolved polymer. If the chains are pulled apart from this coil, the entanglements will keep the coil together. We might call this phenomenon a temporary network. The chains can disentangle only when their elements are free to move and to uncoil. Three means are helpful for this purpose: a solution (as mentioned), thermally induced movement and the absence of intermolecular forces (i.e. a temperature above T_g, section 2.6).

Polyethylene is a flexible material. How can a stiffer material be made? Four approaches are widely used: making the molecular chain longer; preventing disentanglement geometrically; increasing the attractive forces between chains; or making the chain itself from an inherently stiffer repeat unit.

The longer the molecular chain, the more opportunities there are for more entanglements, leading to higher strength (Fig. 2.4). However, the entanglements persist when the material is melted, and the higher viscosity makes it more difficult to shape the melt. The chain length is described by the number of repeat units (called the degree of polymerization) or by the molecular mass. Not all the chains have exactly the same length, so a given polymer mass has a molecular mass distribution and an average molecular mass. The various grades of a particular polymer have different average molecular masses, representing different compromises between solid strength and melt viscosity to suit different needs.

Geometrical entanglement derives from replacing one of the small hydrogens in the monomer by a bulkier group (Fig. 2.5). In styrene the replacement is a massive benzene ring, which makes polystyrene much stiffer than polyethylene because the rings catch on adjacent chains and prevent slip.

There are different kinds of secondary forces: Van der Waals bonds, hydrogen bonds and polar bonds. Van der Waals bonds or dispersion forces are the most common, as they work between hydrocarbon chains, but they are also the weakest.

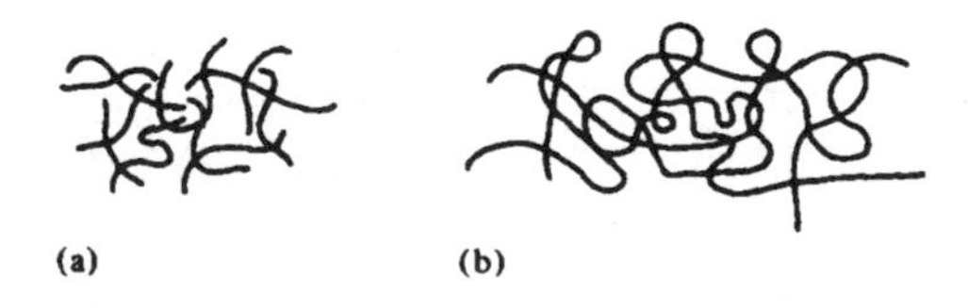

Fig. 2.4 Longer chains have more entanglements per chain

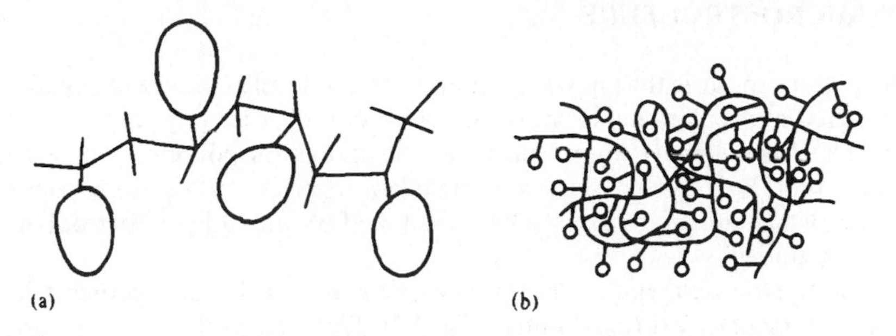

Fig. 2.5 Bulky side groups inhibit disentanglement

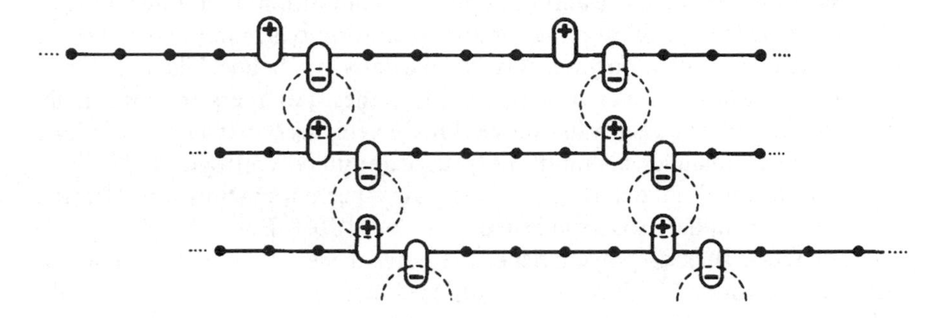

Fig. 2.6 Attractive forces between charged groups in adjacent chains

Hydrogen bonds, such as those acting between specific groups in cellulose or in polyamide chains, are stronger. In polyamides the ethylene repeat unit is separated at regular intervals by amide groups that give rise to hydrogen bonds which hold adjacent chains together. The more closely spaced the amide groups, the stiffer the polyamide.

A further increase in bond strength results from the introduction of polar side groups. In vinyl chloride, one hydrogen atom has been replaced by a chlorine atom. In PVC each repeat unit has a region of positive and negative charge: adjacent chains are attracted (Fig. 2.6). An ultimate increase in bond strength is produced by introducing chemical bonds, which transfers our attention to the rubbers and thermosets: chemical bonds cannot be reversibly broken by temperature increase.

A stiffer repeat unit can be achieved by introducing ring structures or other atoms along the main chain. The repeat unit for polycarbonate of bisphenol A consists of large ring groups, oxygen atoms and carbon atoms with bulky groups rather than just hydrogen atoms: these features make for a stiff chain and also prevent chain slip.

The chemical changes outlined here naturally change the chemical or corrosion resistance and general chemical survival qualities (including high-temperature survival) of the resulting polymers.

2.5 MICROSTRUCTURE

The picture so far is that polymers consist of randomly coiled or kinked and entangled molecular chains showing no structural order (e.g. Figs 2.4b and 2.5b); such a system is **amorphous** and a clean mass of polymer is inherently transparent. Polystyrene is an example taken from the 40% or so of thermoplastics which are amorphous. The crosslinked plastics and rubber polymers are also amorphous.

Some narrow and regular linear polymer chains can cluster together in local axial alignment to form crystallites (Fig. 2.7). The close packing confers on the crystallite a higher density, higher axial and transverse stiffnesses and strengths, and lower toughness. The traditional picture is of randomly oriented crystallites suspended in and reinforcing a matrix of amorphous material of the same polymeric type. Such a polymer is called **partially crystalline**. The crystalline aggregates are large enough to scatter light, so these polymers are not usually transparent, except possibly in thin sections. Examples of partially crystalline thermoplastics include polyamide, polyethylene and polypropylene.

The traditional model (Fig. 2.7) explains behaviour quite well if the amount of crystallized material is less than about 50%. But in highly crystalline polymers, such as polyethylene, the current view is that thin plates or lamellae are formed with molecules ordered *through the thickness* by chain folding (Fig. 2.8). Lamellae grow to form a flat ribbon, with surface imperfections then providing a nucleation source for another layer. These lamellae tend to suck pure polymer from the surrounding amorphous matrix, leaving behind very entangled material and foreign matter. The growing ends of the lamellae twist somewhat, and spread out to reach more crystallizable material. The bundles of ribbon-like lamellae have a spherical growth front, and are called spherulites. Spherulites contain both ordered material and amorphous material. Where two spherulites meet, the spherical character is lost at the plane boundary, which is likely to contain a layer of amorphous material potentially rich in any impurities.

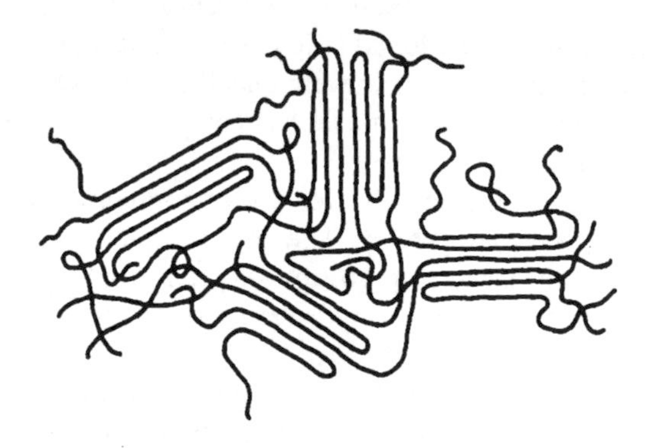

Fig. 2.7 Fringed-micelle model of crystallites in amorphous matrix

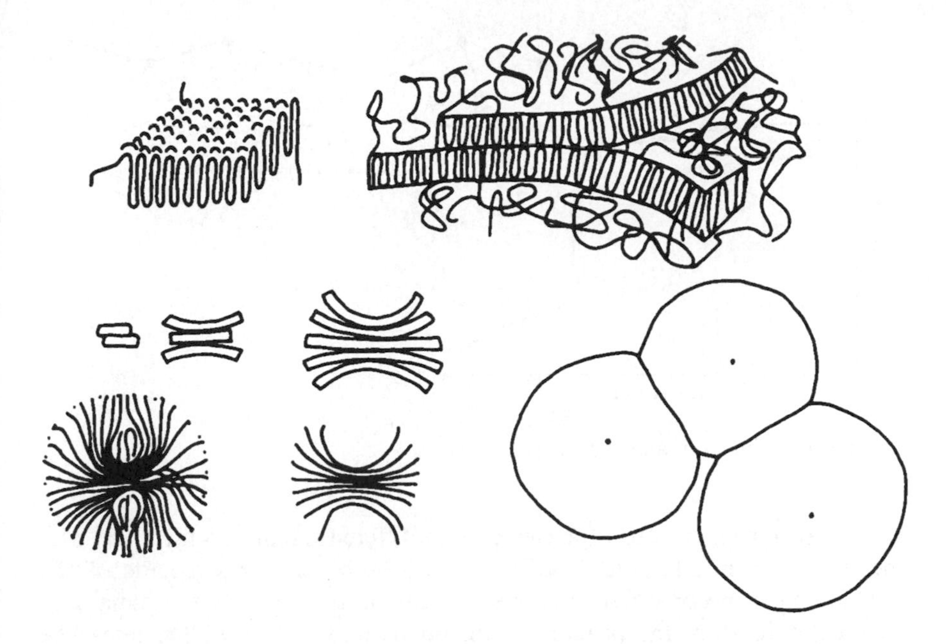

Fig. 2.8 Folded-chain lamellae and the development of spherulites

Linear polyethylene can be up to about 90% crystalline, and the modulus is then about 1 GN/m^2 because of the stiffening effect of the spherulites, in spite of the inherent flexibility of the polyethylene molecule. The greater the proportion of crystallized material, the higher the density and the modulus.

2.6 MOLECULAR MOBILITY

For simplicity, this section describes the molecular mobility of linear thermoplastic polymers. The mobility of crosslinked polymers is described in sections 2.7 and 2.8. It is convenient to consider amorphous and partially crystallizable polymers separately. In the following discussion, it is assumed that the heating is slow and that the temperature in the polymer is uniform.

2.6.1 Amorphous polymers

If the polymer is really cold, say at 0 K, no chains or parts of the molecule can move: the polymer is frozen into a rigid glassy state. On slow heating the secondary or β transition temperature, T_β, is reached, at which side groups or other elements of the molecule have sufficient thermal energy to rotate or to otherwise move: the modulus drops slightly. On further heating, at the glass–rubber transition temperature, T_g, volumetric expansion and increased

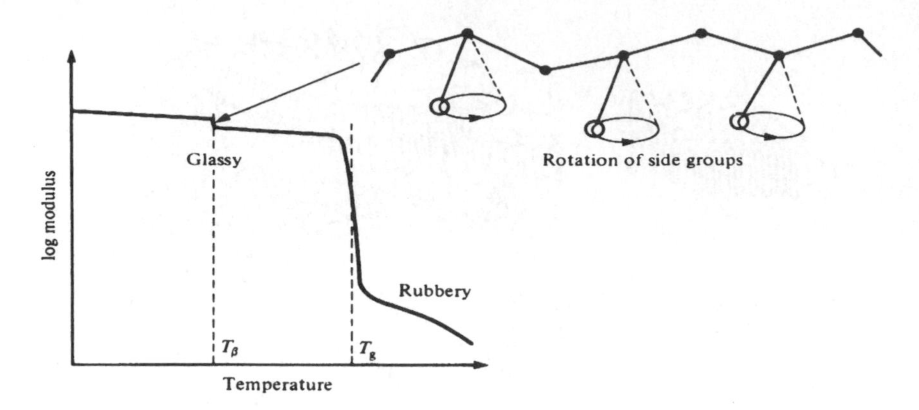

Fig. 2.9 Transitions in amorphous polymers

molecular movement push the elements of different chains so far apart that the secondary bonds fade away and complete blocks or segments of the polymer chain become free to move as local entities. Finally the remaining coherent forces are the primary chain bonds and the entanglements. The modulus now drops by several orders of magnitude (Fig. 2.9). The rise in temperature is accompanied by expansion, typically 60–80 × 10^{-6}/K (linear coefficient; cf. steel 11 × 10^{-6}/K), with an increase at T_g.

The previous paragraph suggests that a sharp transition occurs at T_g. In fact, the transition usually occurs over a temperature interval of some 10 K when measured conventionally under no stress. Moreover, as the temperature approaches T_g, some smaller segments become free to move, aided by any spaces in the polymer mass; there is a slight drift downwards in modulus before the catastrophic drop of three orders.

If molecules are subjected to prolonged stress, then near T_g there is some opportunity for a little rearrangement of chains as small-scale uncoiling occurs; this would not happen appreciably under short-term loading because chains could not uncoil fast enough. The nearer T_g is approached, the more pronounced the uncoiling becomes and hence the more time dependent the stiffness of the polymer. Thus, not only can an amorphous polymer creep at temperatures some way below T_g, but the temperature at which softening becomes apparent depends on how fast the test measuring T_g is conducted (Fig. 2.10). Behaviour above T_g can be glassy too, if the rate of loading is rapid (e.g. impact loading), so that the nominally flexible segments of chains have insufficient time to uncoil; this feature is made worse by the presence of notches or sharp cracks which can further suppress ductility.

Slow cooling essentially reverses the effects described above. Thermoplastics are indeed viscoelastic.

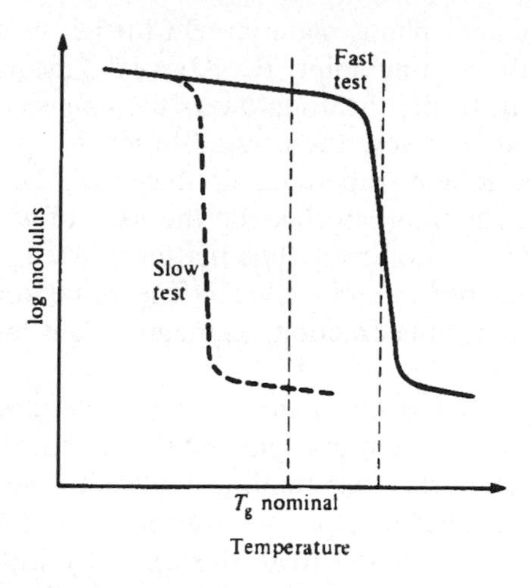

Fig. 2.10 Rate dependence of T_g

2.6.2 Partially crystalline polymers

A partially crystalline polymer is rigid and glassy when at very low temperature. On heating to T_β, the slight increase in mobility of the amorphous fraction of the polymer is swamped by the reinforcing effect of the crystalline fraction. At T_g the amorphous material becomes rubbery, but the crystals can still confer substantial stiffness. When the temperature is further increased above T_g, the material softens, as the matrix becomes more and more rubbery (Fig. 2.11). The mobile matrix confers ductility and extra toughness, and at constant temperature above T_g a constant load will induce creep because of

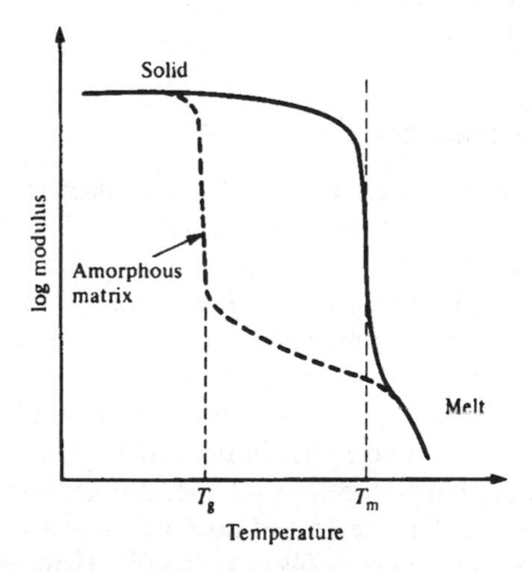

Fig. 2.11 Transitions in partially crystalline polymers

the uncoiling of the amorphous material. On further heating above T_g, the crystallites melt at the melting point, T_m. Above T_m the material is a viscous and amorphous melt; further heating lowers the viscosity.

On cooling a melt to below the crystallization temperature T_m, having passed a metastable zone of supercooling, nuclei will be formed. These are needed to start the growth of lamellae, on the basis of which spherulites are developed. The formation of crystalline matter is a temperature-dependent process, further described in section 9.2.5. The cooled product will always contain a certain amorphous fraction; its magnitude is highly dependent on the cooling conditions.

If cooling occurs under tensile or shear stress, molecules are already partly aligned in the direction of the stress field and this favours the development of many locally aligned nuclei: surrounding molecules are to a considerable extent sympathetically aligned, so crystallization is rapid. Here the molecules do not form spherulites, but so-called 'shish-kebabs' instead. Cooling conditions determine to what extent they will appear in the rigid product.

The transition between the crystalline and the molten phase is associated with a considerable amount of enthalpy [kJ/kg]. This is different from the glass–rubber transition where, at the transition itself, no heat effect can be observed (see Fig. 8.23), but where the thermal capacity [kJ/kg K] above T_g is higher than below T_g. The close packing in crystallites means that partially crystallizable polymers shrink much more than amorphous ones when cooled from the melt state to a solid at room temperature.

The linear expansion coefficients of solid, partially crystalline polymers above T_g are about $100–200 \times 10^{-6}$/K. An important aspect of the properties of partially crystalline polymers is their high impact strength at temperatures between T_g and T_m. The framework of flexible amorphous material is held together by the intermediate crystal structures which consist of parts of the same molecular chains.

2.6.3 Molecular movement below T_g

All molecules of non-polymeric materials are subject to thermally induced movement. In the gas phase they do not interact with each other, except in the case of a collision, which is ruled by the laws of momentum. The same holds for the fluid phase, except for the free surface and the related surface tension which prevents most molecules escaping. The situation is different for the solid crystalline phase, where the molecules are held on their theoretical grid places and are only allowed to oscillate around these places.

Amorphous polymers above T_g are in principle in their fluid phase. When they are cooled from above to below T_g, thermal shrinkage brings parts of adjacent chains so close that they could become subject to secondary attraction forces and these chain parts become 'fixed'. However, the coiled and ever-changing conformation of the molecule chains brings them only locally into contact and so cooling down through T_g builds a kind of network of secondary bonds between the chains with, in between, free mobility of the chain parts in the so-called 'free volume'. This network holds glassy polymers together and is responsible for their cohesive strength: (1) the lower the temperature below T_g, the closer the chain parts and the higher the secondary forces; and (2) the longer the residence below T_g, the more chances, due to thermal motion of chain parts, of increasing the number of contact points in the network (physical aging; see section 5.2.4(b)).

The effect of fast cooling is that the above process is more pronounced. Elements of chains do not have time to reach an equilibrium position. The result is that there remains space within the polymer mass which is capable of being filled by polymer but which is initially vacant. Between T_g and T_β the side groups and short segments of the chain can adjust their position and slowly fill the free volume.

2.7 CROSSLINKED PLASTICS

It was noted in section 2.2 that formaldehyde-based polymers become thermoset on heating, with a high crosslink density and rigid short groups between randomly distributed links. The structure is amorphous and glassy. On heating they may soften slightly at high temperatures, but they start to char before reaching their T_g (Fig. 2.12). Throughout their working temperature range they are essentially elastic with moduli of the order of 1–3 GN/m^2, and rather brittle with breaking strains of only a few per cent. The most rigid of these materials is phenolic, in which the repeat unit consists essentially of a carbon atom plus a large stiff ring. Less rigid is the polymer based on urea–formaldehyde which softens at lower temperatures than phenolic. Other differences, including those relating to corrosion resistance, will be mentioned in Chapter 3.

The other crosslinking plastics are normally two-component systems supplied separately, and adjustments can readily be made by choice and proportion of components to give polymers of the same generic class but having

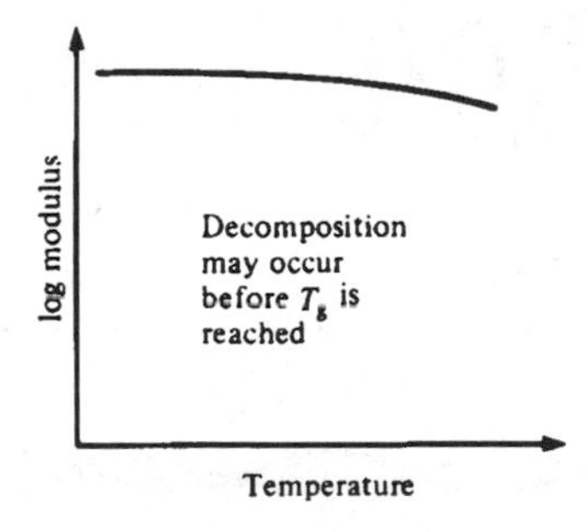

Fig. 2.12 Transition in a highly crosslinked polymer

quite different physical or other characteristics. This leads to a wide variety of crosslinked polymers, and scope for the careless operator to make the wrong polymer or the wrong version of the intended polymer. This category includes the unsaturated polyesters, the polyurethanes, the epoxies and the alkyds. These are usually amorphous and glassy. The exact combination of properties such as stiffness, strength, hardness and toughness depends on the chain length and character of the repeat unit in the polymer or prepolymer, the flexibility and character of the crosslinking agent, and the crosslink density. Not only can the character of the repeat unit be changed using the devices outlined for thermoplastics in section 2.4, but the identity and spacing of active sites for crosslinking can also be varied. The crosslinking agent may contain two or more active sites, so the crosslink density depends on the number of sites per crosslinking unit as well as the proportion used. Some polymer-forming systems, e.g. epoxies and urethanes, will react with different chemical types of crosslinking agent (or hardener) to give polymers having different properties and chemical resistances. The very versatility of these systems, which offer the potential of 'tailor-made' combinations of properties, makes it much more difficult to quote 'typical' values of properties for a given polymer type such as the epoxies.

2.8 CROSSLINKED RUBBER POLYMERS

The flexible rubber linear polymers or copolymers are only lightly crosslinked, so T_g is low, and in normal service use these materials are amorphous and rubbery, with moduli of about 1 MN/m^2 and modest strengths. Between crosslinks, the rubber chains are free to move or extend *locally*, but adjacent molecules are prevented by the links from slipping permanently past each other (Fig. 2.13). Above T_g the rubber is reversibly elastic to high strains. The elasticity may be explained as follows. The chance that a rubber chain is straight is minimal in a properly amorphous material; indeed, there is a very good chance that the molecule is considerably coiled up with the ends or link points close together. If these ends are separated by applying tension, then

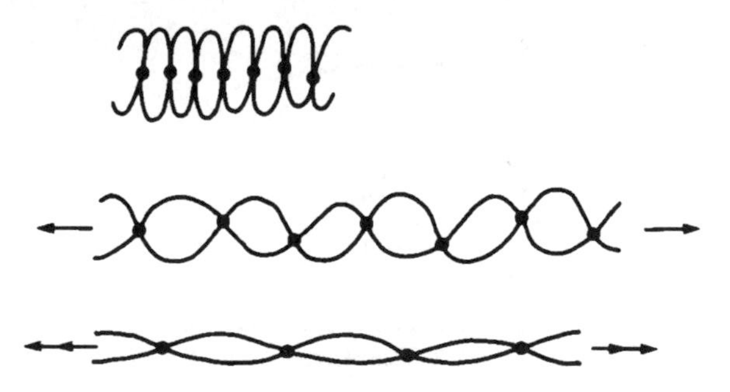

Fig. 2.13 Local extension of chains without link slip

the molecule will crumple up again on removing the external force. So the two ends of the molecule will act as if there were an internal force tending to draw them together.

In some respects, crosslinked rubber above its T_g has the character of an elastic liquid, because the chains are so mobile. On heating, the thermal energy of the chains between links increases and this tends to interfere with the extension of molecules under stress; thus the modulus E (or G) rises slightly with increase in absolute temperature, T. More links per unit length of rubber chain, N, inhibit the maximum flexibility of the rubber, thus also increasing stiffness. Experiment confirms that

$$E \text{ or } G \propto NT \quad \text{for } T > T_g$$

Above T_g the stiffness in tension or shear of crosslinked rubber depends on the easy uncoiling of molecules, hence the low moduli. However, under hydrostatic compression, deformation overcomes forces between molecules, and the quasi-liquid-like character of rubber confers a much higher bulk modulus of the order of 1 GN/m^2. Thus rubber is virtually incompressible with a Poisson's ratio of 0.5. The corollary of this is that if the rubber cannot move out of the way when compressed (for example, if it is prevented from bulging sideways in a simple block under uniaxial compression), then the stiffness will be governed by the bulk modulus and will be up to three orders of magnitude greater than that based on the uniaxial modulus.

Below T_g, molecular motion of the rubber molecules is frozen (following the arguments in section 2.6.1), and rubber becomes a stiff, brittle, glassy solid with a modulus of about 2 or 3 GN/m^2. T_g depends on the character of the rubber repeat unit (or units for copolymers).

Thermoplastic elastomers are in fact thermoplastics, composed of two different incompatible polymers, one that is rubbery (T_g below temperature of use) and the other glassy. Their monomers are copolymerized in order to prevent segregation. When they are cooled down below T_g of the glassy elements, these separate out into small domains. These domains act like large polyfunctional crosslinks, because long, flexible, elastomeric sequences are linked together by them. However, the glassy domains are held together only by secondary bonds and they yield and flow under high stresses or at temperatures near their T_g.

Most rubber molecules will not partially crystallize because of the nature of the chain, often a copolymer or homopolymer with side groups which inhibit alignment. However, natural rubber (*cis*-polyisoprene) and polychloroprene molecules are sufficiently narrow and regular to be able to crystallize partially either at low temperatures (but above T_g) in the absence of stress or at higher temperatures under large tensile strain. The process is reversible: the amorphous state is recovered by increasing the temperature or removing the stress.

At room temperature and above, the thermal energy of natural rubber molecules is too great for any nuclei – local alignment of a few chains – to be

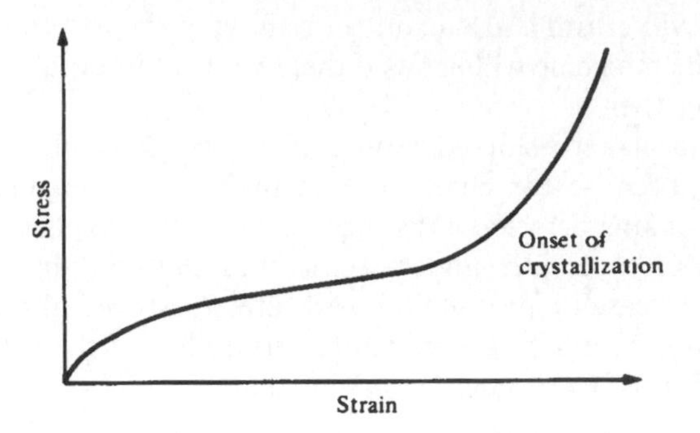

Fig. 2.14 Stress–strain curve for a crystallizable cured rubber polymer

stable. But at 0°C or below, these small aggregates survive long enough to attract more chains and so crystallites begin to grow. The growth rate is proportional to size and increases rapidly until entanglements restrict mobility. The close packing increases the density by about 2 or 3%, and the randomly oriented crystallites which account for up to 30% of the polymer mass cause the modulus to rise to about 100 MN/m^2. This low-temperature crystallization is well documented and is a slower version of what happens when partially crystallizable thermoplastics are cooled down from the melt.

At high tensile strains, typically a few hundred per cent, some localized molecular alignment occurs, leading to the formation of stable nuclei aligned in the direction of extension. Further extension aligns more chains, which are then in a state favouring spontaneous crystallization with crystallites aligned preferentially in the direction of extension. Resistance to extension now has to overcome the covalent forces along chains rather than Van der Waals' forces between chains, leading to a dramatic rise in stiffness in the stretch direction (Fig. 2.14). On removing the stretching force, the polymer recovers its dimensions and its amorphous condition. This phenomenon explains why natural rubber is so tough: when a crack is loaded in the opening mode, the local high stress at the crack tip induces crystallization which acts as a crack stopper. SBR does not crystallize under stress and is less tough.

2.9 MOLECULAR ORIENTATION

The concept of permanent gross orientation is important in plastics technology. So far, the discussion has assumed that a solid thermoplastic made under conditions of no stress is isotropic, whether it is amorphous or partially crystalline. However, if the polymer is stretched severely in the melt (or in the solid form usually above the effective T_g), chains or crystallites can become ordered and aligned in the stretch direction. This alignment or **orientation** can be permanent if, while still under stress, the amorphous material is cooled below T_g or the partially crystallizable material is well below T_m.

The orientation confers additional stiffness and strength in the stretch direction at the expense of planes of weakness. Extreme orientation is possible in some systems under carefully controlled conditions; for example, highly oriented high-density polyethylene 'fibres' can have an axial tensile modulus as high as 50 GN/m^2. In less extreme forms, unidirectional orientation is exploited in fibres, tapes and plastic string, where the weakness in the transverse direction is unimportant. The polymer can also be stressed in two directions in the plane, e.g. tubular film (section 9.4.2), some blown bottles, or sheet, which results in 'biaxially oriented' products. This provides extra strength and stiffness in each stretch direction (but also greater proneness to delamination). It is possible that the cell walls of foam plastics of low density may also be oriented. In injection-moulded parts, a **distribution of orientation** is frozen in; the orientation varies from point to point throughout the product.

The result of orientation is anisotropy, which indicates that different directions may show different magnitudes of the properties. The anisotropy, resulting from molecular orientation, not only refers to modulus and strength, but also to other properties such as heat conduction (which increases in the direction of orientation) and linear thermal expansion (which decreases in this direction).

2.10 SOME BROAD GENERALIZATIONS

The following remarks summarize, in a grossly oversimplified way, the main points discussed in this chapter. There are many exceptions to these statements, but it is hoped that they will help the reader unfamiliar with polymers to develop a general feel for these materials.

In service, rubber polymers are usually lightly crosslinked: they are amorphous and capable of large reversible deformations, with tensile moduli of the order of 1 MN/m^2. Thermoset and chemiset plastics are densely crosslinked, amorphous and usually glassy: they tend to be rather brittle with a modulus of the order of 1 GN/m^2. Amorphous thermoplastic polymers are usually used in their glassy state: they are usually rather brittle with moduli of the order of 1 GN/m^2. Some partially crystalline thermoplastic polymers are used above T_g, some below. If above T_g, the material can be fairly ductile and mechanical behaviour will be rather time and temperature dependent. If used below T_g, the time and temperature dependence will be less pronounced, but the material may be less ductile.

PROBLEMS

1. Why are some glassy polymers stiffer at room temperature than some partially crystallizable polymers?
2. (a) Natural rubber derives from the tree *Hevea brasiliensis*. The polymer is *cis*-polyisoprene and has a molecular weight of 1 000 000 or more. What is its degree of polymerization, if the molecular weight of the mer is 68?
   (b) Estimate the range of degrees of polymerization of nylon-6 having molecular weights in the range 9000–15 000. (The molecular weight of the repeat unit is 113.)
   (c) A linear polyethylene has an average molecular weight of about 160 000. If the distance between adjacent carbon atoms in the main chain is about 0.15 nm and the angle between adjacent carbon–carbon bonds is 108°, estimate the maximum possible separation of the ends of the average molecules. Why is the separation likely to be very much less than this? (The molecular weight of the ethylene mer is 28.)
3. How can the concept of molecular mobility be used to explain why the rate of growth of crystallites is likely to be least near the glass–rubber transition temperature?
4. How does the presence of crosslinks affect molecular mobility and microstructure in polymers?
5. To what extent is molecular weight important in controlling the strength of commercial polymers?
6. To what extent is the glass transition temperature likely to be related to the temperature at which polymers become brittle in service?
7. In injection moulding, a hot melt is injected into a cold cavity to define the shape of the desired product. It is sometimes claimed that an amorphous thermoplastic is cheaper to produce in this way than a partially crystallizable one. What information would you need in order to agree or disagree with this claim?

FURTHER READING

Birley, A.W., Haworth, B. and Batchelor, J. (1992) *Physics of Plastics; Processing, Properties and Materials Engineering*, Hanser, Munich.
Brydson, J.A. (1995) *Plastics Materials*, Butterworth-Heinemann, Oxford.
Gedde, U.W. (1995) *Polymer Physics*, Chapman & Hall, London.
Gent, A.N. (ed.) (1992) *Engineering with Rubber*, Hanser, Munich.
Haward, R.N. and Young, R.J. (1997) *The Physics of Glassy Polymers*, Chapman & Hall, London.
Osswald, T.A. and Menges, G. (1996) *Materials Science of Polymers for Engineers*, Hanser, Munich.

Ward, I.M. (1997) *Structure and Properties of Oriented Polymers*, Chapman & Hall, London.

Young, R.J. and Lovell, P.A. (1991) *Introduction to Polymers*, Chapman & Hall, London.

3 Plastics and Rubber Components and Compounds

3.1 INTRODUCTION

Within the family of polymers there are various subgroups, e.g. thermoplastics, crosslinkable plastics and crosslinkable rubber polymers; within each subgroup there are many individual polymers, each having its own individual portfolio of properties. It is, however, possible to make some general statements about polymers as a class of materials; these statements help the designer to decide in general terms whether polymers are likely to be suitable candidates for a given product or component. If unsuitable, other materials need to be examined; if suitable, then it is necessary to look more closely at the individual polymers in order to narrow down the choice. This chapter introduces the properties of polymers as a class.

Pure polymers are hardly ever used on their own to make articles (the same also applies to many metals), because polymers have a number of limiting features. It is entirely commonplace to use compounds, made from polymers and ingredients ('additives') selected to confer desirable characteristics. The term 'plastics' describes the compound based on a plastics type of polymer and the additives. The term 'rubber' applies to the rubber polymer plus additives. This chapter identifies the common additives, why they are used and what effect they have on the behaviour of a given compound. The discussion provides an appreciation only; it is not intended that the reader becomes a recipe designer.

3.2 POLYMERS AS A CLASS OF MATERIALS

To introduce the properties of polymers as a class, it is helpful to discuss two important applications, pressure pipes and car tyres, and to generalize from there.

3.2.1 Plastics pipes and fittings

About 10% of all pipes and fittings are made from plastics, and about 10% of plastics are used in pipes, mainly thermoplastic pipes. Thermoplastics used in pipes have the following general characteristics.

1. Low density: easy to transport and install.
2. Corrosion resistance: minimal maintenance, negligible build-up of scale, and able to resist aggressive media (by suitable choice of plastic).
3. Insulation: low thermal conductivity or built-in lagging; low electrical conductivity – a possible hazard in pumping non-conducting powders.
4. Easy to make: by extrusion of a polymer melt through a die.
5. Colour coded: some plastics are transparent too.
6. Expansion: thermal expansion must be allowed for in the design of the pipe system.
7. Flammability: the hydrocarbon nature of polymers ensures that all polymers will burn, some more readily than others.
8. Temperature: the service range is from about – 5°C. Most plastics can cope with 50°C, relatively few with 100°C under prolonged pressure, one or two survive 200°C.
9. Stiffness: modulus of the order of a few giganewtons per square metre or less.
10. Strength: yield stress usually less than 20 MN/m^2.
11. Toughness: in the range 1–3 $MN/m^{3/2}$, less under cyclic or prolonged load; able to withstand 'normal abuse'.

3.2.2 Radial tyres for car wheels

Vehicle tyres account for more than half the total use of rubber. Treads are made from SBR, most of the rest from natural rubber. Tyres are produced in heated moulds (to effect crosslinking) following a rather elaborate fabrication procedure. The rubber in the tyre has the following general characteristics.

1. Corrosion resistance: adequate resistance to water, petrol, oil and salt.
2. Insulation: thick-walled tyres tend to get warm, especially if underinflated.
3. Abrasion resistance: excellent.
4. Fatigue resistance: excellent.
5. Toughness: adequately resists crack growth provided the rubber is protected from oxidative degradation.
6. Flexibility: modulus $\simeq$ 1 MN/m^2; grips road and seals to wheel rim.
7. Energy absorption: a smooth, quiet ride over rough surfaces (part of the suspension system).
8. Lubrication: water is a superb lubricant for rubber; road-holding relies on efficient tread design to squeeze water out of the way.

9. Complicated shape: achieved with repeatable precision.
10. Orientation of plies: selected to confer desired road-holding, suspension and steering characteristics.
11. Low density: lightweight construction.

3.2.3 General properties of polymers

In the two examples chosen, there are many common features which can usefully be brought together; a few others which are not dominant in pipes and tyres are also added. First are given those properties applicable to both plastics and rubber materials, then remarks on the special qualities of each group. (It is, of course, possible to find exceptions; these remarks are intended to provide some frameworks for an overview.)

(a) Properties of polymers

1. Density. Typically 800–1500 kg/m^3 for uniform polymers; foamed or cellular polymers down to 10 kg/m^3; heavily filled polymers to about 3000 kg/m^3.
2. Insulation. Outstanding electrical insulation (unless conducting fillers are present), exploited in wire covering and capacitor dielectrics. High surface resistivities may lead to dust pick-up. Polar polymers are, of course, rather lossy in high-frequency electric fields; clean non-polar polymers can have minute dielectric loss angles. Thermal insulation: conductivity of solid polymers is about 0.2 W/m K, i.e. four orders of magnitude lower than copper. Foams lower still by one order.
3. Expansion coefficient. At about room temperature, linear expansion coefficient in the approximate range $60\text{–}200 \times 10^{-6}$/K.
4. Dimensional stability. Expansion already mentioned. A few polymers can absorb some liquids, causing swelling or even dissolution, accompanied by changes in physical properties. For example, natural rubber readily absorbs large quantities of hydrocarbon liquids (use special oil-resistant rubbers instead); nylon absorbs moisture in small quantities.
5. Chemical resistance. Can be very good but depends on the chemical nature of the polymers. A useful concept is that 'like dissolves like'; for example, hydrocarbon polymers, such as polyethylene, are not happy with (hydrocarbon) oils. The corrosion resistance of polymers is generally complementary to that of metals; for example, many polymers resist acids and alkalis which corrode many metals, but some polymers are not oil resistant (beware of sweeping generalizations). The presence of tensile stress can make hostile an otherwise 'safe' environment.
6. Burning. All polymers can be destroyed by flame or excessive heat, although the rate of destruction depends on the type of polymer, the surface-to-volume ratio, the temperature and the duration of exposure to

heat; i.e. some polymers are better than others. Comparable equivalent notions for plastics: cellulose nitrate – explosive; low-density polyethylene – hard wax; PMMA – hardwood; polyethersulphone – only burns if flame is present; asbestos-filled phenolic – used as ablation shield for spacecraft re-entry into Earth's atmosphere.

7. Processing. It is normal to make in one piece three-dimensional articles (which can have a complicated shape) with repeatable precision by mass production techniques (see Chapter 4 for examples).

(*b*) *Some special features of rubber polymers*

1. Reversible high extensibility. For example, up to several hundred per cent in gum natural rubber vulcanizates stretched above T_g.
2. Modulus. Typically, about 10^6 N/m^2.
3. Energy absorption. There is a massive area under the stress–strain curve, even though the modulus is low, which provides a large capacity for storing strain energy.
4. Fatigue resistance. For example, tyre behaviour.
5. Toughness. Good resistance to crack growth under cyclic load if the rubber is protected from oxidative degradation.
6. Temperature range. See Table 2.1.

(*c*) *Some special features of plastics*

1. Modulus. About 10^9 N/m^2 or less.
2. Range of toughness. Some plastics are 'tough', e.g. low-density polyethylene, some fragile, e.g. general-purpose polystyrene.
3. Friction coefficient. Unlubricated, some polymers have coefficients of about 0.3–0.5; PTFE rubbing on itself about 0.02. Some soft plastics just adhere.
4. Temperature range. Amorphous plastics are not used above T_g. Partially crystalline polymers used mainly between T_g and fairly well below T_m, and some are used a little below T_g (see Table 2.1).
5. Appearance. Amorphous plastics can be very transparent; partially crystalline ones can be translucent or opaque. Colour plastics with dyes or pigments.

3.3 DETAILED PROPERTIES OF POLYMERS

It is not appropriate here to discuss in detail the properties of each individual polymer, not least because processing and additives play an important part in determining the useful properties of a given polymer, and play a major role in choice of material. However, the following paragraphs provide thumbnail sketches of the main polymer types, in terms of properties and typical applications.

3.3.1 Thermoplastics

1. General-purpose polystyrene: hard, stiff, transparent, but rather brittle; packaging, light fittings and toys.
2. Toughened polystyrene (rubber-modified): tougher than general-purpose polystyrene; vending cups, dairy produce containers, refrigerator liners, toys – particularly model kits for assembly.
3. ABS: tough, stiff, abrasion resistant; dinghy hulls, telephone handsets, housings for vacuum cleaners and grass mowers.
4. Unplasticized PVC: hard, tough, strong and stiff, good chemical and weathering resistance, self-extinguishing and can be transparent; pipes, pipe fittings, rainwater goods, wall cladding and curtain rails.
5. Plasticized PVC: lower strength and increased flexibility depending on amount and type of plasticizer, compared with unplasticized PVC; insulation of wire for domestic electricity supply, domestic hosepipes, soles of footwear.
6. Polyolefins: distinguished by excellent chemical resistance and electrical insulation properties.
7. Low-density polyethylene (918–935 kg/m^3): exploits toughness at low temperatures, flexibility and chemical resistance in pipes for chemical plant, low-loss electrical wire covering, blow moulded and large rotationally moulded containers, and packaging film.
8. High-density polyethylene (935–965 kg/m^3): much stronger and stiffer than LDPE and is used for dustbins, milk bottle crates and mechanical handling pallets.
9. Polypropylene: has good fatigue resistance and can be used at higher temperatures than polyethylene; the copolymer version is more impact resistant than the homopolymer at low temperatures; pipes and pipe fittings, beer bottle crates, chair shells, capacitor dielectrics and cable insulation, twines and ropes.
10. Acrylic (PMMA): completely transparent, not attacked by ultraviolet light (UV), stiff, strong and does not shatter; domestic baths, lenses and illuminated signs.
11. Polycarbonate: tough, stiff, strong, transparent and good electrical insulation properties: street lamp covers, feeding bottles for babies, safety helmets.
12. Modified PPO: tough, strong and stiff, with good electrical properties; connectors and circuit breakers in electrical equipment, and for business machine housings.
13. Polysulphones: stiff, strong, excellent dimensional stability, transparent, burn only with difficulty and without smoke; passenger service units in

aircraft, components for high-temperature duty in electrical and electronic equipment.

14. Nylons: stiff, strong, tough and abrasion resistant; absorption of moisture increases toughness, but reduces stiffness and dimensional stability; gears, bushes, cams and bearings; glass-filled nylon in power tool housings.
15. Polyacetals: stiff, strong, extremely resilient, and abrasion resistant; taps and pipe fittings, light-duty beam springs, gears and bearings.
16. PTFE: outstanding electrical properties and corrosion resistance, exceptionally low coefficient of friction, tough, can be used continuously at 250°C; bearing surfaces of journal bearings, coatings for cooking utensils, high-frequency high-temperature cable insulation.

3.3.2 Crosslinked plastics

1. Phenolics: strong but rather brittle, good electrical insulators, used at up to 150°C; car distributor heads, electrical plugs, switches and sockets; laminates of paper and phenolic used in high-voltage insulation; laminates of fabric and phenolic used in gear wheels.
2. UF: low cost, hard, abrasion resistant, good chemical resistance, high tracking resistance, heat resistant to about 70°C; buttons, domestic electrical fittings and coloured toilet seats.
3. MF: compared with UF, MF is harder, and has better heat and stain resistance; tableware and picnicware; glass-reinforced MF laminates resist 200°C.
4. Polyesters: fairly good electrical insulation properties, good chemical and UV resistance, usually reinforced with glass fibres to give strengths and stiffnesses 2–10 times those for thermoplastics; roofing and building insulation, car bodies and lorry cabs, boat and ship hulls, and chemical plant.
5. Epoxides: tough, resist alkalis; encapsulation of electronic components, adhesives; glass-reinforced epoxide used in chemical plant; heavily filled grades for moulds for metal shaping and plastics thermoforming.

3.3.3 Vulcanized rubber compounds

1. Natural rubber (natural *cis*-polyisoprene): good fatigue resistance, high strength even when unreinforced, range of hardness by compounding, good low-temperature flexibility; used extensively in commercial vehicle tyres, antivibration mountings and dock fenders.
2. Styrene–butadiene rubber (SBR, styrene–butadiene copolymer): when compounded with reinforcing fillers it resembles natural rubber, but has better abrasion resistance and power fatigue resistance; lorry and passenger car tyres.

3. Chloroprene rubber (polychloroprene): physical properties inferior to those of natural rubber, but better ozone, chemical, oil and high-temperature resistance; used in machine mountings where very oily environment prevents use of natural rubber, and in sponge form in divers' wet suits.
4. Nitrile rubber (NBR, acrylonitrile–butadiene copolymer): increasing acrylonitrile content improves oil resistance at the expense of resilience and low-temperature flexibility; O-rings and oil seals.
5. Ethylene–propylene rubber (EPM, ethylene–propylene copolymer) and EPDM (ethylene–propylene–diene monomer rubber): comparable in general properties with SBR but with much better resistance to atmospheric aging, ozone and oxidation; EPM is peroxide cured and has slightly better weather resistance, but EPDM can also be crosslinked with sulphur; used in sealing strips for car windows, and in water seals in washing machines and car radiator hoses.
6. Butyl rubber (isobutylene–isoprene copolymer): good ozone, weathering, heat and chemical resistance, extremely low gas permeability, high damping capacity at room temperature; inner tubes and liners for tubeless tyres.
7. Butadiene rubber (polybutadiene): low hysteresis, good low-temperature flexibility and abrasion resistance, poor tear resistance; tyres (in blends with SBR and natural rubber).
8. Thermoplastic rubber: similar properties to SBR except that upper temperature for use is around 50°C, due to softening and does not need filler reinforcement; shoe soles.

3.4 THE NEED FOR COMPOUNDS

In general, the usefulness of pure polymers is extremely limited. Many pure polymers cannot be processed into articles having satisfactory properties. The following notes discuss three principal reasons for incorporating additives: processing additives, additives to ensure general survival, and additives to confer useful mechanical and load-bearing properties. The aim is to identify classes of additives and their effects. The principles and possibilities are important here; detailed experience lies in the province of compounding technologists, from whom advice should be sought where necessary. Usable compounds are based on a selection of additives – not all of them!

There are close parallels with the technologies of other materials. Concrete is a mixture based mainly on cement, sand or aggregate, and water, and properties depend on the type and proportions of each and on proper mixing of them. There is an enormous range of steels, based on iron to which carbon and other elements are added to confer the required balance of strength, hardness and other properties; the properties depend on microstructure, which is also affected by fabrication and heat treatment.

3.4.1 Processing into a form-stable shape

1. Heat stabilizers. Prevent thermal degradation at high temperatures during shaping, i.e. prevent molecular chains being broken into shorter lengths.
2. Anti-oxidants. Prevent rapid oxidative degradation at high temperatures during melt processing, or slow oxidative degradation in service.
3. Crosslinking agents in rubber. Sulphur crosslinks promote elasticity; crosslinking rubber with sulphur atoms takes hours, even at high temperature, so add an **activator** to start the reaction, and an **accelerator** to speed up the reaction once it has started. Processors seek a 'safe time' during which they can fill the mould and shape the polymer before it begins to crosslink, hence the use of a delayed-action accelerator. But once crosslinking starts, fast production rates call for a fast crosslinking reaction, so the mix may contain an ultra-fast accelerator.
4. Crosslinking of plastics. For example, polyester plus styrene; the styrene crosslinks may take months to form at 20°C. A **catalyst** speeds up the reaction and an **initiator** is used to start the reaction.
5. Lubricants. Used mainly in PVC to stop molten polymer sticking to metal surfaces.
6. Nucleating agents. Prevent supercooling in partially crystallizable thermoplastics, and give a uniform microstructure with small spherulites.

3.4.2 General survival properties

1. UV absorbers. PMMA, PVC and unsaturated polyester resist UV, but most other polymers degrade quickly in sunlight unless protected by UV absorbers. Most efficient are carbon black or titanium dioxide, but other UV absorbers can be used if coloured products are required.
2. Anti-ozonants. Protect rubber against degradation and thus prevent cracks growing under stress.
3. Burning. As mentioned above, all carbon-based polymers will burn, some more readily than others. Ease of flame initiation is delayed by certain additives but these often give noxious fumes when the polymer actually burns.
4. Inert fillers. Used for cheapness in non-critical products, e.g. chalk or clay in 'PVC' hosepipe and in 'natural rubber' hot water bottles, to make a harder, stiffer and heavier product. Oil used as an extender in SBR for industrial tyres.
5. Colour. Dyes in transparent polymers (car rear lamp housing) and pigments in all polymers. Surface metallizing of plastics is possible.

3.4.3 Mechanical and load-bearing properties

1. Flexibility. Some polymers are inherently flexible and extensible. Other stiffer polymers can be made more flexible as follows:

(a) Add **plasticizer** (especially used in PVC) – the liquid plasticizer separates adjacent PVC molecules, thus permitting slip. Unplasticized PVC, $E \simeq 3\,\mathrm{GN/m^2}$; plasticized PVC, E as low as $3\,\mathrm{MN/m^2}$ (depends on amount used). Nylon is plasticized somewhat by water which reduces E, yield stress and dimensional stability but confers greater toughness.
(b) Incorporate **foaming agents** to reduce modulus and strength by making a cellular structure.

2. Hardness and resilience. Mainly a preoccupation in rubber compounds, **reinforcing** fillers such as carbon black particles in natural rubber form part-physical, part-chemical links which rubber technologists call 'structure'. Carbon black therefore increases strength and stiffness of the rubber compound. The use of lossy fillers can reduce resilience.
3. Stiffness. Incorporate stiff strong fillers to improve stiffness of plastics. Glass fibres are most widely used because they are cheap, having $E \simeq 70\,\mathrm{GN/m^2}$ and tensile strengths of $\simeq 1500\,\mathrm{MN/m^2}$. Increasing the glass content gives a stiffer stronger material. The upper limit to fibre volume fraction V_f is normally dictated by the processing method:
(a) continuous fibres or filament winding: 70% V_f common;
(b) hand-laid-up woven cloth laminates: 40–50% V_f;
(c) short-fibre composites: 10–20% V_f.
The major problem is that fibres emphasize directionality. There are several ways round the problem:
(a) reduce directionality by randomly orienting the fibres (makes inefficient use of the glass but makes stressing calculations easier);
(b) align fibres to directions of principal stresses (efficient use of fibres but awkward to make, and need special involved stressing techniques).
Note: tyre cord and steel wires are used to reinforce rubber in the special case of vehicle tyres having a 'cross-ply' or 'radial-ply' laminate construction.
4. Toughness. Incorporation of rubber particles can increase the toughness of an otherwise brittle polymer; for example, at 20°C, for general-purpose polystyrene, the fracture toughness $K_{Ic} \simeq 1\,\mathrm{MN/m^{3/2}}$, but for a typical rubber-toughened polystyrene $K_{Ic} \simeq 1.7\,\mathrm{MN/m^{3/2}}$ (effect depends on amount and type of rubber present). The rubber particles act as craze stoppers and deform readily, thus absorbing energy. Note the compromise – the toughened polymer is more flexible and is translucent. It is also possible to improve toughness by incorporating tough fibres with the polymer, e.g. cotton fibres in phenolic resin ('Tufnol').
5. Stiffness per unit weight. Plastics can have high modulus/density ratios, and incorporation of glass fibres can increase this ratio by an order of magnitude or so. In bending situations, the good designer exploits second moment of area effects – this is why plastics products often bristle with

ribs. Foams exploit second moment of area effects; foamed products have higher bending stiffness per unit weight (but are more chunky and burn more readily).

6. Combinations of fillers. For example, a floor tile in a 'vinyl' composition – up to 400 phr chalk or clay or whiting for cheapness, hardness and wear resistance; plasticizer to prevent cracking and embrittlement; copolymer (vinyl chloride–vinyl acetate) for greater flexibility compared with PVC.
7. Adventitious or accidental additives. Seldom desirable and therefore best avoided. The usual cause is poor housekeeping during processing, or poor choice of processing conditions or equipment. Examples include blobs of foreign polymer, blobs of degraded polymer, rust or metal particles, brick dust, human hair, cotton and other fibres. The effect may be visual displeasure, or stress concentrations which encourage premature failure.

3.4.4 Examples of compounds

(*a*) *Rubber*

Consider a rubber band. Natural rubber polymer is tacky, fragile, not very form-stable and certainly not elastic – much of the deformation on stressing is permanent, because molecules slip past each other. Elasticity and non-tackiness are conferred by sulphur crosslinking, typically about 2.5 phr (parts of ingredient by weight per hundred parts of rubber or resin). The vulcanization (crosslinking) takes many hours, even at say 140°C. The accelerator could be CBS (0.6 phr), with an initiator and activator consisting of 5 phr zinc oxide and 2 phr stearic acid. Such a mix would take about 40 min to cure at 140°C or 20 min at 150°C; the vulcanizate would have a modulus of about 2 MN/m^2 with an elongation-to-break of some 700%.

Rubber is attacked by atmospheric oxygen which causes degradation on the exposed surface: these faults aggregate to form minute cracks (the surface is 'perished'). Under opening forces the cracks would grow, leading to brittle fracture. For technical applications, the rubber band mix would also contain an anti-oxidant, say about 1 phr.

To make a pencil eraser, a mild abrasive powder such as talc would be added: this makes the eraser stiffer, harder, more leathery and less extensible.

For an antivibration mounting pad or shock absorber, what is usually needed is a much stiffer compound which can be shaped more readily and cured more quickly. To achieve a higher modulus, the rubber is reinforced with perhaps 35 phr of carbon black; 2 phr of a processing aid would be typical; accelerator, initiator system, sulphur and anti-oxidant would still be necessary. Such a mixture might take 20 min to cure at 140°C, giving a modulus of 4.5 MN/m^2 and a maximum elongation of about 500%.

The old-fashioned ebonite car battery boxes derived their stiffness from a very high crosslink density, with 35–40 phr of sulphur, and would be 'rigid' up to 80°C; with such a high T_g, it is obviously untypical of normal rubber compounds.

If extra flexibility is required of a rubber compound, it is quite feasible to make a cellular structure by incorporating a gas, leading to moduli down to about 0.01 MN/m^2.

It is not possible here to go into the detailed formulation of the different rubber polymers, and their compounds, used to make up a car tyre. The following is not a recipe, but it is interesting to note that overall a typical tyre consists of (as percentages by weight):

Rubber hydrocarbon	42 wt%
Carbon black	28 wt%
Steel wire	14 wt%
Oil	9 wt%
'Rubber chemicals'	5 wt%
Textile fibres	2 wt%

(*b*) *Plastics*

PVC is widely used for pipes for cold water services. The pure PVC is all too ready to degrade in the melt state, while it is being extruded; to prevent this, 6–8 phr of heat stabilizer is added. To facilitate processing, 2 phr of a lubricant is often incorporated. To improve the robustness of the pipe, up to 10 phr of an impact modifier may also be present in the pipe compound.

If, for land drainage purposes, a PVC pipe was stiff enough but too heavy, it would be possible to resist buckling collapse by extruding a foam sandwich pipe with stiff, strong skins and a foamed PVC core. This would require the use of a blowing agent to provide the gas in the foam, reducing the core density by some 50% compared with that of the solid polymer.

PVC is also used in a flexible form to convey cool liquids under modest pressures. The tube is often transparent (PVC is amorphous). The flexibility is achieved by a liquid plasticizer, which reduces the strength. Reinforcement is provided by a nylon-fibre braid: an inner wall of plasticized PVC is extruded, the braid is wrapped around this, and then the braided tube is covered by an outer layer of plasticized PVC extruded onto it.

An unsaturated polyester system might be selected for making a pipe or a tank. The recipe for the final plastic could be 100 phr unsaturated polyester, 35 phr styrene, 1 phr initiator and 1 or 2 phr catalyst; this system would be a mobile liquid. The cured resin would be too weak and flexible for such a purpose, but this can be put right by incorporating glass fibres. By filament winding, a volume fraction of 70% can be achieved, leading to a modulus of some 50 GN/m^2 in a layer of the composite along the glass direction – not a simple design value, as will be discussed later.

A lorry cab could be moulded from a fibre-reinforced plastic, typically using a sheet moulding compound. Here, the aim is to have a cheaper mix which can be easily handled. The raw sheet contains short fibres randomly oriented in the plane and some 50 mm long, with a weight fraction of 25% (giving a final modulus of about 5 MN/m^2). The resin system would be cheapened and thickened to a doughy consistency (to permit handling) by mixing in about 35% of chalk powder. The sheet would be shaped by pressure and cured by heat.

In both the above examples, the fibres confer extra stiffness and strength, and reduce thermal expansion. The fibre–resin composite is tougher than the resin alone because of the enhanced opportunities for energy absorption.

PROBLEMS

1. Why use 'plastics in general' to make the following: (a) gear wheels; (b) telephone handsets; (c) rear lamp housings for cars; (d) car distributor point assemblies?
2. Why use a vulcanized rubber compound to make the following: (a) O-ring seals; (b) sealing strips for car windscreens; (c) dock fenders; (d) part of a railway rolling-stock buffer system?
3. A plastics journal bearing is to be run unlubricated under load at high speed. What would be the main design problem?
4. A concentric pipe of 50 mm outside diameter with a 2.5 mm wall is rigidly clamped on 1.7 m centres. What temperature rise would just start to trigger distortion if the coefficient of linear thermal expansion were taken as 10^{-4}/K?

FURTHER READING

Plastics and rubber materials

Bashford, D.P. (1996) *Thermoplastics Directory and Databook*, Chapman & Hall, London.
Birley, A.W., Heath, R.J. and Scott, M.J. (1988) *Plastics Materials Properties and Applications*, Blackie/Chapman & Hall, London.
Brydson, J.A. (1995) *Plastics Materials*, Butterworth-Heinemann, Oxford.
Dyson, R.W. (1997) *Specialty Polymers*, Blackie, London.
Freakley, P.K. and Payne, A.R. (1978) *Theory and Practice of Engineering With Rubber*, Applied Science, London, ch. 1.
Harper, C.A. (ed.) (1975) *Handbook of Plastics and Elastomers*, McGraw-Hill, New York.
Hepburn, C. (1992) *Polyurethane Elastomers*, Applied Science, London.
Hilyard, N.C. (ed.) (1982) *Mechanics of Cellular Plastics*, Applied Science, London.
Monk, J.F. (ed.) (1981) *Thermosetting Plastics*, Godwin, Harlow.
Oswin, C.R. (1975) *Plastics Films and Packaging*, Applied Science, London.
Potter, W.G. (1975) *Uses of Epoxy Resins*, Butterworths, London.

Powell, P.C. (1977) *Selection and Use of Thermoplastics*, Oxford University Press, London.
Saechtling, H. (1995) *International Plastics Handbook*, Hanser, Munich.
Young, R.J. and Lovell, P.A. (1991) *Introduction to Polymers*, Chapman & Hall, London.

Compounds

Ahmed, M. (1979) *Colouring of Plastics: Theory and Practice*, Van Nostrand Reinhold, New York.
Blow, C.M. and Hepburn, C. (eds) (1982) *Rubber Technology and Manufacture*, Butterworths, London.
Eirich, F.R. (1978) *Science and Technology of Rubber*, Academic Press, London.
Katz, H.S. and Milewski, J.V. (eds) (1978) *Handbook of Fillers and Reinforcements for Plastics*, Van Nostrand Reinhold, New York.
Mascia, L. (1974) *The Role of Additives in Plastics*, Edward Arnold, London.
Pritchard, G. (ed.) (1997) *Plastics Additives*, Chapman & Hall, London.
Shakespeare, W. (1623) *Macbeth*, Act 4, Scene 1.
Wake, W.C. (1971) *Fillers for Plastics*, Iliffe, London.

4 Important Polymer Processing Methods

A key attraction of polymer compounds is that they can be readily converted into useful shapes having desirable attributes. Polymer processing embraces the technology of making articles from polymer compounds.

It is possible to buy stock shapes of some plastics in the form of block, sheet, rod or profile for subsequent machining and joining to achieve a product with the desired configurations. However, almost universally the philosophy is to put the plastic or rubber where it is needed, rather than remove after processing what is not needed by cutting away. This calls for polymer processes notionally equivalent in concept (but not necessarily in scale) to die casting, rolling and extrusion of metals, blowing of glass containers and casting of concrete, with the added advantage in mass production that plastic and rubber parts are made in one operation and that substantial post-forming operations are seldom required.

Three common concepts underlie most of the common methods. The first is to make from the raw polymer and other ingredients the required compound in a physical state (usually a fluid) which can be subsequently shaped. The second is to deform the compound into the required shape and size of the product, and the third is to ensure that the product holds its shape and dimensions after it has been shaped.

This chapter outlines the salient features of the most important processes and identifies which of the above themes or subthemes are the controlling factors. When the reader has reached the end of this chapter, he or she should be able to identify the type of process used to make many commonplace articles from plastic or rubber, and should also be aware of some of the implications of choosing a particular process in terms of economics of production and design opportunities.

4.1 THE MAIN CONCEPTS

The first main concept is mixing and making fluid. This involves dispersion of stabilizers, pigments and other additives into the polymer, exclusion of air and volatiles, transformation of solid polymer feedstock (if powder or granules) into a melt, and achieving a fluid state which is uniform in composition, temperature and deformation history. This implies considerable movement of the mix, and 'uniformity of deformation history' is a shorthand way of saying that all elemental volumes of the final compound should have suffered to the same extent.

The second concept is shaping and implies more movement of the mixed fluid composition. The product may be defined in three dimensions by (batch) flow within a split mould, or by being pressed onto the surface of an open mould. Alternatively it may be defined in two dimensions by (continuous) flow through an extrusion die to confer a primary shape such as a tube or rod; there may then follow a secondary shaping operation by slitting or by applying tension to the still-fluid polymer mass.

The third concept is stabilizing the required shape. The two main ways of doing this are to cool a thermoplastic melt until it solidifies, or to crosslink a thermosetting polymer (either hot or cold, depending on the nature of the compound).

The fourth concept is to realize that a shaped product is not a shape filled with a material with continuous and idealized properties, but a product subject to shrinkage and other after-effects introduced by each processing method. Chapter 9 deals with some of these effects. Moreover, materials are seldom fully homogeneous or free from internal defects such as (micro-) cracks.

A small proportion of polymer compounds are fairly mobile liquids at room temperature and pressure. Examples of these include rubber latex, some casting resins such as epoxides, the unsaturated polyesters for making glass fibre-reinforced plastics structures, and some polyurethane systems. These compounds are easy to shape at low pressures and the amount of energy involved is modest.

In contrast, most polymers are solids or extremely viscous liquids at ambient temperature. The majority of polymer products are made by heating the polymer until it is a (still) viscous melt and shaping that melt to achieve the required result. Most conventional polymer processing is therefore fairly energy intensive; energy is needed to heat the polymer, to move the melt and then to stabilize the final shape of the product. The high viscosities, typically 10^2–10^5 N s/m^2 at processing temperatures, call for high pressures, which require rugged and robust, capital-intensive equipment. The low thermal diffusivities suggest that heating and cooling of thick sections based on conduction alone is likely to be slow as well as expensive.

The solid polymer is melted by conduction from a hot surface, often augmented by heat generation from mechanical work by deliberately shearing the polymer. There are two methods of causing the polymer to move. The first is to apply pressure: if applied between fixed boundaries, and assuming no slip at the boundary, shear flow occurs in channels of constant section, and the shear is accompanied by extensional deformation in channels where the cross-section changes along the flow direction(s). The second method is to use the high viscosity to effect drag flow between two moving surfaces; this also develops a pressure gradient, of course. Polymer molecules are long and thin, and it is possible for some of the deformations occurring during shaping movements to be 'locked in', leading to products showing anisotropy of properties; sometimes this is welcome, sometimes not.

The choice of processing temperature is usually a compromise between many factors, and depends on whether the polymer is thermoplastic or crosslinkable. For thermoplastics a low melt temperature implies rapid production rates, high pressures and low thermal energy usage at the expense of some loss of robustness. High temperatures reduce the viscosity and thus reduce the pressure needed for a given flow rate, although the effectiveness of drag flows at constant speed is impaired; on the other hand the longer cooling times reduce the production rate in batch processes. In crosslinkable polymers a high temperature reduces the time needed to achieve a satisfactory crosslinked state, but there may be an increased risk of premature crosslinking during movement or shaping. For either type of material the risk of degradation or other changes makes it undesirable for the polymer to spend long periods at high temperature. Common ways of reducing these risks include ensuring that the compound can flow smoothly through the equipment (e.g. avoiding deadspots such as sharp internal corners in the flow channels where polymer can stagnate) and adding suitable stabilizers which make the compound more risk tolerant (and usually more expensive).

Mixing is essential for nearly all polymer processing, since nearly all plastics products have to contain additives, either to ensure proper processing or optimal product properties or both. It is, however, difficult to incorporate additives in high-viscosity melts and fluids. Various systems have been developed, based on shear flow and/or (more effective) extensional flow. The incorporation can be performed in a separate process (often performed in batch) or as a part of the melt making and processing operation (e.g. in single- or double-screw extruders). It has to perform dispersion of the additives and their homogeneous distribution in the melt (section 8.5).

4.2 BATCH MIXING PROCESSES

Compounds based on polymers are often thoroughly mixed in a batch operation which is separate from the process used to shape them. The two main processes for fine-scale dispersion, used particularly in the rubber industry, are the internal mixer and the two-roll mill, in both of which the compound is an extremely viscous fluid. Where the polymer is available as a free-flowing powder or as granules, other ingredients can be distributed on a coarse scale by tumble blending: this is satisfactory if later processing happens to involve a further melt mixing stage. Liquid polymers which have a low viscosity can be mixed with other ingredients by simple stirring; care may need to be taken to minimize the inclusion of air.

4.2.1 Internal mixer

The internal mixer consists of a robust casing defining two connecting cylindrical cavities side-by-side, within which two rotors counter-rotate (Fig. 4.1). There is only a small clearance between the tips of the rotors and the casing. Ingredients are fed through an aperture in the top of the chamber: this aperture can be closed to apply pressure to the contents of the full chamber to ensure contact between polymer and mixer. The fully mixed compound is often removed through a port in the underside of the casing.

The rotating blades moving past the stationary casing induce drag flow of the polymer and other ingredients: there is a high degree of shear in the clearance between the blade tips and the casing, and much less shear elsewhere. Rotor blades can have different shapes according to the mixing action to be performed and the recommendations of the designer. Some pairs of rotors are represented in Fig. 4.1(b). The mixture is therefore usually pumped along the axis of the rotors as well as becoming circumferentially and radially mixed by the interaction of the blades.

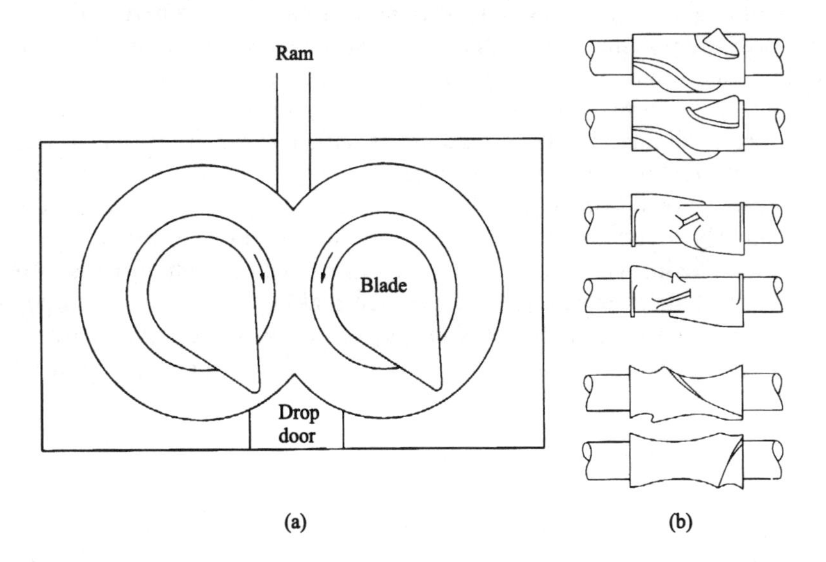

Fig. 4.1 Internal mixer: (a) mixing chamber cross-section, (b) some typical rotor shapes (courtesy Rapra Technology Limited)

Because of the high viscosity of the polymer, the power needed to drive the blades is considerable: this power is dissipated as heat, causing the polymer to melt. Excessive and prolonged mixing could cause premature crosslinking in crosslinkable polymers such as rubber, so in practice the actual vulcanizing ingredients are often not added at this stage, even if the casing is water cooled.

The range of capacities of internal mixers spans 1 to 500 kg of polymer compound.

4.2.2 Two-roll mill

The two-roll mill consists of two robust, horizontal, cylindrical polished rolls fitted with water-cooling channels (Fig. 4.2). The rolls counter-rotate and are separated by a narrow gap called the 'nip'.

The feed, either raw rubber or the part-mix taken directly from the internal mixer, is dropped in between the two rolls. The sticky mass is dragged down from the rolling reservoir or 'bank' into the nip and adheres as a band onto one of the rolls. There is therefore a high but laminar shear flow through the nip: this flow develops a substantial pressure profile which causes the bank well upstream of the nip to roll, thus promoting radial mixing; the pressure may also cause the rolls to bend. There is no mechanism for axial mixing, so the band must be cut on a helix at an angle to the roll axis and it is simply folded over – this seems crude but is often acceptably effective.

Because of the high viscosity, the power input to drive the rolls is high, but the water-cooled rolls have a large surface area in contact with a thin band of polymer, so this prevents excessive temperature rise. It is therefore safe enough to add the vulcanizing ingredients to the mix.

The range of capacities of two-roll mills spans 1 to 500 kg of polymer compound. The homogeneous mix from the mill may then be shaped, usually by one of the following processes: calendering, extrusion, and compression or injection moulding.

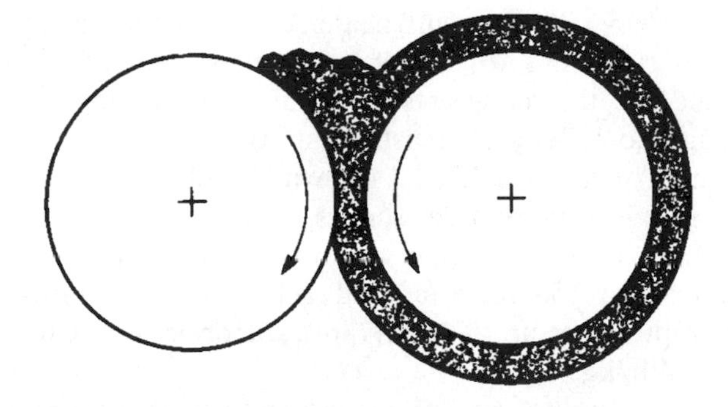

Fig. 4.2 The principles of a two-roll mill

4.3 EXTRUSION PROCESSES

Extrusion processes are used to make continuous products of constant cross-section, usually from thermoplastics or rubber compounds but seldom from crosslinkable plastics. There are two basic types of equipment: the screw extruder and the calender (closely related to the two-roll mill).

4.3.1 Screw extrusion

The purpose of the extrusion line is to make a continuous uniform product of constant cross-sectional shape along its length. Typical extruded products include pipe (1–1500 mm diameter), and profiles such as curtain rail, film, sheet, fibre, tape, rod, wire coverings and cable sheathing. The wall thickness can range from a few micrometres to 75 mm depending on the product.

The essential parts of the extrusion line are the extruder, the die, the haul-off arrangements and provision for stabilizing the required shape of the extrudate. Sometimes the extrudate undergoes a secondary shaping stage before stabilization: examples include extrusion blow moulding and tubular film blowing (both described below) and tape and string manufacture.

(*a*) *Extruder*

The extruder (Fig. 4.3) consists of one or sometimes two screws which rotate inside a close-fitting barrel of constant diameter. The screw is held at one end in a robust axial thrust bearing and is connected to the drive motor by a reduction gearbox. In a single-screw extruder, the helical channel is typically of the single-start type and of constant width; its depth is larger but uniform at the feed end (the feed section), tapers in the compression section, and is again uniform but shallower in the metering section. The screw is sometimes cored for circulation of coolant. The barrel, usually horizontal, can be externally heated by collar or cuff heaters. Part of the barrel may also be cooled. The extruder has to deliver a steady stream of uniform molten polymer to the die; the screw acts as a pump developing drag flow, contributes strongly to melting or 'plasticating' the feed, and its action usually ensures adequate mixing too. This is known as a plasticating extruder and is widely used. Plasticating extruder screws range from 25 to 150 mm in diameter, with screw lengths typically of 25 to 30 diameters, rotate at speeds ranging from 50 to 150 rev/min, and deliver melt at rates from 10 to 1000 kg/h with pressures up to 40 MN/m^2; the drive motor would consume some 0.1–0.2 kWh/kg.

Raw materials suppliers sometimes use large melt-fed extruders, i.e. the feed is melted before it reaches the extruder, which then acts as a mixing

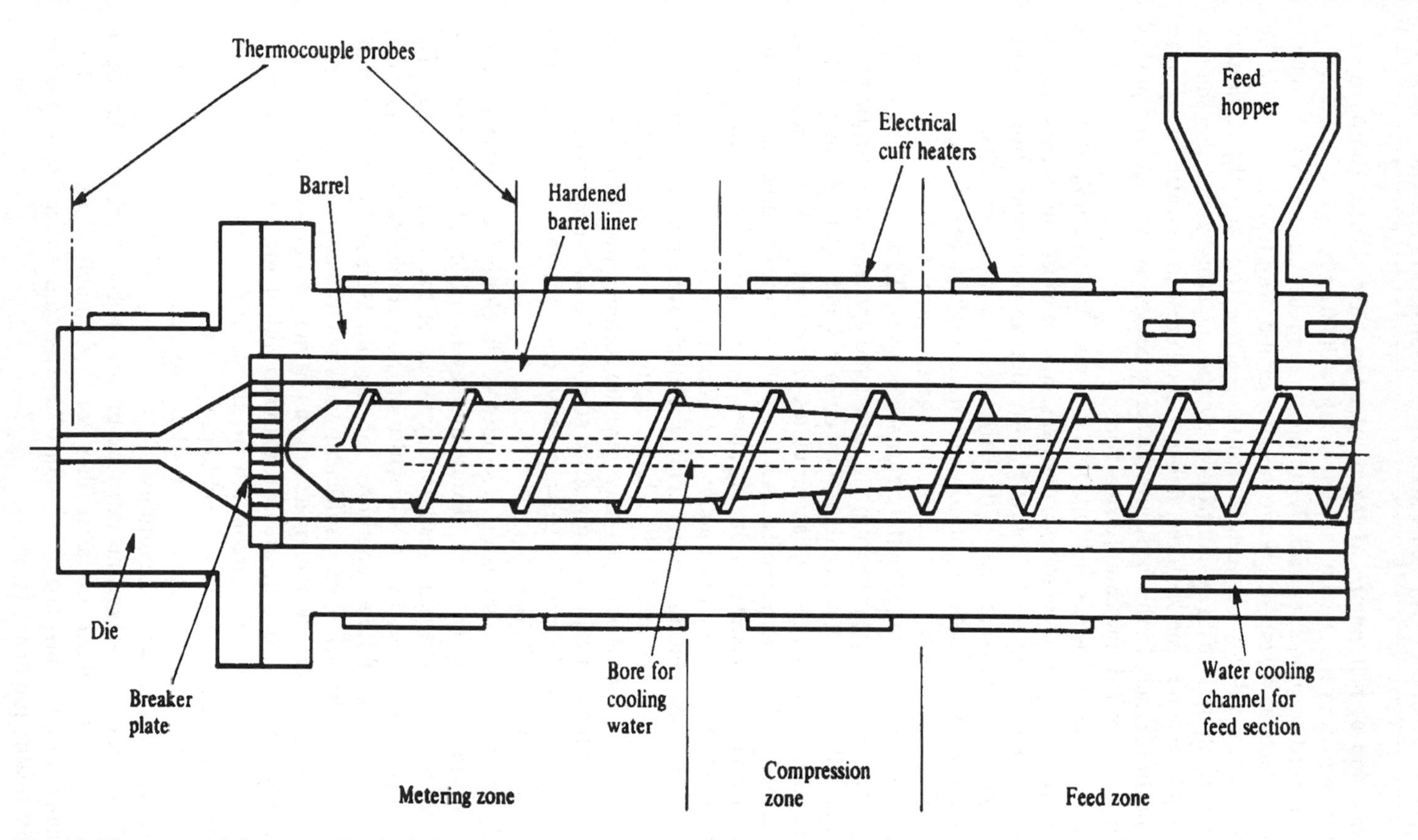

Fig. 4.3 The essential features of a single-screw extruder

device and a melt pump. Melt-fed extruders have shorter screws up to 300 mm in diameter, which can deliver up to 5000 kg/h.

It is worth commenting briefly on the three main functions of the screw. First, it acts as a viscosity pump because of the drag flow induced by the difference in velocity between the rotating screw root and the stationary barrel. Output is usually controlled by the metering zone: the shallow channel generates the required delivery pressures and ensures a uniform supply of the melt. In the feed zone the screw compacts the solid and pushes it forward towards the compression zone.

The second function is melting. This is a much more complicated process than conveying a single-phase material, and it starts in earnest in the compression section of the screw by two mechanisms: conduction of heat from hot metal surfaces such as the barrel and the screw, and dissipation of heat caused by the shearing of polymer in the flight of the screw. A film of melt forms on the hot surface of the barrel which is above the polymer melting point. Ignoring here the clearance between flight tip and barrel, the flight scrapes away the melt from the barrel, and under pressure the first melt formed fills the interstices in the solid bed of polymer in the channel. When these interstices are full, melt accumulates in the melt pool upstream of the solid bed and circulates there. As the polymer moves downstream, the proportion of the channel width occupied by the melt increases and that of the solid bed decreases. Eventually – and often some way down the metering zone – the solid bed disappears. This is a much oversimplified account, of course. The drag induced by the relative motion between metal surfaces (and also the solid bed of polymer) may generate so much heat as to dictate the cooling of both barrel and screw after start-up, to avoid an excessive temperature rise which might otherwise cause degradation, premature cross-linking or just too fluid a melt.

The third function is mixing. The idea of developing the screw channel and assuming drag flow only along the developed length will give some mixing in the downstream direction by virtue of the velocity profile. The component of velocity across the width of the channel leads to a circulating component of flow superposed on the downstream flow. For many mixing purposes this is still insufficient and stream lines of additives can be seen in the product. The mixing action can be much improved by increasing the resistance to the output stream (combined with an increased screw speed) in order to increase the circulation (at the cost of a much higher power consumption and an increased melt temperature).

A variety of screw designs is available to meet the needs of different materials and operating conditions.

Before the melt leaves the extruder proper, it usually passes through a 'breaker plate' which acts as a filter for any unmolten polymer and for accidental contaminants which might otherwise damage the die or produce poor-quality product.

(*b*) *Mixing and mixing elements*

The material passing the screw tip is expected to be uniform (well distributed) in composition and temperature. In addition, agglomerates of additives are expected to have been broken down (dispersed) to an appropriately small size. As stated above, this will not happen automatically in a standard screw. A high shear, as required for dispersion, may appear in the clearance between the flight tip and the barrel, but only a limited fraction of the melt will pass this clearance.

Mixing elements can be added to the screw with the accent on dispersive or distributive mixing action. For dispersive action, usually one or more forced passages of all of the melt between the barrel and a ridge on the element are used; the high shear in this clearance will break the agglomerates of the suspension or the droplets of the emulsion. For distributive action an appropriate element could prescribe a forced pattern of splitting up, circulation and recombination of flow, thus imitating turbulence, which is otherwise not possible because of the high viscosity of the melt. A well-designed distributive mixing element will consume a relatively small amount of energy, because shear forces can be kept low. It will therefore contribute not only to homogenizing the temperature but also to preventing the melt from heating up any further.

(*c*) *Extrusion dies*

To acquire the desired cross-sectional shape, the melt passes through an externally heated die which is firmly connected to the delivery end of the extruder barrel. The flow channel in the die will change its cross-section smoothly from that of the end of the extruder barrel to that of the shape of the product. The polymer swells on leaving the die, so allowance must be made for this when deciding the size of the flow channel and its exit dimensions.

The simplest die is in line with the barrel and consists of a conical taper leading to a relatively short land of constant diameter, and is used for making thin rod. In-line dies are also used to make tape of rectangular cross-section, and profiles (in which the changes of cross-section can be difficult to design). In dies for making sheet up to 10 mm thick and a metre or so wide, the melt passes to a manifold which distributes it to the die land and exit so that after swelling the polymer has a uniform velocity over all its cross-section. Within in-line tube dies the melt flows between two metal surfaces defining an annular gap. It is necessary to be able to hold the core of the die in place and adjust its position to ensure a concentric flow channel. This means of location interrupts the flow of polymer, and care is needed in design and operation to ensure that the polymer welds together downstream, in order to avoid lines of unacceptable weakness in the product.

To facilitate subsequent operations it is sometimes necessary to change the direction of melt flow by using a crosshead die. Thin-walled tubes are extruded vertically upwards with elaborate precautions taken to eliminate weld-lines, and then inflated to make tubular film. Thick-walled tubes are extruded vertically downwards for blow moulding. Wire is covered by pulling it axially along the centreline of the slightly larger exit of a crosshead die to which molten polymer is fed under pressure. Two or more previously covered wires can be sheathed by a subsequent extrusion-covering operation, taking care not to melt the original coverings.

On emerging from the die exit, the extruded polymer inherently tries to swell in cross-sectional dimensions (and contract in the direction of extrusion to conserve constant volume of material). Post-extrusion die swell is not well understood as yet: it relates to the elastic properties of the polymer melt, depends on the velocity distributions within the die, and provides a real but unsought challenge to die designers.

The commercial objective of extrusion is to produce a defined cross-section of acceptable quality at an acceptable rate. The extrudate must be taken away from the die. Profile, rod, tube and sheet are hauled off under a slight tension, often applied using coated rollers or belts. At the same time it is necessary to ensure that the shape and dimensions of the product are as required and then stabilized; a variety of operations (including combinations) is possible.

In pipe extrusion, for example, the soft tube is inflated under slight internal pressure – using a plug attached by a light chain to the die core through which air is supplied – and fed through a separate close-fitting jacket called a calibration die. Small variations in wall thickness at a given output rate can be achieved by small adjustment of haul-off tension. In profile extrusion for sealing strips or window frames, the separation between adjacent webs can be maintained by inserting jig templates in the extrusion line.

In the tubular-film process the thin-walled tube of molten polymer is inflated by air carefully fed through the core of the die at a very closely controlled pressure; downstream the cold tube is sealed by two nip rolls which counter-rotate to draw the molten tube under axial tension. The result is a film which is biaxially drawn to a considerable extent (with extension ratios of up to three in the hoop and axial directions): the orientation is frozen in and strengthens the product considerably. Tubular film is widely used for making carrier bags and refuse bags, and thicker gauges when slit can be used for building and horticultural purposes. Partially crystallizable polymers such as polyethylene and polypropylene are most commonly used. Lay-flat widths of tubular film range from about 100 mm to many metres.

4.3.2 Calendering

The calender (Fig. 4.4) consists of four horizontal robust counter-rotating steel rolls (sometimes called 'bowls'). The top two bowls look just like a two-

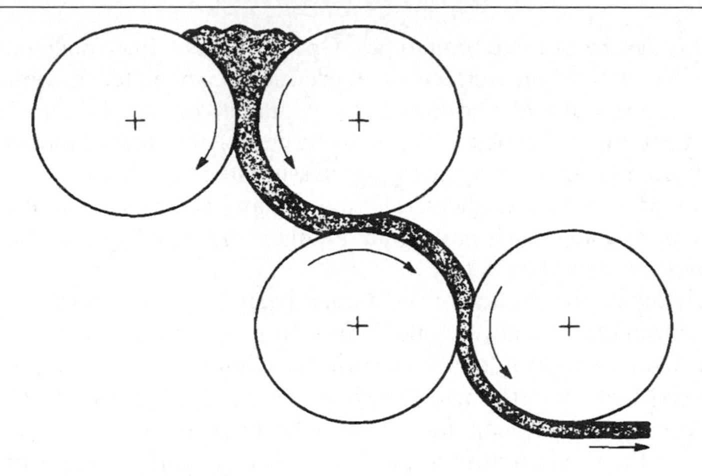

Fig. 4.4 Typical arrangement of a four-roll calender

roll mill and are charged with a uniform polymer compound. The compound is drawn through the first nip and, adhering to one bowl, is transferred to the next nip where it is reduced in thickness again. The thickness reduction in the third nip provides fine control of the sheet and confers the required surface finish, either plain or embossed.

Sheet up to a few millimetres thick and a metre or so wide can be made this way. It is also possible to use the calender to make a sandwich of substrate and polymer, such as the fibre-reinforced plies for subsequent manufacturing into car tyres. The thermoplastic foil or sheet has to be cooled after calendering. Rubber sheet must be crosslinked in a heating chamber if it is required as a finished product; more frequently the rubber sheet is left in the uncured or 'green' state because it will later be fabricated into such products as hose or tyres before being crosslinked.

The feed has to be premixed and preheated. The early operations provide some shaping but essentially deliver the melt to the final nip, which is essentially a moving die. The rolls and their bearings must be sufficiently robust to withstand, without unacceptable distortion, the high pressures arising in the melt from the drag flow of a high-viscosity melt.

4.4 MOULDING PROCESSES

Moulding is a batch process, that is, products are made one at a time (or a few at a time) on a cycle which can be repeated as often as necessary. The essential steps in the moulding process are delivery of a shapable compound to a mould, so that a three-dimensional product can be made, and stabilizing its shape either by cooling or by curing. Prior to shaping the compound may be an unshaped liquid or melt, a sheet or a tube.

Moulds can be of three basic types. Open moulds (either male or female) are used to define one surface of a product. Two matched female half-moulds are used to make hollow items. A matched male and female pair is used to form an impression which is to be completely filled. Shaping in the mould is usually achieved by applying pressure to cause the fluid to conform to one or both mould surfaces. There are many combinations of delivery, mould and shaping, each having advantages and disadvantages. Only the main ones are described here.

The choice of process and of the mould depends greatly on the number of pieces to be made, bearing in mind the lowest possible price per product, and on the tolerances to be attained. In turn, the choice of the process will affect the choice of the material and the shape of the product, which should be producible and functional. In Fig. 4.5 the price of a tray is given as a function of the total number of products, as made with different processes. The price consists of costing elements such as raw material, mould, machine and personnel. As each of these changes, the positions of the curves will change, including their crossover points.

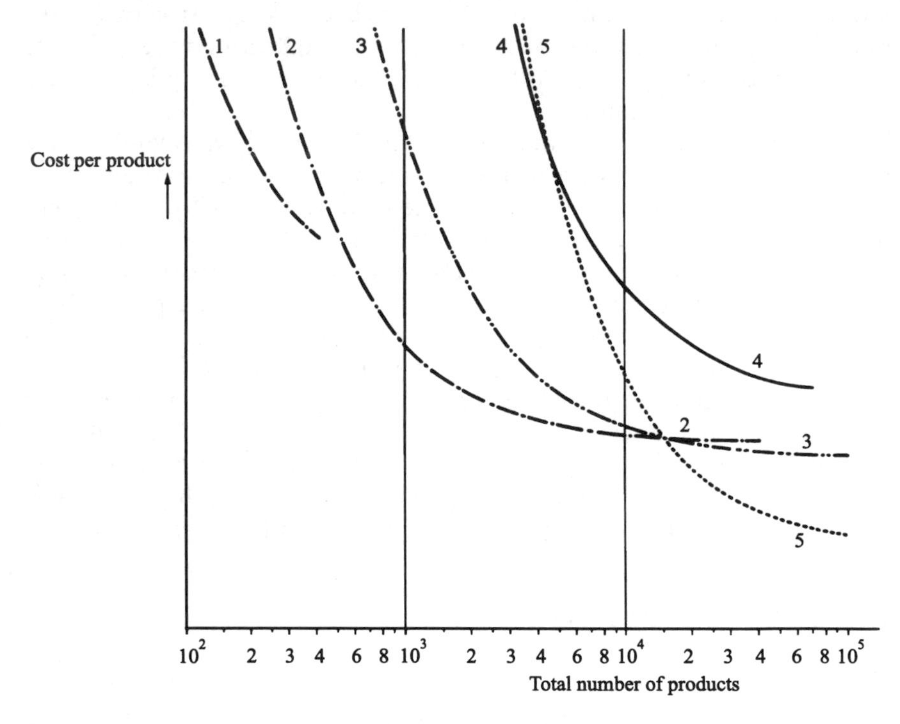

Fig. 4.5 Production costs per product of a tray made using different processes: 1, thermoforming with wooden mould; 2, thermoforming with double aluminium mould; 3, extrusion blow moulding with single mould; 4, hot mould compression moulding with sheet moulding compound (SMC); 5, injection moulding (Courtesy F. Irrgang, Breda)

4.4.1 Injection moulding

The essence of injection moulding is the injection of polymer melt into an impression within a closed split mould which completely defines product dimensions (Fig. 4.6). Thermoplastics, crosslinkable plastics and rubber compounds can all be injection moulded. The main parts of the hardware are the mould, the injection unit, and the means of opening and closing the mould and keeping it closed.

Many aspects of the construction of simple compression and injection moulds are similar in principle. The main differences are that in injection moulds the parting line is vertical (i.e. with horizontal movement of the moving half), there is no deliberate clearance for flash because it is not needed, there must be a runner to permit the passage of melt from the injection unit to the impression, and water or oil cooling channels are needed (for thermoplastics). For efficient control of mould temperatures the coolant

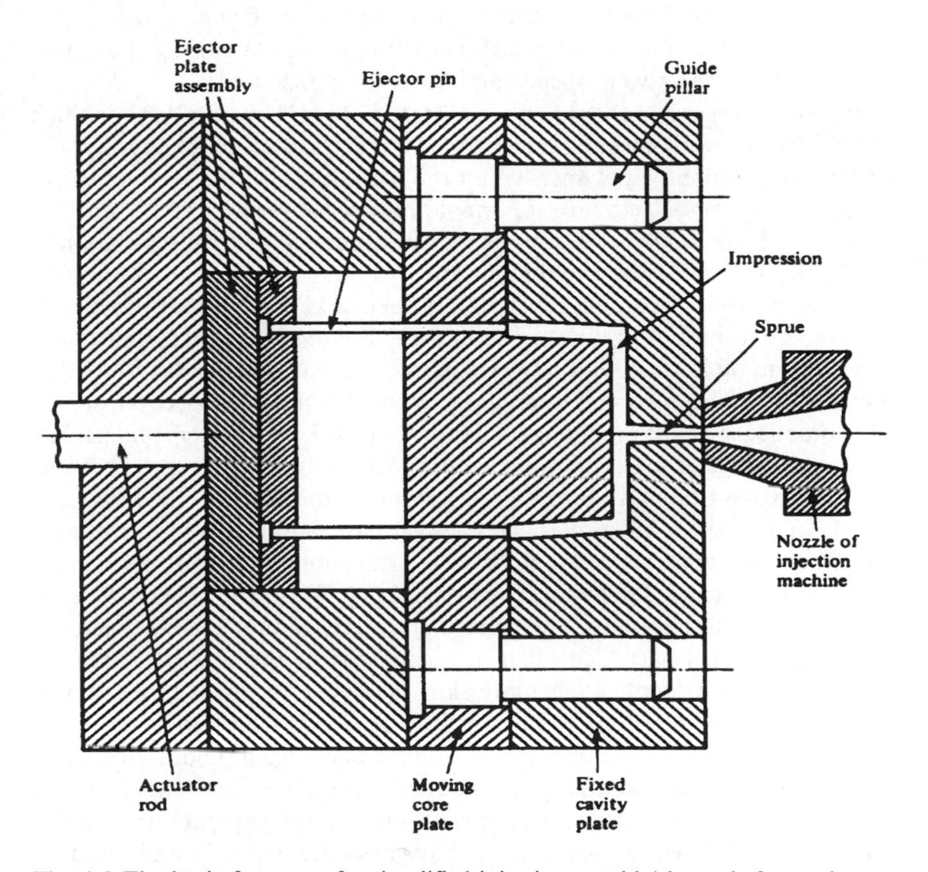

Fig. 4.6 The basic features of a simplified injection mould (channels for coolant or heaters not shown)

flow should be turbulent, i.e. Reynolds number greater than 2500. Where re-entrant shapes and holes are required, moulds incorporating devices such as sliding cores and collapsible side cheeks become necessary and they increase the cost of the mould.

The injection unit looks rather like an extruder. A single screw rotates within a close-fitting heated barrel and plasticates the granular powder or rubber strip feed into a uniform melt. The difference is that it is a batch rather than a continuous process. A nozzle valve is closed to prevent the exit of melt from the barrel, so that as hot melt is pumped past the screw tip, the rotating screw moves back axially under the developed pressure until a sufficient volume of melt has accumulated. The valve is then opened and the screw (no longer turning) is used as a piston to inject rapidly a pre-determined quantity of melt into, and completely fill, the closed mould. While the moulding is solidifying (by cooling or crosslinking), the nozzle valve is shut off and the screw prepares more melt ready for the next injection sequence. Eventually the product becomes sufficiently form-stable to be ejected from the mould, which is then opened. On closing the empty mould, the nozzle valve is opened and the cycle is repeated.

Injection pressures are commonly in excess of 100 MN/m^2 and can be as high as 200 MN/m^2, with injection rates of 3×10^{-4} m^3/s or more, generating substantial orientation of polymer within the moulding. Mould locking forces based on the projected area of moulding are obviously high, and for large mouldings 1000 tonnes or more may be needed to prevent separation of mould faces, which would make unwanted flash. Opening and closing the mould is controlled hydraulically; the mould can be kept closed either hydraulically or by using a toggle-clamp action which moves just past dead centre when the mould is just shut. It is worth noting that the time needed to solidify polymer in the impression is dominated by the thickest cross-section, so it is good practice to design mouldings of uniform wall thickness where practicable.

The size range of injection-moulded products includes computer parts the size of a match head, ball-bearing cages from 5 mm diameter upwards, gear wheels, pipe fittings, shoe soles and complete waterproof boots, car battery cases, car rear lamp housings, chair shells, stacking containers, mechanical handling pallets and dinghy hulls up to 4 m long. These and other articles of simple or complicated shape are made in one piece with repeatable precision. Injection-moulded parts can often be recognized by the scar left when the gate (where the melt enters the impression) is removed – this is easily seen on a moulded shoe sole, but in other articles it is often located in a hidden position for aesthetic reasons. Many smaller injection moulding machines operate on a fully automatic cycle, with cycle times from less than a second (for some thin-walled dairy produce containers in polystyrene) up to a few minutes for thick-walled items. Small items are often made in multi-impression moulds, and the general layout of the runners and impressions can be seen in the plastics model assembly kits (e.g. for aircraft).

The capital cost of the injection moulding machine can be high compared with other moulding machinery, because the heated barrel must withstand high internal pressures and the mould locking force must be adequate. Mould costs can be high compared with those for other moulding processes because the mould must be sufficiently robust to withstand the high injection pressures without undue distortion, and precision engineering is involved to obtain the required cavity shape and to prevent material escaping from the mating surfaces.

4.4.2 Reaction injection moulding (RIM)

Reaction injection moulding (RIM) is a process for making products by direct polymerization of two highly reactive, low molecular mass components in a closed mould. The process is performed within one integrated machine with the mixing element for the components playing the central role.

In many cases these components are a polyol (containing one or more –OH groups) and a diisocyanate (containing two –NCO groups), which react to form either a linear chain or a branched chain or a crosslinked polyurethane (characterized by its –NH–CO–O– groups). The choice of monomers and additives yields a polymer with properties very much tailored to the function of the product, from flexible (elastomeric) to rigid or foam. As the process is essentially of a chemical nature, the knowledge of the chemist and control of the chemistry is essential.

The components have low viscosity (also directly after having been mixed). They therefore flow easily and only need low pressure. The sensitive element is the mixing device, which usually consists of two metering pumps and two injection nozzles placed face-to-face for impingement of the flows to promote turbulent mixing. The impingement space provides to the mould but also has a cleaning device in order to make space for the next shot. The injection pressure can be low (0.35–0.5 MPa – compare with data for injection moulding) because of the low viscosity of the fluid. Therefore only a low mould clamping force is required, so the clamping unit and the mould can be cheap and large product dimensions will be possible.

The mould cavity should be filled from the bottom and vented at the top. With foaming resins a distribution of density through the wall thickness can be realized. The mould temperature should be controlled within close margins in order to control the chemical hardening reaction and to obtain a repeatable production cycle. In order to conduct away the rather high reaction heat the mould should be made of metal.

Cycle times are normally in the range of 2–4 min, depending on the part geometry and the formulation used. Typical products are bumpers and dashboards for cars, and housings.

Polyurethanes are not the only materials suitable for RIM; epoxy and nylon based on caprolactam are also used.

4.4.3 Compression moulding

The outside surface of the moulding is formed by a complementarily-shaped recess (the cavity) machined in the cavity plate. The inside surface is formed by a projecting core in the core plate. When the cavity and core plates are brought together within the platens of a press, the space created (the 'impression') defines precisely the shape and all dimensions of the product. To ensure accurate mating of the two plates, projecting guide pillars are fitted to one mould plate and guide bushes to the other. Considerable shrinkage may occur during moulding, and ejector pins are used to push the moulding from the core as the mould opens; generous tapers ease extraction.

Crosslinkable polymers are compression moulded in the sequence shown in Fig. 4.7. A charge of preheated moulding powder or a preform, or a rubber compound from a mixer, is placed in the cavity and the mould is closed vertically under pressure in the range 5–50 MN/m^2. Both mould surfaces are heated electrically or by steam and maintained at a suitable temperature, so that by conduction alone the charge melts and spreads to fill the impression. A small excess of that melt is allowed to flash out through small clearances between the mould plates. The charge must be premixed because there is no mixing during flow in the mould, and there is no opportunity to generate heat by mechanical working. Continued heating promotes crosslinking, and when this is sufficiently complete the mould is

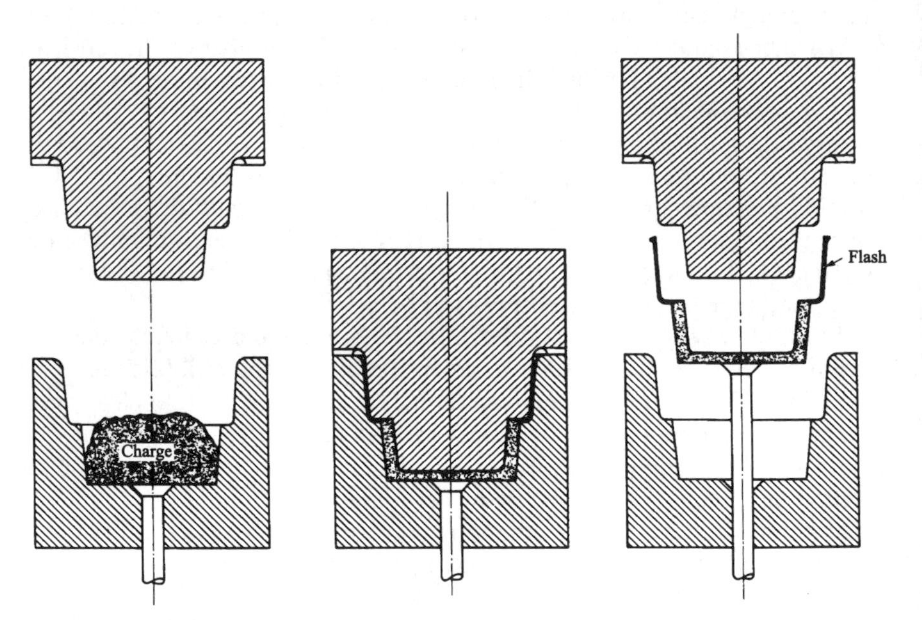

Fig. 4.7 The compression moulding operation

opened, the moulding removed hot and the flash removed. The mould is then cleaned and recharged, and the process is repeated.

With small mouldings, multi-impression moulds are used to make the most economic use of press capacity. Functional features such as holes, bosses, threads and robust metal inserts can be moulded in. Both the mould and the press are of rugged construction to withstand the high pressures involved.

Typical compression moulded plastics products include electrical plugs and sockets, commutators for small electric motors and automotive distributor heads. Rubber products include O-ring seals, springs and antivibration mounting pads.

Fibre-reinforced crosslinkable polymer compounds can also be compression moulded.

On the plastics side, compounds have a dough-like consistency based on, for example, liquid polyester resins mixed with styrene, together with glass fibres and thickeners. The charge may be a dough moulding compound (DMC) or a sheet moulding compound (SMC). DMC products range from power tool cases, electricity and gas meter boxes and car heater cases to electrical distribution equipment. SMC is well suited to large shell-like structures such as car body parts, lorry cabs and fascias to structures up to the size of a double bed. For moulding products with prescribed fibre directions and laminates (Chapter 7) prepregs can be used, consisting of parallel fibres impregnated with (usually epoxy) resin or rubber, packed between film on rolls and stored cold in order to prevent premature hardening. They can be used as plies and placed in the mould according to a prescribed stacking order.

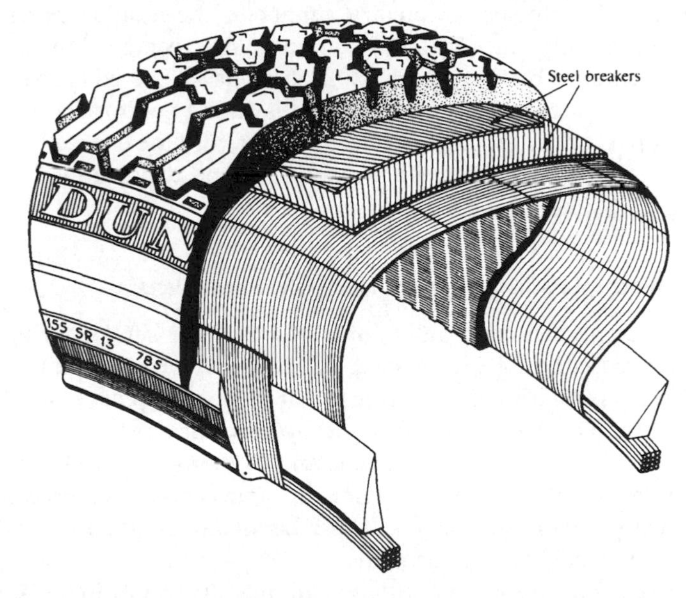

Fig. 4.8 The anatomy of a radial car tyre (courtesy Dunlop)

On the rubber side, tyres are fibre-reinforced composites and the completion of the tyre-making process is by compression moulding. Much of a tyre's performance depends on three parts: the casing, the tread and side walls, and the beads (Fig. 4.8). The casing is a series of plies each calendered in 'green' rubber with steel wires or synthetic unidirectionally aligned fibres. The beads consist of high-tensile steel wire which is extrusion-covered with uncured rubber. Tread and side-wall components are also extruded in uncured rubber. The tyre is then built up from these components on a drum. The unit is then transferred to a heated split steel mould which defines an open cavity providing the tread pattern and the outside of the tyre wall. The inside surface of the tyre is provided by an inflatable (cured!) rubber bag which can be pressurized into the correct shape using hot water or steam. The outer steel mould contains many small vent holes to enable air to escape as the rubber flows in the mould under heat and pressure; new tyres often still have the short spikes on them where the rubber has flowed a little way into the vent holes, and it is also possible to see witness lines of mating surfaces of adjacent sections of the mould. Flash round the rim of the tyre is trimmed off after ejection from the mould. Car tyres typically weigh about 6 or 7 kg; tyres for earth-moving equipment can weigh over 1 tonne.

4.4.4 Transfer moulding

Transfer moulding is a variation on compression moulding in which the rubber or thermosetting plastic is melted in a separate chamber and then transferred under pressure through a runner to the mould impression for shaping and crosslinking (Fig. 4.9). The mould is vented, but there is no deliberate clearance between faces for flash to pass. The advantages of this procedure are that the charge to the mould is already uniformly hot, so that the heating part of the moulding cycle is shorter, and the charge entering the impression is fluid so it flows more readily around delicate inserts without damaging or displacing them.

4.4.5 Moulding glass mat thermoplastic (GMT) products

Along with compression moulding of SMC, a rather similar technique has been developed, starting with a sheet of glass mat or glass fabric that has been impregnated with thermoplastics such as polypropylene or polyetherimide (GMT). The sheets can be made by extruding molten film between layers of preheated glass reinforcement and bringing this package (with an extra film on top and underneath) under compression in a double belt press provided with moving steel bands. These bands are heated first and cooled later in order to consolidate the plate.

Having been cut to size and heated again (e.g. in an infrared oven) above T_g or T_m, this material can be deformed easily in a mould under pressure.

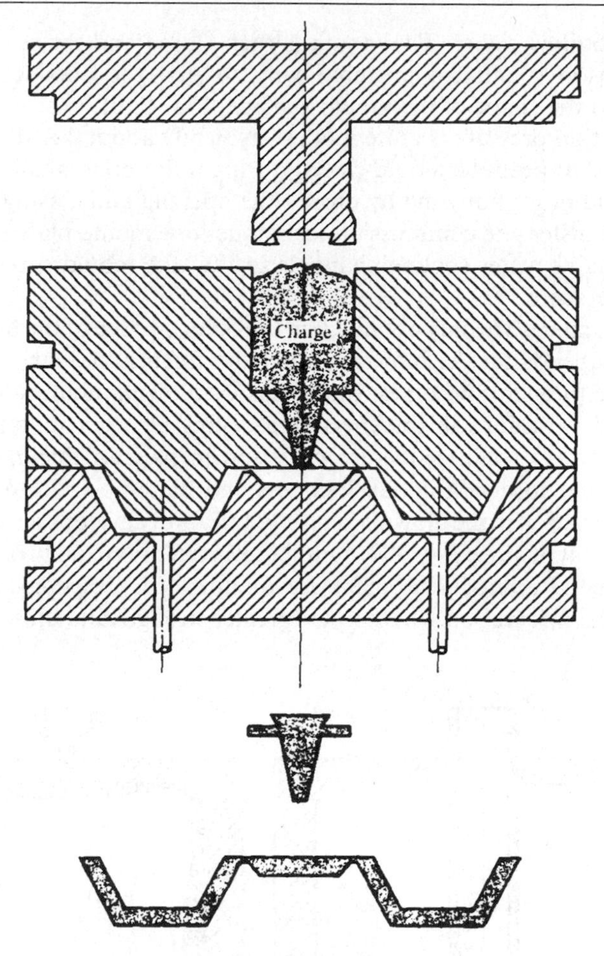

Fig. 4.9 Essential features of transfer moulding

After the product has cooled it can be taken out. A reasonable deformation of the GMT in the mould is possible, but ribs cannot be shaped. Shaping of the fabric in the sheet is limited as too much deformation would cause wrinkles. The surface finish is not perfect because some of the glass fibres tend to become slightly visible: they do not shrink as much as the thermoplastic when the product cools in the mould.

A typical application is in large-area thin-walled panels used in cars.

4.4.6 Extrusion blow moulding

As the name suggests, extrusion blow moulding is based on extrusion. A thick-walled tube of uniformly molten thermoplastic (called a 'parison') is extruded vertically downwards between the open faces of a cold split mould defining a hollow cavity. The mould is then closed and the sealed (but still warm) parison is inflated pneumatically so that the outside of the tube conforms to the shape of the mould surface (Fig. 4.10).

The inflation pressure is quite modest, typically about $0.2\,MN/m^2$, so the force needed to keep the mould closed during inflation is small. Both halves of the mould move horizontally during the opening and closing (whereas in injection, transfer and compression techniques one mould plate is fixed). The mould contains water-cooling channels so that the mould can be kept at a fixed temperature.

Close tolerances are only obtained on the outer surface of the product. Wall thickness depends on the blow-up ratio: the further the parison has to be inflated, the thinner it becomes. It is usual for blow mouldings to have quite pronounced variations in thickness, especially in corner features. Modern machines control some deliberate variation in parison thickness along its length in order to match initial thickness and blow-up ratio to reduce wall thickness variation in the blown moulding. Inflation develops tensile deformation in the melt of some 200–300%, either uniaxial or biaxial in different parts of the product. Recent developments have controlled, preserved and exploited the molecular orientation. This confers greater strength and thus thinner walls

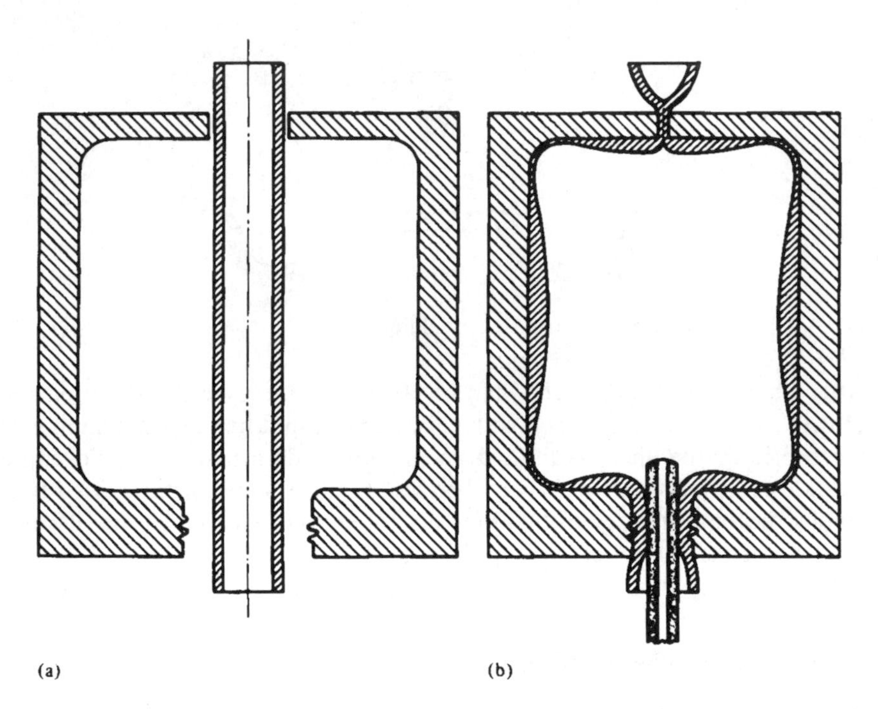

Fig. 4.10 Essential features of basic extrusion blow moulding: (a) the molten hot parison is extruded between the open halves of a cold split mould, (b) the mould is closed and air is blown through the spigot to inflate the parison

capable of withstanding the stresses in clear bottles containing carbonated beverages.

Blown products range in mass from a few grams to more than 20 kg, and in size from hollow spheres 5 mm in diameter (to reduce evaporation loss from liquid surfaces) through bottles and jerrycans to dustbins (blown in pairs and then divided) to 200 l drums and 1 m^3 oil storage containers. With ingenuity, complicated shapes can be achieved, notably in heater ductings and petrol tanks for cars.

Not all the parison is used to make the product: the ends which protrude from the closed mould are trimmed off, regranulated, kept clean and recycled. Where the parison is pinched-off or sealed, there will be a long scar where the end of the parison has been removed from the blown product. This will be found on bottles on the base and is in line with the witness mark where the two mould faces meet.

To achieve a satisfactory wall thickness at the thinnest points, the thickness elsewhere may be greater than it needs to be. This limitation of the process is compounded by the moulding being cooled from the outside only (by the cool mould) so that for a given part thickness the cooling cycle is much longer for blow moulding than for injection moulding.

4.4.7 Thermoforming

Thermoforming produces hollow or shaped articles from thermoplastics sheet and is analogous to sheet metal shaping. Amorphous thermoplastics are preferred because they have a wider range of softening temperatures than partially crystalline ones. The feed is of uniform composition and more expensive than granules or powder because it has already been processed once – by calendering, extrusion or casting. Typical thermoformings include vending cups, chocolate box liners, refrigerator liners and dinghy hulls.

The basic process requires an open mould or tool, a means of heating and softening (but not melting!) the sheet in, say, a thermostatically controlled oven, and a means of making the softened sheet conform to the surface of the mould. The necessary shaping pressure can be applied using either a simple plunger or by introducing a modest air pressure difference by means of a vacuum or compressed air supply. In vacuum forming, the sheet is clamped over a mould, heated *in situ*, and then shaped by evacuating the air under the sheet in the mould. Shaping results from tensile deformations.

As with blow moulding, the wall thickness of a thermoformed article is not defined precisely by the mould: the dimensions which can be accurately held relate to the surface of the article which is in contact with the mould. For a female mould these are the outside dimensions, and for a male mould the inside ones. The best possible thickness distribution in the final shaping may result from a judicious combination of pressure and vacuum forming, with or without the use of a plunger.

The main attractions are that the equipment is inexpensive, moulds are generally inexpensive (because they do not have to withstand high temperatures and pressures), and large thin-walled parts can be made which could not be produced economically or technically by moulding polymer melts.

There are several disadvantages in thermoforming. Thinning occurs, particularly at corners: this orients material in the direction of draw, and the amount of thinning depends on the drawdown. After shaping and cooling, trimming operations are usually required: this produces scrap which may sometimes be reprocessable. Heating is by conduction or radiation only and is slow for thick sheet – so is cooling. Cycle times can be long and hence labour charges may be high. Thin-walled articles are often produced in multicavity moulds in line with an extruder and sheet die line, and operated automatically.

4.5 CONTACT MOULDING TECHNIQUES

Contact moulding techniques are widely used in the manufacture of articles from crosslinkable plastics reinforced with long or continuous fibres. The fibres are usually glass but specialized applications may involve fibres based on carbon or aromatic polymers. For convenience only, the following remarks describe the use of glass fibres; technology for other fibres is broadly similar. The basic plastic, often called a resin, is frequently supplied as a mobile liquid which can be readily shaped and which will wet the fibres easily as well as the binding agent used to hold the fibres together. When the product has been shaped, the shape is stabilized by crosslinking: this can occur either hot or cold depending on the chemistry of the system.

4.5.1 Hand lay-up

The mould defines the surface of the product: in its simplest form it can be either a male or a female type, and made from wood, plaster, metal or even reinforced plastic. Because the resins used are good adhesives too, the mould surface must be carefully coated with a wax or spray release agent.

Liquid resin, initiator, catalyst and crosslinking agents are thoroughly mixed in the correct proportions. A reinforcement-free gel coat is applied to the mould. This gel coat is then covered with successive layers of resin and reinforcement typically having the form of woven cloth or of randomly oriented chopped strand mat. This sandwich is compacted with brushes or hand-rollers to expel air and ensure that all the glass fibres are properly wetted with resin. The laminate is built up until the correct thickness, i.e. the required amount of glass, is present, after which the resin is left to gel and harden.

Hand-laid-up fibre-reinforced plastics products usually have a plate- or shell-like form, and include cladding panels for buildings, swimming pools and large liquid storage tanks, pipe bends and branches, and boat and ship hulls up to 50 m long made in one piece from 65 tonnes of resin and 65 tonnes of glass fibres.

The advantages of this technique are that the capital cost of equipment is low, moulds are inexpensive (the lightweight constructions must be form-stable and resist hand-rolling loads and general handling), reinforcing struts and stringers can be readily incorporated into the product, and the main upper limit on size is dictated by transporting the finished product.

Some disadvantages are inherent in such a process. The tolerance in fibre distribution in the as-supplied glass cloth is typically ±10% in accordance with national standards. It is not too difficult to double this spread in operations involving handling, cutting and stretching the cloth over doubly curved mould surfaces and other awkward configurations. Moreover it is frequently necessary to overlap the cloth. These difficulties can be appreciated in the construction of, say, a pipe bend or branch. Thus the disadvantages are that uniformity of composition depends largely on operator skill, wall thicknesses are difficult to control exactly, only one surface of the moulding has a good finish, and the process is slow, labour intensive and difficult to mechanize.

The most commonly used resin is unsaturated polyester (UP). It is used (and made liquid) with styrene, which acts as a solvent and as a crosslinking agent at the same time. Styrene vapour is a health hazard even in very low concentrations. During build-up of the laminate and thereafter during hardening, the free surface is exposed to the air, giving rise to styrene emission. Since this health hazard is well known there is a strong tendency to avoid these free-surface methods (including the next mentioned, spray-up) and to give preference to resin transfer moulding.

4.5.2 Spray-up

Reinforcement (chopped strands of glass) and the resin mixture are sprayed onto the mould surface using a twin-nozzle compressed-air gun, and then compacted with rollers. Compared with hand lay-up, increased production rates are possible but operator skill is still important and styrene emission has to be taken care of.

4.5.3 Resin transfer moulding (RTM)

Dry fibre material (glass fibre mat, fabric and/or rovings) is brought into an open two-part mould and the mould is closed. Reactive resin is injected until the mould is filled. When the resin has cured the product can be taken out.

The flow of the resin is subject to flow resistance during its passage through the fibre material. The driving force for the flow is usually a vacuum, which at the same time evacuates the air from the cavity and

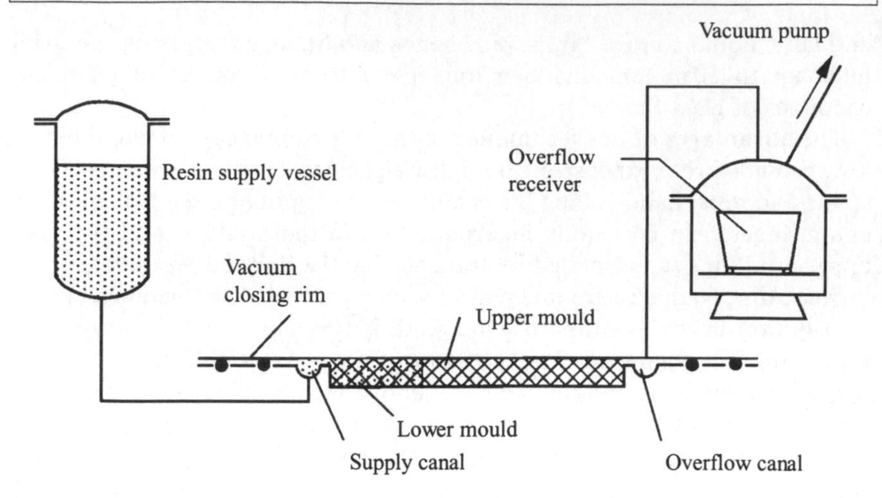

Fig. 4.11 Schematic diagram of vacuum-driven resin transfer moulding (courtesy Technische Universiteit Delft)

helps to keep the mould halves together (Fig. 4.11). The flow speed may be increased by injecting the resin under pressure. If, however, the mould is flexible, this will affect the shape of the cavity and in consequence the suggested flow path of the resin. The local speed of the resin will be determined by the local permeability of the fibre material, the viscosity of the resin and the local pressure decay. By summation these factors determine the total filling time. Local speed differences of the flow front may lead to a situation where not all of the product is impregnated. The shape of the product, together with the arrangement of channels supplying the resin, can also lead to 'dead spots'.

The permeability of the fibre material is low if its density is high (as is the case of fabric compared with mat), so a high fibre volume fraction can only be realized at the cost of a low production rate or of a stiff mould construction.

The vacuum will help to make a void-free product by evacuating air from the dry fibre material. On the other hand, it will help dissolved air in the resin to evaporate, so that the resin flow will contain little bubbles that can be trapped between the fibres.

The RTM technique was primarily developed to avoid styrene emission in the workshop. It has proved to be very flexible and economic, and is most suitable for rather large products in smaller series.

4.5.4 Matched-tool moulding

The cold matched-tool moulding process consists of charging the female tool with reinforcement and resin, closing the mould and forcing the resin through the reinforcement under light pressure. Curing occurs at room temperature. This is used where contact moulding is too slow, where close tolerances are needed, where larger quantities of mouldings are needed and where a good surface finish is required on both sides of the product. The hot matched-tool moulding process is a form of compression moulding used particularly for SMC and DMC, as described in section 4.4.3.

4.6 CONTINUOUS FIBRE TECHNIQUES

4.6.1 Filament winding

Continuous glass-fibre reinforcement is wound onto a rotating mandrel (Fig. 4.12). The fibres are coated with resin immediately prior to winding. The direction of reinforcement can be aligned with the directions of principal stresses which occur in the moulding under service conditions. The mandrel can be dismantled or collapsed after the resin has been cured. This process is limited mainly to the production of hollow cylindrical or spherical mouldings such as pipes, large tanks, pressure vessels, rocket engines and aerospace components which have to meet exacting performance requirements.

4.6.2 Pultrusion

Continuous glass-fibre reinforcement, together with fabrics or chopped strand mat, is pulled through a heated die of certain length. The die is shaped to form the glass structure into a profile. The glass is impregnated with reactive resin before entering the die or, alternatively, the resin is injected into the glass structure at the beginning of the die. The die is heated and the resin will cure during the passage through the die. At the exit of the die a transport mechanism pulls the profile, and a saw will cut it to length. Different types of profiles can be made, including hollow ones.

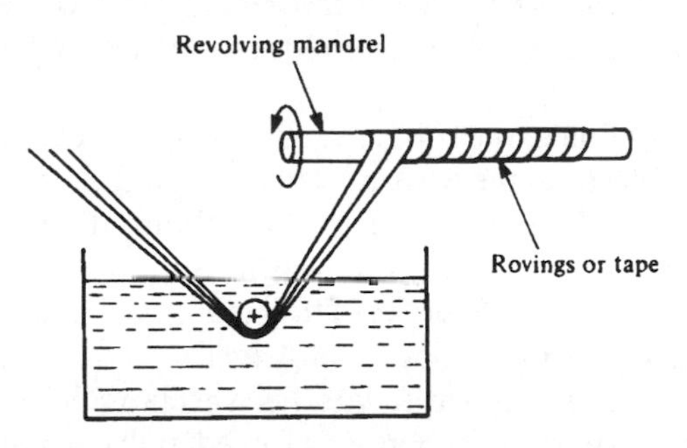

Fig. 4.12 The filament winding process for FRP

PROBLEMS

1. What is the case for making the wing of a sailplane ('glider') from plastics? What method of construction would you use if the overall wingspan was 30 m? Be as specific as possible.
2. What are the main similarities and differences between the processes of compression moulding and extrusion blow moulding?
3. Two thermoset plastic articles have similar shape, appearance and 'feel'. One has been compression moulded and the other injection moulded. How would you distinguish which article had been injected?
4. Dustbins for domestic use can be made from a propylene–ethylene copolymer either by injection moulding or by extrusion blow moulding. Suppose (for the sake of this question only) the dustbins are moulded by both processes from the same polymer at the same melt and mould temperatures. Which dustbin is likely to be the heavier? Describe the molecular orientation you would expect at a point halfway up the side of the bin. Would you expect the crystalline texture to be the same in each bin at that point? Would you expect bins made by the two processes to use the same or different grades of the same type of polymer?
5. Why are thermoplastics usually not compression moulded?
6. Sketches for eight cylindrical plastics products are reproduced in Fig. 4.13. For each sketch suggest a method of manufacture, and indicate clearly how the design would need to be modified to conform to good practice and to ensure as far as possible that manufacture would be trouble-free. For some shapes more than one method is possible, depending on the shape modifications suggested.

   If the length/diameter ratio of the cylindrical section were (a) decreased by a factor of 2, and (b) increased by a factor of 2, and other features were unchanged, would this alter any of the suggestions you have made? If the wall thickness of the original cylindrical wall were reduced by a factor of 2, would this affect your recommendations?
7. What would be one of your worst technical fears if you were in charge of the injection moulding of thermosetting plastics or hard rubber compounds?
8. How would you make the rubber sealing strip used for car windscreens?
9. A sketch produced on the back of an envelope (Fig. 4.14) is for a box to be made in polyethylene by injection moulding. If the tool were made, how would you expect the moulding to appear (a) immediately after demoulding, (b) a week later? (c) How might you improve the design if no critical load-bearing duties are involved?
10. A sketch has been submitted to you for a rectangular box lid (Fig. 4.15) to be made in short-glass-fibre-reinforced polycarbonate by injection moulding. Identify the major inadequacies of the proposed design.

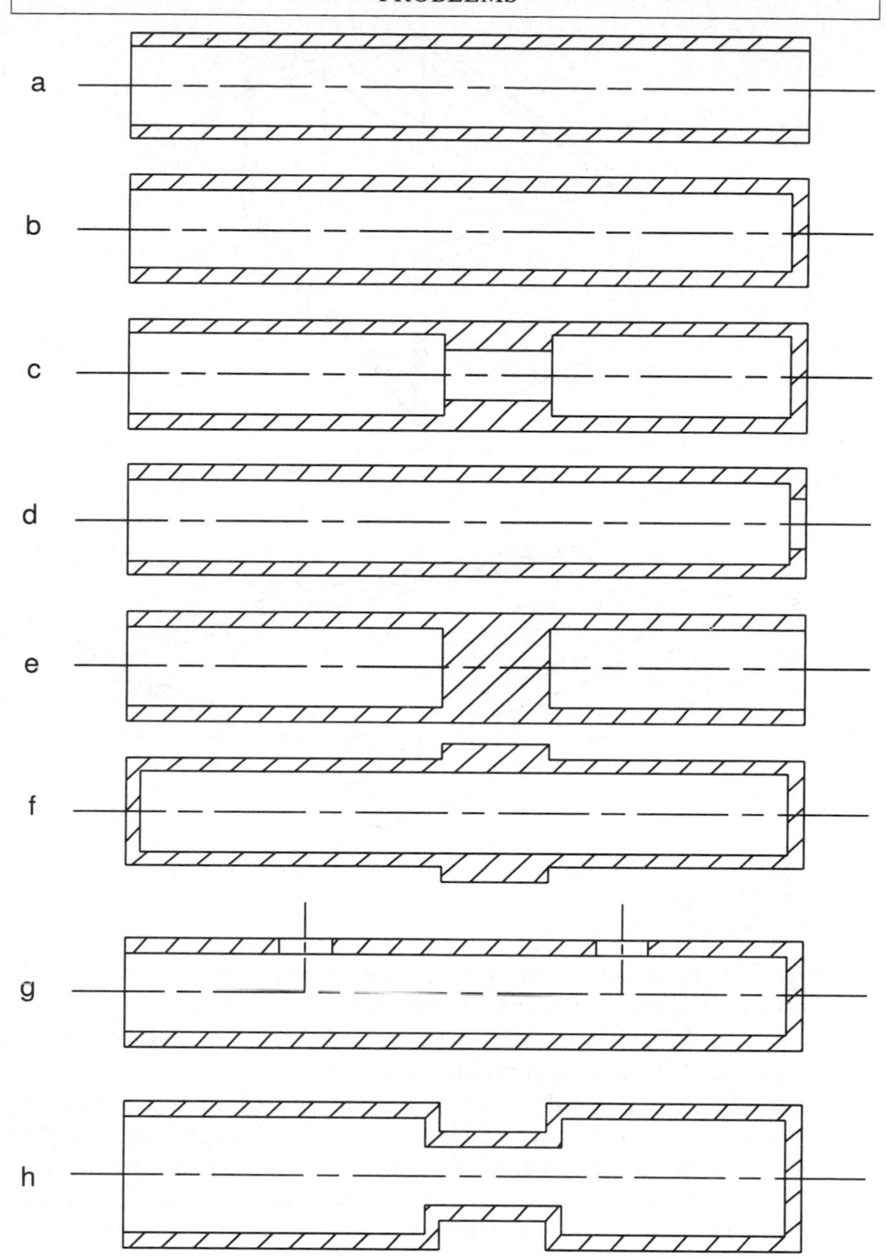

Fig. 4.13

11. A general-purpose bottle suitable for, say, orange squash is to be extrusion blow moulded in an unplasticized PVC compound. The designer is not a (polymer) engineer and has proposed the shape sketched in

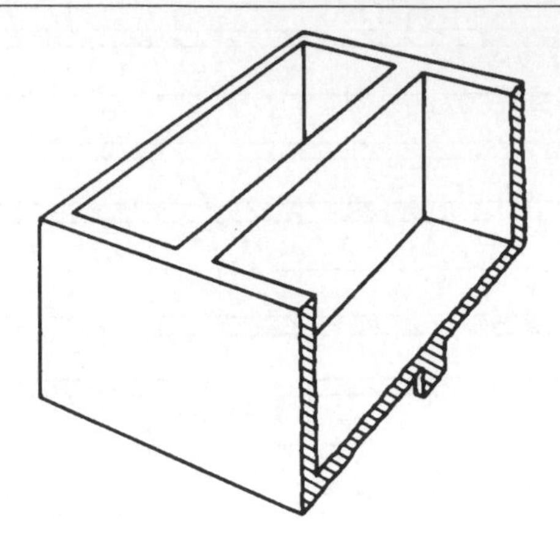

Fig. 4.14

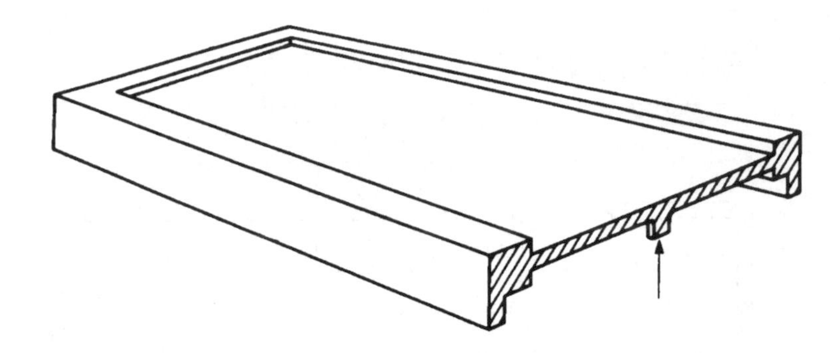

Fig. 4.15

Fig. 4.16. How would you want to improve the general design of the bottle by suggesting minor amendments?

12. Why is it difficult to achieve high volume fractions of glass fibre in articles made by a contact moulding process?
13. ABS tube is to be extruded through a die having an annular gap. During start-up the pin or mandrel which defines the inside of the tube is slightly off-centre. What effect would you expect this to have on the extrudate? If such a tube were extruded vertically downwards as a parison for blow moulding, how would you detect the fault in, say, a cylindrical blown bottle?
14. Why is it good practice to design flow channels in polymer processing equipment with generously radiused rather than sharp internal corners? (Think of the polymer rather than the hardware.)
15. An integral mesh netting is made from polyethylene in many configurations and gauges. Applications for this netting include packaging fruit

Fig. 4.16

and vegetables, fencing, and stabilizing reclaimed land. How would you make one such mesh?

FURTHER READING

Birley, A.W., Haworth, B. and Batchelor, J. (1992) *Physics of Plastics, Processing, Properties and Materials Engineering*, Hanser, Munich.

Blow, C.M. (1971) *Rubber Technology and Manufacture*, Butterworths, London.

Grossman, R.F. (1997) *The Mixing of Rubber*, Chapman & Hall, London.

Isayev, A.I. (1987) *Injection and Compression Molding Fundamentals*, Marcel Dekker, New York.

Janssen, L.P.B.M. (1978) *Twin Screw Extrusion*, Elsevier, Amsterdam.

Manas-Zloczower, I. and Tadmor, Z. (1994) *Mixing and Compounding of Polymers*, Hanser, Munich.

Osswald, T.A. and Menges, G. (1995) *Materials Science of Polymers for Engineers*, Hanser, Munich.

Potter, K. (1997) *Resin Transfer Moulding*, Chapman & Hall, London.

Rauwendaal, C.J. (1990) *Polymer Extrusion*, Hanser, Munich.

Rosato, D.V. (1997) *Plastics Processing Data Handbook*, Chapman & Hall, London.

Stevens, M.J. and Covas, J.A. (1995) *Extruder Principles and Operation*, Chapman & Hall, London.

Stevenson, J.F. (ed.) (1996) *Innovation in Polymer Processing: Molding*, Hanser, Munich.

Tadmor, Z. and Gogos, C.G. (1979) *Principles of Polymer Processing*, Wiley, New York.

5 Stiffness of Polymer Products

Polymers are widely used because of, or in spite of, their low moduli, which normally lie in the range $1\ MN/m^2$ to a few giganewtons per square metre. Applications relying on the prespecified stiffness of polymer products range from springs (in bridge bearings, antivibration mountings, flexible couplings and suspension units) through shoe soles and inflated tyres, to window sealing strips and main window frames, pipes, tanks and vessels, not to mention lorry cabs, helicopter rotor blades and ship hulls.

Another major application of plastics is packaging. With the exception of films, the design of much packaging is based on maximizing the stiffness and strength per unit mass and minimizing the cost per unit mass. This also involves use of the principles discussed in this and the next chapter, even if the product seems 'trivial'.

This chapter discusses the stiffness of nominally uniform and isotropic polymers and ways of predicting the stiffness of products made from them. The material is discussed in three main sections. The first discusses the behaviour of plastics where loads are applied over times ranging from a few seconds to a few years: this introduces the idea of creep under constant load, and how to allow for this in design. The second section discusses creep under non-constant loads, and then discusses the more generalized relationships between stress and strain for different models of material behaviour, including stress relaxation and the deformation response to cyclic loading. The third section examines the properties of rubber polymers and discusses aspects of the analysis of rubber spring units under common types of loading. This chapter does not discuss the stiffness of fibre–polymer composites; this is the subject of Chapter 7.

5.1 STIFFNESS OF PLASTICS: ELEMENTARY CONCEPTS

5.1.1 Modulus from the conventional tensile test

The test is a standard one for metals, usually done at constant rate of crosshead separation. Within the elastic range, the stress–strain relationship

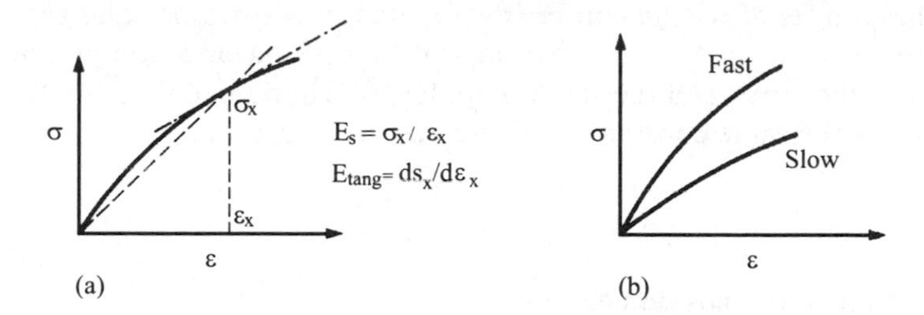

Fig. 5.1 Stress versus strain at constant crosshead rate

is a straight line of slope *E*, the elastic or Young's modulus. This value of modulus is sensibly constant over a wide range of test speeds and temperatures. Modulus is widely used both to characterize the stiffness of a material and as the basis for elastic design. It is understandable to call for the modulus of a polymer tested under similar conditions. Such tests on plastics reveal the following points.

1. The stress–strain relationship can be somewhat curved, as shown in an exaggerated form in Fig. 5.1a. As the curvature is usually fairly small, designers retain the concept of linearity (but not elasticity) by approximating the curve to a secant straight-line construction at a defined strain, ε_X, typically 0.2% or 1% for plastics. In order to make its exact meaning clear, $\sigma_X/\varepsilon_X = E_s$ is defined as the secant modulus for the strain ε_X. Such an approach can permit the use of elasticity theory, as will be described in section 5.1.4(a). The tangent to the curve at ε_X, with its defined tangent modulus $d\sigma_X/d\varepsilon_X = E_{tang}$, has an even lower slope. E_{tang} is of importance for buckling calculations in non-slender cases. At the origin ($\varepsilon_X = 0$) E_{tang} and E_s coincide and are called E_0.
2. Modulus depends on test speed. A slow speed permits more disentanglement of molecular chains, and hence a lower modulus; a fast test speed gives insufficient time for chains to uncoil and disentangle (Fig. 5.1b).
3. Modulus depends on test temperature, as already explained in section 2.6 and shown in Fig. 2.10.
4. There are potential interactions between speed and temperature. Well below T_g, a change in rate has little effect (Fig. 2.10). Well above T_g, rate has some effect but it is not usually too troublesome. But in the region of T_g, say within some 50°C or so (above or below), rate has a big effect; for example, very slow (i.e. long-term) tests can drop T_g by several tens of degrees to the region where the plastic may be used in service, with a corresponding drop in modulus.
5. Poisson's ratio, ν, sometimes called lateral contraction ratio, for glassy polymers (i.e. $T < T_g$) is $\nu \simeq 0.33$. For amorphous polymers above T_g, $\nu \simeq 0.5$ and for partially crystalline polymers above T_g, $\nu \simeq 0.4$.

6. Other modes of testing can be investigated; for isotropic solid plastics at small strains (up to 1 or 2%), moduli in tension and compression are almost the same, and the shear modulus, G, can be calculated from E and ν under the same conditions of rate and temperature as

$$G = E/2(1+\nu) \tag{5.1}$$

5.1.2 Effect of composition on modulus

The effect depends on the relative stiffness of the plastic and the dominant ingredient. For a stiff polymer, a flexible additive reduces modulus, and the following examples refer to tests carried out at 20°C. Incorporating a small amount of small rubber particles ($E \simeq 1\,\text{MN/m}^2$) into polystyrene ($E \simeq 3\,\text{GN/m}^2$) reduces the modulus by a factor of about 2. Incorporation of plasticizer can reduce the modulus of PVC by up to three orders of magnitude, and reduce the T_g to as low as −35°C, i.e. the highly plasticized PVC is quite rubbery at 20°C. A cellular structure formed from gas inclusions increases flexibility, as expected: injection mouldings made from foamed polystyrene would have a volume fraction of polymer of about 0.5, with a modulus about one-quarter that of the unfoamed polystyrene; expanded polystyrene made by a quite different process for packaging or thermal insulation has a volume fraction of polymer of the order of 0.05 or less and a modulus of only $50\,\text{kN/m}^2$. For a stiff polymer, an even stiffer reinforcement, e.g. glass or carbon fibres, will raise the modulus of the composition.

If molecules become lined up during processing and are frozen in the lined-up configuration (e.g. by shear in mould filling, or by inflation of parisons or drawing of films), then the material becomes stiffer along the direction(s) of alignment and more flexible in the transverse direction(s); the material becomes anisotropic, that is, properties vary with direction. Fibrous or plate-like reinforcement may also confer anisotropy of modulus – either deliberately or accidentally. If the anisotropy becomes significant, then analysis and design will have to rely on the techniques of Chapter 7 rather than those given in this chapter.

5.1.3 Long-term loading of plastics

Tests at constant crosshead speed are normally quite quick, typically lasting only a few minutes at most, where stiffness data are required. They are useful for quality control comparisons. But many plastics products in service carry loads for far longer times, t, up to a few years being commonplace; often the loads or deformations are applied and maintained reasonably constant. The corresponding behaviour of the plastics material is measured in a creep or stress relaxation test, and long-term load-bearing design must be based on these data, not the short-term data (section 5.1.4(a)).

(a) Creep tests and data for plastics

Tests are carried out on parallel-sided specimens at constant temperature and constant load. Creep is defined as the strain, which is time dependent, resulting from the applied load – hence, a concurrent control test is done on a similar sample not under load, to detect changes caused by other factors. The curve of linear strain versus linear time flattens out and is not very meaningful (Fig. 5.2a). Therefore data are usually presented as linear strain versus log time (Fig. 5.2b; BS 4618) or on double log coordinates (Fig. 5.2c). Note that strain is time dependent from the instant the load is applied: there is no instantaneous strain to which a time-dependent strain must be added, and claims that there is an instantaneous strain usually reveal that measurements at sufficiently short times have not been taken.

As expected, an increase in stress increases strain at a given time (although doubling the stress may not double the strain except at short times) as shown in Fig. 5.3a. At a given stress and time under load, the higher the temperature, the higher the strain (Fig. 5.3b). Creep at some temperature above ambient temperature does not include expansion caused by the difference in temperature: the datum point is zero strain at the test temperature, not ambient temperature.

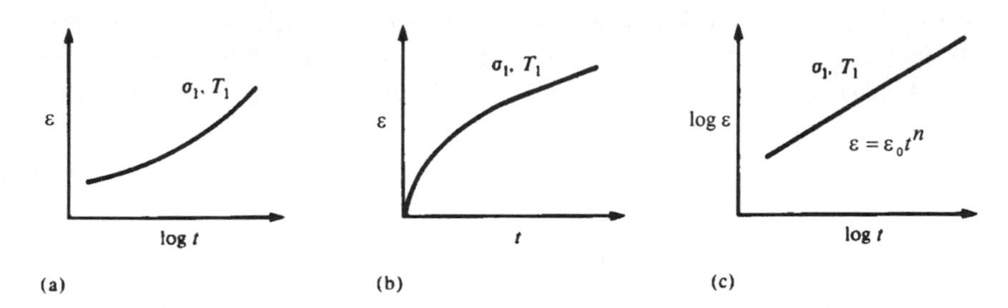

Fig. 5.2 Presentation of creep data at the same stress σ_1 and the same temperature T_1

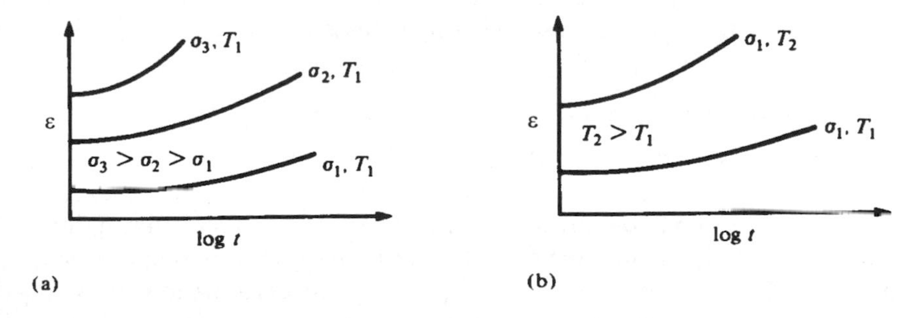

Fig. 5.3 Effect of creep on different σ or T

Almost all published data refer to tensile creep. For practical purposes, creep in compression is almost the same as that in tension. The lateral contraction ratio (Poisson's ratio) varies only slightly with time (and the time dependence can often be ignored). Creep in shear seems to relate to tensile creep using the lateral contraction ratio. There is some evidence to suggest that creep under combined stress can be related to uniaxial data using the concepts of equivalent stress and strain.

Creep data are expensive to generate. Interpolation between sets of data has to be practicable, for example between creep curves using the isochronous curves to provide a proportioning factor (and if no isochronous curve is available, assume a straight line for interpolation), or between families of creep curves at two different temperatures using a short-term modulus–temperature curve and assuming the long-term behaviour is similar to the short-term behaviour when devising the proportioning factor.

There is some evidence to suggest that the effects of stress and temperature can be regarded as equivalent in their effect. A similar equivalence between time and temperature is also apparent. This could suggest that a short-term test at high stress or high temperature could be used to predict long-term behaviour at low stress and lower temperature. It is generally considered unwise to extrapolate experimental data points by more than a factor of 10 on the time scale. Guessing creep data is best left to experts.

Other factors, outside the scope of this section, may affect creep; these include free volume effects in amorphous plastics (section 5.2.4(b)), secondary crystallization in partially crystallizable plastics, and any absorption of liquids.

(*b*) *Representation of creep data*

The family of creep curves can be represented by cross-plotting to show isochronous stress–strain curves, isometric stress–log time, and creep modulus–log time curves, as shown in Fig. 5.4. Creep modulus is the stress divided by the desired strain at a given time and temperature, and represents a secant modulus constructed on the relevant isochronous curve. All the cross-plots convey the same information, so the designer uses whichever is available.

(*c*) *Stress relaxation in plastics*

Stress relaxation is defined as the stress, which is time dependent, resulting from the applied deformation (compare the definition for creep). Figure 5.5 gives an impression of the stress response to a suddenly applied strain. Stress relaxation tests are more expensive than creep tests and are seldom carried out. For strains up to 1% or so, and provided the deformation stays constant, stress relaxation can, with reasonable accuracy, be described by the relevant isometric stress–time curve generated from creep tests.

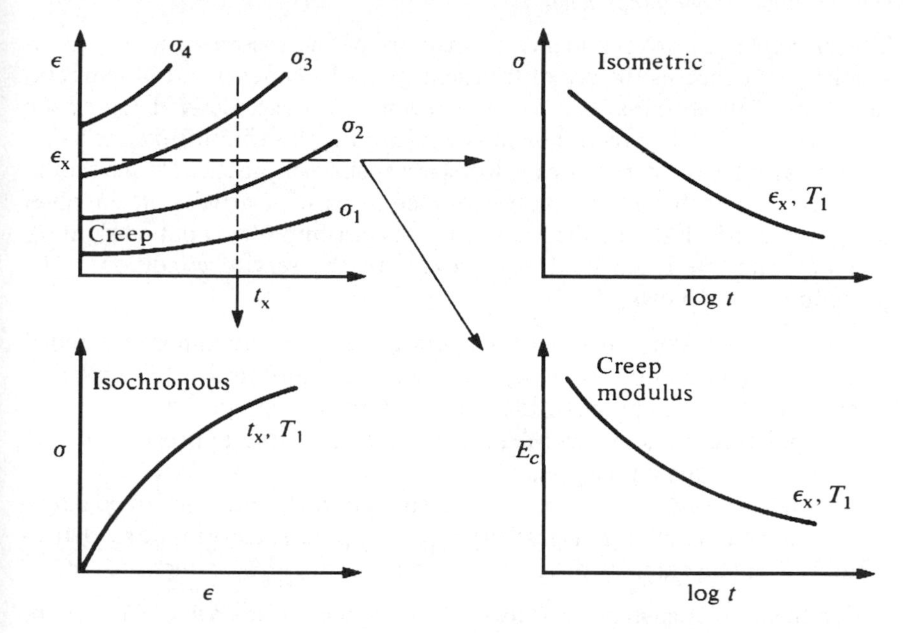

Fig. 5.4 Representation of creep data

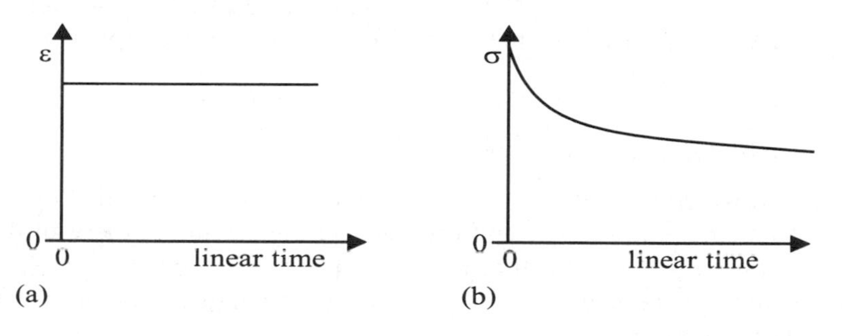

Fig. 5.5 Relaxation: (a) imposed strain and (b) resulting stress as a function of time

5.1.4 Prediction of long-term stiffness of plastics products under constant load or deformation

This is a special case of the more general problem of designing for acceptable stiffness. The discussion centres on reasonably well-designed and properly manufactured articles made from unreinforced plastics which are in equilibrium with their safe operating environment at constant temperature.

(*a*) *Principle of correspondence*

The principle of correspondence forms part of the theory of linear viscoelasticity. It concerns the use of transient (time-dependent) and of temperature-dependent stiffness data in the relevant formulae, derived for elastic deformation. For the practice of constant loading, this simplified conclusion is of great interest. It provides a worst-case, simple, practicable procedure for predicting deformation in homogeneous plastics articles. It involves using standard elasticity theory (taking its assumptions on board, too), together with the creep modulus relevant to the service conditions. The procedure is as follows:

1. identify the maximum service temperature and the maximum (assumed continuous) duration for which the maximum constant load is applied;
2. calculate the maximum stress(es) in the proposed design;
3. interpolating as necessary, read the strain from the appropriate creep curve for the maximum stress;
4. calculate creep modulus = maximum stress/(strain read from creep data).
5. use this creep modulus in elasticity theory to predict deformations, deflection, buckling stability or relaxation in the proposed product.

For the special case of bending only, in which strains are small, a convenient but slightly approximate short-cut is to assume the isochronous stress–strain curve is actually linear to, say, 0.5 or 1%; steps 2 to 4 above are then replaced by reading the creep modulus at 0.5 or 1% from curves of creep modulus versus log time. This does not, of course, apply to stress systems other than bending.

(*b*) *Sample stiffness calculations*

A polypropylene component is bolted onto a metal frame at 20°C. The section under the washers is uniform and, during assembly, the wall thickness is reduced by 2%. If the torque applied in securing the part were T, what torque would be needed to just (further) begin to compress the plastics part after (1) 1 day and (2) 1 year?

This is a stress relaxation problem. From the isometric creep curve at 2% strain (Fig. 5.6a), 20°C, the initial stress at, say, 10 s is 22 MN/m^2; after 1 day, the stress has fallen to 11.5 MN/m^2 and after 1 year to 6.5 MN/m^2. Thus, after 1 day, the torque needed to just compress the part is $(11.5/22)T = 0.52T$ and after 1 year is $(6.5/22)T = 0.3T$.

A straight, rectangular cross-section polyacetal cantilever is 100 mm long, 6 mm wide, and carries a constant transverse load of 10 N at the free end. What minimum depth is needed if the deflection at the free end should not exceed 2 mm over 2 years at 20°C?

The engineer's theory of bending gives the required second moment of inertia as

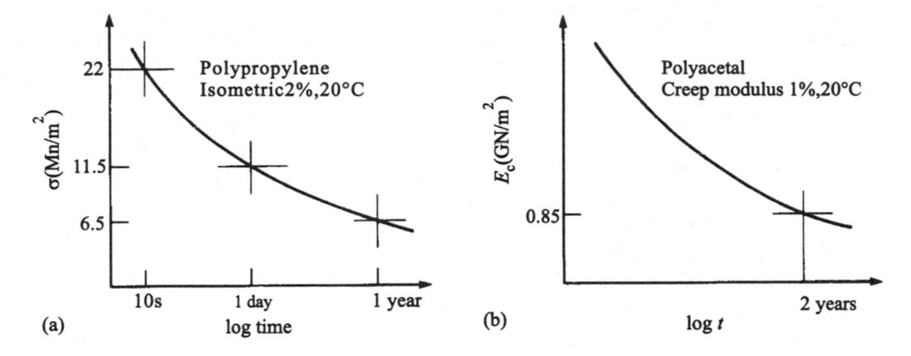

Fig. 5.6 Data for problems in schematic form

$$I = \frac{bh^3}{12} = \frac{PL^3}{3E_c\delta} \tag{5.2}$$

where P = load, L = span, δ = free end deflection, and E_c = creep modulus at 20°C, 2 years. Assuming the isochronous stress–strain curve is linear at 2 years up to 1% strain, the approximate method is to identify the creep modulus from the E_c versus log t plot for 1% strain (Fig. 5.6b), i.e. $E_c = 0.85\,\text{GN/m}^2$, giving the depth as 15.8 mm. The creep data clearly show that the beam would seem to be overstiff in the short term in order to be adequately stiff after 2 years.

The more precise method involves calculating the maximum bending stress, $\sigma = (PLh)/2I$, then reading the strain from the creep curve at that stress at 20°C, and hence calculating the creep modulus. Given a family of creep curves, how would you actually calculate the beam depth?

(c) *Need for improved stiffness*

The low modulus of polymers results in high deformation under load. This may lead to problems for many applications in which deformations have to stay within strict limits. In section 10.3 indications are given of how to deal with this problem. In general it has to be noted that, in designing with materials such as metals and concrete, attention is firstly directed to calculation of strength. If that requirement is fulfilled, stiffness would normally be satisfactory. With polymers attention to stiffness should be primary. If that criterion is met, strength may be no problem, but mind the many traps that high or low temperature, environment or impact load may offer!

PROBLEMS

1. Estimate a value of secant creep modulus (in GN/m^2) under constant load under the following conditions:

Material	*Stress* (MN/m²)	*Strain* (%)	*Time*	*Temperature* (°C)
(a) Acetal copolymer	2.5	–	100 s	20
(b) Acetal copolymer	–	1.0	100 s	20
(c) Acetal copolymer	–	1.0	100 s	60
(d) Acetal copolymer	–	0.5	1 year	40
(e) Propylene homopolymer	4.0	–	2 weeks	20
(f) Propylene homopolymer	3.0	–	2 weeks	50

Creep data are provided in Figs 5.7–5.11.

2. A tube made from cast acrylic is 3 mm thick, of 50 mm bore and 300 mm long with sealed ends. The pipe, in a test room at 20°C, is pressurized internally with air to 1 MN/m² gauge. Estimate the outside diameter midway along the pipe after 1 year under this constant pressure. Relevant creep data are given in Fig. 5.12.

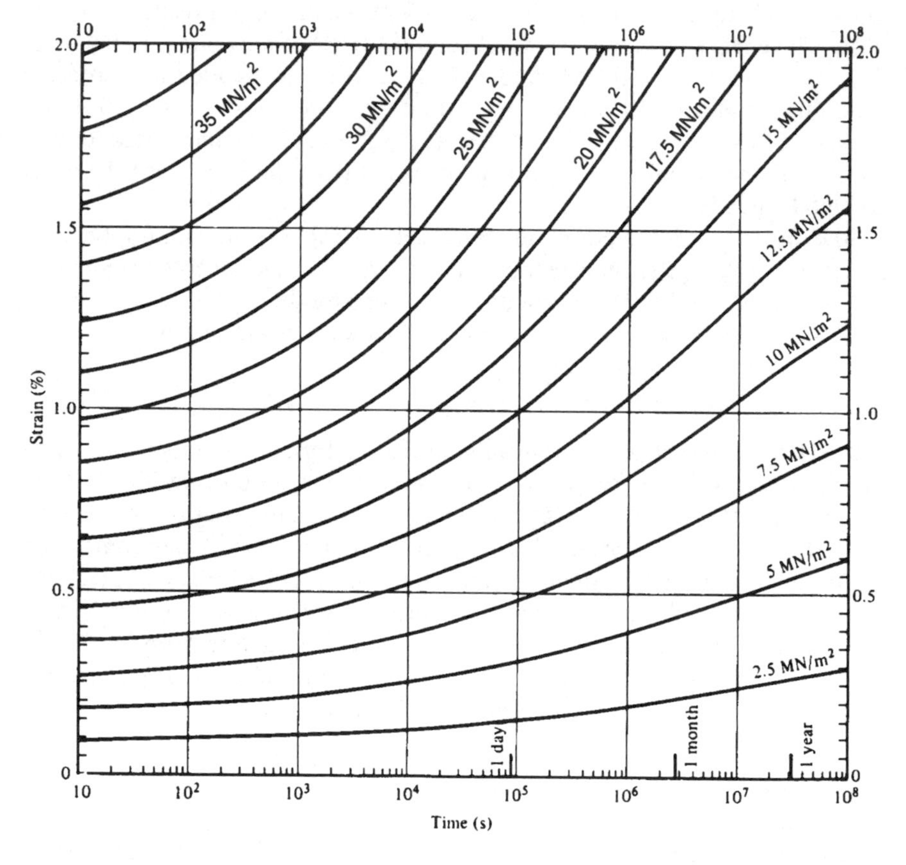

Fig. 5.7 Creep in tension; acetal copolymer, 20°C, 65% r.h. (data courtesy Amcel)

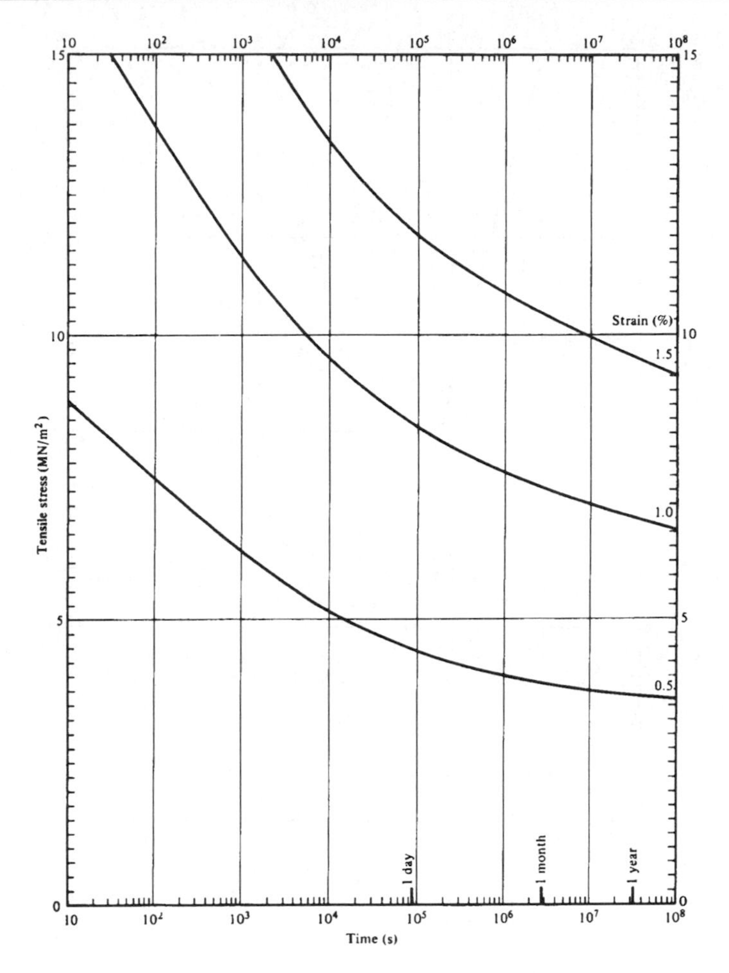

Fig. 5.8 Isometric stress versus log time; acetal copolymer, 60°C (data courtesy Amcel)

3. Acrylic sheet is available in the following thicknesses: 6, 8, 10, 12, 15, 20 and 25 mm. A bookshelf with simply supported ends, of 600 mm effective length, has to carry, at 20°C for 3 years, paperback books 120 mm wide which have a mass of 11 kg/m. What shelf thickness would you use if the maximum sag in the shelf should not exceed 3 mm?
4. By comparison with the maximum bending stress approach, what is the likely percentage error if the modulus is taken at 1% strain in calculation of the deflection of a polyacetal beam 150 mm long, 10 mm deep and 2.5 mm wide under a constant load of 2.5 N at 20°C after (a) 10 s, (b) 3 years?

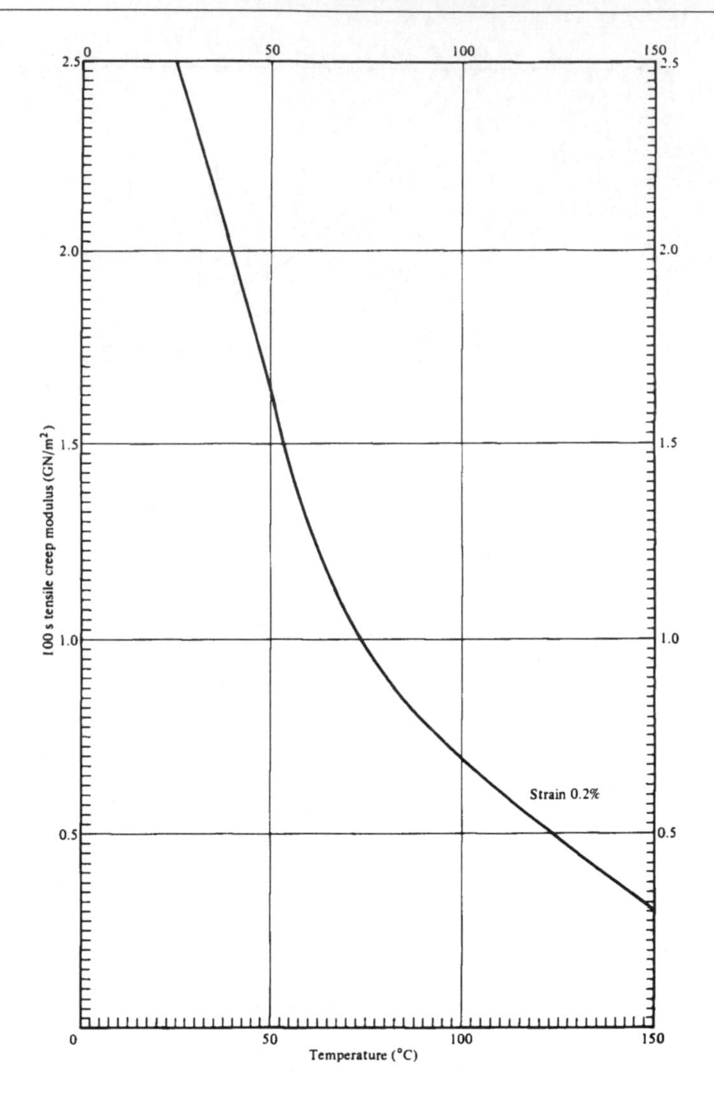

Fig. 5.9 100 s creep modulus (at 0.2% strain) versus temperature; acetal copolymer (data courtesy Amcel)

5. Part of a mechanism operating at 20°C consists of a straight cantilever spring 200 mm long made from a propylene–ethylene copolymer and which must pass through a rectangular aperture 12 × 20 mm. The proposed cross-sectional shape is shown in Fig. 5.13. The free end of the cantilever is transversely deformed by 5 mm in the direction of the arrow, and is required to exert a thrust within the load range 2.5–12.5 N for a period not exceeding 1 year.

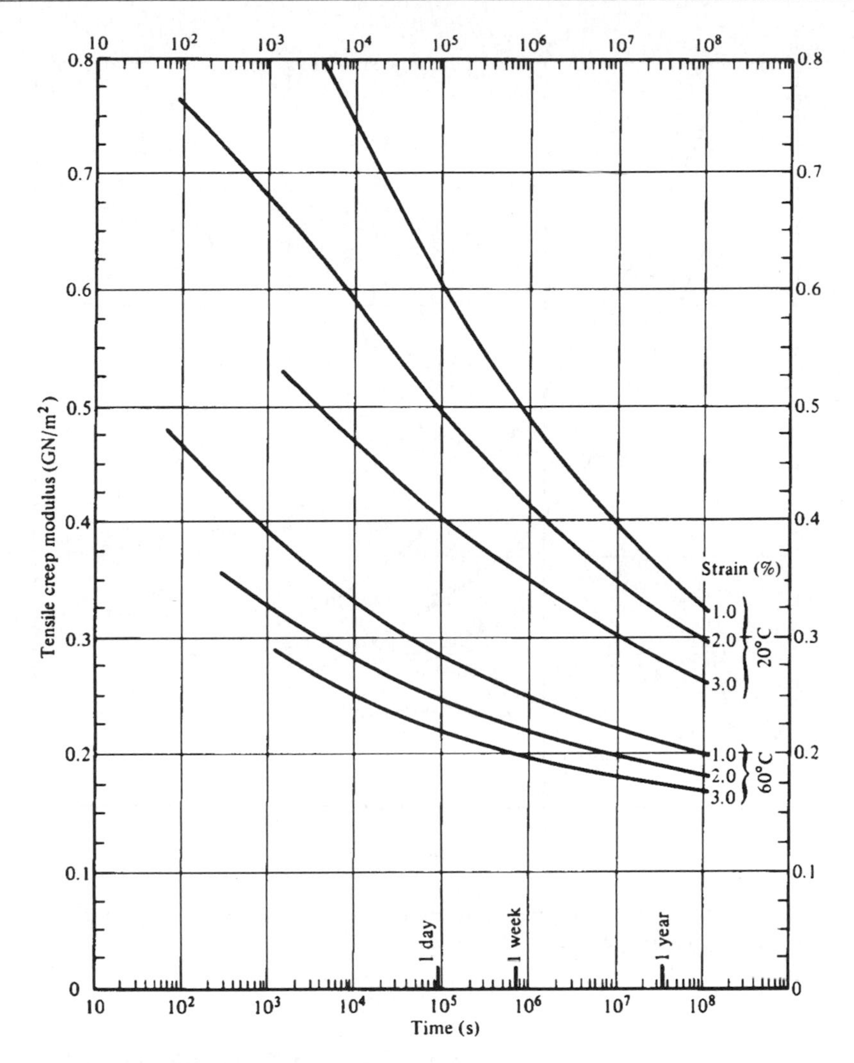

Fig. 5.10 Propylene homopolymer: tensile creep modulus versus time. Specimens annealed to a density of 0.909 g/cm^3 (data courtesy ICI)

Discuss the inadequacies of the proposed design. Recommend a more suitable design, supporting your recommendation with a dimensioned sketch, with all necessary calculations, and outlining any assumptions you make.

6. Part of a moulded mechanism consists of an end-loaded cantilever 3 mm wide, 12 mm deep and 120 mm long. It is made from acetal copolymer and operates at 20°C. The free end is deflected and held in position by a catch mechanism for 1 year. What is the minimum force exerted on the catch?

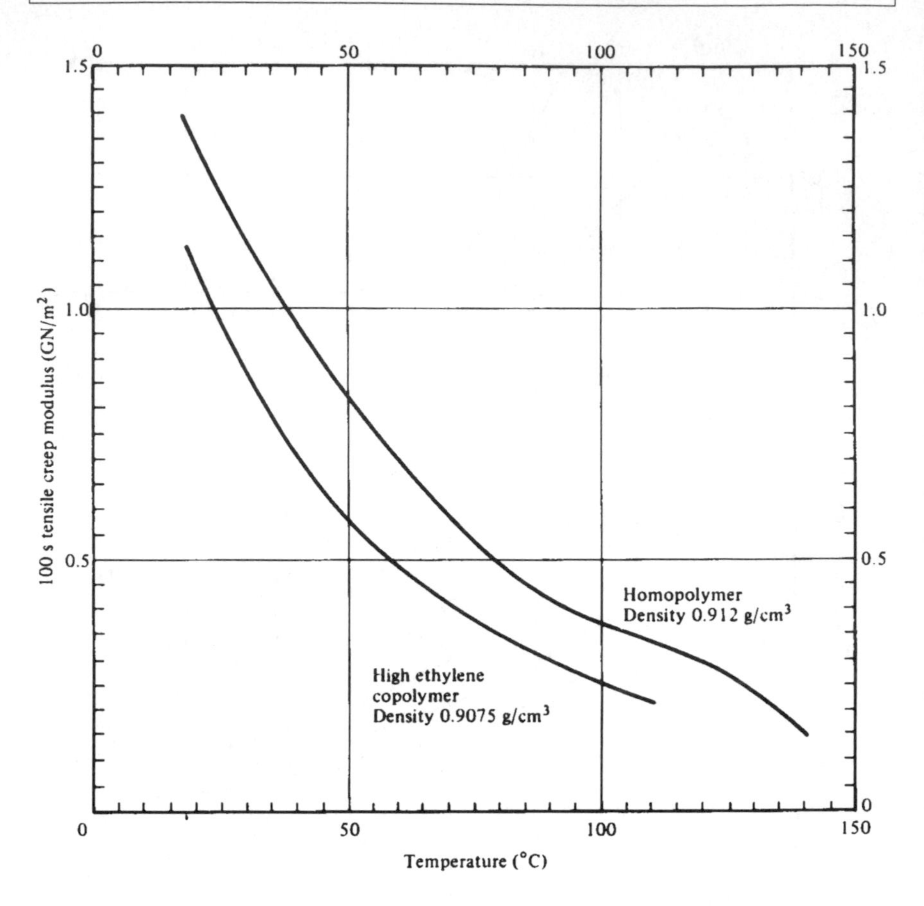

Fig. 5.11 Propylene homopolymer and copolymer: 100 s tensile creep modulus (at 0.2% strain) versus temperature (data courtesy ICI)

5.2 STIFFNESS OF PLASTICS: A MORE GENERAL APPROACH

The previous section discussed two special cases. The first was the stiffness of plastics described by modulus as measured at constant crosshead speed and constant temperature in a conventional tensile test. The second was creep under constant load, together with a design procedure for predicting deformation under constant load; in effect, this procedure amounts to time- and temperature-dependent elastic design.

This section extends the argument to more general cases; in particular, it asks, and attempts to answer, questions of the following kinds: Can we predict the strain response to a stress which varies with time? Can we generalize relationships between stress, strain and time? Can we predict a

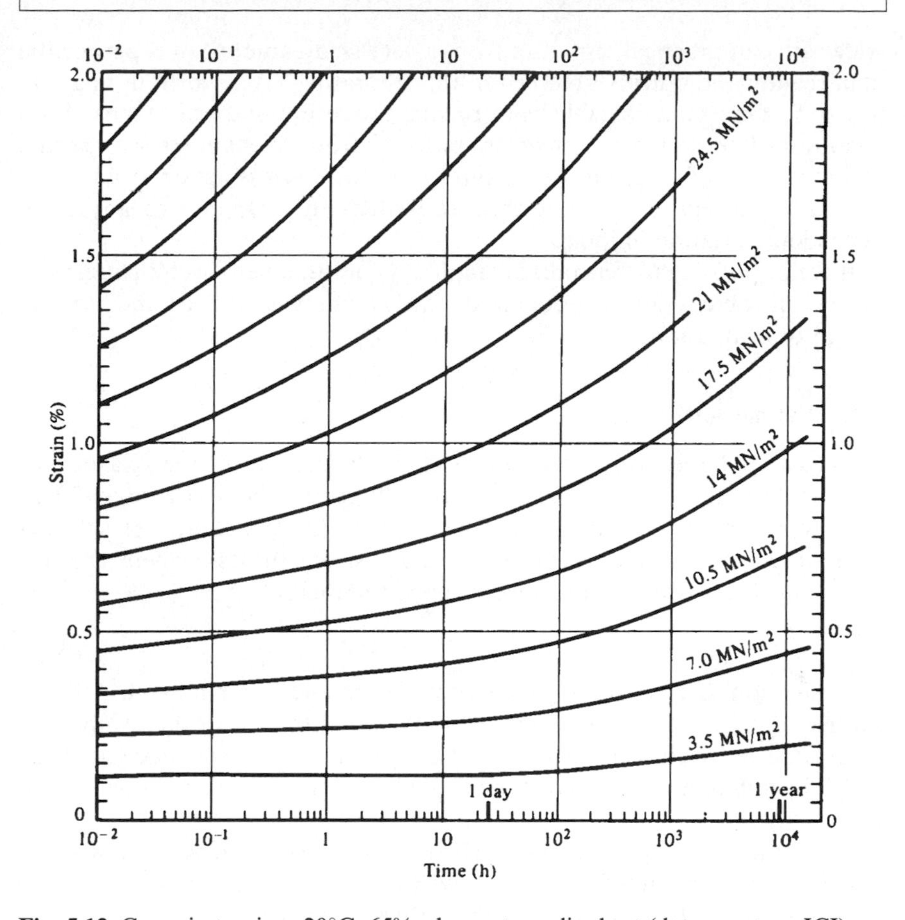

Fig. 5.12 Creep in tension; 20°C, 65% r.h., cast acrylic sheet (data courtesy ICI)

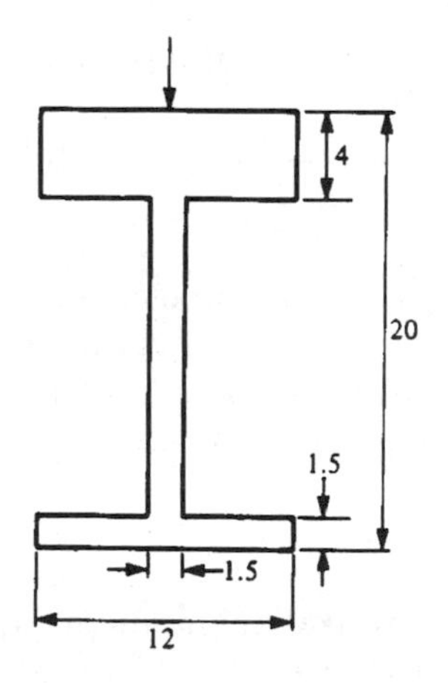

Fig. 5.13

wide range of practical responses from a very small amount of experimental data readily and quickly obtained in the laboratory? It is worth noting that any procedure should enable us to predict the response to the standard test conditions (constant stress or strain, recovery from constant stress or strain, constant rate of stressing or straining, impulse stressing or strains and cyclically varying stress or strain), as well as the response to any time-dependent variation in input.

It is not possible to generalize completely, but there are useful patterns of behaviour which can be recognized, fairly readily analysed, and used in design or evaluation.

5.2.1 Viscoelasticity

A perusal of creep behaviour shows that plastics are neither elastic nor viscous, but seem to be, at least superficially, somewhere in between. This time-dependent behaviour is termed viscoelastic. Of special interest is the behaviour at small strains, typically at 0.5% strain or less, where the real behaviour (at a given temperature) is linear viscoelastic, i.e.

$$\varepsilon = \sigma C(t) \tag{5.3}$$

The function $C(t)$ is called creep compliance and its unit is $[\mathrm{m}^2/\mathrm{N}]$. Its inverse, $1/C(t) = E_c(t)$, is the creep modulus but the use of the compliance is formally more correct; it indicates that creep under a given load (equation (5.3)) is different from relaxation following a given deformation (equation (5.5)).

In the theory of linear viscoelastic behaviour, $C(t)$ is independent of strain and only a function of time. This well-documented but rather involved theory shows relations between the different transient (time-dependent) functions, which describe the deformation behaviour of a polymer (e.g. Ferry, 1980). We will outline the relationship between static and dynamic compliance in section 5.2.2(d).

At higher strains, the behaviour becomes non-linear, i.e.

$$\varepsilon = \sigma f_1(\sigma, t) \tag{5.4}$$

and the correlation between measured behaviour and the predictions based on the theory of linear viscoelasticity become less acceptable. Non-linear viscoelasticity is not well understood and lies outside the scope of this book.

Stress relaxation is a manifestation of the same viscoelastic behaviour. It is expressed with the formula

$$\sigma = \varepsilon E_r(t) \tag{5.5}$$

where $E_r(t)$ is called the relaxation modulus with unit $[\mathrm{N}/\mathrm{m}^2]$. Theory teaches that the product $C(t) \times E_r(t) \leqslant 1$; in most cases, however, it is less than 5% smaller, depending on the magnitude of t and on the polymer. On the basis of models we will return in section 5.2.3 to some aspects of the difference between $E_c(t)$ and $E_r(t)$.

The principle of correspondence can now be specified somewhat more precisely in that the modulus used should be either the creep modulus, $E_c(t) = 1/C(t)$ (section 5.1.4(a)), if the load is prescribed; or the relaxation modulus, $E_r(t)$, if the deformation is prescribed. Each of these should take the place of Young's modulus E in the relevant elasticity formula.

5.2.2 Superposition

The simplest approach to predicting the strain response (creep) to a time-dependent load (or the stress response (relaxation) to a time-dependent deformation) uses the concept of superposition. According to Boltzmann, superposition assumes that the strains from different stress inputs (and vice versa) at different times are independent and additive. At time zero, the stresses and strains are all assumed to be zero. Superposition assumes that the response to a sequence of loadings is the sum of the individual isolated responses made at the appropriate time intervals. Correlation between theory and practice is quite good, provided the material response at any time is within the limits of linear viscoelasticity (we discuss these limits in section 5.2.4(a)). The following examples illustrate the procedure.

(a) Superimposed load

Suppose a stress σ_1 is applied at $t = 0$ and then at time t_1 the stress is increased to σ_2 (Fig. 5.14). The strain at some later time $t > t_1$ is the superposition of the strain ε_1 caused by σ_1 and of the strain ε_2 caused by the increase of σ_1 to σ_2, applied at $t = t_1$:

$$\varepsilon = \varepsilon_1 + \varepsilon_2 = \sigma_1 C(t) + (\sigma_2 - \sigma_1)C(t - t_1) \tag{5.6}$$

Because of the nature of the function C, the strain response to this stress sequence for $t \gg t_1$ is essentially the same as if σ_2 had been applied (alone) from $t = 0$, and also as if σ_1 had never been applied.

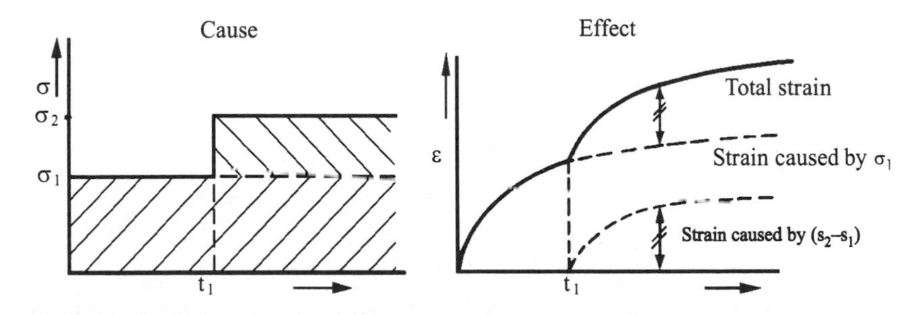

Fig. 5.14 Superimposed load and related deformation response

(b) Recovery from creep

At the end of a deformation test on mild steel within the elastic limit, on removing the load the steel returns immediately to its original unstrained state. At the end of a creep test on a metal, only the elastic deformation is recoverable, not the plastic deformation.

Plastics behave rather differently. On removing the load at the end of a test, the plastics material can recover most if not all the strain, provided it has not yielded, but the recovery process takes time. Recovery can be rapid and complete under short-term low-stress (linear) creep, but is overdelayed and may not realize completion under long-term creep or high-stress (non-linear) creep (or both).

The calculation based on superposition is as follows. Suppose a constant stress, σ_1, is applied at time $t = 0$ [giving a strain response $\varepsilon_1(t)$] and removed at time t_1. The strain for $0 < t < t_1$ will be

$$\varepsilon_1(t) = \sigma_1 C(t)$$

The removal of load at t_1 can be modelled as continuation of the stress σ_1, for $t > t_1$, to which is added a negative stress, $-\sigma_1$, applied at $t = t_1$ (Fig. 5.15). [The response to $-\sigma_1$ acting on its own from time t_1 is simply $\varepsilon_2(t - t_1) = -\sigma_1 C(t - t_1)$.] By adding the two responses, the resultant strain, ε, for $t > t_1$ is given by

$$\varepsilon(t) = \varepsilon_1(t) + \varepsilon_2(t - t_1) = \sigma_1 C(t) - \sigma_1 C(t - t_1) \tag{5.7}$$

(c) Integral form of superposition

So far, the emphasis has been on discrete step-function changes of stress at 'long' time intervals (e.g. longer than 1 min). If the strain response has the form $\varepsilon = \sigma C(t)$ and the change in stress at time t_1 is small, e.g. $\delta\sigma$, then the strain at $t > t_1$ may be adapted from equation (5.6):

$$\varepsilon(t) = \sigma C(t) + C(t - t_1)\delta\sigma \tag{5.8}$$

If the stress history consists of a series of steps $\delta\sigma$ applied at times t_n, then the strain response may be written as

$$\varepsilon(t) = \sum_{n=0}^{N} C(t - t_n)\delta\sigma_n \tag{5.9}$$

In the limit this reduces to the convolution integral

$$\varepsilon(t) = \int_{-\infty}^{t} C(t - t')\frac{\mathrm{d}\sigma}{\mathrm{d}t'}\mathrm{d}t' \tag{5.10}$$

which says that the strain at any time t depends on the previous loading history, $\sigma(t')$ (introduced as a series of steps of stress in time $\{\mathrm{d}\sigma(t')/\mathrm{d}t'\}\mathrm{d}t'$),

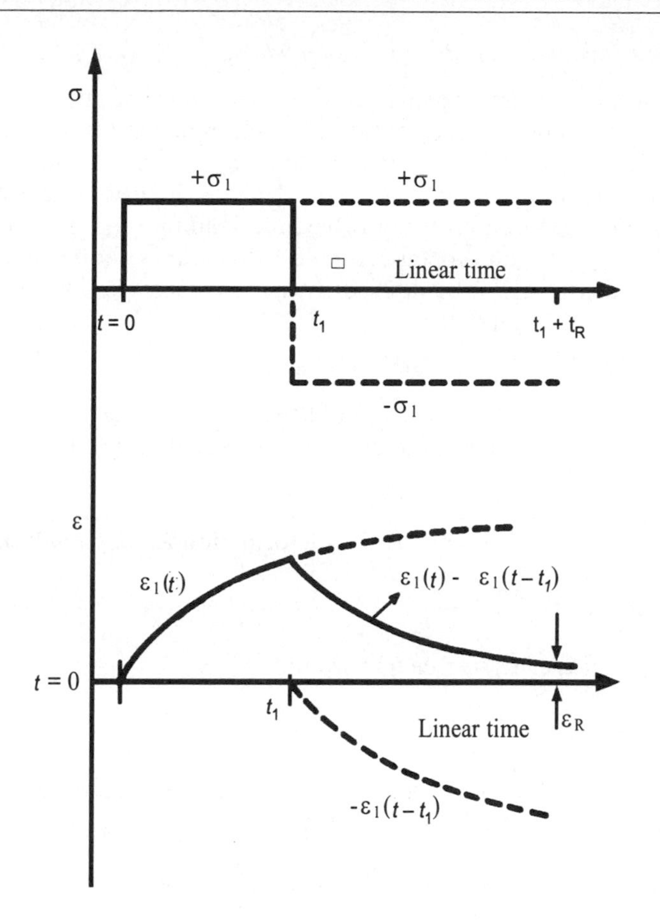

Fig. 5.15 Schematic creep recovery

and the material compliance $C(t)$. Note the meaning of t, the elapsed time between a stress signal and the moment of observation of the resulting strain (this is a kind of looking back), and t', the running time which determines the actual value of σ (a kind of looking forward from the first application of σ).

The convolution integral (equation (5.10)) forms the basis of linear visco-elasticity theory, which permits the prediction of strain from any arbitrary form of stress input. Analysis usually involves Laplace transforms. Relationships between different types of inputs and responses can be compared, e.g. the similarity between creep and stress relaxation. This theory contains the basic principle of correspondence, which provides justification of the use of time-dependent elasticity (section 5.1.4) for problems involving long-term constant stress. These aspects of analysis will not be explored further using the convolution integral approach, but the interested reader will find amplification of these remarks in Ferry (1980), Williams (1980) or Lockett (1981).

(d) Dynamic modulus; relation to creep modulus

The convolution integral provides means to calculate the effect of changing load. One of the most common forms is a cyclic stress input equation (5.11). This introduces us into the area of short loading times and dynamic loading, which cannot be studied with creep tests. There is, however, a clear theoretical connection between creep and dynamic loading, and one of the outcomes of this connection for viscoelastic behaviour is the phenomenon of damping, which first will be defined as the loss factor $\tan\delta$. Suppose the stress input has the cyclic form

$$\sigma(t') = \sigma_0 \sin \omega t' \tag{5.11}$$

(we choose t' to indicate the input and keep t for the time of observation of the output). ω is a constant. The derivative of equation (5.11) is

$$\mathrm{d}\sigma(t')/\mathrm{d}t' = \omega\sigma_0 \cos \omega t' \tag{5.12}$$

Choosing $s = t - t' \Rightarrow \mathrm{d}s = -\mathrm{d}t'$, the deformation at time t will be

$$\varepsilon(t) = \int_0^\infty C(s)\, \omega\sigma_0 \cos \omega(t-s)\mathrm{d}s \tag{5.13}$$

By working out the cosine function we find

$$\varepsilon(t) = \sigma_0\omega \sin \omega t \int_0^\infty C(s) \sin \omega s\, \mathrm{d}s + \sigma_0\omega \cos \omega t \int_0^\infty C(s) \cos \omega s\, \mathrm{d}s$$

The deformation ε has the same frequency ω but not the same phase as σ. It is usual to introduce the following values:

$$\text{Storage compliance } C'(\omega) = \omega \int_0^\infty C(s) \sin \omega s\, \mathrm{d}s \tag{5.14}$$

$$\text{Loss compliance } C''(\omega) = -\omega \int_0^\infty C(s) \cos \omega s\, \mathrm{d}s \tag{5.15}$$

Both of these are fully determined when ω and the full range $(0 < t < \infty)$ of $C(t)$ are given. The minus sign in C'' is chosen in order that C'' would be positive. Equation (5.13) can now be written as

$$\varepsilon(t) = \sigma_0(C' \sin \omega t - C'' \cos \omega t) \tag{5.16}$$

The maximum amplitude is

$$\varepsilon_0 = \sigma_0\sqrt{(C'^2 + C''^2)} \tag{5.17}$$

The loss angle δ is defined as

$$\tan\delta = C''/C' \tag{5.18}$$

We can now rewrite equation (5.16):

$$\varepsilon(t) = \varepsilon_0(\sin \omega t \cos\delta - \cos \omega t \sin\delta) \tag{5.19}$$

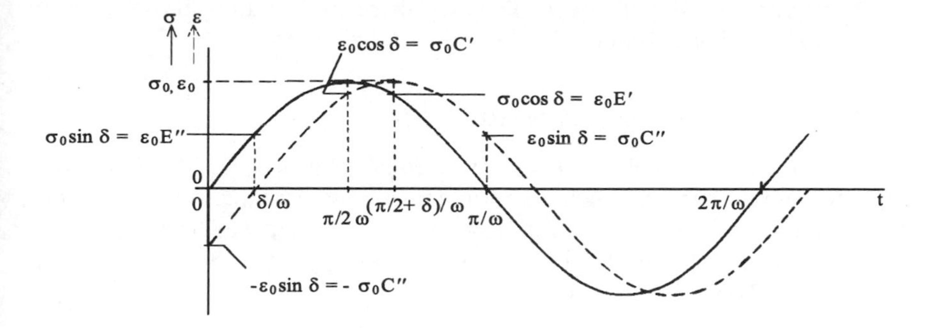

Fig. 5.16 Retarded strain (---) response to sinusoidal stress (——) input

which is similar to

$$\varepsilon(t) = \varepsilon_0 \sin(\omega t - \delta) \tag{5.20}$$

If $C(t)$ is fully known and if ω is given, δ can be calculated as well as the dynamic response to the cyclic input (Fig. 5.16).

Using equations (5.14) and (5.15) it is now possible to calculate the dynamic compliance $C'(\omega)$ and $C''(\omega)$ for sinusoidal load as a function of frequency from the creep compliance $C(t)$, which is measured at constant load as a function of time, and vice versa. The numerical calculation, however, can be troublesome. In principle it is only possible if measurements are available for the full time (frequency) spectrum from 0 to ∞. If the measuring points are chosen carefully, e.g. in a logarithmic row, calculations can be simplified considerably. Because of experimental limitations, both dynamic and creep measurements are necessary in order to cover the full time (frequency) spectrum from 0 to ∞. The results can be recorded on one scale with equivalent t and $1/\omega$. They will show that C decreases (and stiffness increases) as the loading time t decreases or frequency ω increases.

From Fig. 5.16 it can be understood that for the dynamic case relaxation and creep are closely related; it is a matter of experimental definition whether load $\sigma(t) = \sigma_0 \sin \omega t'$ or deformation $\varepsilon(t) = \varepsilon_0 \sin(\omega t' - \delta)$ is given. If deformation were given, calculation of the load would proceed via storage modulus $E'(\omega)$ and loss modulus $E''(\omega)$ (without the minus sign) and define the same loss angle $\tan \delta = E''/E'$. This leads to a clear relation between E' and E'' on the one hand and C' and C'' on the other, for which we refer the reader to the books mentioned in section 5.2.2(c).

Tan δ is often recorded as a function of temperature and, as such, an indicator of the presence of a relaxation mechanism that causes a transition between levels of stiffness of the polymer. Tan δ is usually freely translated as damping and as such it plays a role in the damping of vibrations (section 5.3.1(d)). With dynamic loading a high value of tan δ indicates conversion of mechanical into thermal energy and thus heat build-up, for example in rubber tyres, but also in other vibrating elements including those used in ultrasonic welding assembly.

5.2.3 Models for viscoelastic behaviour

Models can help us to obtain a better notion of the behaviour of the material they represent. They may help us to interpret and sometimes even to extrapolate the behaviour. We should, however, be very cautious with the latter, as this may exceed the limit of their validity, into an area where other mechanisms become dominant.

(a) Spring and dashpot models

The precursor to viscoelasticity theory took the view that the behaviour of a material could be described by models consisting of combinations of springs and dashpots.

For example, the elastic behaviour of metals can be represented by a linear spring (Fig. 5.17a). The stiffness of the spring, K, is related to the applied load, P, and the deformation, x, by

$$K = P/x$$

Thus, for the special and convenient case of a spring having unit length and unit cross-sectional area, the stiffness becomes the tensile modulus, E:

$$\sigma/\varepsilon = E \tag{5.21}$$

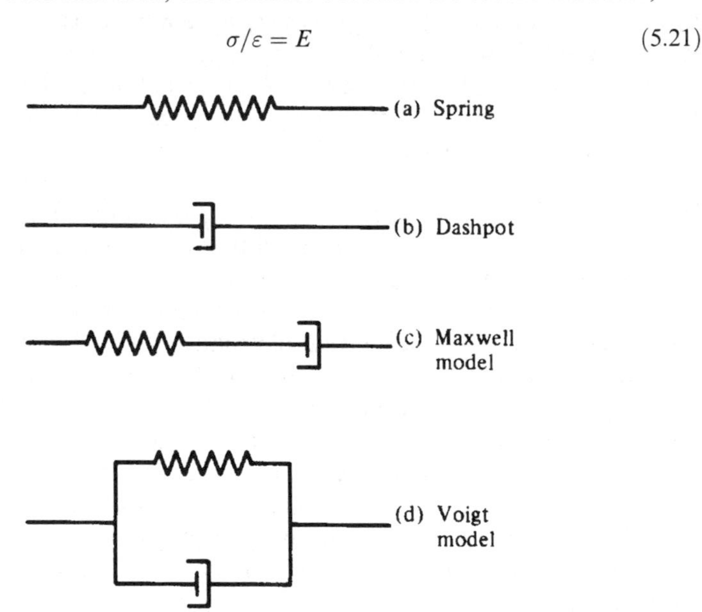

Fig. 5.17 Simple spring and dashpot models

The behaviour of a fluid may be described by a dashpot (Fig. 5.17b). The viscosity (here assumed constant) of the fluid controlling the dashpot behaviour is related to the applied axial force, P, and the rate of extension, x, by

$$P = \eta \frac{\mathrm{d}x}{\mathrm{d}t}$$

If the dashpot has unit length and cross-sectional area, the dashpot behaviour is described by

$$\sigma = \eta \frac{\mathrm{d}\varepsilon}{\mathrm{d}t} \tag{5.22}$$

The spring and the dashpot elements may now be combined together in different ways. Our objective is to ascertain if any simple combination can be useful in describing the qualitative form of the stress–strain relationship for linear viscoelastic behaviour. At this stage, we are not trying to define quantitative relationships. The point in seeking a simple model of the correct form is that, if it describes known experimental behaviour, it is hoped that it would also work for an unknown stress or strain input: the predictive capacity of the model for the new experience can then be tested by experiment.

The spring on its own is not useful in that it does not describe time-dependent behaviour. The dashpot alone cannot describe recovery behaviour. The Maxwell model (Fig. 5.17c) consists of a spring and a dashpot in series. This alone does not describe complete recovery because of the dashpot. The Voigt model (Fig. 5.17d) consists of a spring and a dashpot in parallel. This will not describe stress relaxation under constant deformation.

The next simplest model is the **standard linear solid** (Fig. 5.18) which consists of a spring in series with the Voigt element. Because there are two springs, E_1 to represent the glassy and E_2 to represent the rubbery behaviour, this seems to work quite well as a model of polymer behaviour. The analysis consists of developing the linear differential equation relating the overall stress, σ, and the strain, ε, in the model, and then solving the equation for particular inputs of stress or strain.

Equilibrium and compatibility give the following relationships:

$$\begin{gathered} \sigma = \sigma_1 = \sigma_2 + \sigma_3 \\ \varepsilon = \varepsilon_1 + \varepsilon_2 \qquad \varepsilon_2 = \varepsilon_3 \end{gathered} \tag{5.23}$$

The subscripted terms can be expressed in terms of σ and ε by the following steps:

$$\varepsilon_1 = \sigma_1/E_1 = \sigma/E_1 \Rightarrow \qquad \varepsilon_2 = \varepsilon - \sigma/E_1 \tag{5.24}$$

$$\sigma_2 = E_2\varepsilon_2 = E_2\varepsilon - E_2\sigma/E_1 \tag{5.25}$$

$$\sigma_3 = \sigma - \sigma_2 = \sigma - E_2\varepsilon + E_2\sigma/E_1 \tag{5.26}$$

$$\frac{\mathrm{d}\varepsilon_3}{\mathrm{d}t} = \frac{\sigma(1 + E_2/E_1) - \varepsilon E_2}{\eta_3} \tag{5.27}$$

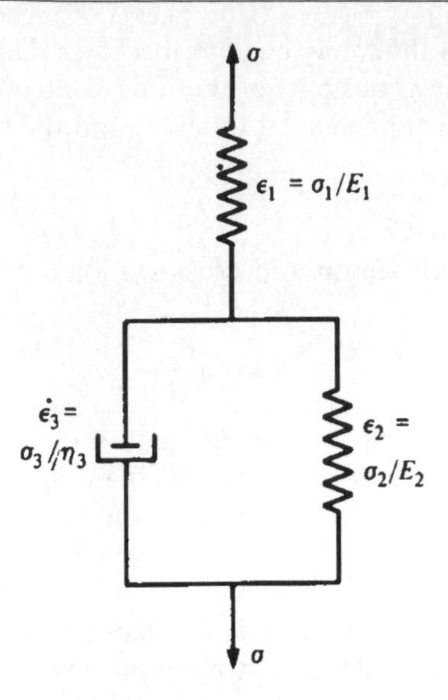

Fig. 5.18 The standard linear solid

If the Voigt subsystem is to keep together, $\dot{\varepsilon}_3 = \dot{\varepsilon}_2$, i.e. from equation (5.24):

$$\frac{\mathrm{d}\varepsilon_3}{\mathrm{d}t} = \frac{\mathrm{d}\varepsilon_2}{\mathrm{d}t} = \frac{\mathrm{d}\varepsilon}{\mathrm{d}t} - \frac{1}{E_1}\frac{\mathrm{d}\sigma}{\mathrm{d}t} \tag{5.28}$$

Equating equations (5.27) and (5.28), and rearranging, gives the general relationship between the applied stress and the overall strain:

$$\frac{\mathrm{d}\varepsilon}{\mathrm{d}t} + \frac{E_2}{\eta_3}\varepsilon = \frac{E_1 + E_2}{\eta_3 E_1} + \frac{1}{E_1}\frac{\mathrm{d}\sigma}{\mathrm{d}t} \tag{5.29}$$

This equation can now be solved for particular inputs of stress or strain as the following examples show.

Creep
Putting $\sigma = \sigma_0 =$ constant, $\mathrm{d}\sigma/\mathrm{d}t = 0$, so equation (5.29) becomes

$$\frac{\mathrm{d}\varepsilon}{\mathrm{d}t} + \frac{E_2}{\eta_3}\varepsilon = \frac{E_1 + E_2}{\eta_3 E_1} \tag{5.30}$$

This may be solved to give

$$\varepsilon(t) = \frac{\sigma_0}{E_1} + \frac{\sigma_0}{E_2}\left[1 - \exp\left(-\frac{E_2 t}{\eta}\right)\right] \tag{5.31}$$

Equation (5.29) may be expressed in terms of the creep compliance:

$$C(t) = \varepsilon(t)/\sigma_0 \tag{5.32}$$

giving

$$C(t) = \frac{1}{E_1} + \frac{1}{E_2}\left[1 - \exp\left(-\frac{E_2 t}{\eta}\right)\right] \tag{5.33}$$

At very short time, $t \to 0$, the exponential term tends to unity, and hence

$$C(t) \to 1/E_1$$

At the other end of the timescale, as $t \sim \infty$, the exponential term disappears and hence

$$C(t) \to \frac{1}{E_1} + \frac{1}{E_2}$$

The creep modulus, $E_c(t)$, is the inverse of the creep compliance, so the limits of creep modulus described by the standard linear solid are E_1 and $E_1 E_2/(E_1 + E_2)$, as shown in Fig. 5.19. The shape of the transition region depends on $E_2 t/\eta$ and most of the change occurs around the time denoted as

$$\tau_c \simeq \eta/E_2 \tag{5.34}$$

This time is called the **retardation time** (the suffix 'c' denoting creep).

The plot of creep modulus versus dimensionless time certainly has the right qualitative form, and parts of the sigmoidal shape can be seen in the creep data for polypropylene, particularly the right-hand end (Fig. 5.10). The major practical problem is that, for many real materials, the change from the short-term, high-modulus value to the long-term, lower bound

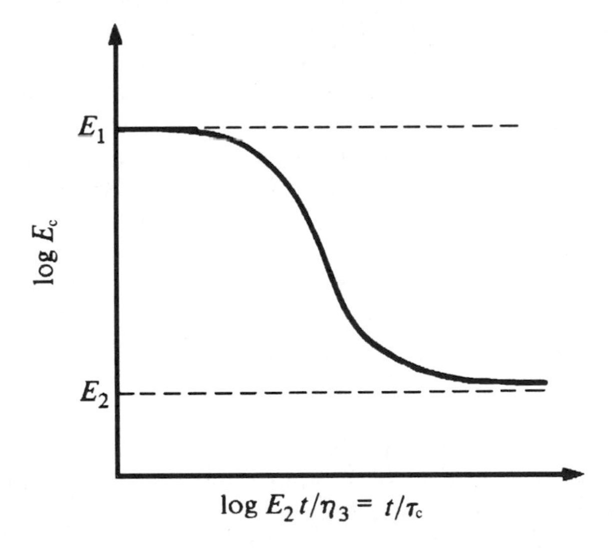

Fig. 5.19 Creep modulus versus time

occurs over a wide time range, e.g. 6–10 orders of magnitude, which often spans at least part of the timescale of normal use. This is a criticism of the way polymers behave rather than of the model.

It is interesting to note that already the concept of the standard linear solid has enabled us to see a general case – the upper and lower bounds to stiffness – which a preoccupation with detailed data such as creep had not so far revealed.

Stress relaxation

A similar analysis can be made of the response of this model to stress relaxation by putting $\varepsilon = \varepsilon_0$, a constant. The basic differential equation now becomes

$$\frac{1}{E_1}\frac{\mathrm{d}\sigma}{\mathrm{d}t} + \frac{E_1 + E_2}{\eta_3 E_1}\sigma = \frac{E_2}{\eta_3}\varepsilon \tag{5.35}$$

On solving this, and noting that the relaxation modulus, $E_\mathrm{r}(t) = \sigma(t)/\varepsilon_0$ we find:

$$E_\mathrm{r}(t) = E_1 - \frac{E_1^2}{E_1 + E_2}\left[1 - \exp\left(\frac{t}{\tau_\mathrm{r}}\right)\right] \tag{5.36}$$

where the **relaxation time**,

$$\tau_\mathrm{r} = \eta/(E_1 + E_2) \tag{5.37}$$

the suffix ‘r’ denoting stress relaxation.

The following points should be noted.

1. Relaxation has two meanings in this context. First, in ‘relaxation modulus’, relaxation is short for ‘stress relaxation’, i.e. stress decays at constant deformation. Second, relaxation in ‘relaxation time’ means the gradual decay of modulus from the short-term to the long-term value. (It is semantically correct but confusing to describe τ_c as the creep relaxation time, i.e. measured under constant load, as modelled, and τ_r as the stress-relaxation relaxation time because it is measured in a stress relaxation experiment.)
2. τ_r (equation (5.37)) is always less than τ_c (equation (5.34)): this is obvious from the algebra.
3. $E_\mathrm{r}(t)$ is always rather less than $E_\mathrm{c}(t)$ – in practice, the discrepancy is usually only 1 or 2% over the range of practical interest.

(*b*) *Exponential function of time*

Upon presentation of creep data, it has been observed in a number of cases that these give straight lines on a double log scale (Fig. 5.2c). The mathematical representation for such lines is

$$\varepsilon = ct^n \tag{5.38}$$

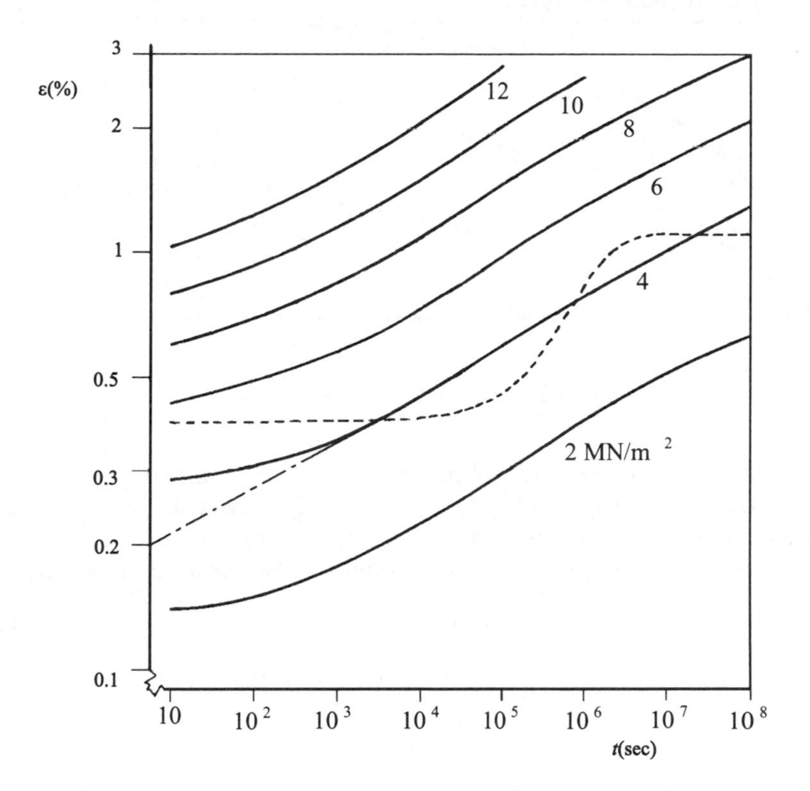

Fig. 5.20 Creep data for a polypropylene, 20°C on a double logarithmic scale. The data for $\sigma = 4\ \mathrm{MN/m^2}$ can be approximated with $C_1 = 0.425 \times 10^{-3}\ \mathrm{m^2/MN}$, $n = 0.11$ (see its continuation as ----- line). The dotted line gives the least squares approximation of the data for $\sigma = 4\ \mathrm{MN/m^2}$ according to the standard linear solid with $E_1 = 1030\ \mathrm{MN/m^2}$, $E_2 = 577\ \mathrm{MN/m^2}$, $\tau_c = 1.05 = 1.05 \times 10^6$ s.

The related model for the creep compliance is

$$C(t) = C_1 t^n \quad \text{or more precisely} \quad C(t) = C_1 (t/t_1)^n \tag{5.39}$$

where C_1 is the compliance for unit time ($t = 1$ s or $t = t_1$).

In Fig. 5.20 creep data for a polypropylene are represented in such a way. This indicates that, *for a certain region of time* (see the right-hand, rather straight parts of the curves) equation (5.38) gives a good approximation. For a number of thermoplastics in a relevant region this approximation has proven to be very useful.

The general shape of creep curves with small and large values of n respectively is indicated in Fig. 5.21; $n = 0$ represents a purely elastic material, $n = 1$ a purely viscous one. A material with a high value of n has a

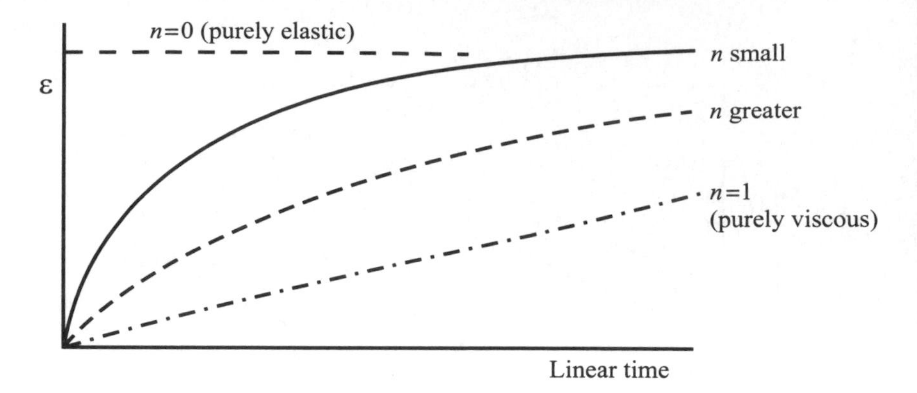

Fig. 5.21 Graph representing $\varepsilon = ct^n$ for different values of n

rather viscous behaviour and will not be suitable for engineering purposes. Plastics suitable for engineering show values of $0.1 < n < 0.2$; n and C_1 can be considered to be materials constants within certain limitations.

The same approximation gives as a model for the relaxation modulus:

$$E_r(t) = E_1 t^{-n} \tag{5.40}$$

Theory of the viscoelastic behaviour shows that in this model:

$$C(t) \times E_r(t) = \frac{\sin n\pi}{n\pi} \tag{5.41}$$

For the mentioned values $0.1 < n < 0.2$ we find $0.98 < \sin n\pi/n\pi < 0.94$. This demonstrates that the error is not great if we confuse E_c and E_r. (compare item 3 of list in section 5.2.3(a) above, and the start of section 5.2.1).

Theory further shows that

$$\tan\delta = \tan\left(\frac{1}{2}n\pi\right) \tag{5.42}$$

The consequence is that $\tan\delta$ would be a constant for the full frequency–time spectrum. This is obviously not true and it shows the limitations of this kind of model.

The model is, however, quite often used, but one should be well aware of its limitations. One of its uses is for the extrapolation of creep measurements to higher values of time. This is usually allowable over a decade, or even further if the materials data do not suggest the close proximity of a maximum of $\tan\delta$ (see section 5.2.6(b)).

In comparing both models, the standard linear solid and an exponential function of time, it is apparent that the first emphasizes more the extremes (e.g. rubbery and glassy behaviour), whereas the second points more to the transition between the two.

5.2.4 Deviations from viscoelastic behaviour

(*a*) *Non-linear stress–strain curve*

The stress–strain relationship is usually curved. This applies not only to the one obtained in the conventional tensile test (Fig. 5.1), but also to the isochronous stress–strain relationship as obtained from creep data (Fig. 5.4c). In section 5.1.1 we indicated that the lower left part of the curve may be considered to be a straight line, allowing linear elasticity to be used according to the principle of correspondence. This is allowable as long as the secant modulus E_s does not deviate from the modulus at the origin E_0. This means that there is a limit to proportionality, namely ε_p, and $\sigma_p = \varepsilon_p E_0$, beyond which linear elasticity calculations are not valid.

For normal practice this limit is often too low (it may be very close to zero) and a higher limit is chosen according to the required accuracy of the calculations. In view of other sources of inaccuracy, it is often set at $E_s/E_0 \geqslant 0.85$. These values can be found in the isochronous stress–strain diagram by drawing the tangent at the origin (which defines E_0), drawing the line for $E_s = 0.85 \times E_0$ through the origin, and defining the section with the curve as ε_u, then $\sigma_u = \varepsilon_u E_s$ (Fig. 5.22). Even beyond this limit the material

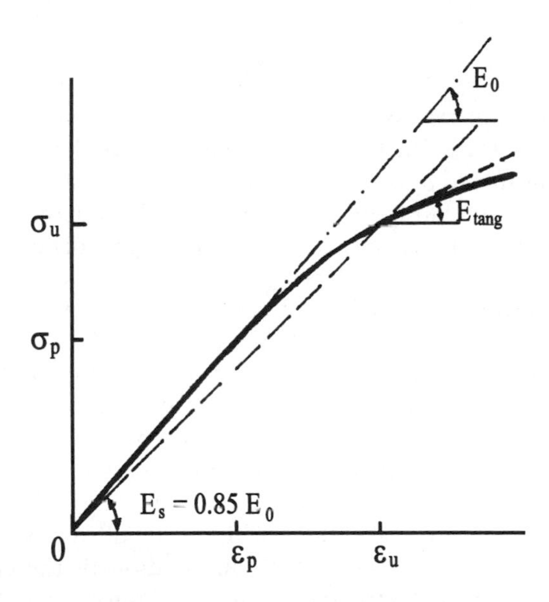

Fig. 5.22 Proportional and 'ultimate elastic' limits indicated in an isochronous σ–ε curve

may still behave properly, but elasticity calculations to prove this will not be valid, nor the principle of superposition. Before or after an elasticity calculation, validation that $\sigma_{max} < \sigma_u$ or $\varepsilon_{max} < \varepsilon_u$ has to be carried out.

Non-linear behaviour plays an important role with bending. Here, in contrast to pure tensile load, the strain in a cross-section changes; in fact it is proportional to the distance from the neutral axis (maximum distance termed c). As long as the strain stays below ε_p, the stress will also have this proportionality, but not beyond ε_p. This means that elastic formulae such as $f = ML^2/(2EI)$ and $\sigma_B = Mc/I$ are not valid there. However, since the strain near the neutral axis is small and likely to be smaller than ε_p, part of the cross-section still obeys proportionality. Compared with pure tensile stress, where ε is equal throughout the section, deviation of deformation from the elastic analogue will be smaller in bending.

Measurement of the modulus in tensile and in bending should give the same value as long as the strain is within the proportionality limit ε_p. In many standard modulus tests a maximum strain of $\varepsilon = 0.25\%$ is prescribed. Insofar as even this would exceed ε_p, it should be clear from the above that a lower modulus would be found for tensile testing than for bending testing with the same nominal strain. Therefore in bending, somewhat higher ε is allowed than in the case of tensile load, and in fact the magnitude of ε_u proposed above is best suited for bending rather than for tensile load.

For bending tests a higher strain (up to 3.5%) is sometimes allowed. The non-linear behaviour will certainly then result in a lower bending modulus.

(*b*) *Physical aging*

Above it was supposed that the properties $C(t)$ and $E(t)$ were independent of time, i.e. that these functions of a polymer should be the same regardless of the moment in real time that t (at first loading) was set to zero. In simple terms, a product should have the same properties whether it is new or old. This, apparently, is not exactly true. A polymer product cannot be regarded as fully stable.

The normal temperature of use of polymers is not as far below the transition temperatures T_m (melting temperature of crystallites) or T_g (glass transition temperature) as is the case with most metals, which are used well below their melting points. The molecular structure of polymers is rather open, leaving space for the movement of chain parts. These parts will therefore continue to move according to thermally induced movement, until they are constrained by surrounding forces (section 2.6.3). This process, which leads to an increase in density and in stiffness, is called physical aging.

A similar source of change is the continuation of the curing (chemical crosslinking) process in thermosets. Of a different nature, but also leading to fundamental changes in properties, are secondary crystallization, dynamic fatigue, loss of additives, breakdown of chains caused by chemical or physical influences, etc. In contrast to physical aging, most of these processes are irreversible.

Physical aging is a property of the amorphous phase, including the amorphous regions in partially crystalline polymers. It is caused by the mobility of molecular chain parts in the so-called free volume. Above T_g the chains are normally too far apart and their thermal motion is too great to be able to develop stable contacts with their neighbours. Below T_g (and with partially crystalline polymers even above T_g) the situation is different. If and when a chain part is caught in the attraction field of another part, its free movement ceases, leading to a higher local density and a higher stiffness, and impeding the free mobility of surrounding chain parts. So the free volume decreases but the rate of aging slows down; this process is self-retarding.

A fairly close packing can be reached in a very slow cooling process, but industrial processes never allow time for this thermodynamic equilibrium. Normal quick cooling (quenching) brings the thermal movement down to a low level and impedes the packing process, resulting in a relatively high amount of free volume (and related low density). However, thermal movement is still present and according to temperature and free volume the aging process will continue, unless a secondary transition (an immobilization of chain elements, chain sections or side groups; section 2.6.1) blocks all motion.

Physical aging can be undone by heating above T_g (or close to T_m with partially crystalline polymers) in order to restore full mobility of all chain parts.

In order to demonstrate physical aging, creep experiments were performed as presented in Fig. 5.23. It shows that, during the storage between production and testing of the sample, its stiffness increases. The various curves can be described with the formula

$$C(t) = C_0 e^{(t/t_0)^m} \tag{5.43}$$

This represents a very general type of behaviour of amorphous materials not too far below their T_g, at low stress and for creep times that were clearly shorter than the preceding storage time after quenching. Quenching, storage and creep temperatures should be the same. A magnitude of m of about $\frac{1}{3}$ seems to apply to a broad spectrum of materials.

The storage time t_e determines the magnitude of t_0:

$$t_0 = t_2 \left(\frac{t_e}{t_1}\right)^{\mu} \tag{5.44}$$

where t_1, t_2 and μ are parameters dependent on the material and on the conditions of testing such as temperature. The shift rate μ is zero above T_g; just below T_g it rapidly increases to about unity, remaining at unity over a small or wide (depending on the polymer) temperature range below T_g. Finally at lower temperatures, due to a secondary transition, the aging process slows down and μ decreases.

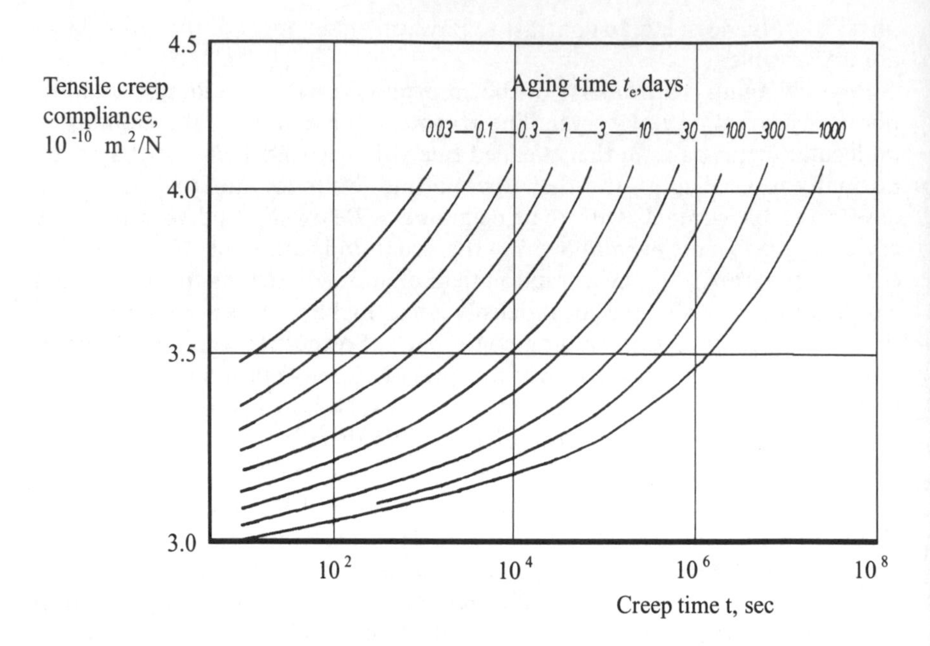

Fig. 5.23 Creep compliance of rigid PVC as measured at 20°C after quenching from 90°C to 20°C and storing at 20°C. The period of storage before loading for creep is indicated as t_e. (courtesy L.C.E. Struik)

If we now consider creep experiments, which exceed the storage time, it is clear that the aging process will also continue during creep under load. Thus creep and aging will act simultaneously on the sample, which leads to deviation from relation (5.43). This is shown in Fig. 5.24. The shape of the measured curves fits with one that could be calculated on the basis of a storage time $t + t_e$, which progresses with creep time t.

The above clearly shows that the magnitude of the creep modulus and even the shape of the creep curve is determined by the history of the sample. This makes it clear that storage of samples should be part of any testing procedure. It also demonstrates that production, cooling and storage will influence the stiffness of products. Besides, the changes brought about by physical aging may also affect yield stress and brittleness.

5.2.5 Time-dependent buckling

Buckling of viscoelastic structures is a time-dependent phenomenon, like creep and relaxation. Its manifestation, however, is different; under certain circumstances of shape, loading, etc., buckling (instability) will appear after

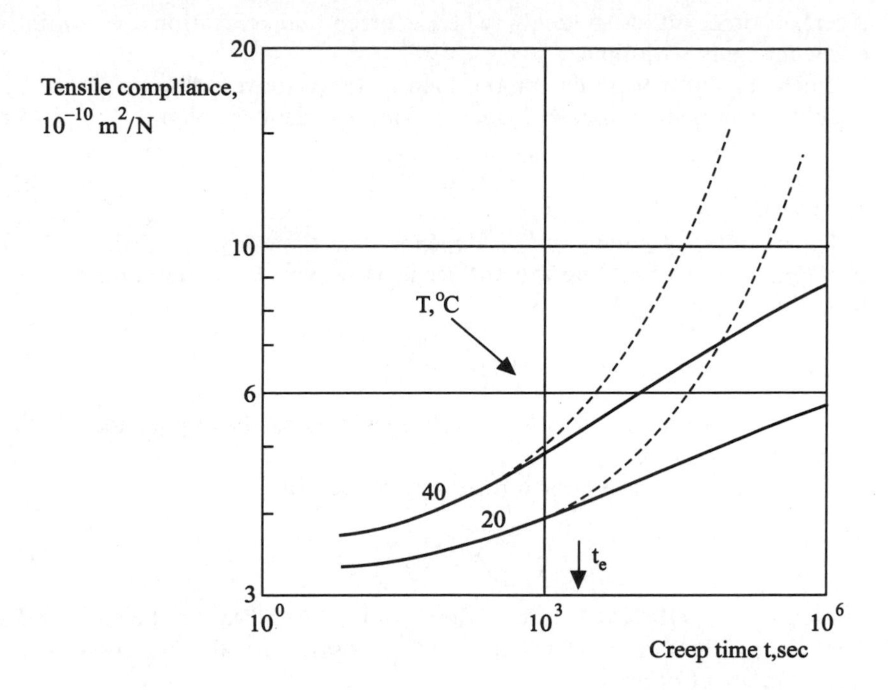

Fig. 5.24 Tensile creep of rigid PVC: long-term tests with samples stored for $t_e = 1800$ s. The dashed lines conform with equation (5.43) (courtesy L.C.E. Struik)

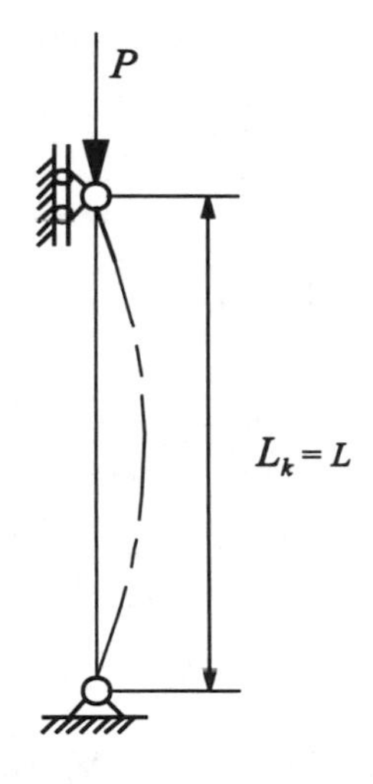

Fig. 5.25 Buckling

a certain time, rather suddenly, whereas creep and relaxation are continuously changing with time.

Euler derived a formula for the load at (and above) which collapse by buckling occurs in an axially loaded pin-jointed slender column (Fig. 5.25):

$$P_c = \frac{\pi^2 EI}{L^2} \tag{5.45}$$

The modulus of polymers E_c is dependent on time (Fig 5.4). When a load $P < P_c$ is applied, buckling will not occur. However, P_c is now subject to the formula

$$P_c(t) = \frac{\pi^2 E_c(t) I}{L^2} \tag{5.46}$$

and so it will decrease. As soon as P_c decreases below the applied load P, the column will collapse.

For columns the slenderness ratio λ is defined by

$$\lambda^2 = \frac{L_k^2 A}{I} \tag{5.47}$$

where L_k is the effective length of the column, depending on the end conditions (Figs 5.25 and 5.26), where A is the cross-sectional area, and I is the second moment of area.

Now equation (5.46) can be rewritten as

$$\varepsilon_c = \frac{P_c(t)}{AE_c(t)} = \frac{\pi^2}{\lambda^2} \tag{5.48}$$

Apparently buckling strain is independent of creep modulus and loading time.

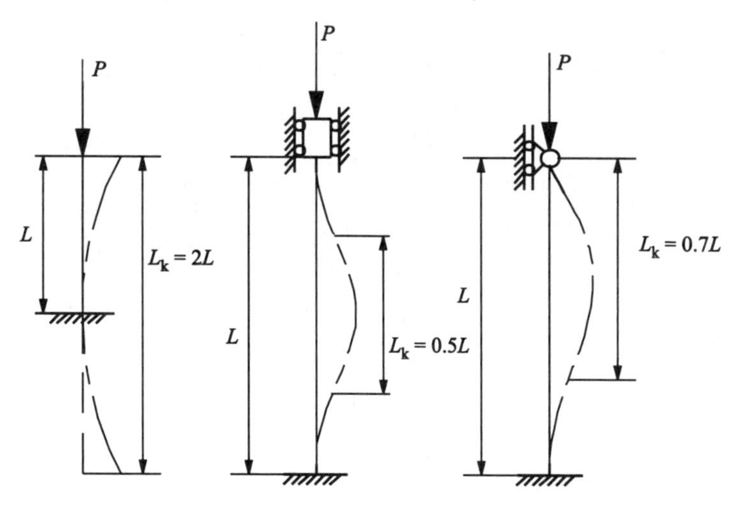

Fig. 5.26 Effective length for bucking L_k depending on end conditions

Euler's formula was derived for perfect axial loading. Buckling loads, however, are much reduced when load is applied offset (introducing an extra bending moment), which may be the case when the column is not perfectly straight or when internal stress or orientation have no symmetry around the axis of loading, so that they tend to induce a preferred curvature.

In many cases of buckling the slenderness is so high, and therefore ε_c so low, that it falls below the limit of proportionality ε_p, allowing E_0 as creep modulus. In short columns this may not be the case. In order to understand the situation it is important to realize that bending, on top of compression, plays a central role in Euler's formula: the bending strain ε_b should be superimposed on the actual compression strain ε_c. If ε_c is outside the proportional domain, additional bending strain ε_b will be along the local tangent of the isochronous $\sigma-\varepsilon$ curve (Fig. 5.22). It is important to note that E_{tang} is smaller than E_s. Table 5.1 provides an indication of some limiting values of ε_p for practical use. It should be clear that above these values the magnitude of ε_c in equation (5.48) loses its physical significance.

Table 5.1 Maximum strain ε and minimum slenderness ratio λ for which linear deformation behaviour can be assumed

	ε_{limit} (%)	λ_{limit}
Rigid PE	0.15	81
PA	0.2	70
POM	0.28	59
Rigid PVC	0.3	57

If the slenderness is much smaller than the indicated values, buckling will not occur at all and the limiting load will be determined by compression creep strength. In thin elements there will be a risk of local buckling. In the case of ribs and webs this will be dealt with briefly in section 10.3.4(c).

5.2.6 Temperature-dependent stiffness

Viscoelastic processes like creep and relaxation usually run faster if the temperature is higher. The same stiffness is reached either by loading for long times at low temperature or by loading for short times at high temperature. This raises the idea that temperature could be equivalent to (log) time.

(*a*) *Time–temperature equivalence*

This equivalence is already suspected from the concepts of upper and lower bounds on modulus (compare Fig. 2.10) and the notion that molecular mobility can be described in terms of temperature (section 2.6) and time or frequency (section 5.2.2). Tests can be carried out to cover both temperature and frequency; tests are usually done at (approximately) constant frequency over a range of temperatures or (more difficult) at constant temperature over a range of times or frequencies. The experimental results can be plotted as shown in Fig. 5.27a. Within certain limits, determined by, for example, the presence of other transitions nearby or by non-linear effects, the individual curves, obtained at different temperatures, show the same shape so that they can be combined by a suitable shift factor a_T to give a master curve (Fig. 5.27b). This, however, is theoretically only allowed provided that only one molecular mechanism is involved ('thermorheologically simple' behaviour) and that no other effects (e.g. physical or chemical aging, changes in crystallinity, relaxation of internal stress, or environmental interactions) occur over the complete timescale (compare section 5.2.4(b)).

The principle of time–temperature equivalence can be used to calculate the stiffness under given conditions of time and temperature. For this purpose we

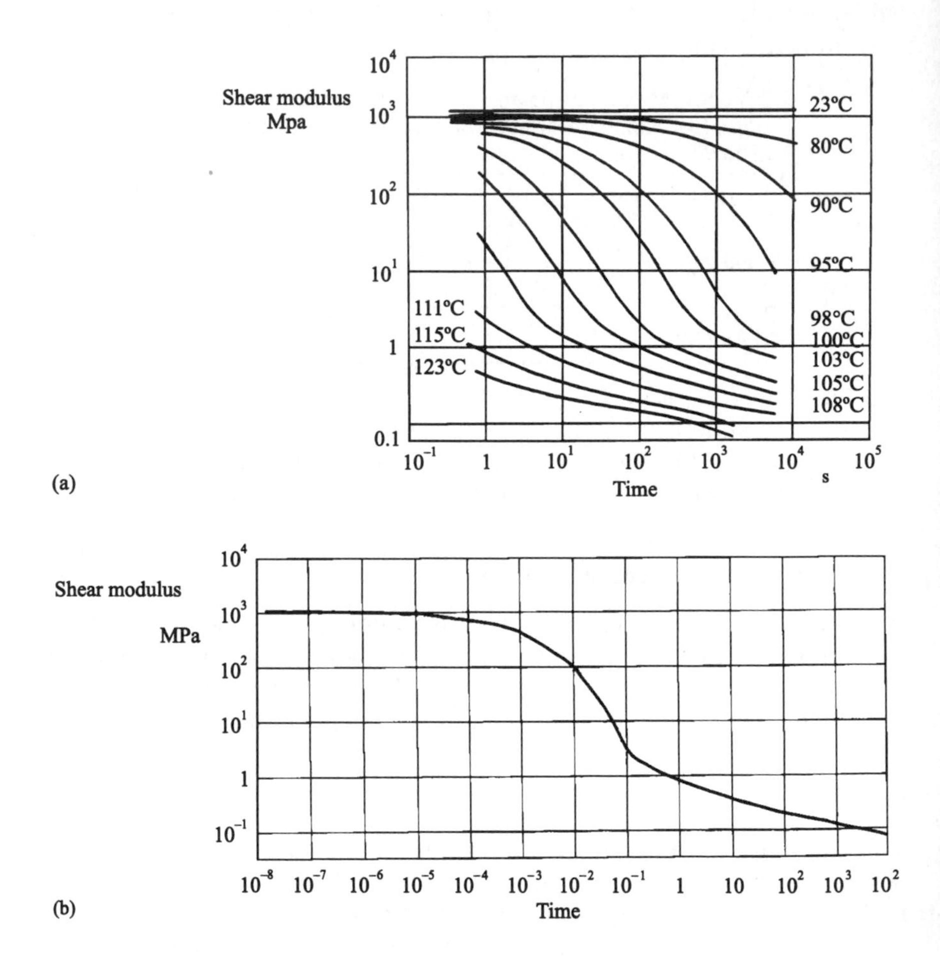

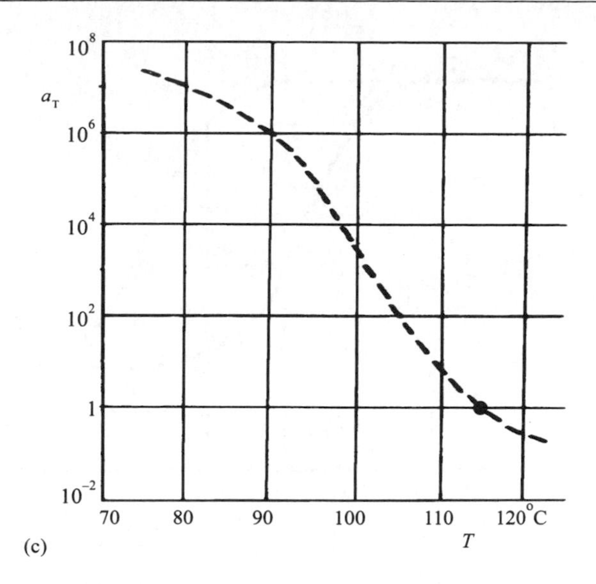

Fig. 5.27 Condensation of test data of polystyrene using a temperature-dependent shift factor: (a) shear modulus, measured at various temperatures; (b) master curves at 115°C; (c) shift factor a_T ((a) and (b) courtesy W. Knappe, VDI-Z **107**, 853/1085)

need to know the master curve of the modulus E at a reference temperature T_{ref} and the magnitude of the shift factor a_T. We will illustrate the application of this principle for stress relaxation, but it applies to creep and dynamic load in a similar way.

The relaxation time τ_r is characteristic for a particular molecular mechanism and its contribution to the stiffness of the polymer; in fact it determines the position of the modulus curve along the (log) time axis. If an external load is present for a shorter time than τ_r ($t/\tau_r < 1$) the mechanism contributes to the load transfer; if it lasts longer the mechanism ceases to contribute and the transition to a lower stiffness has taken place. The Deborah number, $De = t/\tau_r$, indicates to what extent a relaxation mechanism has been addressed thus far by the load.

When the temperature rises, this relaxation process will take place earlier, and the relaxation time will be shorter. So with a given load, a given deformation will be reached earlier. We now define the shift factor a_T as

$$a_T = \tau(T)/\tau(T_{ref}) \tag{5.49}$$

where $\tau(T)$ is the relaxation time τ_r of the mechanism at temperature T, $\tau(T_{ref})$, later on referred to as τ_0, is the relaxation time of the mechanism at the reference temperature T_{ref}.

This formula states that relaxation time, and therefore the shift in the position of the modulus curve along the (log) time axis, at a certain temperature $\tau(T)$, is determined by $a_T \times \tau_0$.

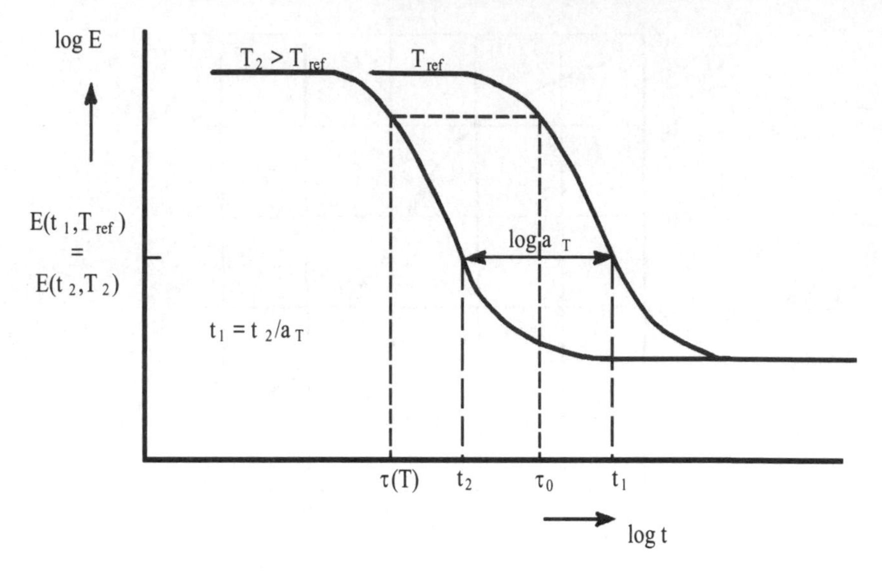

Fig. 5.28 Temperature of a modulus curve containing one relaxation mechanism

If we load a polymer (with a given τ_0) at its reference temperature T_{ref} with a certain deformation pulse ε_1, then a certain stress $\sigma(t_1, T_{\text{ref}})$ is reached after a time t_1. The modulus is $E(t_1, T_{\text{ref}}) = \sigma/\varepsilon_1$ and the Deborah number $De_1 = t_1/\tau_0$ (Fig. 5.28). If the temperature is increased to T_2, the relaxation time falls to $\tau(T_2) = a_T \times \tau_0$ and the value of De_1 (and its associated value of E) is reached earlier, i.e. after a loading time $t_2 = a_T \times t_1(\log t_2 = \log a_T + \log t_1$; note the negative value of $\log a_T$ for $T > T_{\text{ref}}$).

If we now look for the stress that is reached after the same time t_1 but at the increased temperature T_2, the Deborah number is apparently different:

$$De_2 = t_1/\tau(T_2) = t_1/(a_T \times \tau_0) = De_1/a_T$$

The modulus for this case is

$$E(t_1, T_2) = E(t_1/a_T, T_{\text{ref}}) \tag{5.50}$$

which we can read from the known E–t graph for T_{ref} at

$$t = t_1/a_T \tag{5.51}$$

The problem is now reduced to the question of whether we know the magnitude of a_T. For the case of Fig. 5.27 we have added its value in Fig. 5.27c as an illustration, with the master curve as reference ($T_{\text{ref}} = 115°\text{C}$).

In general, one can suppose that the process of relaxation, like many other time and temperature dependent processes, can be described with the Arrhenius equation:

$$\tau = Ce^{A/RT} \tag{5.52}$$

where A is the activation energy of the relaxation process [J/mol], R is the gas constant (8.3 J/K mol) and T is the absolute temperature [K]. The shift factor will then be, according to definition (5.49),

$$a_T = \exp\left[\frac{A}{R}\left(\frac{1}{T} - \frac{1}{T_{ref}}\right)\right] \tag{5.53}$$

If, as is often the case, the magnitude of A is unknown, it is best to measure (log) a_T as the shift of two (log) modulus versus log time curves, as measured at two different temperatures (e.g. 20 and 60°C) according to the scheme of Fig. 5.28. The value of A/R can be calculated from the magnitude of a_T (compare to formula (5.50)) and the values of $T_{ref} = (20 + 273)$ K and $T = (60 + 273)$ K. Now for every other value of T the related value of a_T can be calculated. If the curve at 20°C is chosen as the reference, then a curve at higher temperature (e.g. 60°C) has a shorter value of τ_r (shift to the left) so $a_T < 1$ and $\log a_T < 0$. It may be more practical to choose as T_{ref} the highest temperature for which a good modulus curve is available.

Williams, Landel and Ferry made a classic study of temperature–time equivalence. Their best-known result, known as the WLF formula, is given for our case as

$$\log_{10} a_T = \frac{-17.44(T - T_g)}{51.6 + T - T_g} \tag{5.54}$$

It requires that the reference curve of modulus versus log time should be known at T_g. The validity of the WLF formula is restricted to $T_g < T < (T_g + 100)$, i.e. the domain of rubbery behaviour. Thus it does not apply to most engineering applications of plastics.

To reduce the complexity of equation (5.53), we can make use of a linear approximation of it, bearing in mind that aspects mentioned at the beginning of this section would restrict the validity of equation (5.53) to a limited area in any case. This approximation reads:

$$a_T = c^{\frac{T_{ref} - T}{10}} \tag{5.55}$$

where c is a constant, to be determined in a similar way to that described for A/R above.

Knowledge of a reference curve for E versus $\log t$ and of the function a_T enables the engineer to calculate E for given requirements of temperature and time. Temperature may not always be constant, but equation (5.51) can be supposed to be an additive function:

$$t = \sum_i \frac{t_i}{a_{T_i}} \tag{5.56}$$

If temperature is known as a function of time, the 'effective time' t at reference temperature T_{ref} can now be calculated, so that E can be read from the reference curve.

(b) *Modulus as a function of temperature*

The principle of time–temperature equivalence made it obvious that a good way of exploring the full time spectrum could be to explore the temperature range. This produces more information and is experimentally more accessible. Caution is, however, necessary, first because of the aspects mentioned at the beginning of section 5.2.6(a) and second because every molecular mechanism, e.g. the glass transition, will have its own activation energy A, which is different from other mechanisms, e.g. a secondary transition, in the same polymer. Nevertheless the thermal spectrum can provide very useful information.

The most widely used method measures the frequency and the damping of free torsional vibrations at a frequency of about 1 Hz. From these the shear modulus G and the damping $\Lambda \simeq \pi \tan \delta$ can be calculated. These 'mechanical spectra', as presented in Figs 5.29 and 5.30, give a good impression of the useful temperature range (the glass transition temperature shows up as a maximum of $\tan \delta$ or Λ), the level of stiffness to be expected (e.g. the glass transition for polyvinylchloride at about 80°C at 1 Hz, which does not allow much stiffness for a reasonable period above 60°C) and possible hazards of brittleness (e.g. polypropylene below 0°C) of the polymer.

PROBLEMS

1. Distinguish between (a) 'relaxation' and 'recovery'; (b) residual strain and recovered strain.
2. At 20°C, a hollow sphere made from cast acrylic has an internal diameter of 400 mm with a uniform wall 10 mm thick. The sphere is connected to a reservoir A of inert gas at 20°C at 1.03 MN/m^2 above atmospheric pressure for 10 days; it is then vented to the atmosphere for 1 day, and finally connected to reservoir B of inert gas at 700 kN/m^2 above atmospheric pressure. Calculate the volume of gas enclosed by the hollow sphere 4 days after being connected to reservoir B. The value of lateral contraction ratio may be taken as 0.4. Creep data are given in Fig. 5.12.
3. A tensile bar, made of polypropylene homopolymer (Fig. 5.10), was loaded with $\sigma = 1\ MN/m^2$ at 20°C for 10^5 s. At that moment the load was increased to 2 MN/m^2 for a further 10^4 s. (a) What is the final strain ε? (b) What would the final strain be if the second period had lasted not 10^4 but 10^6 s? (c) What is the final strain of a fresh bar, loaded with $\sigma = 2\ MN/m^2$ at 20°C for 1.1×10^6 s? Compare this result with (b).
4. Calculate the remaining strain ε_R (Fig. 5.15) for a material obeying the exponential function of time, as a fraction of the maximum strain $\varepsilon_1(t_1)$ which was present at the end of the loading cycle, for the case that the recovery time t_R is equal to the original loading time t_1. Choose $m = 0.1$ and 0.2 respectively.

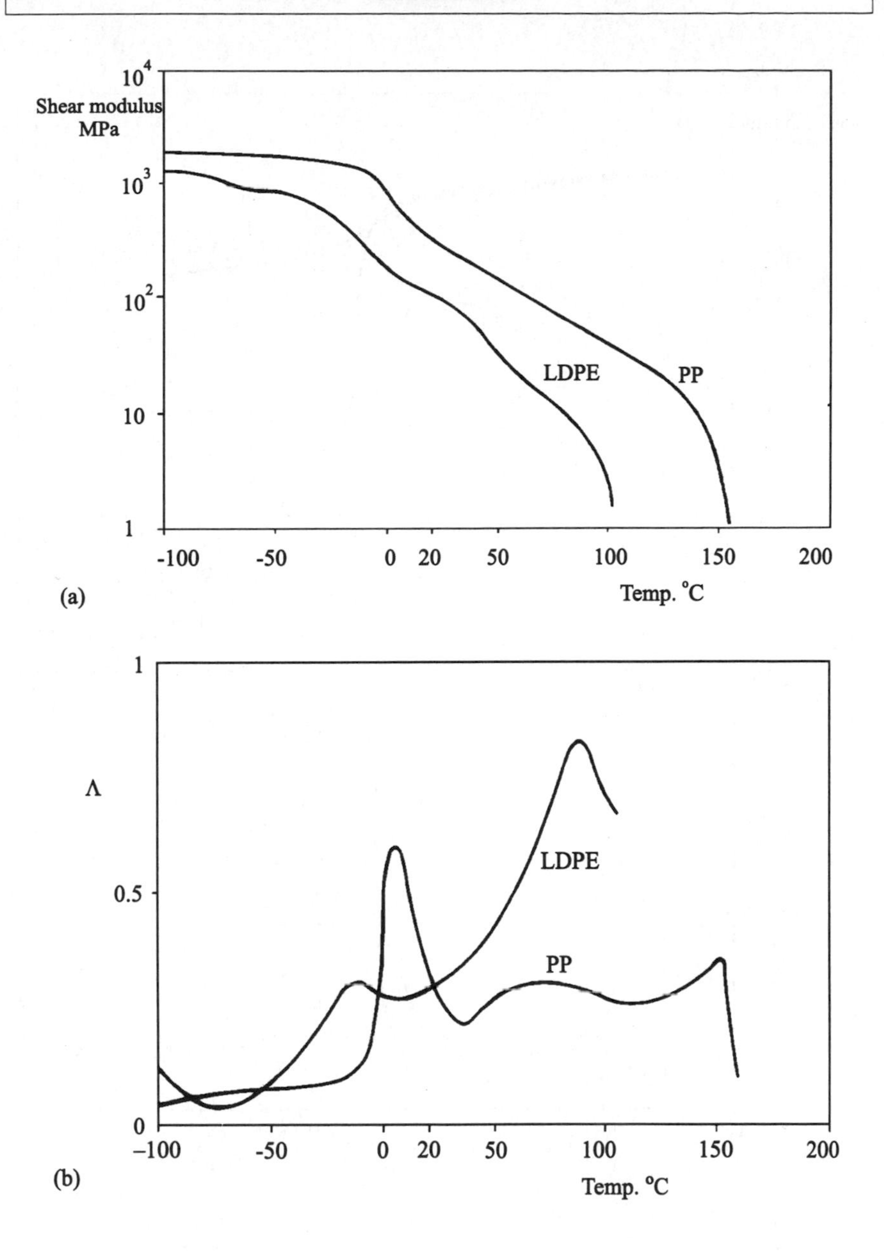

Fig. 5.29 Shear modulus G (a) and damping $\Lambda \simeq \pi \tan \delta$ (b) of low-density polyethylene (LDPE) and polypropylene (PP)

5. If the isochronous stress–strain curve (derived from creep measurements) of a viscoelastic material were a straight line, would the curve obtained

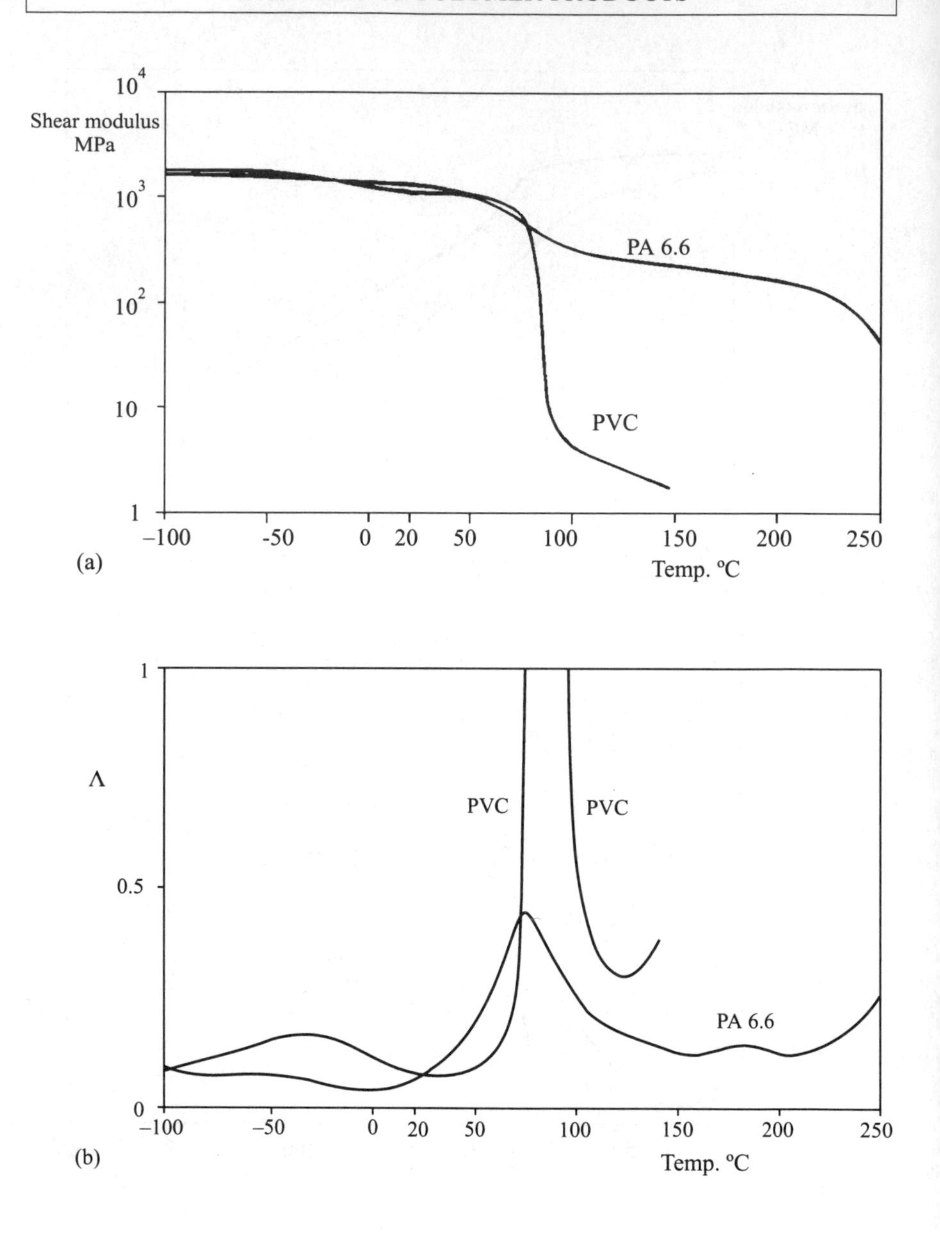

Fig. 5.30 Shear modulus G (a) and damping $\Lambda \simeq \pi \tan \delta$ (b) of polyvinyl chloride (PVC) and polyamide 6.6 (PA 6,6). Note the secondary transition of polyvinyl chloride at about $-40°C$

with a tensile test, performed at constant crosshead rate, be straight as well?

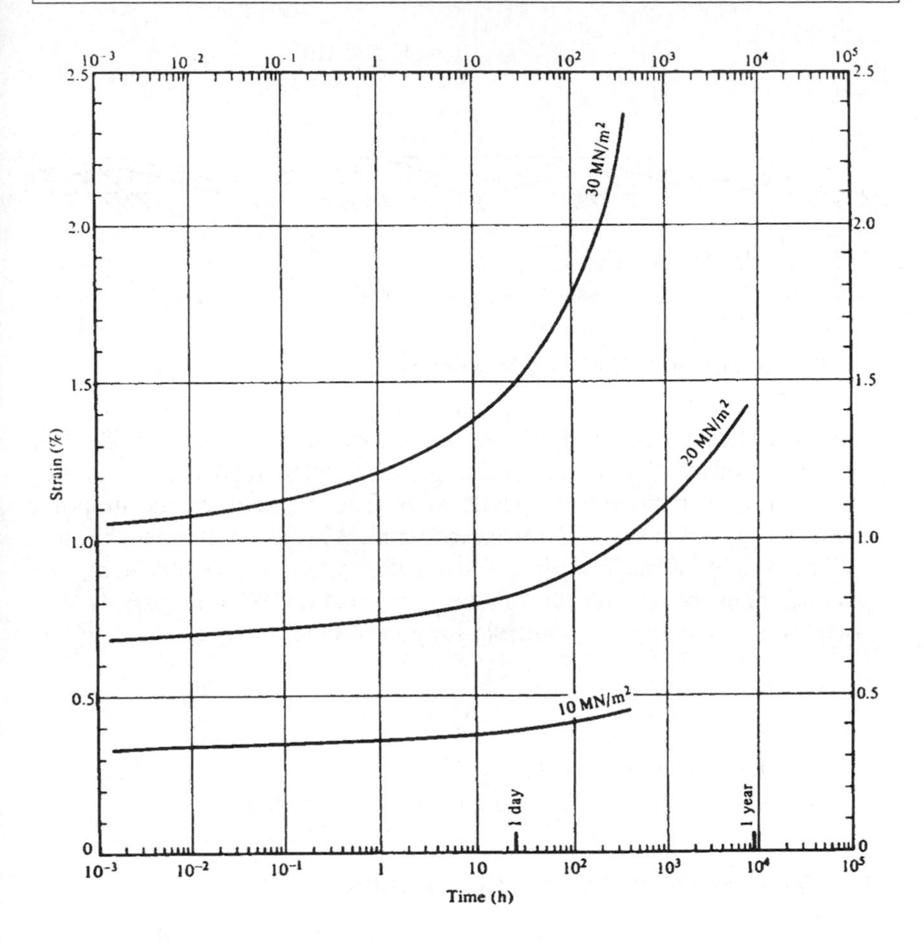

Fig. 5.31 Creep curves in tension; 20°C; unplasticized PVC (data courtesy ICI)

6. Express the creep curve for acetal copolymer at 20°C and 10 MN/m^2 in the form $E = E_0t^n$, with E in GN/m^2 and t in seconds. Is this a satisfactory representation of the data over the range of times given?
7. Given the creep data for unplasticized PVC at 20°C in Fig. 5.31 make an order of magnitude estimate of the viscosity of the dashpot element if a single standard linear solid model were used to describe creep at a stress of 20 MN/m^2. Would you expect the viscosity of the element to vary with the magnitude of the applied stress?
8. What is the shape of the curve of the local bending stress versus the distance to the neutral axis in a cross-section of a bar, which is loaded in pure bending, 1 h after the load was applied?
9. A hot water pipe for domestic use is fed from a boiler at 60°C. The pipe, made of polypropylene, outer diameter 25 mm, wall thickness 2.3 mm, is

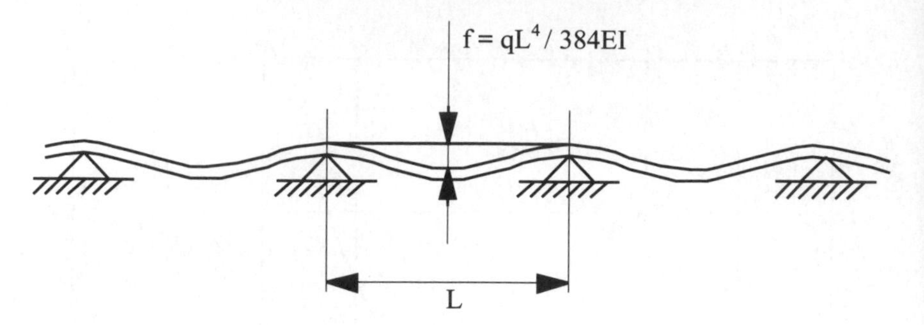

Fig. 5.32 Hot water pipe with multiple supports

supported horizontally at 1 m intervals in a room at 20°C (Fig. 5.32). The pipe is tapped for 1 h daily, so for that period it will be at 60°C, and will cool down to room temperature shortly after. The sag between the supports under its own weight should not exceed 5 mm. Will this requirement be met?

The weight of such a pipe, filled with water, is $q = 4.96\,\mathrm{N/m}$; the second moment of inertia of the cross-section of the pipe is $I = 10\,600\,\mathrm{mm}^4$. The sag of a multiple-supported pipe is

$$f = \frac{qL^4}{384EI}$$

The modulus of polypropylene at 20 and at 60°C is given in Fig. 5.10; using the curves for $\varepsilon = 1\%$, explain why this is permitted.

5.3 STIFFNESS OF VULCANIZED RUBBER

Vulcanized rubber is a flexible elastic material in the temperature range of normal use (i.e. $T > T_g$), with a modulus of about 1–8 MN/m^2, and is capable of storing large amounts of energy per unit volume. This section discusses the stiffness of vulcanized rubber compounds (amplifying remarks made in section 2.8) and some factors affecting stiffness, and introduces simplified methods of analysis of the stiffness of rubber springs under static loads.

5.3.1 Mechanical properties

(a) Stress–strain relationships for short-term loading

Tests at a constant rate of shear deformation show that vulcanized rubber compounds have a linear stress–strain relationship up to almost 100% shear strain, and therefore exhibit Hookean behaviour, i.e.

$$G = \tau/\tan\gamma \tag{5.57}$$

It is necessary to denote shear strain by $\tan\gamma = x/H$ (Fig. 5.33) because displacements are often large; small strain elasticity usually relies on the identity $\gamma \simeq \tan\gamma$. The design of springs under shear loads is based on Hooke's law.

In contrast, the tensile load–deformation relationship is markedly non-linear. As the load is increased, the rubber chains between crosslink points uncoil and provide less obstruction to other nearby molecules; the local slope of the load deformation curve decreases markedly. Eventually, the rubber molecules between crosslinks either become fully stretched (leading to fracture in rubbers which cannot crystallize) or the alignment of molecules in crystallizable rubbers leads to pronounced crystallization accompanied by a dramatic increase in stiffness (followed by fracture).

Because of the large increase in length during the tensile test, typically several hundred per cent, data are usually expressed in terms of the extension ratio λ, which for a parallel-sided specimen is defined as the gauge length under load divided by the gauge length under no load. The relationship between engineering stress σ and extension ratio λ (Fig. 5.34) can then be described over a range up to about 50% extension ($\lambda \approx 1.5$) by

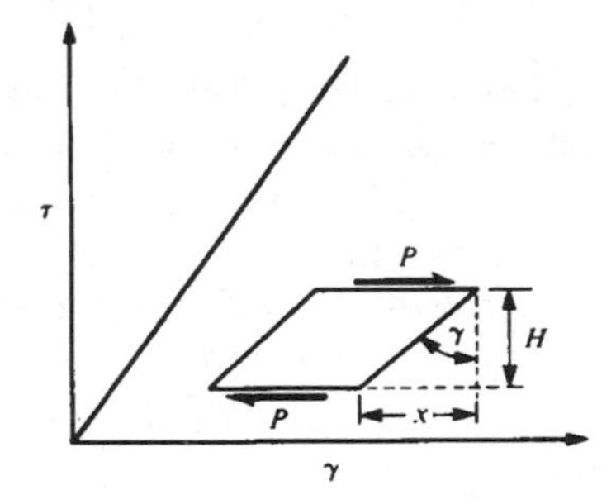

Fig. 5.33 Shear stress versus strain for rubber

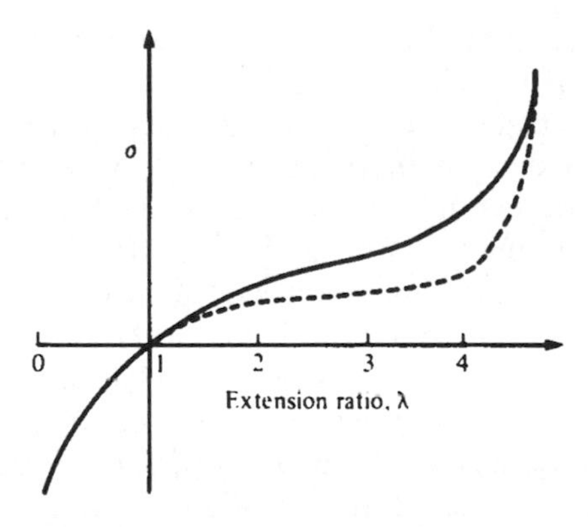

Fig. 5.34 Direct stress versus extension ratio for a crystallizable rubber

$$\sigma = E_0(\lambda - \lambda^{-2})/3 \tag{5.58}$$

where E_0 is the tensile modulus at zero strain ($\lambda = 1$).

This equation reduces to Hooke's law at very small strains. Fortunately, many – but certainly not all – technical applications impose direct strains of less than 15% and it is often acceptable to use Hooke's law in place of equation (5.45):

$$\sigma = E_0\varepsilon \tag{5.59}$$

Rubber is known to be incompressible; this expression, however, is not fully exact. The correct statement is that the bulk modulus $E_B = -p/(\Delta V/V) \approx 1000\text{–}2000\,\text{MN/m}^2$ is two to three orders of magnitude greater than the tensile modulus (note that p is given as pressure and ΔV as the increase in volume). The Poisson's ratio

$$\nu = \frac{1}{2}\left(1 - \frac{E}{3E_B}\right)$$

is therefore only slightly lower than 0.5 and $G = E_0/3$ is a fairly accurate approximation.

After loading, the crosslinked gum vulcanizate shows (visco) elastic recovery on removing the load: the rubber is elastic (but not linear elastic), even after large elongations of several hundred per cent and even where stress-induced crystallization has occurred.

The temperature dependence of the modulus (e.g. E_0 or G) of the normal lightly crosslinked rubber compounds above T_g is different from that of plastics: the modulus actually increases slightly with increases in temperature, for reasons discussed in section 2.8.

(*b*) *Effect of reinforcing fillers*

The modulus of a rubber compound can be significantly increased by incorporation (before crosslinking) of reinforcing fillers, of which the most common is carbon black having a small particle size of the order of a micrometre. It would seem that the black provides extra physical links between adjacent rubber molecules (over and above those provided by crosslinks). The more black that is used, the stiffer the compound.

The modulus of a rubber compound is related to the (elastic) ball-indentation hardness of the compound, which is usually described in terms of 'international rubber hardness degrees' (IRHD) or Shore A hardness. Hardness values range from 0 degrees when the modulus is 0, up to 100 degrees when the modulus is infinite. Corresponding values of shear modulus G for these hardness scales are given in Fig. 5.35. Because the hardness is so easy to measure, it is used as a major characteristic of the rubber compound.

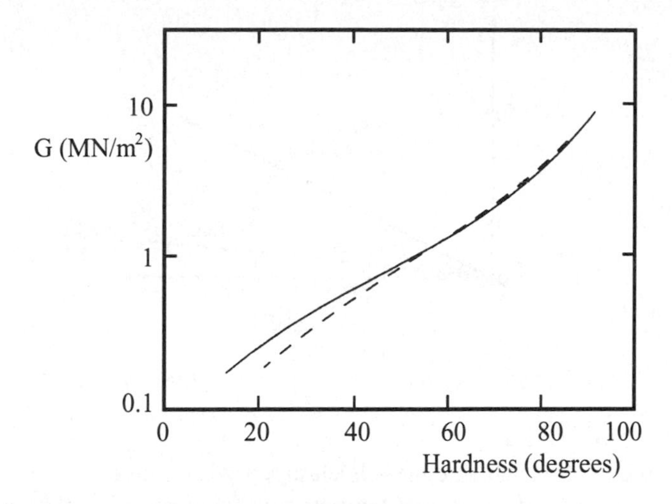

Fig. 5.35 Relation between shear modulus *G* and hardness (degrees): solid curve IRHD, dashed curve Shore A degrees (from Gent, 1992)

Some of the physical links between rubber and black are unsatisfactory and break during the first stressing; rubber technologists talk of 'structure breakdown'. To take account of this, it is normal practice to stretch the rubber testpiece or the component to the maximum expected service stress and then remove the load. Properties are then measured on the 'scragged' (prestretched) rubber, not on the as-moulded material.

(*c*) *Creep and stress relaxation*

Although elastic under short-term loads, the stiffness of rubber is slightly time dependent. The amount of creep or stress relaxation varies approximately linearly with logarithmic time, and depends on the chemical nature of the crosslinks in the rubber. It is current practice in rubber testing to ignore creep up to 1 min, and to express creep as the change in dimensions in one decade as a percentage of the initial elastic strain under the applied load. A plot of creep strain versus log time passes through zero creep strain at 1 min (Fig. 5.36). Creep is evidently stress dependent. Typically, a gum natural rubber might creep at about 1.5% per decade and a hard rubber containing a lot of reinforcing carbon black might creep as much as 6% per decade.

Similar to creep, rubber is also subject to stress relaxation. This property is particularly important for seals. If a tightening stress has relaxed too much, the seal may start to leak. For rubber it is usually quantified with the notion of **permanent set**, i.e. remaining strain after a cycle of relaxation or creep. In many cases the temperature is increased (sometimes decreased) during the cycle. The usual measuring set-up is as follows:

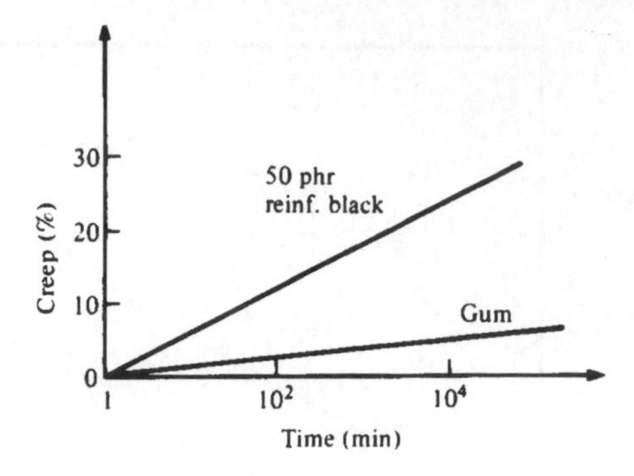

Fig. 5.36 Effect of reinforcing black on creep for natural rubber, 20°C (courtesy MRPRA)

1. a specimen of known thickness is clamped with a given deformation, δ_c;
2. the unit is exposed to the test temperature for a specified time (e.g. 24 h);
3. the unit is released to allow recovery at room temperature;
4. the residual deformation δ_r is measured a fixed time (e.g. 30 min) later;
5. δ_r/δ_c, together with temperature and time, is recorded as permanent set.

(d) Dynamic properties of rubber

Section 5.2.2(d) introduced a discussion of the stiffness behaviour of polymers under cyclic loading. That analysis also applies to rubber compounds. Applications where this is relevant include car tyres, conveyor belts, anti-vibration mountings, and pulsed pressure flows in rubber hose.

Even above the glass transition temperature, the modulus under cyclic loading (frequencies greater than, say, 0.1 Hz) will be greater than that under static loading, because under cyclic loading the molecules have less time to rearrange themselves during the cycle. This argument suggests that the higher the frequency, the higher the modulus for the same applied stress. Reinforcing fillers confer an even higher modulus under static and dynamic loads.

Associated with the stiffness is the loss factor, tan δ, which is responsible for hysteresis, i.e. the amount of energy lost or not recovered during a load–unload cycle. This was defined in equation (5.18). Tan δ reaches a peak at the glass transition where large amounts of energy are 'lost', i.e. transformed from mechanical to thermal energy. If dissipation of this thermal energy is prevented, then the temperature rises: this is most likely in thick sections of components under high stress at high frequencies. If the temperature rises too much, the component may become unserviceable because of, for example, oxidation or degradation of the rubber, or breakdown of the bond between rubber and a metal housing. Well above T_g, i.e. in the normal service range, the value of tan δ is much lower. Gum natural rubber in its

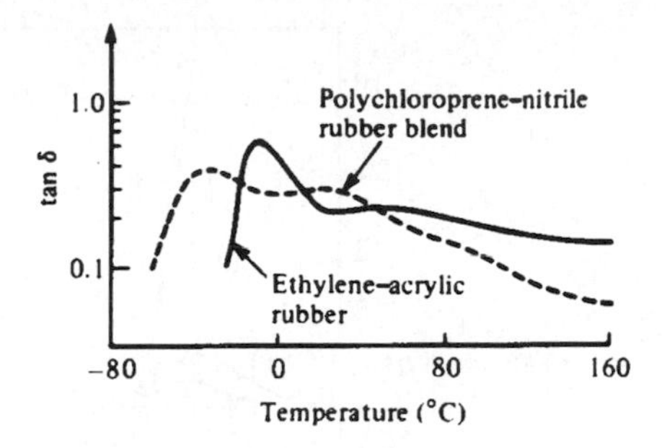

Fig. 5.37 Data for two 'lossy' rubber compounds

working temperature range is not a 'lossy' material, i.e. it is a low-damping rubber, with tan δ typically of the order of 0.02. (A high-damping rubber, such as a polysulphide, has a much higher value of tan δ by one or two orders of magnitude.) Typical data are given in Fig. 5.37 for two materials which are contenders for front engine mounts for cars.

The existence of tan δ suggests that this property of rubbers can be used to damp vibrations. However, this is only true for systems used in the range of their natural frequency. In order to make this clear we have to look to a damped vibrating system (Fig. 5.38a). The equation of motion of such a system is

$$m\left(\frac{\mathrm{d}^2x}{\mathrm{d}t^2}\right) + \eta\left(\frac{\mathrm{d}x}{\mathrm{d}t}\right) + Kx = P_0 \sin \omega t \tag{5.60}$$

where $K = P'/x$ and $\eta = P''/\dot{x}$. The steady state solution is

$$x(t) = \frac{P_0 \sin(\omega t - \psi)}{[(K - m\omega^2)^2 + (\omega\eta)^2]^{1/2}} \tag{5.61}$$

where tan $\psi = \omega\eta/(K - m\omega^2)$ which (due to the added mass m) is different from the tan $\delta = \omega\eta/K$ of a rubber with the same characteristics as the Voigt element in Fig. 5.38a. From equation (5.61) follows the magnitude of the excitation (deflection) of the mass:

$$x_0 = \frac{P_0}{[(K - m\omega^2)^2 + (\omega\eta)^2]^{1/2}} \tag{5.62}$$

which is shown in Fig. 5.38b as a function of ω for various values of tan δ. The graph is normalized with respect to the natural (or resonance) frequency of

$$\omega_\mathrm{r} = \sqrt{\frac{K}{m} - \frac{\eta^2}{2m^2}} \tag{5.63}$$

For frequencies ω in the order of ω_r it might be of advantage to use a high-damping rubber, e.g. butyl rubber, in order to avoid high excitation of the

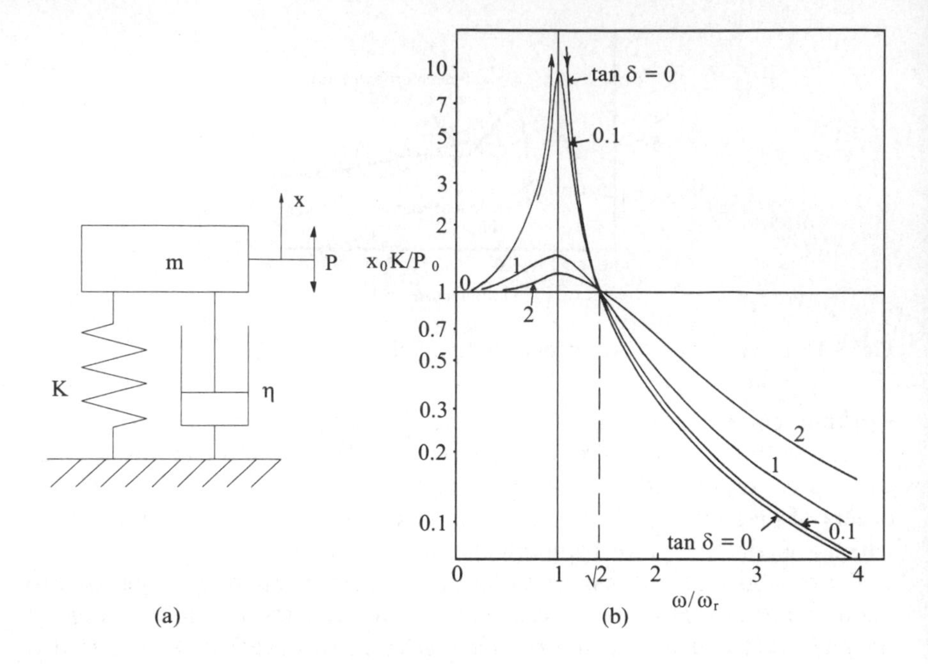

Fig. 5.38 (a) Damped vibration system. (b) Amplitude of oscillation for a suspended mass as a function of frequency (courtesy Gent, 1992)

system; for supercritical frequencies ($\omega > \omega_r\sqrt{2}$) a low value of tan δ, e.g. natural rubber, would be advantageous to keep excitation as low as possible. Many vibrating systems are (designed to be) in the range of their natural frequency ω_r only for short periods and normally perform high above it. In such cases a low tan δ rubber will give the lowest transmission of external vibrations to the system. This applies as much to vehicles and engine supports as to supports of buildings for protection against the vibrations of underground railways or earthquakes.

5.3.2 Rubber springs

With a modulus of only a few meganewtons per square metre, rubber seems an unlikely candidate for serious load-bearing springs. Why, then, is rubber used for dock fenders, bridge bearings, vibration isolation mountings for heavy machinery and for car engines, flexible couplings, and suspension springs for railway rolling-stock? Compared with steel springs, rubber spring units can be extremely compact because they can store large amounts of

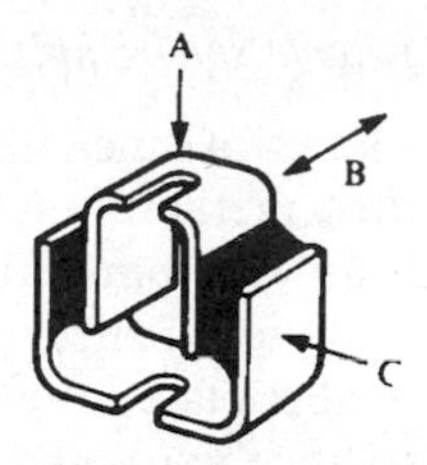

Fig. 5.39 A simple bonded rubber spring unit (courtesy Dunlop)

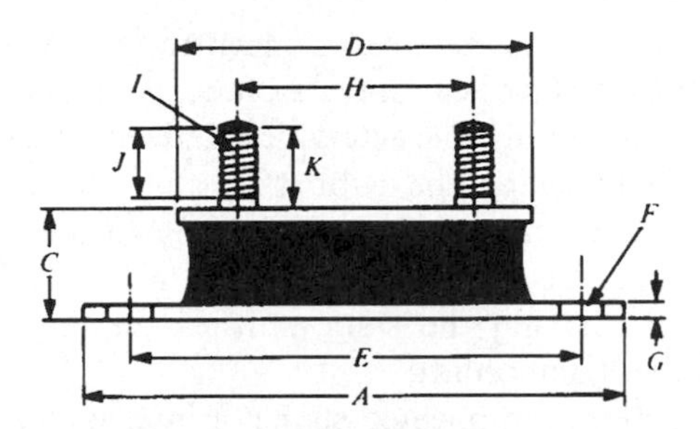

Fig. 5.40 Provision in simple spring for fastening to adjacent equipment (courtesy Dunlop)

energy per unit volume, they can be (and usually are) readily designed to provide markedly different stiffness in different directions, they can accommodate some misalignment, they need no maintenance, and a small amount of hysteresis helps to dampen vibrations in conditions near to resonance. Of the rubber types available, natural rubber is usually the preferred material for springs because it is highly and elastically extensible, it has a wide range of operating temperatures, and it has excellent resistance to fatigue and defect growth, together with low hysteresis. In addition, rubber bonds readily and effectively to metals, so that the bonded surfaces provide load transfer to the rubber, and the metal plates or housings facilitate compact assembly to adjacent equipment, as shown in Figs 5.39 and 5.40.

Rubber springs can be stressed in tension, compression, shear or in any combination of these three modes. There are three basic types of spring. Bonded rubber springs have metal parts or plates chemically bonded to them to provide load transfer and restraint; examples include bridge bearings and flexible couplings for power transmission. Loose rubber springs have no additional parts joined to them; examples include rubber bands, O-ring seals, foam seat cushions and carpet underlay. Encased rubber springs are also made in which the bond is achieved mechanically by friction caused by compressive stresses set up during assembly. The following discussion centres on bonded rubber springs.

(*a*) *Shear stiffness of bonded rubber spring units*

The simplest units consist of blocks of vulcanized natural rubber bonded to parallel metal plates or concentric metal tubes. Discs can carry parallel-plate shear or rotational shear and can accommodate axial misalignment. Annular bushes can carry axial or rotational shear and axial misalignment. For units of simple shape under parallel plate, rotational or axial shear, analysis is based on conventional elasticity theory and assumes (not too unreasonably) that the rubber vulcanizate has a linear stress–strain relationship up to a strain of about 0.35.

Subject always to minimizing stress concentrations at the rubber–metal bond, the free rubber surfaces can have a plane, conical or concave surface to confer desirable operating characteristics. Indeed, it is quite practicable and often desirable to shape the rubber so that the unit operates in a particular mode at constant shear stress, leading to the use of less rubber and a faster moulding cycle (compared with a unit of constant cross-sectional dimensions, having the same stiffness in that mode and made from the same rubber compound).

In units having a more complicated shape or internal load distribution, or where larger strains are involved, it is usually necessary to use approximate or numerical methods to predict performance.

Parallel shear of bonded rubber blocks

Consider a rubber block (Fig. 5.41) of thickness H and plane area A under a shear load P which causes a relative displacement x. The shear stiffness of the unit is

$$K_s = \frac{P}{x} = \frac{GA}{H} \ (\mathrm{N/m}) \tag{5.64}$$

This expression applies to rectangular blocks as well as to circular discs, giving a linear spring characteristic up to $x/H = \tan\gamma = 0.35$.

Bonded rubber disc in rotational shear

The disc (Fig. 5.42) has a thickness H, inside and outside radii R_1 and R_2, and carries a torque T applied to the end-plates. At the point P on a radius R, which has rotated through an angle θ, the shear strain is

$$\tan\gamma = \frac{R\theta}{H} = \frac{\tau}{G} \tag{5.65}$$

The torque $\mathrm{d}T$ acting on a thin annular element bounded by radii R and $R + \mathrm{d}R$ is

$$\mathrm{d}T = R\tau \mathrm{d}A = 2\pi R^2 \tau \mathrm{d}R \tag{5.66}$$

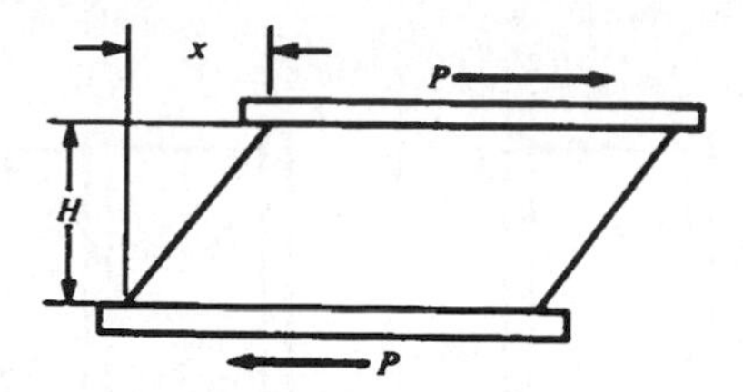

Fig. 5.41 Shear of bonded rubber block

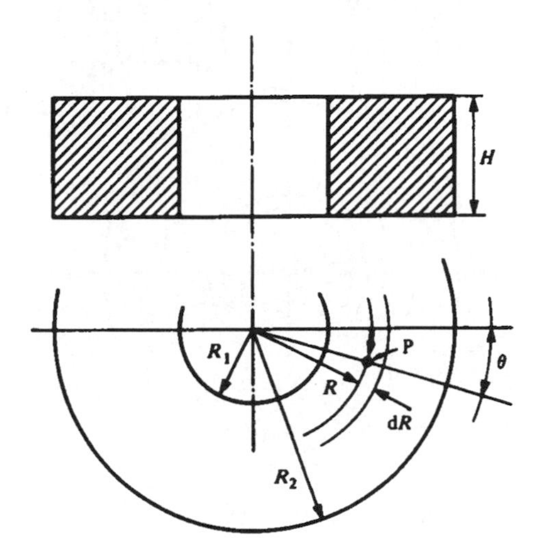

Fig. 5.42 Leading dimensions for torsion of bonded disc

Combining equations (5.65) and (5.66), the total torque T is found by integration as

$$T = \pi G\theta(R_2^4 - R_1^4)/2H \tag{5.67}$$

and hence the rotational stiffness K_r is

$$K_r = \frac{T}{\theta} = \pi G(R_2^4 - R_1^4)/2H \ (\text{N m/rad}) \tag{5.68}$$

Sleeved rubber annulus in axial shear

The annulus (Fig. 5.43) has a length L_1, inside and outside radii R_1 and R_2, and carries an axial load P, which displaces the inner metal sleeve an axial distance x. At any radius R, the shear strain in the axial direction is

$$\tan\gamma = \frac{\tau}{G} = \frac{P}{2\pi RLG} \tag{5.69}$$

The axial shear strain in a thin annular element bounded by R and $R + dR$ is

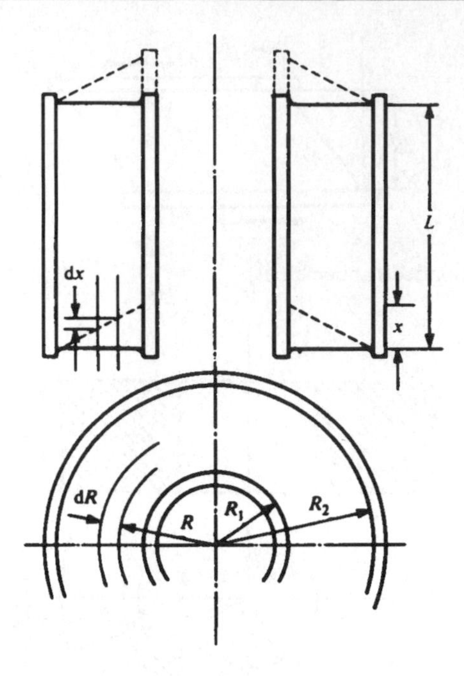

Fig. 5.43 Axial shear of sleeved rubber spring

$$\tan \gamma = \frac{dx}{dR} \tag{5.70}$$

Combining equations (5.69) and (5.70) and integrating gives the total displacement:

$$x = \frac{P \ln(R_2/R_1)}{2\pi LG} \tag{5.71}$$

and hence the axial stiffness of the unit is

$$K_a = \frac{P}{x} = \frac{2\pi LG}{\ln(R_2/R_1)} \text{ (N/m)} \tag{5.72}$$

Sleeved rubber annulus in rotational shear

With reference to Fig. 5.44, the shear strain at a radius R is

$$\tan \gamma = \frac{P}{AG} = \frac{T/R}{2\pi RLG} \tag{5.73}$$

The shear strain in the thin annular element is

$$\tan \gamma = R\frac{d\theta}{dR} \tag{5.74}$$

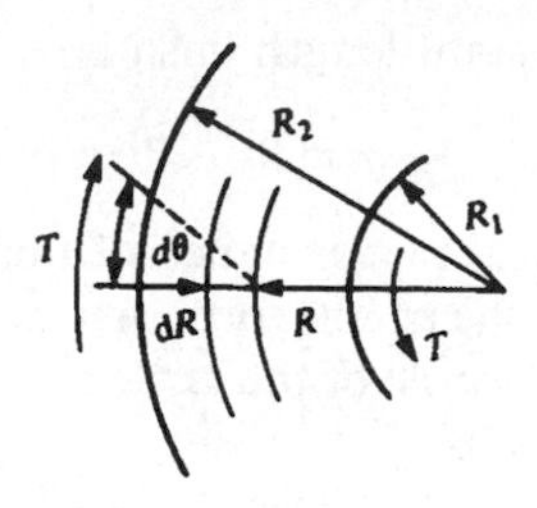

Fig. 5.44 Rotary shear of sleeved rubber spring

and hence the total rotational displacement θ is

$$\theta = \frac{T(R_1^{-2} - R_2^{-2})}{4\pi LG} \tag{5.75}$$

giving the rotational stiffness of this unit as

$$K_r = \frac{4\pi LGR_1^2R_2^2}{R_2^2 - R_1^2} \tag{5.76}$$

Constant stress forms

So far, units have had constant cross-sectional shapes. Apart from the parallel shear of a bonded rubber block, the shear stress is non-uniform and hence most of the rubber is not being used efficiently. One way of making better use of the rubber is to adopt a constant shear stress profile within the unit. This can lead to a dramatic reduction in the volume of rubber of a given hardness needed to achieve the required stiffness in the given mode of deformation. The shapes and lengths or heights may be unusual, but these can still be readily moulded and, for large quantities, the slightly increased cost of making the moulds is more than offset by the faster moulding cycles resulting from thinner cross-sections.

The approach is illustrated for a bonded-sleeve annular rubber unit having internal rubber dimensions R_3 and L_3, and outer dimensions R_4 and L_4. The condition for the unit to operate at constant shear stress in rotational shear is given by equation (5.73), i.e.

$$R^2L = R_3^2L_3 = R_4^2L_4 = C \quad \text{a constant} \tag{5.77}$$

Using the inner dimensions in equation (5.73) and combining it with the general expression leads to the stiffness K_r' of this constant stress bush as

$$K_r' = \frac{2\pi R_3^2L_3G}{\ln(R_4/R_3)} \tag{5.78}$$

The volume of rubber in this constant stress bush is

$$V_2 = \pi R_3^2L_3\ln(R_4^2/R_3^2) \tag{5.79}$$

and the volume of the constant length bush is

$$V_1 = \pi(R_2^2 - R_1^2)L \tag{5.80}$$

Given that the mechanical performance of the constant length bush was satisfactory in rotational shear, the proportion of rubber saved, WS, by adopting a constant stress profile of the same rotational stiffness, and having the same inside dimensions, is

$$WS = 1 - V_2/V_1 \equiv 1 - \frac{2R_3^2 \ln(R_4/R_3)}{R_2^2 - R_3^2} \tag{5.81}$$

The saving can evidently be substantial: this is achieved in the slightly slimmer unit at the expense of a reduction in axial stiffness. The weight of rubber saved will depend on the criteria of comparison used, which need not be the same as the ones used here.

Constant shear stress profiles for the disc in rotational shear and the sleeved annulus in axial shear can be derived using the same principle.

(*b*) *Compressive stiffness of bonded rubber spring units*

Many rubber springs have to sustain compressive loads, including sealing strips, bridge bearings, antivibration mountings and general spring units. Sometimes the loads are applied normal to the load-bearing faces: this will be discussed in this section for the case where the bonded faces are parallel. Sometimes the loads are applied at an angle to the plane of the bonded parallel faces: these loads can be resolved into direct compression and shear components in the unit, and this approach is described in section 10.5.2(b).

The design for compression is complicated by the lateral restraint of the bonded metal faces on the rubber: the rubber is not free to expand in the plane of the unit. This means that the stiffness of the unit depends markedly and non-linearly on the thickness of rubber between restraints. Current design practice is based on taking this restraint into account using a shape factor. This section outlines the analysis for a long slender strip of rubber and describes current practice for a range of cross-sectional shapes.

Long strip with load-bearing faces perfectly lubricated

It is necessary to examine this artificial example first, in order to understand current design practice.

The strip is of rectangular cross-section and infinitely long, and the lubricated faces permit uniform lateral expansion. A useful analysis can be carried out using small-strain elasticity theory. Referring to Fig. 5.45, the compressive load P_z is uniformly distributed. The strip is compressed by an amount δ_z so that $\varepsilon_z = \delta_z/H_0$. The infinite length prevents longitudinal deformation and hence $\varepsilon_y = 0$ and $\gamma_{xz} = \gamma_{yz} = 0$. Perfect lubrication gives

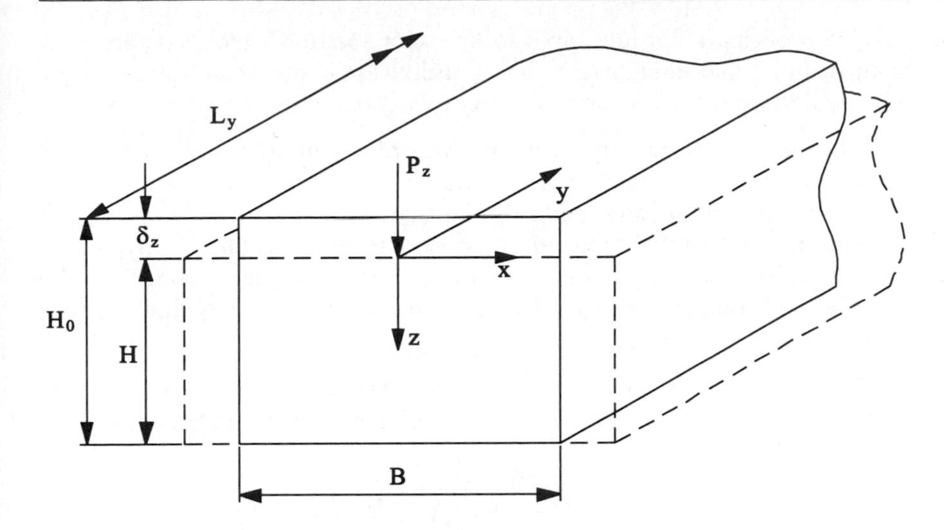

Fig. 5.45 Compression of strip with lubricated faces

$\sigma_x = 0$, and incompressibility leads to $\varepsilon_x + \varepsilon_y + \varepsilon_z = 0$, i.e. $\varepsilon_x = -\varepsilon_z$. Thus, from Hooke's law, the compressive stress in the rubber is

$$\sigma_z = \frac{\varepsilon_z E_0}{1 - \nu^2} \tag{5.82}$$

This leads to the compressive stiffness K_z given by

$$K_z \equiv \frac{P_z}{\delta_z} = \frac{4E_0 B L_y}{3H} \tag{5.83}$$

which is satisfactory if compressions are small ($\varepsilon_z < 0.15$) and provided the loaded faces are lubricated.

Long strip with load-bearing faces perfectly restrained

In practice, mctal plates are bonded to the load-bearing faces of the rubber. It is assumed that the metal plates are rigid. When such a strip is compressed, the unloaded faces bulge out (Fig. 5.46) laterally, except where prevented by the metal plates. How does this restraint affect compressive stiffness?

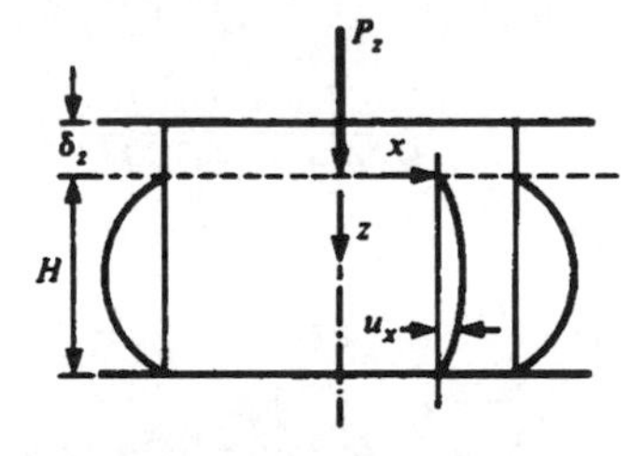

Fig. 5.46 Compression of strip with bonded faces

This problem is not amenable to an exact solution which satisfies both compatibility and equilibrium, but a useful approximate solution can be readily derived from first principles. The essential steps are as follows.

1. Assume the bulging only occurs in the x direction, u_x.
2. Calculate the direct and shear strains.
3. Hence calculate stresses using Hooke's law.
4. Find that σ_z is a function of z as well as x, which cannot satisfy equilibrium because P is also a function of z. However, a practically useful solution is obtained by calculating the mean value of P_z within the strip between the plates.

The results of this analysis suggest that the axial compressive force P_z needed to compress the block by a small amount δ is given approximately by

$$P_z = -\frac{4}{3}\frac{E_0 \delta L_y b}{H}\left(1 + \frac{B^2}{4H^2}\right) \tag{5.84}$$

The first term is identical with the load on a long strip having loaded faces which are perfectly lubricated, provided $H \simeq H_0$. The second term indicates the important restraining effect of the bonded plates and depends on the aspect ratio of the strip. If the strip is thin ($2H \ll B$), compressive stiffness is dominated by the restraint. If the strip is thick and narrow ($2H \gg B$), then stiffness is hardly affected by the bonded plates, but the design may need to take into account the risk of buckling.

In design, it is convenient to work in terms of the apparent modulus E' of the actual spring unit (including restraint). Thus the force P_z required to compress the unit by an amount δ_z may be written as

$$P_z = -E' L_y b \delta_z / H \tag{5.85}$$

Equating (5.84) and (5.85), the apparent modulus is given by

$$E' = \frac{4}{3} E_0' \left(1 + \frac{B^2}{4H^2}\right) \tag{5.86}$$

The degree of restraint is commonly described in terms of the shape factor S. S is defined as the ratio of the area of one loaded face to the total area of the force-free faces of the block. For the infinitely long block discussed here,

$$S = \frac{B L_y}{2H(B + L_y)} \simeq \frac{B}{2H} \tag{5.87}$$

i.e.

$$E' = \frac{4}{3} E_0' (1 + S^2) \tag{5.88}$$

This expression works quite well for gum vulcanizates up to compressions of about 15%. At higher compression and as E' increases with increasing values of S, it needs correction as bulk compression will no longer be negligible. Correction may be found by introducing the corrected compression modulus E'_c from the equation

$$\frac{1}{E'_c} = \frac{1}{E'} + \frac{1}{E_B} \tag{5.89}$$

Block with load-bearing faces perfectly restrained

Similar to strips, analyses can be made of the compressive stiffness of blocks of different plan shape. Of particular interest are cylindrical blocks and rectangular blocks of finite length, in both of which the restraint is obviously more severe than in the very long strip, because bulging can take place in both the in-plane directions. The general form of equation (5.86) is

$$E' = \alpha E_0(1 + \beta S^2) \tag{5.90}$$

For a cylindrical block and for a rectangular block, $\alpha = 1$ and $\beta = 2$, whereas for the long strip, $\alpha = \frac{4}{3}$ and $\beta = 1$.

The compressive stiffness K_c of one layer of rubber of initial thickness H is given by

$$K_c \equiv \frac{P_z}{\delta_z} = \frac{E'A}{H} \tag{5.91}$$

where A is the cross-sectional area and E' is the shape factor corrected modulus from equation (5.90). For a stack of n interleaved identical rubber blocks, each of thickness h, the stiffness of each block will be

$$K_c(h) = E'A/h$$

and the compressive stiffness of the stack will be

$$K_c = \frac{E'A}{nh} \tag{5.92}$$

PROBLEMS

1. (a) Show that equation (5.58) reduces to (5.59) at infinitesimal uniaxial strain. Compare the stresses calculated by these two equations if the same rubber compound is compressed by 15%. Compare the extension ratios calculated by the two equations when the same load is applied so that the strain is 0.15.
   (b) Show that if (small) deformations occur at constant volume, Poisson's ratio $\nu = 0.5$.
   (c) A rubber compound is to be used in the temperature range 0–50°C. By how much would you expect its elastic shear modulus to vary?
   (d) Estimate the amount of energy per unit volume stored in a rubber block deformed to a shear strain of 0.5.

(e) Distinguish the definitions of creep used for plastics and for rubber compounds.
(f) If the compressive creep rate of a given rubber compound is 2% per decade, how much creep strain would you expect to occur in a bearing pad over a 50-year period?
(g) What is the typical maximum frequency of radial loading of a car tyre running along a straight, smooth road?

2. Calculate the shape factors for compression of the following vulcanized natural rubber blocks bonded to metal plates:
   (a) 100 mm × 100 mm × 25 mm thick.
   (b) 100 mm × 100 mm × 10 mm thick.
   (c) 300 mm × 10 mm × 6 mm thick.
   (d) 100 mm diameter × 10 mm thick.
   (e) 100 mm diameter × 18 mm thick with two equispaced metal interleaves.
3. A rubber sandwich compression spring is made from an assembly of five steel plates, each 2 mm thick, equally separated by and bonded to identical isotropic vulcanized natural rubber blocks having a plan area of 200 mm × 100 mm. The overall thickness of the units is 50 mm. In a static test at 20°C, an axial load of 100 kN in the principal direction compresses the unit by 6 mm. Estimate the hardness of the rubber used.
4. The dimensions of the vulcanized natural rubber in a sleeved annular bush are 50 mm ID, 70 mm OD and 40 mm long. If the rubber hardness is 63 IRHD, what is the displacement caused by an axial load of 2.5 kN? To save rubber, a constant stress profile is to be used which has the same axial stiffness, internal length and ID. What are the new external dimensions of the rubber, and how much rubber is saved by using the new shape?
5. An annular bush with bonded sleeves is required to have a torsional stiffness of 35 N m/rad and an axial shear stiffness of 150 kN/m. The vulcanized natural rubber annulus is to be 25 mm long and has an internal diameter of 20 mm. Recommend a suitable and economical design.

FURTHER READING

Stiffness of plastics

Birley, A.W., Haworth, B. and Batchelor, J. (1992) *Physics of Plastics; Processing, Properties and Materials Engineering*, Hanser, Munich.
Ferry, J.D. (1980) *Viscoelastic Properties of Polymers*, Wiley, New York.
Lockett, F.J. (1981) *Engineering Design Basis for Plastics Products*, HMSO, London.
Turner, S. (1983) *Mechanical Testing of Plastics*, Godwin Harlow, Essex.
Ward, I.M. and Hadley, D.W. (1993) *An Introduction to the Mechanical Properties of Solid Polymers*, Wiley, Chichester.
Williams, J.G. (1980) *Stress Analysis of Polymers*, Ellis Horwood, Chichester.

Stiffness of vulcanized rubber

Freakley, P.K. and Payne, A.R. (1978) *Theory and Practice of Engineering with Rubber*, Applied Science, London.
Gent, A.N. (1992) *Engineering with Rubber*, Carl Hanser Verlag, Munich.
Williams, J.G. (1980) *Stress Analysis of Polymers*, Ellis Horwood, Chichester.

6 Strength of Polymer Products

6.1 INTRODUCTION

Failure is depressing and usually unwelcome. Many polymer products are used in applications where failure would be disastrous in terms of health, safety, loss of services or production; pipes and pneumatic tyres are important examples taken from the area of pressure containment, and there are many others. The economics of installing polymer articles or components dictates strongly that the design deploys the minimum amount of the polymer commensurate with satisfactory performance. Satisfactory performance implies that the product survive but without too much overdesign. Survival depends on an awareness of the causes of failure and ways of overcoming them, and is the subject of this chapter.

It is sometimes helpful to envisage a polymer mass as being made up of many very long molecular 'worms' or 'caterpillars'. Taking the analogy further, caterpillars can turn into something else, or can fail by starving, freezing, dehydrating, drowning, fright, or by being chopped into pieces, or being eaten or becoming diseased. Alertness to the causes of death can promote survival. In synthetic polymers there are also many potential causes of failure. It therefore follows that designing to avoid failure is more tricky, and less cut-and-dried, than designing to achieve the required stiffness – but of course this applies to all materials, not just polymers. The basis of avoiding failure involves stress analysis, a failure criterion such as yield or brittle fracture, and a safety factor.

Most engineering load-bearing applications use polymers which are expected to show substantial ductility – examples include many rubber compounds, polyethylene, polycarbonate and PVC. Designing to avoid ductile failure usually relies on the relevant yield stress data. Yield criteria are well established for matching uniaxial yield data to the multiaxial stresses in a proposed product. If the service temperature is well above the glass transition temperature, the simple model of molecular mobility given in section 2.6 surely suggests that ductile failure would be likely to occur. It is indeed fortunate that the limited range of conditions within which a given polymer is reliably ductile does overlap well for some polymer compounds with service conditions – the rubber car tyre is a good example. Yet most readers could cite examples where nominally ductile polymers (as described in some sales literature) actually fail in a brittle manner – even domestic products such as the rubber band snapping at low elongation, the cracked polyurethane shoe sole, the old polyethylene road traffic control cone by the roadside, or the plastic fertilizer bag left in the field. More seriously, brittle failure in pressure pipelines is also possible if the design is inappropriate. There is then a growing suspicion that brittle failure may be the natural mode under searching circumstances, with ductility a welcome respite and a special case.

Much of the literature of polymer testing, like that of many materials of construction, describes the results of tensile or bending tests mainly at 20°C with exploration of such effects as temperature, rate or duration of loading, and composition and effects of processing. This has provided an invaluable wealth of experience, information and data on yield stress, brittle strength, and factors which promote transitions from ductile to brittle behaviour, with elongation-at-break thrown in for good measure. This all provides one conventional basis for checking whether a polymer product is likely to be safe in service, and it is entirely appropriate that this significant background is described here.

In addition, a major cause for concern is the presence within materials of real or apparent crack-like defects. Under suitable conditions these defects can grow and result in brittle fracture. Crack growth or fracture mechanics cannot account for the origin of crack-like defects, but it can describe successfully a wide variety of brittle failure phenomena encountered both in the laboratory and in service. This provides a basis for describing the toughness of polymers and for designing to avoid failure by crack growth or for predicting that an article containing known defects will be safe under particular conditions. This neatly complements the established approach, and it is appropriate to introduce in this chapter the role of fracture mechanics in describing polymer behaviour. The significance of the aphorism 'nature sides with the hidden flaw' will become increasingly apparent as this chapter unfolds.

6.2 THE TRADITIONAL APPROACH

6.2.1 Short-term tensile strength

Most short-term strength data derive from uniaxial tests in air on waisted specimens of uniform cross-section in the gauge length using a tensile testing machine operating at a constant rate of crosshead separation. The output from the machine is a load–extension curve, which is normally transformed into either a stress–percentage elongation curve or (more usefully) a stress–strain curve, where stress is based on the original cross-sectional area.

This section describes some common modes of failure for polymers, with the emphasis on behaviour at about 20°C.

(*a*) *Brittle failure*

Short-term brittle failure is characterized by the creation of new surfaces and fragmentation after the polymer has deformed to only a small extent, typically about 1% strain or less. It is a common form of failure in many polymers at temperatures below their glass transition temperature, T_g, particularly where large bulky side groups or a high density of crosslinks are present to inhibit disentanglement in amorphous polymers. Molecules held together by secondary forces can hardly slip past each other, so the load has to pass through the network of primary and secondary forces in search of a weak spot that would serve as a crack starter.

Examples of this type of failure at 20°C include acrylic, general-purpose polystyrene, and any epoxy and crosslinked polyester resins. The stress–strain curve is very nearly linear (compare Fig. 6.1). The network is not very extensible; the typical modulus for polymers below T_g is of the order of 1 GN/m^2 or higher.

The appearance of the fracture surface is characteristic too. There is usually a mirror-like region where a crack-like defect has grown slowly, surrounded by a region having sharp but coarser features indicative of fast crack growth, which still show no sign of permanent large-scale damage such as yielding.

During the tensile test, many materials of this type show signs of crack-like defects well before the failure strain is reached. These defects are useful indicators of impending failure in products in service: they impair clarity, reflect light and are particularly obvious in transparent materials. At first glance, these defects look like cracks, i.e. two parallel separated surfaces

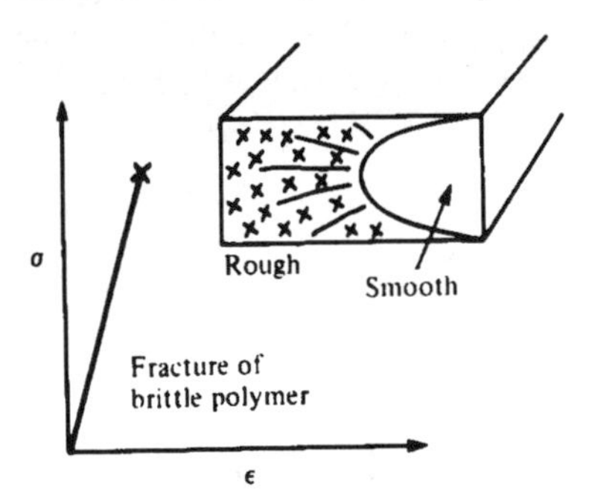

Fig. 6.1 Fracture of a brittle polymer

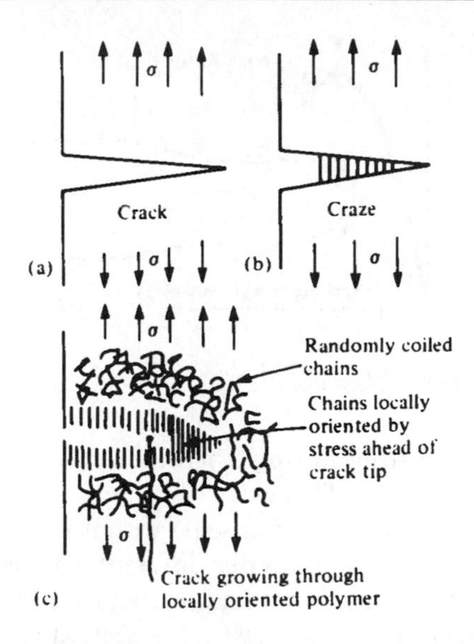

Fig. 6.2 Cracks and crazes in a brittle polymer

normal to the applied stress (Fig. 6.2a). Because the surfaces are separate, no load can be transferred across the crack so the load, together with an associated bending moment, concentrates itself beyond the crack tip in the uncracked material: as a result of the stress concentration at the tip, the crack grows and the specimen fails.

Closer examination reveals that in many amorphous polymers below T_g the two new surfaces are bridged by load-bearing fibrils. The material between the mirror-like surfaces is in fact porous with a density about half that of the unstressed polymer: it constitutes a 'craze'. A craze is initially less dangerous than a crack because at least some load can be transmitted across it (Fig. 6.2b), and most helpfully some energy is absorbed in making it. However, a craze will degenerate into a crack – the voids coalesce – if the load is held constant or increased, as the locally oriented material at the craze tip fails (Fig. 6.2c), and so this eventually leads to brittle failure.

It is worth noting that in uniaxial compression these polymers often show ductile failure by yielding; there is no tensile stress to open up crack-like defects, except at highly deformed free surfaces.

(*b*) *Ductile failure*

Short-term ductile failure under uniaxial tension is characterized by gross and permanent deformation surrounding the region of separation of the specimen into pieces. Some, but not all, ductile failures are preceded by the onset of necking at the yield point. In many polymers there can be seen an

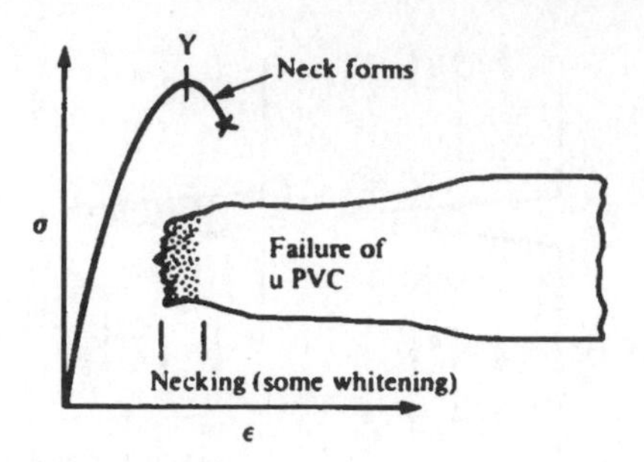

Fig. 6.3 Ductile failure in unplasticized PVC

internal cloudiness occurring during a tensile test at some load well below yield or the onset of necking. This early warning is caused by microvoiding, i.e. localized yielding on a very small scale within the bulk of the material. The minute voids have a refractive index different from that of the surrounding material, thus scattering light and appearing cloudy. At yield, molecules slip past one another in a non-elastic manner, and the microstructure changes – especially for partially crystalline polymers.

Below yield the molecular chains slip past each other, normally to a small extent which is largely, if not completely, recoverable – even if the recovery process takes time as described in section 5.2. In partially crystalline materials it is the amorphous matrix which is deforming most, rather than the crystallites; this occurs most readily above T_g, e.g. at 20°C in low-density polyethylene. At yield molecules slip past each other to a large and non-elastic extent. There is some attempt to straighten out parts of the molecular chain until entanglement prevents this. In partially crystalline polymers there can even be some reordering of the microstructure. Testing a range of polymers reveals a range of effects which are worth describing.

Some polymers just manage to exhibit a yield point before breaking: unplasticized PVC exemplifies this necking rupture quite well, as sketched in Fig. 6.3. The absence of a pronounced neck is partly attributable to the difficulty of melting and mixing the PVC particles with other ingredients during the processing; the better the processing, the less the risk of stress-raising defects acting as crack-like sources, and the more ductile the behaviour.

Another class of polymers shows uniform extension without forming a neck, as shown in Fig. 6.4. The stress-strain curve does show a yield point, beyond which the material forms a porous structure and there is no appreciable necking. Rubber-modified polystyrene shows this pattern of behaviour at about 20°C, and may fail at up to 40% strain compared with 2% for general-purpose polystyrene. The effect of the rubber particles is to reduce the glass transition temperature of the composition and prevent the multiple crazes within the polystyrene from early degeneration into long cracks. So

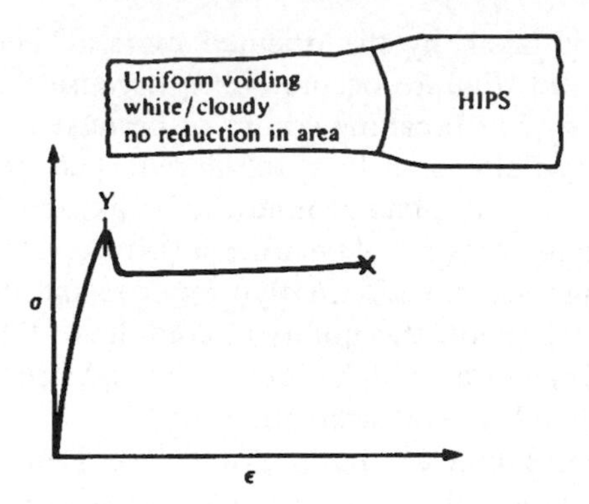

Fig. 6.4 Ductile failure in the waist of a specimen of high-impact polystyrene

beyond yield the material deforms permanently into a spongy structure which looks cloudy because the crazes scatter light. Elongation proceeds until the crazed material can no longer hold together, and the polymer breaks fairly cleanly but with permanently deformed material on the white fracture surface.

A third pattern of ductile behaviour involves cold drawing, and occurs above T_g in polymers based on narrow molecules which can crystallize. The polymer yields (and may show microvoiding before doing so), and then deforms locally to develop a neck of voided whitened polymer (Fig. 6.5). The molecules are now beginning to align themselves considerably in the direction of the applied load: the alignment leads to strain-induced crystallization (or even recrystallization) and the material in the neck therefore becomes much stiffer in the direction of loading than is the unnecked material. The neck is therefore stable and the shoulder of the neck travels along the specimen for as long as the load is applied, or until the neck reaches the wider ends of the specimen where the cross-section is too large for yield to occur without a dramatic increase in load. Very large elongations, up to 10-fold, can be achieved in this way. As the drawing proceeds,

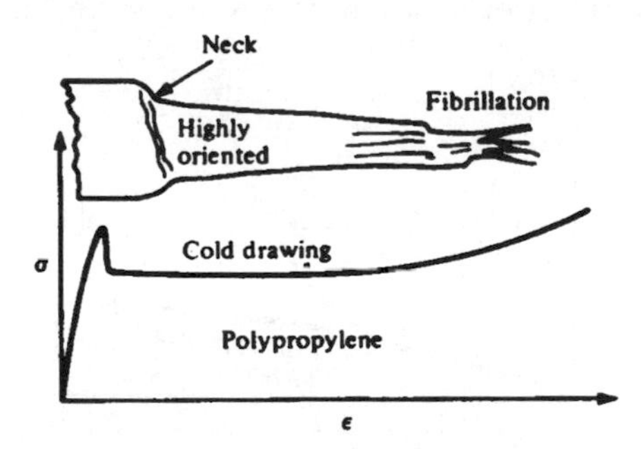

Fig. 6.5 Failure following cold drawing in polypropylene

some of the load is taken by the oriented material and so some strain-hardening is observed. Failure occurs either because many molecules (or parts of them) reach their breaking strain, or because local defects start to grow in the transverse direction. In a well-drawn specimen it is not uncommon for the polymer to fibrillate around the break; the forces between the molecules in the highly drawn polymer are insufficient to hold the material together in the transverse direction. Polypropylene and most partially crystallizable polymers can show this ability to cold draw. Polypropylene string is sold in a highly drawn state and has excellent axial strength (which is what is wanted) but negligible lateral strength.

A fourth pattern has some attributes of both ductile and brittle behaviour. Polymers in this category are well above their T_g and can undergo very large extensions without forming a neck, and the cross-sectional area drops uniformly along the gauge length as the extension increases. Examples include lightly crosslinked rubber and well-plasticized PVC compounds, where the molecules, or parts of them, are evidently free to uncoil to a large extent; on removing the stress, the molecules coil up again. The stress–strain curve shows no peak, but rises continuously as extension proceeds (Fig. 6.6). Failure occurs when a crack-like defect manages to grow across the width of the specimen. The fracture surface then looks shiny – especially in some rubber compounds – with no signs on it of any permanent deformation – just as if brittle fracture had occurred. Because of the shape of the stress–strain curve, non-linear fracture mechanics is needed to handle this sort of behaviour. Away from the crack, much of the enormous strain developed during the test to failure is recovered, although not always completely because of some breakdown of bonds and crosslinks. Plasticized PVC can also show signs of permanent deformation, usually evidence that the plasticizer and PVC are inadequately mixed.

If the lightly crosslinked polymer is able to crystallize, because it is a narrow molecule, then the orientation along all the gauge length results in a considerable increase in stiffness, as indicated in Fig. 6.6 for a gum natural-rubber vulcanizate. Copolymer rubbers do not crystalize in this way and show only a modest amount of orientation. But on removing the stress, the parts

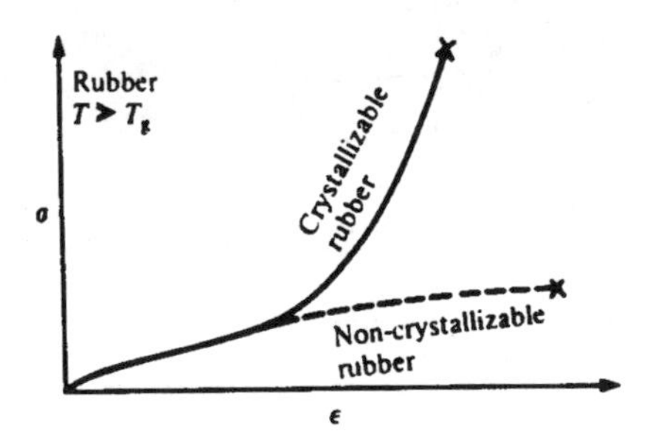

Fig. 6.6 Stress–strain curves for two kinds of crosslinked rubber

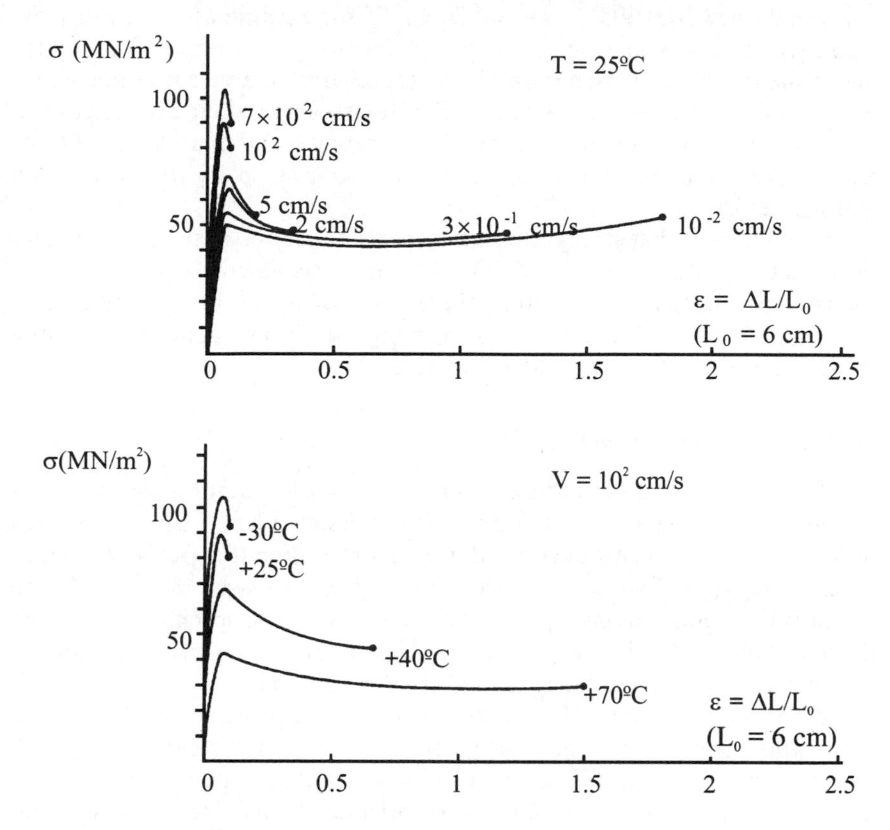

Fig. 6.7 Tensile test curves of PVC at different crosshead speeds *v* and temperatures *T* (from W. Retting, *Rheol. Acta*, **8**, 259, 1969)

of the chain between the fixed points or crosslinks coil up again and cannot 'remember' or accommodate the previous order. Thus the crystallization in rubber is 'elastic' whereas the cold drawing in, say, the polyolefin plastics is irreversible and permanent because of the lack of crosslinks to prevent wholesale molecular motion.

(*c*) *Influence of tensile speed and temperature*

The test results described above were supposed to have been measured under standard conditions in a climatized laboratory, i.e. at 20°C and with a tensile speed that would lead to fracture within 1 min. Tests with polymers under different conditions show that these may strongly influence the results. Figure 6.7 shows the results of such tests with PVC.

These results show a great analogy of the influence of tensile speed and of temperature on yield strength and on strain at fracture. It should be borne in mind that the T_g of PVC is at about 80°C. T_g is the temperature at which the secondary forces fade away. At lower temperatures the tensile force introduces the energy that is necessary for the chains to overcome secondary forces and to slip past each other. The figure shows that an increase of temperature means an increase of chain mobility and that a decrease of tensile speed means an adaptation of the process speed to the existing mobility at 25°C.

This illustrates that strength parameters are functions of temperature and time (and of other influences that determine secondary forces, such as solvents or plasticizers) and that these functions may bring a transition between ductile (low yield stress σ_Y, and high strain at fracture) and brittle (high σ_Y, and hardly any strain at fracture).

(*d*) *Failure criteria for multiaxial stress*

Loading a product may induce a multiaxial stress situation in any point considered to be critical. This means that the designer has to assess the stress tensor of that point in terms of allowable stress. For this purpose various criteria can be used which compare the actual stress tensor $\{\sigma_1, \sigma_2, \sigma_3\}$ with an equivalent uniaxial one σ_{eq}. A criterion is in principle specific for a certain mode of failure: yielding ($\sigma_{eq} = \sigma_Y$), brittle fracture ($\sigma_{eq} = \sigma_{br}$, etc.), crazing or otherwise. The best studied are criteria for yielding.

Any stress situation can be expressed by its three principal stresses, called σ_1, σ_2 and σ_3 in order of decreasing magnitude. This is meaningful for an isotropic material and will be discussed later in this chapter. For an anisotropic material the axes of symmetry (or orientation) have to be chosen, and the situation is much more complex, as is made clear in Chapter 7.

The yield criterion of Tresca can be written as

$$\sigma_1 - \sigma_3 = \sigma_Y \tag{6.1}$$

which states that the yield stress is reached when the difference between the two extreme elements of the principal stress tensor has done so.

The yield criterion of Von Mises is somewhat more complex:

$$\frac{1}{\sqrt{2}}\sqrt{[(\sigma_1 - \sigma_2)^2 + (\sigma_2 - \sigma_3)^2 + (\sigma_3 - \sigma_1)^2]} = \sigma_Y \tag{6.2}$$

These criteria do not account for the fact that the yield stress in compression σ_{Yc} for many materials has a higher absolute value than the yield stress in tension σ_{Yt}. Nor do they account for the fact that hydrostatic pressure is of influence on the magnitude of σ_Y.

For crazing other criteria have been found.

Many long-term yield (and brittle fracture) data have been measured with pressurized pipe experiments (compare Fig. 6.10). The wall stress in such pipes is mainly biaxial with stresses in the hoop or tangential direction

($\sigma_1 = \sigma_{tang}$) and in the axial direction ($\sigma_2 = \sigma_{ax}$); the stress through the thickness has the value of the internal pressure ($\sigma_3 = -p$) and may be ignored in thin-walled pipes. Thus $\sigma_{tang} = 2 \times \sigma_{ax}$ applies. In principle the uniaxial yield stresses had to be derived from these on the basis of a failure criterion. The Tresca criterion states for this case that at yielding $\sigma_{tang} = \sigma_Y$ if the pipe is thin-walled (or more precisely $\sigma_{tang} + p = \sigma_Y$). In section 10.4.1(c) an example of the application of the Tresca criterion is given.

6.2.2 Failure under long-term load

The previous discussion mainly centred on behaviour under short-term loading. In practice many products operate under loads applied for prolonged periods. The loads may be notionally held constant or may be cyclic in nature, and these two patterns of loading are described separately.

(a) Constant load

Products under constant load may fail after a certain period, which indicates that the material's strength $\hat{\sigma}$ decreases under load: if a working stress σ is applied, this decreasing strength is the cause of a restricted time to failure (Fig. 6.8). The higher the stress σ the shorter the time at which failure occurs. The course of the curve for $\hat{\sigma}$ is subject to temperature and many other influences, as described in section 6.2.3, and therefore it is not always easy to make a good estimate of the time to failure. In a similar way it is not always easy to indicate the stress that the material can sustain for a required service life under service conditions.

Some authors use the term 'static fatigue', static referring to constant load, and fatigue unhelpfully referring to the time duration of the loading and not describing cyclic loading as in common engineering usage.

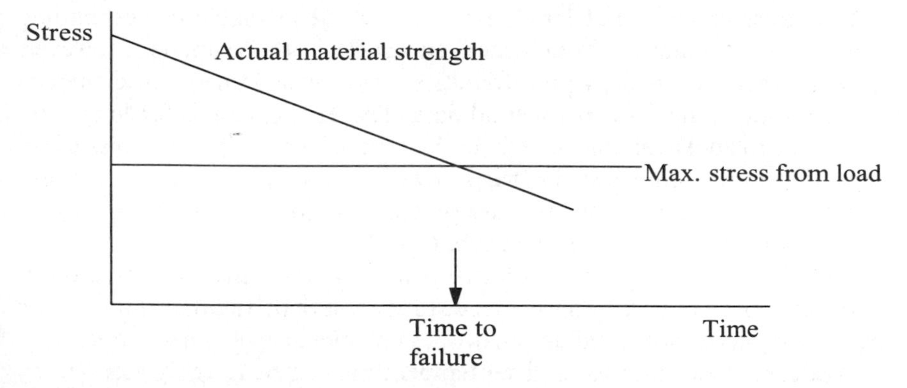

Fig. 6.8 Decreasing materials strength controlling the time to failure after a certain load is applied

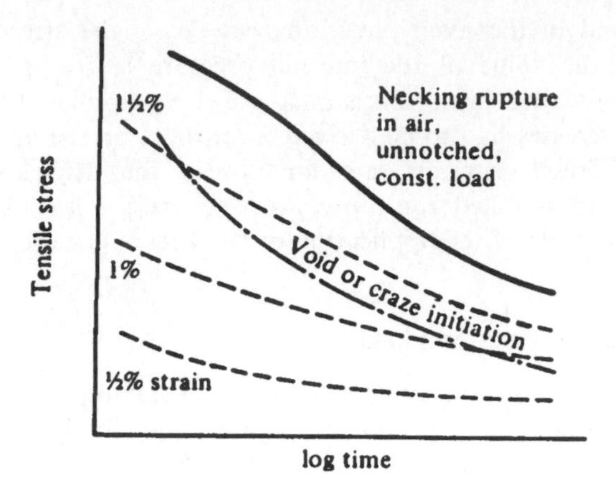

Fig. 6.9 Necking rupture data superimposed on isometric creep data

Tests under constant load in air at about 20°C on unnotched specimens of many plastics show failure in rather a ductile manner in medium-term tests – even for polymers well below T_g. This failure is often called 'creep rupture', which describes the stress at which the specimen either breaks or begins to form a neck. The creep rupture data superpose neatly on the isometric stress–log time curves from the corresponding creep tests, as shown schematically in Fig. 6.9. Optical failure (crazing and voiding) sometimes precedes mechanical failure by a considerable period of time under constant stress. Increasing the temperature of the test reduces the failure stress at any given loading period.

It should be noted that the representation of curves like Figs 6.9 and 6.10 deviates from the standard mathematical ones: the y-axis contains the given load, and the x-axis the measured time to failure. The same holds for fatigue data as in Fig. 6.14.

It is most unlikely that long-term data will be available corresponding exactly to the material from which an article is to be made. However, guessing these is fraught with difficulties – real or imagined – and there is really no substitute for experimental data. The best scheme is for long-term data to be (made) available, mostly in the form of the lifespan of pressurized pipes. Pipes are usually tested as suspended from one end in a room or a bath of a given temperature. The data are presented as log life versus log tangential stress for the relevant temperature (Fig. 6.10).

Equally, the designer cannot wait 10 years (say) while the relevant data are being obtained. There is some evidence to suggest that, all other things being equal, long-term brittle failure also obeys a principle of time–temperature equivalence. Thus data for a given temperature can be roughly assessed by carrying out shorter-term tests at a higher temperature. This has at least proven to be successful for polyolefins as shown in Fig. 6.10. In view of the

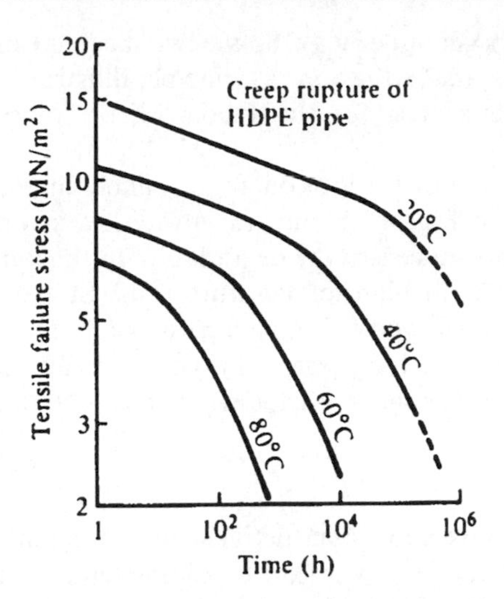

Fig. 6.10 Creep rupture data for high-density polyethylene pipe. The upper left-hand parts of the curves represent ductile failure mode, the lower right-hand parts brittle failure by crack growth

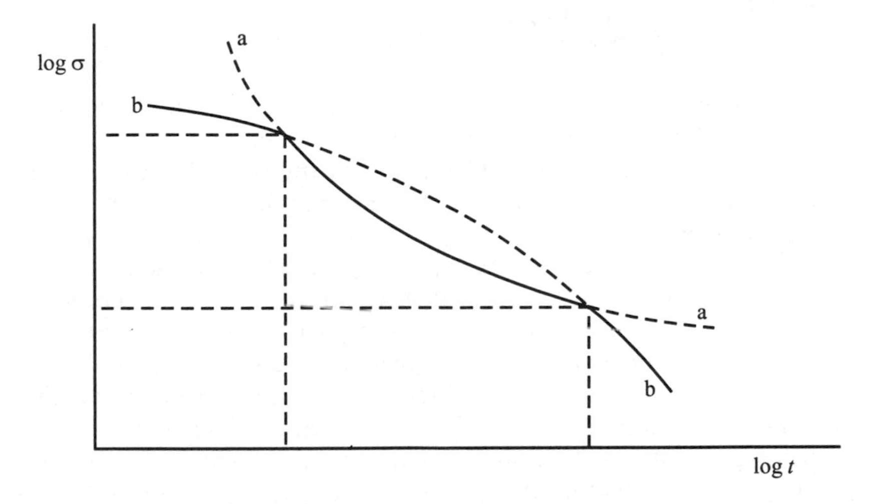

Fig. 6.11 Two types of fracture each with their own envelope, (a) ductile (b) brittle, each 'fighting' for preference on the timescale (courtesy A.K. van der Vegt)

many uncertainties connected with failure it is, however, sensible to perform all extrapolations with great caution.

Figure 6.10 illustrates that long-term failures can be ductile, but also brittle depending on the material, the magnitude of the stress and the temperature. In effect, experience shows that in the long term some of those polymers thought to be ductile show a tendency to become embrittled as crack-like defects get under way. So the two mechanisms fight for a 'which can enforce failure first' situation, a principle illustrated in Fig. 6.11. This figure also suggests a basis for the ductile–brittle transition in the impact speed region.

The designer must be on the lookout for any likely change of mode of failure – all too obvious in Fig. 6.10, but not always so in other circumstances. Weathering may introduce surface degradation in the long term, which does not occur in short-term high-temperature tests. Stresses locked in during manufacture may be relieved at high temperature, but not to the same extent at low temperature in the long term. Many factors can exert an effect on the long-term strength of polymers – including all those mentioned in section 6.2.3.

(b) Cyclic (fatigue) load

Because polymers have a low conductivity, there is a real risk that hysteresis (tan δ) in cyclic fatigue tests will lead to a large temperature rise in polymer articles, which would not occur in a high-conductivity metal product subject to comparable drastic circumstances.

Heat development W is determined by the formula

$$W = n \int_0^{2\pi} \sigma \mathrm{d}\varepsilon = \pi \varepsilon_0 \sigma_0 n \sin \delta \ (\mathrm{W/m^3}) \tag{6.3}$$

as can be derived from equations (5.11) and (5.21). Thus both stress and strain amplitudes σ_0 and ε_0 and frequency n contribute.

This heat has to be conducted to the surface and from the surface transferred to the surrounding medium. The temperature rise mentioned stems from the low conduction and in its turn it will influence the actual value of tan δ. Particularly if the temperature is near T_g or another transition temperature, the increasing tan δ will boost heat development, which cannot be met by increased conduction and transfer. The resulting temperature rise could lead to considerable softening within the polymer mass – especially if it is thick – if the hysteresis is large and the frequency is high. Stress concentrations will encourage the development of local hotspots under cyclic loading.

When a gum rubber vulcanizate is taken through a complete load–unload cycle in a standard tensile test, the energy absorbed (the area under the stress–strain curve) is modest at low strains, but becomes large at high strains because of crystallization, as shown in Fig. 6.12. Where the rubber compound contains reinforcing carbon black, some breakdown of the reinforcement (called breakdown of the 'structure') occurs during the first cycle, thus making the compound rather weaker and more extensible, as shown in Fig. 6.13. In subsequent cycles the hysteresis is much reduced, as shown in Fig. 6.13 for the second and tenth cycles. In practice, rubber compounds are mainly used at much lower strains where the hysteresis is small (but not necessarily insignificant).

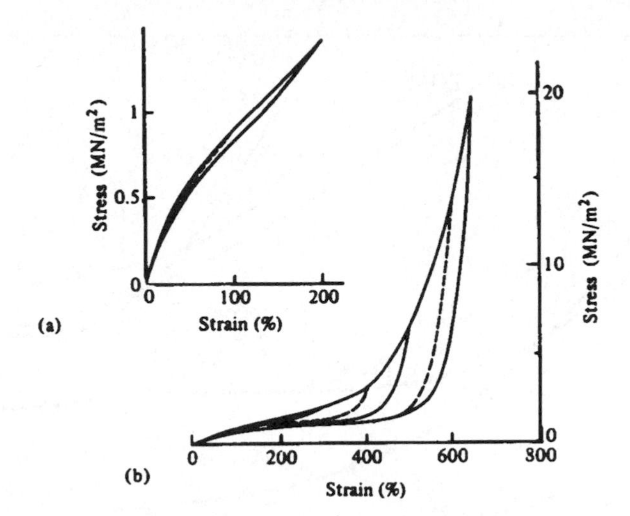

Fig. 6.12 Hysteresis in a natural rubber gum vulcanizate (courtesy MRPRA): first cycle hysteresis loops for gum vulcanizate extended to various strains. At low strains (enlarged in (a)) there is little hysteresis, but it is very much greater at high strains (b) owing to crystallization

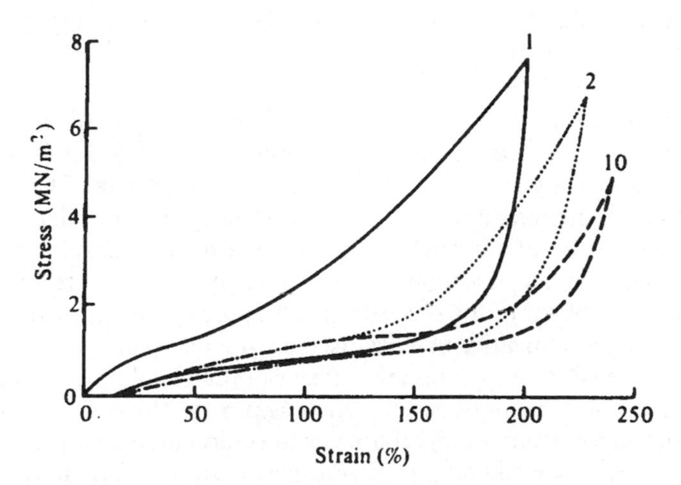

Fig. 6.13 Hysteresis in a reinforced natural rubber vulcanizate (courtesy MRPRA): first, second and tenth tensile stress–strain loops of a vulcanizate containing 50 phr reinforcing black. Most of the structure breakdown (and also set) occurs on the first cycle

Experimental studies of the behaviour of polymers tend to be undertaken at low frequencies, typically about 1 Hz, to avoid excessive temperature rise. Additional tests may be made at higher frequencies but at lower stress or strain amplitudes. The main emphasis has been on square-wave and sinusoidal wave

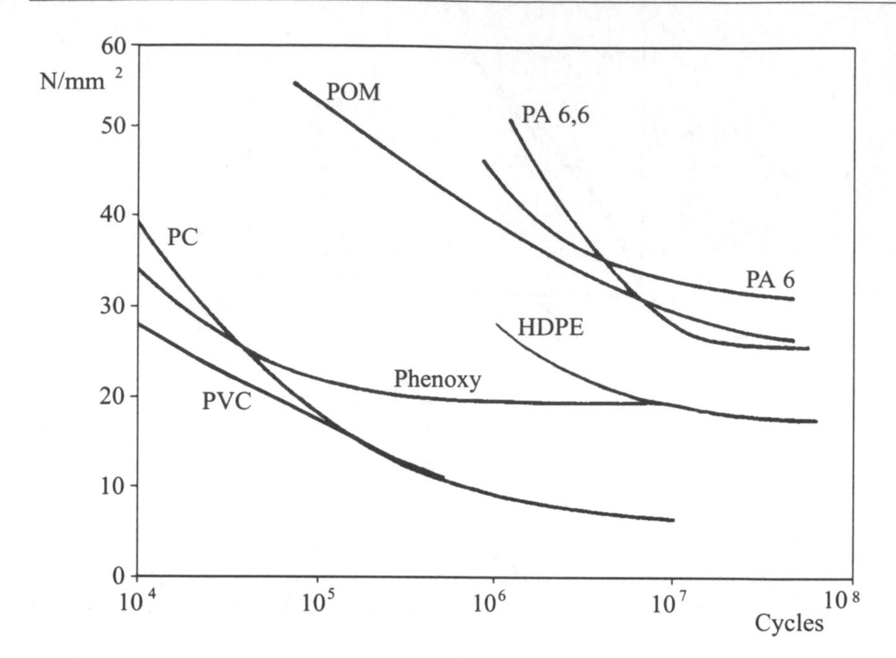

Fig. 6.14 Fatigue life (S–N curves) of several compounds. Frequency 30 Hz, mean stress = 0 (from Sächtling, 1977)

forms, in tension and compression, or in bending. Bending tests the surface of the polymer, whereas push–pull tests the whole cross-section uniformly and is more sensitive to any internal defects which may be present. The data from these tests are sometimes described as 'dynamic fatigue' data – dynamic refers to the cyclic nature of the loading, and fatigue to the duration of loading (compare static fatigue in section 6.2.2(a)). Some typical data are given in Fig. 6.14 – note the tendency for a sharp drop in the fatigue strength after a modest number of cycles, indicative of a change from ductile to brittle fracture.

There are few data illustrating the effect of cyclic loading at amplitude σ_a superimposed on a mean stress σ_m. One might expect that if $\sigma_a \ll \sigma_m$, then in the short and medium term, failure would be dominated by creep rupture, and if $\sigma_a \gg \sigma_m$, then the fatigue at zero mean stress would dominate. The behaviour is complicated by the transitions from ductile to brittle behaviour, by the presence of any orientation and by the risk of thermal failure. Failure might be better understood from a fracture mechanics standpoint as crack growth (section 6.3.4(b)).

6.2.3 Factors promoting a ductile–brittle transition

Many factors conspire to make a nominally ductile polymer behave in a brittle manner. This section lists some of the main 'offenders'. The list looks depressingly long. It is. This account rather artificially attempts to identify the responsibility as if failures had only one cause; thus section 6.2.3(a) lies within the province of the designer, section 6.2.3(b) addresses the compounder and technologist, section 6.2.3(c) outlines the processor's contribution (an important theme developed in Chapter 9), and section 6.2.3(d) blames the environment. However, many of the factors interact as in section 6.2.3(e), so that in practice failures result from a combination of factors. Apportionment of blame among them then becomes a major preoccupation, although there is only space here to introduce the factors rather than to explore them in detail.

(*a*) *Stress and design features*

Polymers show several features of behaviour in common with other materials, albeit some in a more exaggerated form. For example, faster stressing, cyclic loading (fatigue) and triaxial tension all tend to make a nominally ductile polymer more brittle.

A faster rate of stressing inhibits disentanglement, as mentioned in section 2.6. Cyclic loading at the same stress leads to embrittlement because of the higher rate of loading during a cycle compared with that under constant load. A square-wave input of stress or strain is more severe than a sinusoid of the same amplitude and frequency for the same reason. Thick sections of a polymer containing crack-like defects are likely to be more brittle than thin sections because plane strain encourages the development of a triaxial stress system which suppresses shear.

Stress-concentrating features such as holes and internal corners are a major – even the major – source of failure of products in service. The local stress σ_L is larger than the applied stress remote from the region σ by the linear elastic stress concentration factor k. In a circular hole in a flat plate the maximum value of k is 3. In an elliptical hole of major axis a and minimum radius of curvature r, the stress concentration factor is

$$k = 1 + 2\sqrt{(a/r)} \tag{6.4}$$

from which it is seen that k rises dramatically as the sharpness of the defect increases. Failure then occurs when σ_L reaches the brittle strength of the polymer. If the material yields, then perhaps the stress concentration will become blunter and therefore safer because the stress at the tip of the defect is spread out. But at the root the stress is triaxial and this inhibits shear, so the polymer tends to be more brittle anyway: a stress concentration is most efficient at searching out whether a polymer (or any other material) will embrittle.

The moral is quite clear: the designer who insists on introducing features such as notches, embossed surfaces, moulded-in lettering and sharp internal corners (as shown in Fig. 6.15) is courting disaster. It would be wise to avoid

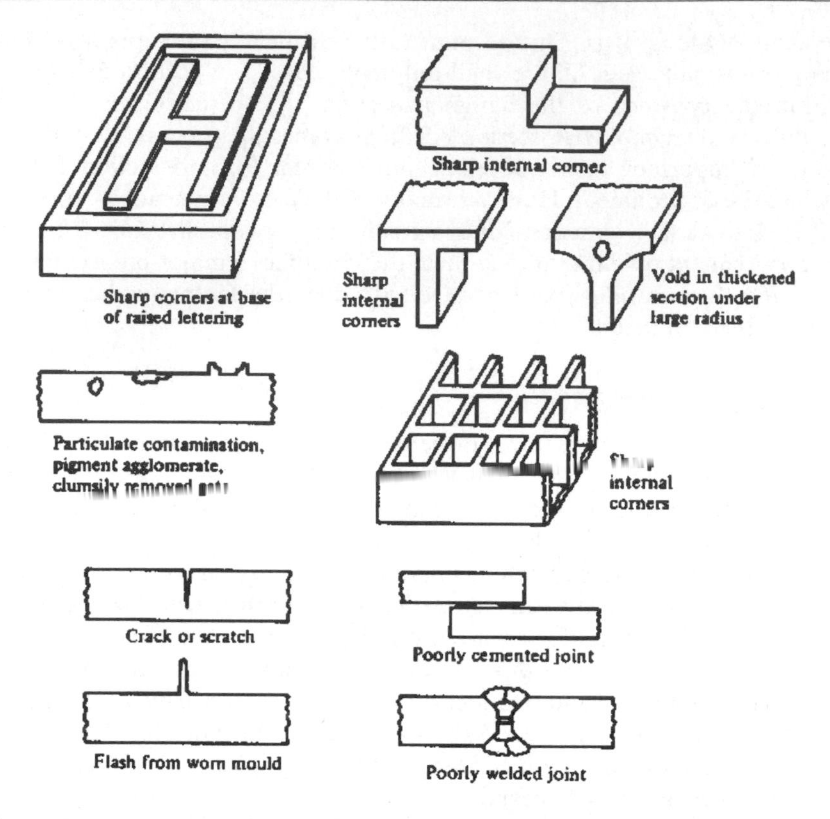

Fig. 6.15 Some common forms of stress concentration

these features or at least put as generous a radius on internal corners as possible. One should be alert to sources of stress concentrations deriving from processing – worn tools can give flashed parts where material escapes under pressure through ill-matched mating metal surfaces, and the clumsy removal of gates can leave sharp scars. More difficult to eliminate are pigment or filler aggregates (which arise from inadequate mixing technology) and small voids. It must be recognized that some designs have to incorporate keyways and thread or tooth forms and may need to provide for self-tapping screw forms, the latter being particularly aggressive, the more so as time passes: here the designer should choose polymers which are tough under these circumstances, particularly if significant loads are applied to the component in service.

The factors may combine to give more severe effects. For example, natural rubber is tough. But under cyclic loading, cracks can grow from sources of stress concentrations. The sketches in Fig. 6.16 indicate the dramatic shortening of fatigue life of rubber parts initially having severe rather than mild stress concentrations.

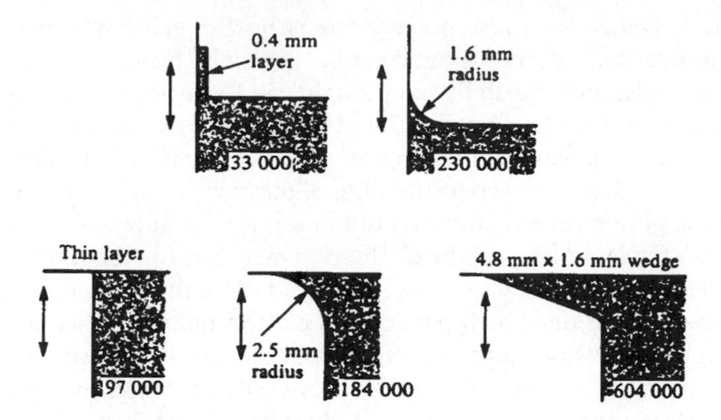

Fig. 6.16 The effect of reducing stress concentrations on the fatigue life of natural rubber: 25.4 mm square × 6.3 mm thick testpieces subjected to cyclic shear deformations of ±100% strain at 2 Hz. Large flaws visible at the number of cycles shown (courtesy MRPRA)

(b) Polymer and compound features

A wide, not to say bewildering, range of polymers is available. Each polymer system has its own portfolio of variants in either the character of the given polymer or characteristics imposed by compounding with other ingredients. The following remarks are indicative rather than comprehensive and indicate that care is needed in choosing and then actually obtaining the most suitable compound: embrittlement can result from incorrect choice or supply of the wrong formulation. Data for the effects mentioned here are given in more specialist texts.

There would seem to be an easy answer to the question 'Is a high molecular weight grade of a given polymer likely to be more brittle than a low molecular weight one?' Thc high molecular weight one would have many more entanglements per chain and hence would be more stiff and brittle. There is evidence to support this: cast acrylic has a high molecular weight and is more robust than articles made from the much lower molecular weight acrylic moulding powders. However, this evidence is far from complete and merits further discussion in section 6.2.3(c), and is elaborated upon in Chapter 9.

An increase in crosslink density restricts molecular mobility and hence makes a polymer more likely to be brittle. It is quite easy to increase the quality of crosslinks by adding more of the crosslinking agent (e.g. sulphur in rubber) or changing the chemistry of the prepolymer in unsaturated polyester to permit more crosslinking per unit volume.

Copolymers and blends are often used to achieve a balance of properties: a change in proportions can lead to a change in the balance, and toughness may suffer. General-purpose polystyrene is brittle, being well below T_g at room temperature; incorporation of rubber (either by blending or copolymerization) provides more flexibility and toughness. Reducing the rubber content makes the compound less tough. A similar argument can be used for rubber-toughened epoxy polymers. Plasticizers are used to increase flexibility, often dramatically; a decrease in concentration of plasticizer – either deliberately or because the plasticizer has migrated out in service – reduces ductility.

Hard particulate fillers reinforce the polymer, but for the most part they impair the toughness of the composition and reduce the elongation at break, e.g. natural rubber filled with particles of carbon black or silica or whiting. Stiff fibrous fillers also impair ductility and usually introduce some anisotropy of strength: unless the fibres are carefully matched to directions of principal stress, there is a real risk of failure across the fibres, analogous to wood splitting readily along the grain but not across it. This point will be explored further in Chapter 7.

Poor housekeeping in processing may lead to contamination, especially where a range of additives is used in different compositions or where thermoplastics scrap is reground and mixed with virgin polymer. Bad mixing may leave agglomerates of additives unbroken, introducing domain boundaries that may serve as crack starters.

(*c*) *Processing features*

Shaping the polymer by melt processing introduces several features which can promote embrittlement. Fundamental among these are the development of **anisotropy** by alignment of molecules or fibrous fillers, **inhomogeneity** giving a distribution of microstructure throughout the wall thickness of a product, and **residual stress** caused by solidification from the melt during cooling. These ideas are introduced here and developed in Chapter 9.

On a more practical note, processing conditions may deviate from the optimal ones.

Melt processing at low melt temperatures can lead to poorly knit weld lines. Running too small an injection machine flat out with too large a cavity can even cause injection of a polymer which is incompletely melted – the partly fused granules then guarantee later disappointment in service because the boundaries of the unfused grains are sources of crack-like defects. At high melt temperatures there is an increasing tendency to form voids and flash.

With rubber compounds and thermosetting plastics, too short a cycle time will give insufficient crosslinking – even a crumbly core to the product – while too long a cycle time may cause the polymer to become thermally degraded. The polymer technologist will need to choose ingredients such as accelerators and heat stabilizers with care so that these effects can be minimized within the constraints of the inherently low thermal diffusivity of polymer compounds.

(d) Environmental factors

A decrease in temperature leads to less mobility and hence to a more brittle pattern of behaviour; the transition is most marked in amorphous polymers near their T_g. Partially crystallizable polymers tend to become more brittle more gradually as the temperature drops towards T_g.

Exposure of most polymers on their own to oxygen, ozone or UV radiation can lead to degradation, resulting in cloudiness on the surface of plastics and perishing of rubber; the result is a sharply cracked surface, with the defects only too willing to grow under tensile stress to give a brittle fracture. The effect is worse in thin products such as film than in thick ones such as conveyor belting because of the larger ratio of surface area to volume of polymer. The 'remedy in advance' is to incorporate small quantities of suitable anti-oxidants, anti-ozonants and UV absorbers into the recipe to inhibit the onset of degradation.

Articles made from polymer compounds frequently have to withstand a range of chemical, usually liquid, environments. Some chemicals have no effect, others readily dissolve the unstressed polymer or cause swelling. The simple concept that 'like dissolves like' is reflected in the data in Table 6.1 for the solubility parameters of liquids and fairly 'non-polar' polymers: where the differences between solubility parameters for liquid and polymer are small, chemical attack is likely – this is a good way to identify solvents for making adhesive joints. However, aggressive chemical reactions such as attack by polar media and strong or oxidizing acids are not covered by this approach. Chemical reactions are diffusion controlled, and proceed all the more vigorously if the temperature is raised, or if exposure is prolonged. Do seek good advice when checking against chemical attack.

(e) Environmental stress cracking

Exposure to chemicals and solvents may seem rather unconnected with the theme of promoting embrittlement. But there is a nasty phenomenon known as environmental stress cracking. A polymer may be ductile when stressed in air, and may resist a given liquid when exposed unstressed. However, if that liquid is a stress-cracking hazard, the exposed stressed polymer may fail in a brittle manner at a disconcertingly low strain after only a short time. There is perhaps no unique explanation for the phenomenon, but the following may be helpful in obtaining a feel for what seems to happen: as the stress increases, crazes or microvoids develop. These regions are extremely porous and hence the hostile liquid is rapidly absorbed. The swollen polymer in the crazed or voided region is surrounded by undamaged polymer which has not swollen, so the pressure exerted by the swollen polymer acts as a wedge to grow more crazes. Eventually the swollen region grows so fast that solvent cannot reach the damage tip, and brittle fracture results. It would seem that

Table 6.1 Solubility parameters at 20°C
(*a*) *Solubility parameters of liquids*

Liquid	*Solubility parameter* $((\mathrm{cal/cm^3})^{1/2})$
Dodecamethyl pentasiloxane	5.35
Hexamethyl disiloxane	6.0
n-Pentane	7.05
n-Hexane	7.3
n-Octane	7.6
n-Decane	7.7
Methyl isobutyl ketone	8.4
Carbon tetrachloride	8.6
Methyl ethyl ketone	9.3
Styrene	9.3
Acetone	9.6
n-Dodecanol	9.8
Isobutanol	10.5
Acrylonitrile	10.5
n-Butanol	11.4
Isopropanol	11.5
n-Propanol	12.0
Ethanol	12.7
Propylene glycol	13.7
Methanol	14.5
Glycerol	16.5
Ethylene glycol	17.1
Formamide	19.2
Water	25.0

(*b*) *Estimated solubility parameters of polymers*

Polymer	*Solubility parameter* $((\mathrm{cal/cm^3})^{1/2})$
Polybutadiene	8.4
Poly(2,6-dimethyl-1,4-phenylene oxide)	8.6
Butadiene/styrene (75/25) copolymer	8.7
Polystyrene	9.1
Polyvinyl chloride	9.5
Polymethylmethacrylate	9.5
Butadiene/acrylonitrile (75/25) copolymer	9.5
Styrene/acrylonitrile (75/25) copolymer	9.6
Polycarbonate of bisphenol A	9.8
Polysulphone of bisphenol A	10.8
Polyacrylonitrile	15.4

the effect of the tensile stress is to widen the range of differences in solubility parameter which cause attack.

After being loaded in tensile stress with a constant load, the unexposed polymer will start to yield or develop and grow a stress-induced crack in due course, but the presence of a stress-cracking fluid (or gas, like ozone for

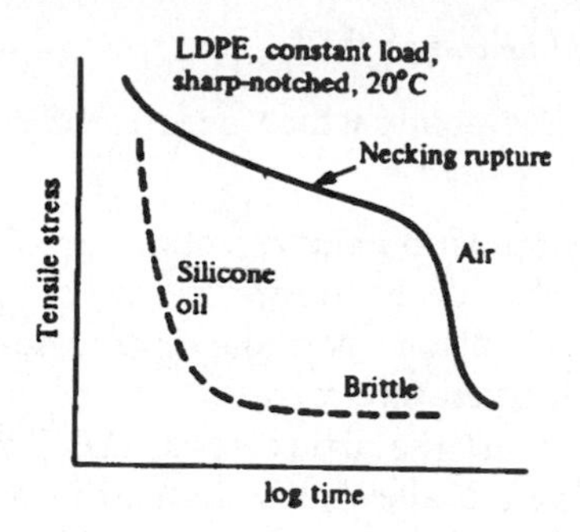

Fig. 6.17 Environmental stress-cracking in sharp-notched low-density polyethylene

rubber) will speed up the process dramatically, as indicated in Fig. 6.17. The growth of crazes and cracks in air or in hostile environments can be described by fracture mechanics (section 6.3.4(a)).

Examples of stress-cracking hazards for polymers include the following: polyethylenes – detergents; low-density polyethylenes – silicone oil and some alcohols; polystyrene – acetone, vegetable oil and white spirit; acrylic – alcohols and some solvents; ABS – glacial acetic acid; polysulphone – acetone, ethyl acetate, petrol and trichloroethylene; polycarbonate – chloroform, petrol.

To conclude this section we have seen that failure by embrittlement is promoted by any of the following circumstances taken singly or in combination:

- long-term loading;
- sharp notches (beware stress concentrations!);
- a hostile stress-cracking environment in contact with the product; the effects described under short-term loading also apply in the longer term and often to a greater extent;
- a cyclic load (fatigue) rather than a constant load.

6.2.4 Designing to avoid failure

The overall procedure is analogous to that described for stiffness in section 5.1.4(a). It is assumed that a stress analysis of the product can be made, or an appropriate structural idealization of it. The polymer is assumed to have a stress–strain curve which has an unambiguous yield point or ultimate strength. The design problem then devolves on a full design brief for load-bearing duties (together with the other aspects of the product function as indicated in section 1.3), selection of the most suitable failure data (by no means always straightforward), and choice of a suitable safety factor. When either impact energy or cracks of known size or both are involved, the approach to avoiding brittle failure, based on fracture mechanics, outlined in section 6.3, should be addressed.

(a) Design brief for load bearing

The following indicate the points which need to be covered, before seeking actual failure data.

- How big, and in what position and direction, are the applied loads in the product or its structural idealization?
- Are the applied loads constant, periodic or occasional, and for how long?
- What is the service temperature range?
- What environments must the article resist: UV, moisture, oils, greases, acids, alkalis, alcohols, brake fluids, industrial and domestic solvents, detergents, 'other chemicals'? Are there any hidden plasticizing or embrittling effects? Are there any potential stress cracking hazards?
- What are the acceptance test schedules for the product? The product must satisfy these as well as actual service conditions.

From the design brief it is possible, at least in principle, to identify the nature of a possible failure and perhaps to find a value of failure stress relevant to the service conditions, if the data exist. The next step is to decide on the value of the design stress. Two complementary approaches are possible: one is based on the value of allowable stress of any kind, the other on a maximum strain criterion.

(b) Basis for the calculation

If we restrict ourselves to a statically determined design, a strength calculation should have the following scheme (Fig. 6.18):

1. Structural aspects. From shape and load the stress tensors can be evaluated for all load–cross-section combinations considered to be critical.

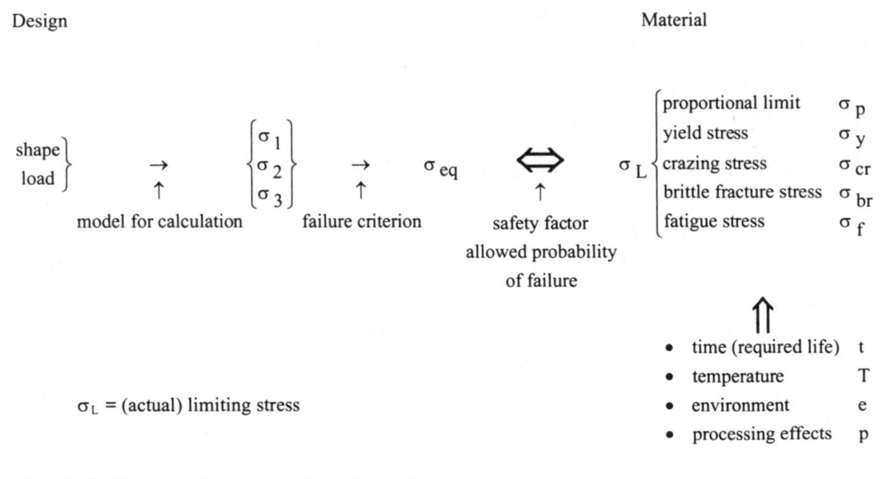

Fig. 6.18 Scheme for strength calculations

This results in one or more critical design stress tensors for any of the working or testing loads.

2. Materials aspects. In view of the polymer (as compounded for the purpose) and the application of the product, the (uniaxial) limiting stress σ_L is determined. This implies that the choice made from proportional limit σ_P, yield stress σ_Y, crazing stress, brittle fracture stress, etc., is valid for creep rupture, fatigue or according to any other loading scheme. The relevant stress figure $\sigma_L(t, T, \text{etc.})$ should be the valid one for the operating conditions, such as the required life, temperature and environment, and for the conditions under which the product was processed (resulting in possible molecular orientation, etc.). It is clear that determination of this stress requires great knowledge and experience in the field of polymer engineering, rather than drawing a stress figure from a database.
3. Actual to permissible. The critical design stress tensor should now be translated into an equivalent uniaxial stress σ_{eq} through the appropriate failure criterion. Each of the σ_{eq} values should now fulfil

$$\sigma_{eq} \leq \sigma_L(t, T, \text{etc.})/F \tag{6.5}$$

where F is the safety factor as discussed in the next section.

(*c*) *Safety factor*

The safety factor F should consider different aspects of design and service of the product.

1. The uncertainty in the calculations. This may vary from rounding-off errors to the use of a calculating model, which is an erroneous interpretation of reality. An example is failure by buckling instead of by compression.
2. Inaccuracy in the choice of the materials data. The many factors that determine the strength of a compound or grade as shaped into a product in service, and our limited knowledge regarding the precise effect of each factor, may often result in a good guess rather than in an exact value of σ_L. In addition we have to consider that all strength data are subject to a statistical spread.
3. The chance that the load is higher or applied differently from the one supposed at the strength calculation, e.g. caused by inexpert use of the product or an unexpected testing load. Also, the loading of commercial products can be considered to be of a statistical nature (a typical example is a vacuum cleaner).
4. The risks that are connected with the possible failure of a product in terms of damage to persons, environment or goods. If failure is announced by yielding or cracking sounds of the material, which enables measures to be taken to minimize the damage, these risks are supposed to be smaller than if only an unannounced failure can be expected.

If the safety factor is defined on this basis – the materials data which are valid for the actual design conditions – it could have a magnitude of 1.3 to 1.5, mainly depending on point (4). This same basis also serves for metals, and to use it systematically would make the comparison of strength calculations for metal and polymer products more transparent. Sometimes σ_L is derived from the strength as measured in a laboratory tensile test, and adapted for the influence of time, temperature, environment or processing by means of a separate reduction factor R. The introduction of such a factor is true to the idea of a comparable strength calculation, but it may violate the proper description of these influences.

This basis, that the materials data are valid for the actual design conditions, is adopted in calculating the necessary wall thicknesses for thermoplastics pressure pipe for cold water services at 20°C. This is possible as the measurements are performed close to the actual application practice. Typical values of F from HDPE pipe standards are 1.3.

When a higher value of F is chosen, this is usually because of a mistrust in the materials data for operating conditions and in the influence of processing. British Gas uses $F = 4$ in the design of pipework systems for gas distribution, with medium-density polyethylene to avoid failure by creep rupture. This factor is greater than that used for cold water pipes because a system is being considered, and because the effects of chemical environment, surface damage, jointing and secondary installation stresses are being judged. From a standpoint of materials engineering it would seem more proper, however, if these factors were taken separately to adapt strength data, to incorporate a jointing factor and to revise the design requirements with regard to possible surface damage and secondary installation stresses.

The effect of processing, such as injection moulding, on strength is, in spite of all specific knowledge developed in recent years, difficult to predict in advance of making the product, and so experience counts.

A study of points (2) and (3) above makes it clear that statistical considerations should form part of a safety procedure. The basis of such a procedure would be to calculate the probability of failure from the combined chances of low strength and high loading. This would require knowledge of their distributions.

(*d*) *Maximum strain approach*

The maximum strain limit is the lower of two values: the limit of deformation allowed in use, and the limit suggested for the polymer.

The deformation in use is really a geometrical effect, because deformation can be transformed into strain. The question then becomes 'By how much can a part deform without impairing its performance or interfering with adjacent equipment?' Such deformation problems may be solved using creep data and the methods outlined in sections 5.1 and 5.2.

Polymer limitations are more difficult to generalize. The following limits are only rough guidelines and are restricted to reasonably well designed and well manufactured articles operating over a long period in air within the temperature range 10–40°C. Reasonably well designed here means a design in which stress concentrations are minimized, and reasonably well made excludes parts made under conditions which introduce excessive molecular orientation. On this basis amorphous plastics should not normally be used at strains exceeding about 0.75% because of crazing or voiding, which are visually disquieting. Welded polyolefins seem to void at about 1% strain. Incorporation of fibrous reinforcing fillers reduces the recommended strain in partially crystalline polymers to about 0.75%. The argument is that crazing or voiding occur long before mechanical failure, so the use of these strain limits does not require the use of an additional safety factor unless the visual defect is unacceptable.

For resin-bonded fibre composites it has been found that for strains from $\varepsilon = 0.2$ to 0.4%, signs of debonding and/or microcracks could be found, on the basis of which $\varepsilon_{max} = 0.2\%$ was introduced as a limit in BS 4994.

All strains mentioned here could form the basis of a strength calculation:

$$\sigma_L = \varepsilon_{max}\, E(t, T, \text{etc.}) \tag{6.5a}$$

PROBLEMS

1. Why might you expect an unnotched tensile bar injection moulded from general-purpose polystyrene and gated at one end to show a fracture surface (Fig. 6.19) which is predominantly smooth over a wide area but totally surrounded by rough material?
2. After a bar made from high-impact (rubber-modified) polystyrene has been broken, you pick it up and try to flex the waisted part. Would you expect it to be stiffer or more flexible than an untested bar? Why? Describe thc difference in 'feel' between the broken and untested bars.
3. Distinguish the behaviour in shear of the highly drawn material in a broken polypropylene tensile bar with that of an undrawn specimen.

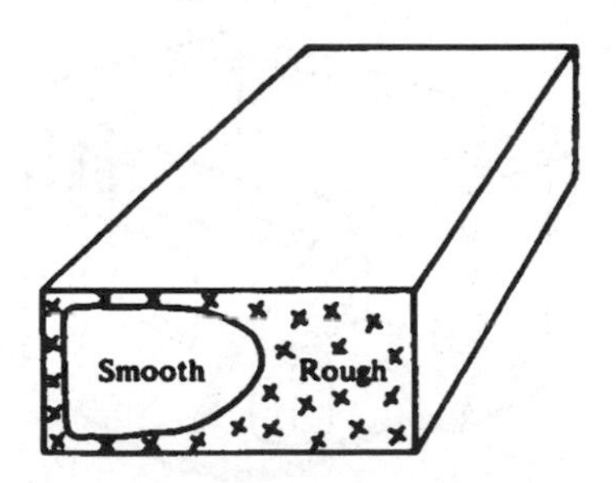

Fig. 6.19 Fracture surface of end-gated injection-moulded bar

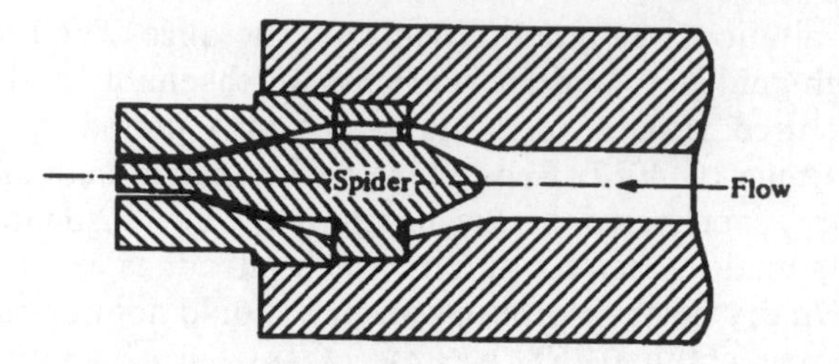

Fig. 6.20 Schematic view of spider within a pipe extrusion die

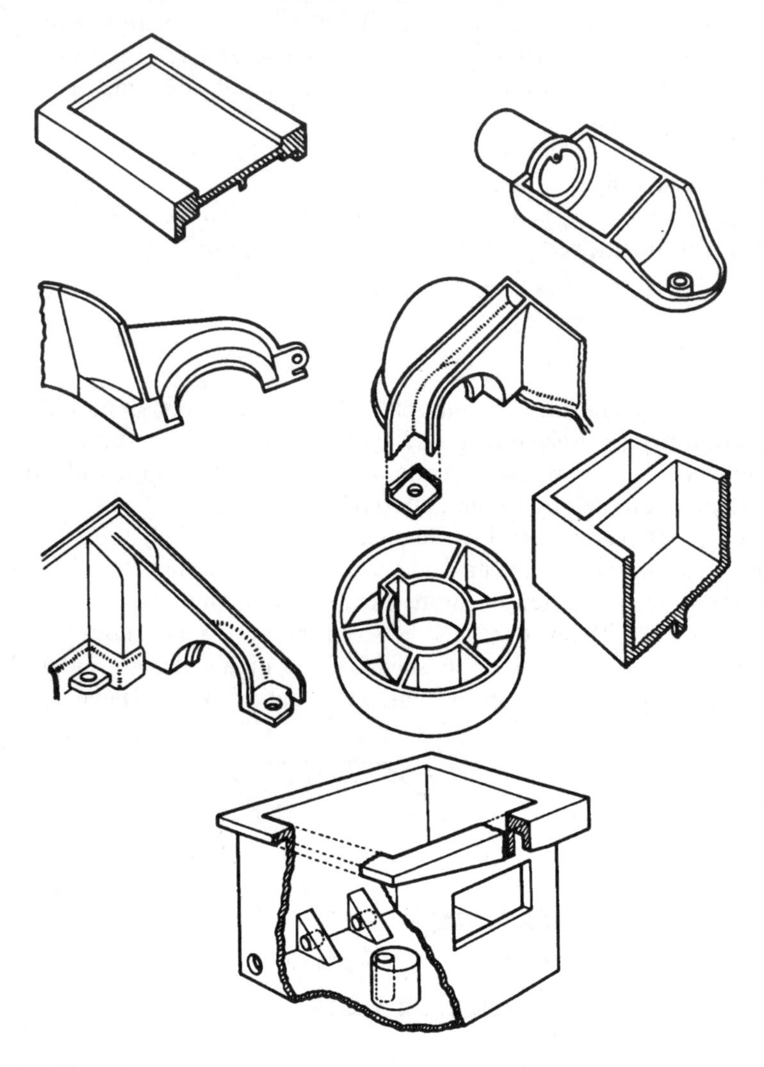

Fig. 6.21 Find the many stress concentrations in these products

4. Natural rubber vulcanizate has the ability to crystallize under suitable conditions, whereas styrene–butadiene copolymer rubber (SBR) has not.Which of the two would you expect to be the more resistant to crack growth?
5. Why do thick rubber bands seem to last longer than thin ones?
6. A plastics pressure pipe is made by extruding it through a die having an annular gap. The mandrel defining the inner wall of the pipe is held in place with respect to the outer part of the die by the spider (Fig. 6.20). What fault or faults might you expect in a pipe extruded from a short-glass-fibre-reinforced plastic?
7. A cube of side 100 mm is to be cast from epoxy resin. The maker knows that the curing reaction is exothermic (gives off heat), but wishes to achieve high production rates. What faults would you look for if the maker mixed too much accelerator in the resin?
8. Which of the following liquids are likely to be effective solvents at 20°C for (a) styrene–butadiene copolymer rubber, (b) polysulphone: water, ethylene glycol, isobutanol, carbon tetrachloride, *n*-pentane?

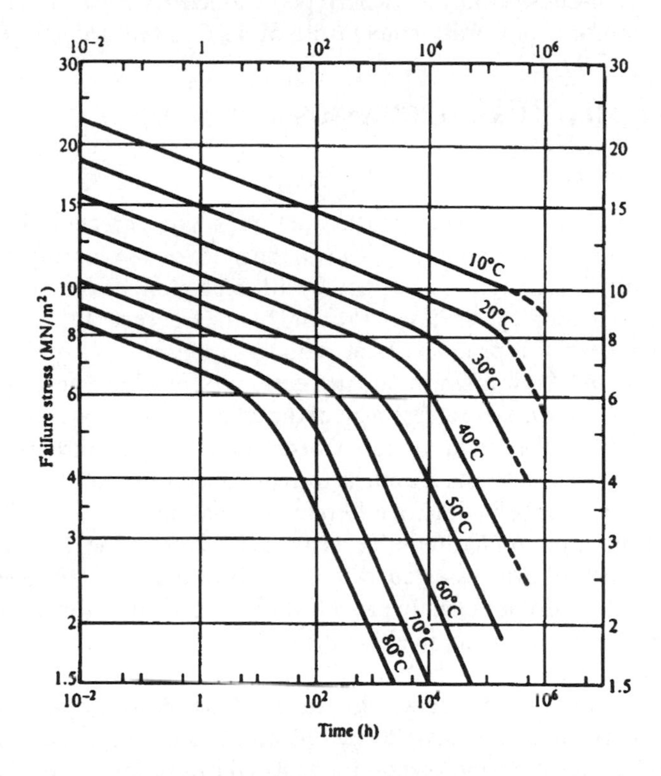

Fig. 6.22 Creep rupture data for high-density polyethylene pipe at 20°C (courtesy Hoechst)

9. Gear wheels are often injection moulded from polyacetal, using a centre-gated cavity. Identify the main factors which contribute to the long-term durability of the gears.
10. Would you expect rubber to perish more readily when stretched or when unstretched?
11. Identify the stress concentrations in the products in Fig. 6.21.
12. (a) A pipe extruded from high-density polyethylene has a mean diameter of 1.5 m and is to contain water at 20°C for 50 years at an internal pressure of 0.35 MN/m^2. What is the minimum recommended wall thickness for this duty? Creep rupture data are given in Fig. 6.22.
    (b) What would be the consequence of foolishly ignoring the ductile–brittle transition in high-density polyethylene pipe destined for a 50-year duty at 20°C?
    (c) If only a 1-year life were required of a high-density polyethylene pipe at 20°C, by how much could you increase the pressure rating quoted for a 50-year life?
    (d) By how much would you downrate the safe working pressure for a high-density polyethylene pipe destined to work for 50 years at a uniform pipe wall temperature of 40°C rather than 20°C?

6.3 THE FRACTURE MECHANICS APPROACH

6.3.1 Introduction

'It just snapped.' Even when brittle fracture is half-expected, the actual incidence of failure is disappointing, and all the more disappointing when the user expects any failure at all to be ductile with high elongation. What is the best material to use for a given design to avoid brittle fracture? Which factors affect the robustness of a product? How damage-tolerant is the polymer (in end-product form)? How often should inspections of products be made in service?

Traditional (and easy) assessment of the tendency of a material or product to brittle fracture relies on fast loading, often with a stress concentration present to inhibit shear. Such tests are quick and easy to conduct, and can provide data for comparing the behaviour of different materials under the conditions of test. In addition, if manufactured objects are being tested, local stress conditions and the effects of processing conditions can be evaluated. However, the traditional methods do not usually provide data for the quantitative prediction of load-bearing performance in either the short or the long term.

A better basis of quantitative design is to measure the strength of a body containing a crack or crack-like defect. This involves more precise experimental work than the 'easy' tests, but analysis of the results provides the required fundamental background as well as the basis for comparing materials.

Load can be applied as a quasi-static force, leading to a distribution of stresses in the product, which it can either sustain or not. But load can also

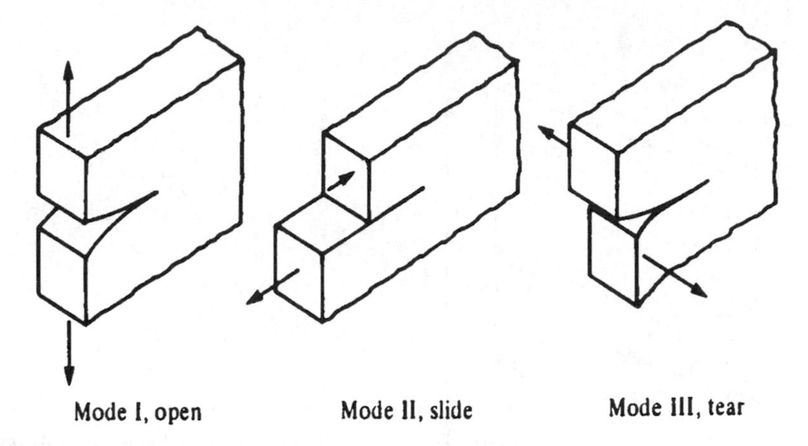

Fig. 6.23 Modes of crack growth

be applied as an impact and the question then is whether the product as a whole can absorb the impact energy without overloading any of the local sustainable stress conditions. The latter vision requires consideration of energy rather than stress, which is less familiar in engineering of products.

Fracture mechanics deals with all kinds of fracture, but with brittle fracture in the first place. This involves failures which occur at low strains with no permanent deformation and therefore in the linear elastic domain. The process involves the creation of new fracture surfaces, and requires a supply of energy. This section states the concepts of linear elastic fracture mechanics (LEFM) embodied in the terms 'strain energy release rate', 'stress intensity factor' and 'fracture toughness'. It identifies factors which affect crack growth in polymers, discusses the time dependence of crack growth, and applies these concepts to simple but representative practical problems. Our discussion centres on the growth of cracks under opening forces (Fig. 6.23) rather than under shear forces.

Yielding is typically non-linear behaviour. Yielding phenomena therefore fall outside LEFM. Fracture mechanics also deals with non-linear phenomena, such as local yielding, making use of other symbols (J instead of G) but that falls outside the scope of this section.

6.3.2 Energy approach to fracture

The earliest analyses of brittle fracture of glass were based on an analysis of energy, and this approach is still preferred for describing crack growth in rubber compounds.

The energy approach, as described by LEFM, considers that two sources are available for the growth of a crack: external energy U_1 (e.g. coming from impact) and stored energy U_2 (the elastic work as exercised by a load on the

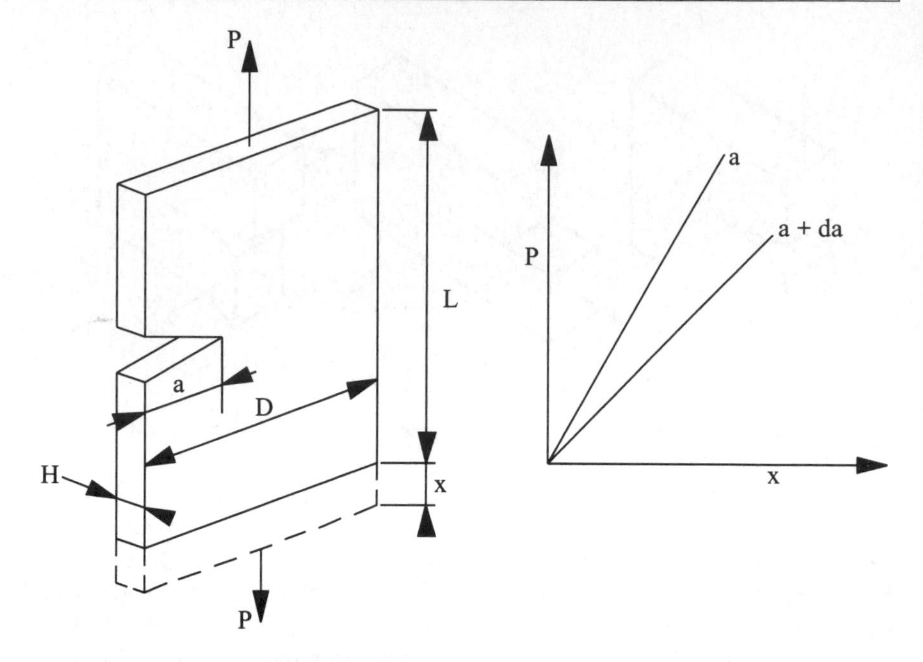

Fig. 6.24 Load–deformation for cracked body

product). When fracture occurs they are (partly) converted into energy to create new surface and internal damage in the material, U_3, and in surplus energy U_4 (in tossing broken pieces around, etc.). The role of stored energy is essential.

First consider an uncracked plate of thickness H and width D and deformed by an amount x under an applied load P (Fig. 6.24, ignoring the crack a). The compliance of the uncracked plate, C_0, is defined as

$$C_0 = x/P \tag{6.6}$$

For this plate, the stress σ and the compliance C_0 are

$$\sigma = P/HD \quad \text{and} \quad C_0 = L/EHD \tag{6.7}$$

and the strain energy U stored in the plate is the area under the force–deformation curve, given by

$$U = \frac{1}{2}Px = \frac{1}{2}P^2C_0 = \frac{1}{2}(\sigma^2/E)(HDL) \tag{6.8}$$

If the plate fails in a brittle manner (with load P_f and connected displacement x_f), the failure stress $\hat{\sigma}$ is

$$\hat{\sigma} = P_f/HD \tag{6.9}$$

and the stored energy at failure is

$$w = \frac{1}{2}P_f x_f = \frac{1}{2}P_f^2 C_0 = \frac{1}{2}(\hat{\sigma}^2/E)(HDL) \tag{6.10}$$

Now consider a plate of the same dimensions, but containing a crack of length a, again deformed by an amount x under an applied load P acting to open the crack (Fig. 6.24). The compliance of the cracked plate, $C(a)$, is defined as

$$C(a) = x/P \tag{6.11}$$

The dependence of C on a is clear: the longer the crack a, the higher the deformation x for a given force P.

The strain energy $U(a)$ stored in the plate is

$$U(a) = \frac{1}{2}Px \tag{6.12}$$

When more energy is applied (by increasing the load) this may result in a change of stored energy, but also in crack growth (to be understood as energy consumed to create a new surface). The change of stored energy could even be negative, in that it contributes to the extension of the crack length from a to $a + da$. In a way, the crack growing phenomenon can be ascribed to the release of stored energy. This has led to the definition of the strain energy release rate G_I per unit thickness as

$$G_I = -\frac{1}{H}\frac{\partial U}{\partial a} \ (\mathrm{J/m^2}) \tag{6.13}$$

Here the word 'rate' denotes surface area and not time and the suffix I denotes the crack opening mode (Fig. 6.23).

We shall see that G_I depends on modulus and thus is affected by all the environmental factors, such as temperature and time under load, which influence the modulus of the polymer. Failure in an elastic material occurs almost instantaneously when G_I reaches the critical value G_{Ic}. G_{Ic} is sometimes called the critical strain energy release rate (but sometimes also the fracture toughness).

Typical values of G_{Ic} at 20°C for a range of materials are shown in Table 6.2. Again, the values of G_{Ic} depend on such factors as test conditions, composition, and on method and conditions of processing. It should be noted that G_{Ic} relates to the generation of a new fracture surface, and that when a crack of width H grows by Δa, the amount of new surface is $2H\Delta a$. Some books use the term fracture energy γ_c which refers to the energy needed to make one surface on the material: $G_{Ic} = 2\gamma_c$.

For a plate containing a crack of length a, the same analysis as for the uncracked plate suggests that the measured failure energy w should be equated to the elastic strain energy:

Table 6.2 Typical values of plane strain critical strain energy release rate in air at 20°C

Material	$G_{\mathrm{Ic}}(\mathrm{kJ/m^2})$
Glass	0.02
Cast acrylic	0.2–1
Polystyrene	1.4
Polyesters	0.4
Natural rubber	25
Aluminium alloy	50–200

$$w = \frac{1}{2}P_{\mathrm{f}}x_{\mathrm{f}} = \frac{1}{2}P_{\mathrm{f}}^2 C(a) \tag{6.14}$$

For different crack depths a, failure energy w will be different because C is a function of a:

$$\frac{\partial w}{\partial a}\left(= \frac{\partial U}{\partial a}\right) = \frac{1}{2}P_{\mathrm{f}}^2\frac{\partial C}{\partial a} = \frac{w}{C}\frac{\mathrm{d}C}{\mathrm{d}a} \tag{6.15}$$

Substituting in equation (6.13):

$$G_{\mathrm{Ic}} = w/(HD\phi) \tag{6.16}$$

where

$$\phi(a) = \frac{C}{\mathrm{d}C/\mathrm{d}(a/D)} = \text{a calibration factor} \tag{6.17}$$

If we measure the fracture energy w of a plate with a crack of size a and know the value of $\phi(a)$, equation (6.16) gives us the value of G_{Ic}. In order to be more accurate, a series of G_{Ic} tests are made over a range of crack lengths, $0.04 < a/D < 0.5$, and the crack length is measured after fracture to account for any slow growth before catastrophic fast fracture.

The above procedure requires $C(a)$ and $\phi(a)$ to be known functions for the sample to be tested. Analytical determination of these functions is only possible in a limited number of cases (see problem 4 and section 6.3.7(c)). So mostly they have to be determined using FEM analysis or experimentally (with a number of samples containing different crack sizes a, in order to measure $C(a)$ as a graph, from which $\mathrm{d}C/\mathrm{d}a$ should be deduced graphically).

If the plate were loaded in impact, the measured energy w_{m} also contains other components, such as the kinetic energy w_{k} of the pieces that flew away (if the clamping permitted). If w_{k} could be taken in the average case as a constant, equation (6.16) would look like

$$w_{\mathrm{m}} = G_{\mathrm{Ic}} \times (HD\phi) + w_{\mathrm{k}} \tag{6.18}$$

and a plot of w_{m} versus $HD\phi$ would have a slope G_{Ic} (Fig. 6.25).

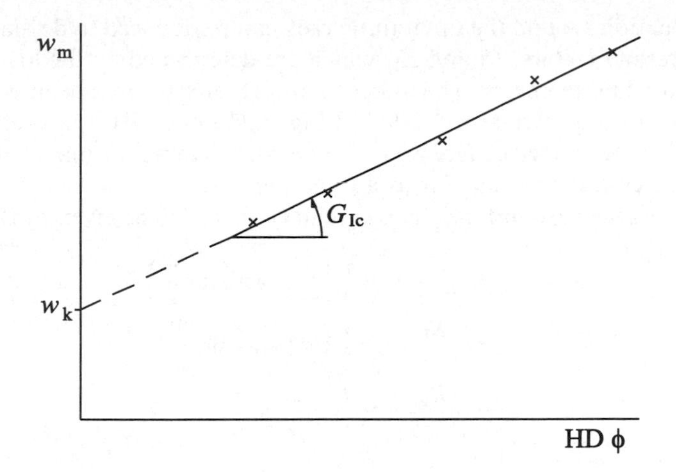

Fig. 6.25 Impact energy measured with specimens containing cracks of different sizes, a, plotted on a graph that allows direct reading of G_k

The results of impact tests, e.g. Charpy and Izod, are exclusively bound to the shape and dimensions of their specimens. The determination of G_{Ic} with the aid of $\phi(a)$ allows us to reduce results of all of these tests to the same characteristic of the material, which describes its resistance to impact independently of the shape and dimensions of the specimen, but dependent on temperature, processing conditions, residual stress, and so on.

6.3.3 Stress intensity approach to fracture

(a) Stress intensity factor (opening mode)

Consider a thin plate, made from a linear elastic material, containing a crack (with a straight front through the thickness) and loaded in its plane. The

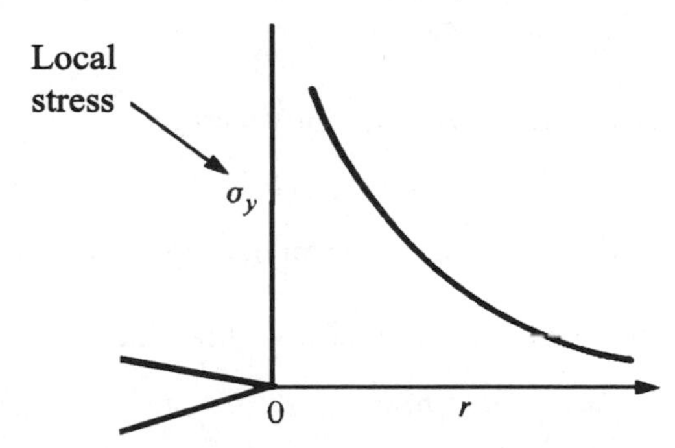

Fig. 6.26 σ_y along x-axis near crack tip

stress situation around the tip of that crack can be deduced to depend on two **stress intensity factors**, K_{I} and K_{II} which are determined by the loads on the plate and by its geometry. The indices I and II refer to the opening and slide modes of crack growth as indicated in Fig. 6.23 (index III is not relevant for the situation described). Here we only consider loading normal to the crack, which makes K_{II} irrelevant for us as well.

The stress field around the crack tip can be shown to be given by (Fig. 6.26)

$$
\begin{aligned}
\sigma_x &= \frac{K_{\mathrm{I}}}{\sqrt{(2\pi r)}}\cos\frac{\theta}{2}\left((1-\sin\frac{\theta}{2}\sin\frac{3\theta}{2}\right) \\
\sigma_y &= \frac{K_{\mathrm{I}}}{\sqrt{(2\pi r)}}\cos\frac{\theta}{2}\left(1+\sin\frac{\theta}{2}\sin\frac{3\theta}{2}\right) \\
\tau_{xy} &= \frac{K_{\mathrm{I}}}{\sqrt{(2\pi r)}}\cos\frac{\theta}{2}\cos\frac{3\theta}{2}\sin\frac{\theta}{2}
\end{aligned}
\qquad (6.19)
$$

σ_x and σ_y clearly tend to infinity at the crack tip.

The local stress, σ_y, along the x-axis near to but not at the crack tip is given by

$$\sigma_y = \frac{K_{\mathrm{I}}}{\sqrt{(2\pi r)}} \qquad (6.20)$$

where r is the distance ahead of the crack tip.

The applied load also gives rise to a local stress σ_x along the x-axis, having the same form as σ_y near the crack tip.

The form of the local stress distribution $\sigma_y(x)$ is the same for all cracks and loads, but the magnitude of $\sigma_y(x)$ depends on the applied load. K_{I} is independent of the material, provided it is isotropic linear elastic, and has the rather unfamiliar units $\mathrm{N/m^{3/2}}$ or $\mathrm{Pa\ m^{1/2}}$.

For a plate having a central crack of length $2a$, and carrying a unidirectional stress σ which is uniform remote from the crack, and which acts to open up the crack, a stress function for σ_y (Fig. 6.27) can be derived:

$$\sigma_y = \sigma\frac{x}{\sqrt{x^2 - a^2}} \qquad (6.21)$$

which for stresses near the crack tip approaches

$$\sigma_y = \sigma\sqrt{(a/(2r))} \qquad (6.22)$$

Clearly the magnitude of $\sigma_y(x)$ depends on the applied load and the length of the crack, $2a$.

Comparing equation (6.22) with (6.20) we find that

$$K_{\mathrm{I}} = \sigma\sqrt{(\pi a)} \qquad (6.23)$$

In general, the value of K_{I} (and of K_{II} and K_{III}) is determined by the distribution of the applied load(s), the size and shape of both the crack and

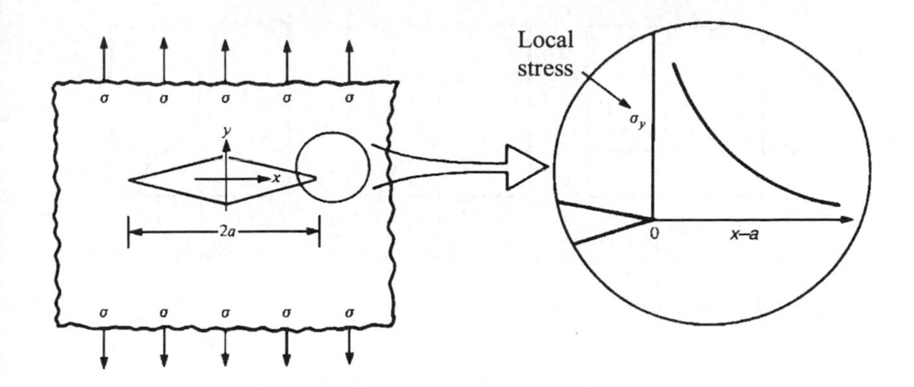

Fig. 6.27 Stresses near crack tip in thin infinite plate containing a crack

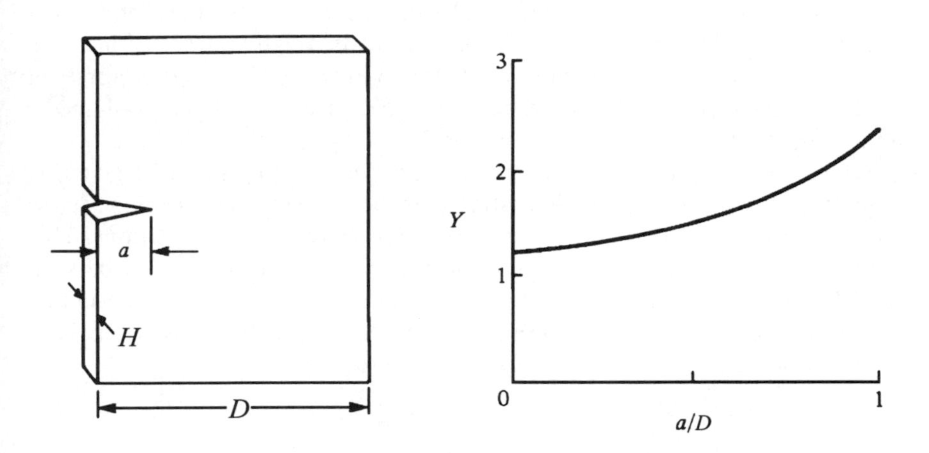

Fig. 6.28 Correction factor Y for single edge crack in sheet of finite size

the body, the location of the crack, and on the boundary conditions. It is customary to compare all of these situations with that of the plate having a central crack of length $2a$, and carrying a unidirectional stress σ, as in Fig. 6.27. This is normally expressed by introducing a dimensionless correction factor Y. Equation (6.23) than becomes

$$K_{\mathrm{I}} = Y\sigma\sqrt{(\pi a)} \tag{6.24}$$

Values of Y are tabulated in the standard handbooks. (Some books use Y, some use $Y' = Y\sqrt{\pi}$; i.e. $K_{\mathrm{I}} = Y'\sigma\sqrt{a}$; the distinction is usually made clear.) Two values are of special interest. For a small crack in a flat plate (Fig. 6.27) $Y = 1$ and for a small crack at one edge of a plate $Y = 1.12$. For the latter case with a bigger crack, see Fig. 6.28.

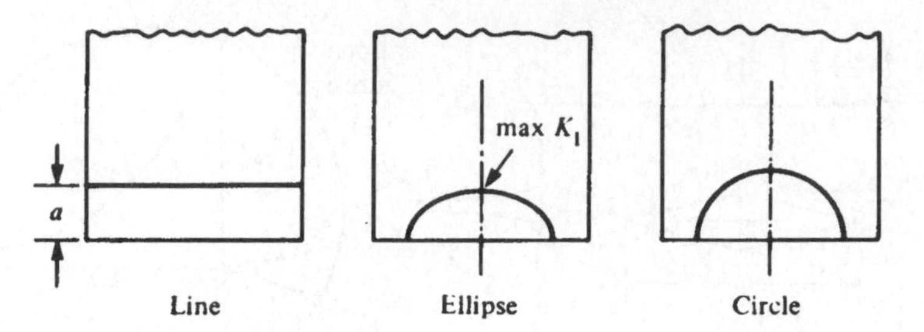

Fig. 6.29 Shapes of plane crack front

The crack front does not have to be straight. A more general case is the elliptical crack front shown in Fig. 6.29. K_I varies with position round the ellipse and is always greatest where the elliptical front intersects the minor axis; nevertheless, K_I for the elliptical front is always less than that for a straight crack front of the same depth a. Two limiting cases are of interest. As the ratio of major and minor axes tends to infinity, $Y \to 1$ (thus agreeing with the result for a straight crack front). For a circular ('penny-shaped') crack front, the maximum value of Y is $2/\pi$.

The stress field $\{\sigma_x, \sigma_y, \sigma_z\}$ around the crack tip in a thin plate is given by equation (6.19) and $\sigma_z = 0$ (plane stress). If the plate is thick the plane stress supposition no longer holds, as the stress shows strong local variations. This implies strong variations in lateral contraction as well, which can be seen clearly near the crack tip as a dimple in the surface of a ductile specimen under tensile load (see Fig. 6.39a). So for thick products the more severe case of plane strain, $\sigma_z = \nu(\sigma_x + \sigma_y)$ should be followed.

(b) Fracture toughness

In a linear elastic material, brittle fracture occurs almost instantaneously when K_I reaches a critical value K_{Ic} corresponding to a combination of applied stress and crack length just prior to fracture. It is not meaningful to talk about a critical applied stress for a part, nor about a critical crack length: only the combination of the two has any meaning. K_{Ic} is called the fracture toughness, and is a material property which depends (like conventional strength) on the polymer type, on test conditions, on composition, and on the method and conditions of processing. Not all of these factors are well documented as yet. Typical values of K_{Ic} for fast fracture at 20°C are given in Table 6.3. A polymer having a high value of K_{Ic} is more resistant to damage by crack growth (Fig. 6.30).

The following examples illustrate the influence of composition on fracture toughness. In unreinforced crosslinkable linear polymers, the toughness (in this case expressed as critical strain energy release rate G_{Ic}) depend on the

Table 6.3 Typical values of plane strain fracture toughness in air at 20°C

Material	K_{Ic} (MN/m$^{3/2}$)
Epoxy	0.6
General-purpose polystyrene	1.0
Cast acrylic sheet	1.6
Rubber-toughened PMMA	4.2
Polycarbonate	2.2
uPVC pipe compound	2.6
HDPE pipe compound	3.0

Table 6.4 Critical strain energy release rate of epoxy resins cured with different hardeners

Resin	*Hardener and amount*	G_{Ic} (J/m^2)
	Diethylene triamine (10 phr)	86
Epikote 828	*m*-Phenylene diamine (14.6 phr)	110
	Tris(dimethylaminomethyl)phenol (4 phr)	180
	Diphenyldiaminomethane (27 phr)	340

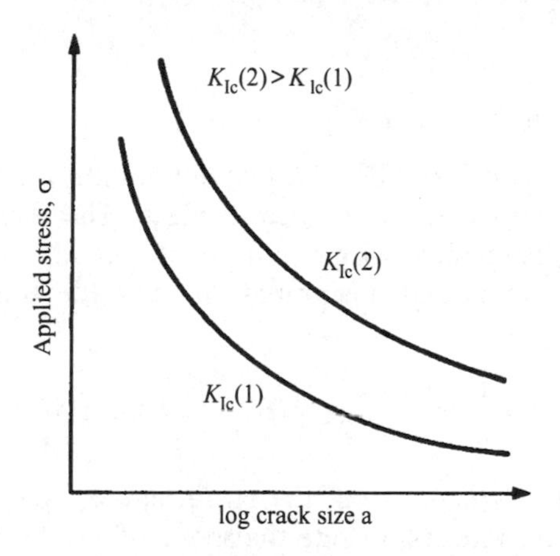

Fig. 6.30 Applied stress versus crack length

type of crosslinking agent used, and on the proportion of crosslinks present in the final product, as shown for an epoxy resin in Table 6.4.

The effect of change in environment on the toughness of several acrylic compositions is given in Table 6.5. Acrylic absorbs a small amount of moisture, which has a plasticizing action, thus increasing toughness. The effect of biaxially stretching a film or sheet of this material in the plane is to align polymer molecules so that they obstruct crack growth (rather like trying to chop wood across rather than along the grain).

Table 6.5 Fracture toughness values for some acrylic materials

Material	$K_{Ic}(MN/m^{3/2})$
Cast acrylic sheet in air, 20°C	1.6
Cast acrylic sheet in water, 20°C	1.7
Acrylic base cement in air, 20°C	1.85
Acrylic base cement in water, 20°C	2.1
Biaxially stretched acrylic sheet in air, 23°C, new	2.9
Biaxially stretched acrylic sheet in air, −18°C, new	1.3
Biaxially stretched acrylic sheet in air, 23°C, after 12 months' weathering in air	2.5

Where rigid particulate fillers are present, much depends on the properties of the filler and the efficiency of the bond at the interface between the filler and the polymer. In epoxy resin, adding rigid silica particles increases the fracture toughness considerably; in acrylic, the 10% barium sulphate particles are responsible for the slight increase in fracture toughness of the base cement compared with that of the unfilled cast sheet. Rubbery particles above their T_g can absorb a great deal of energy and can also allow the matrix to yield (even if it is below its T_g).

(*c*) *Relationship between G and K*

The stress in the loaded plate (Fig. 6.24) is not equally distributed and so the density of the energy varies from place to place. The total energy stored in the plate could (in agreement with equations (6.2) and (6.8)) be expressed in terms of local stress σ_i and a summation over the (differently stressed) elements as

$$U(a) = \frac{1}{2}Px = \sum_i \frac{1}{2}(\sigma_i^2/E) \times (\text{specimen volume})_i \tag{6.25}$$

On the basis of this expression, the strain energy release rate G_I can be shown (but using techniques outside the scope of this book) to be

$$G_I = Y^2\sigma^2\pi a/E \tag{6.26}$$

and hence, by comparison with equation (6.24) for the same situation, the energy and stress intensity approaches are related by

$$K_I^2 = EG_I \tag{6.27}$$

Both their critical values share the same relationship:

$$K_{Ic}^2 = EG_{Ic} \tag{6.28}$$

which makes it confusing when sometimes G_{Ic} is indicated as fracture toughness as well. Note that the value of E (and those of K_c and G_c– see below) are time dependent.

6.3.4 Time dependence of fracture

The analysis so far would seem to suggest that there is no risk of fracture occurring, provided that $K_I < K_{Ic}$. This is not actually so. Under load, a plastic creeps and the yield stress decreases. A cracked plastic still creeps, the yield zones surrounding the crack increase or the crazing zones widen and finally the crack grows, and the bigger the crack, the faster it grows, until catastrophic fracture occurs. What are needed are the understanding and the data relating **crack initiation** and **growth rate** to stress intensity factor. On this basis, it would then be possible to relate applied stress and initial crack size so that, after a given duration of loading, the final crack size would not cause fracture at the applied stress. It would also then be possible to plan inspection schedules so that parts containing cracks which are excessively large for a given load could be removed from service before a disaster occurred.

(a) Crack growth under constant load

Defects do not need to be dangerous; where they are small enough and where the stress is small enough, the formal theory says that they can be tolerated, even if they have the shape of cracks. In fact polymers are crowded with voids and fillers that act as such, but their submicroscopic dimensions make them harmless. However, the properties of polymers, such as modulus and yield stress or crazing stress decrease under load, which can change a harmless situation into one to be cautious about. In fact this statement implies that stress may lead to a situation where a harmless crack a will start to grow and then perhaps become harmful. The time elapsed since the stress σ was applied is called **initiation time**. It is supposed to be determined by the magnitude of $K_I = \sigma\sqrt{(\pi a)}$.

Theory, supported by experiment, shows that, for many polymers under constant load, the crack length a increases at rate $\dot{a} = da/dt$ which is related to the stress intensity factor at time t by

$$\dot{a} = C_I K_I^n \qquad (6.29)$$

where C_I and n are 'constants' which depend on the material and on test conditions. This power law shows that bigger cracks grow faster than smaller ones, because $K_I = Y\sigma\sqrt{(\pi a)}$.

In many polymers it is possible to observe three distinct regions of crack growth rate, as indicated in Fig. 6.31. Slow crack growth (which mostly

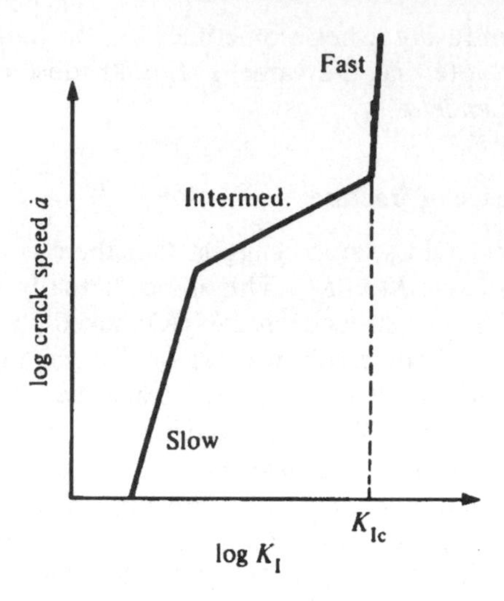

Fig. 6.31 Constant load power-law behaviour

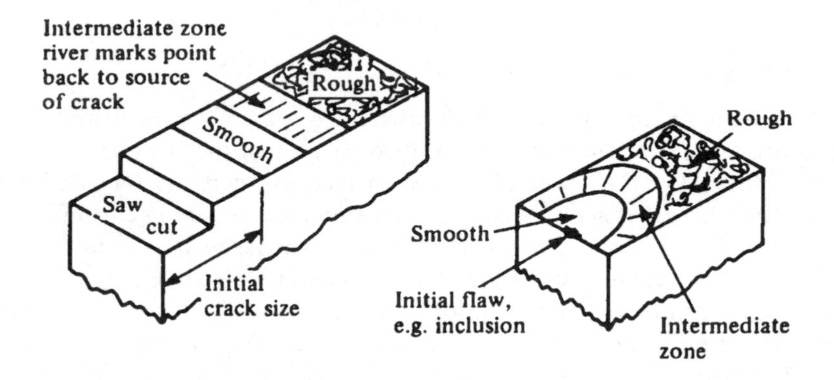

Fig. 6.32 Marks on fracture surface (schematic)

cannot be distinguished from initiation of crack growth) is often characterized by a mirror-smooth fracture surface; intermediate crack growth has a rather rougher surface, often with river lines pointing to a focus from which the crack started to grow (Fig. 6.32); and, finally, fast crack growth corresponds to a very rough fracture surface (but in some polymers the surface corresponding to fast fracture may be very smooth). Finding out where the crack started from is useful in blame analysis – was it the designer, the processor, the user, the environment or just an 'accident'? For most practical purposes, the third stage occurs practically instantaneously at $K_I = K_{Ic}$. In each of the three regions, different values of C_I and n apply. In cast acrylic at

20°C in air, slow growth is described by $\dot{a} = 7 \times 10^{-6} K_{\rm I}^{24}$ and intermediate crack growth by $1.3 \times 10^{-4} K_{\rm I}^{12.5}$.

The time taken to grow a crack from a_1 to a_2 under a constant applied stress a is found by integrating the power-law equation. Thus, for a finite width plate under uniform constant tensile stress σ, with a crack of depth a:

$$K_{\rm I} = Y\sqrt{\sigma}/(\pi a) \tag{6.30}$$

Therefore

$$\frac{{\rm d}a}{{\rm d}t} = C_{\rm I} K_{\rm I}^n = C_{\rm I}(Y\sigma)^n(\pi a)^{n/2} \tag{6.31}$$

If Y is assumed constant over the range a_1 to a_2, the time to grow the crack is given by:

$$t = 2(a_1^{1-n/2} - a_2^{1-n/2})/[(n-2)C_{\rm I}(Y\sigma\sqrt{\pi})^n] \tag{6.32}$$

Experiment shows that, for slow growth, the value of n is large, often in the range 7–25. It follows therefore that the time to grow a crack from a_1 to a_2 is dominated by the initial crack size, because a small crack grows more slowly than a large one.

A general conclusion is that $K_{\rm Ic}$ is a function that decreases with time under constant stress (Fig. 6.33). One has to ascertain that $K_{\rm I} < K_{\rm Ic}(t)$, just as one has to ensure that the design stress σ is smaller than the allowable long-term strength $\hat{\sigma}(t)$.

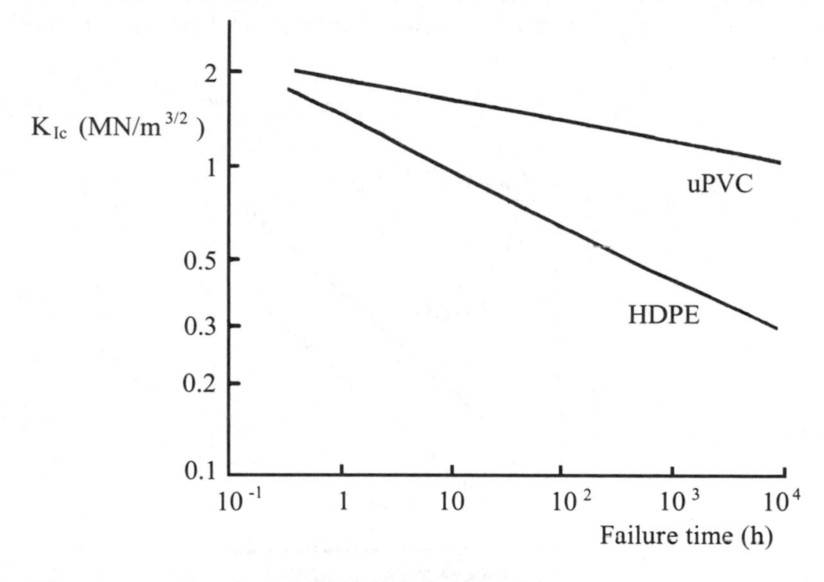

Fig. 6.33 $K_{\rm Ic}$ regression data on different precracked pipe materials as measured with C-ring tests as described in section 6.3.6. uPVC is an unmodified standard pipe grade PVC; HDPE is a 'conventional' high-density polyethylene (courtesy Pipeline Developments, Manchester)

(b) Crack growth under cyclic load

Fatigue damage is most often caused by crack growth under cyclic loading. It is to be expected that, for a given stress amplitude, crack growth under cyclic loading would be much faster than under constant load, because the rate of loading would be faster. Also, fatigue at a higher frequency would be more damaging than at a lower frequency.

Experiments show that the crack growth rate per cycle, $\mathrm{d}a/\mathrm{d}N$, is related to the stress range ΔK by the Paris power law

$$\mathrm{d}a/\mathrm{d}N = C_2(\Delta K)^m \tag{6.33}$$

where m is a 'constant' in the range 4–5 for polymers. The Paris law does not indicate that a higher mean stress is more severe than a lower one, although this is what is observed. Many fatigue data are more usefully correlated by a modified Paris law (Fig. 6.34) of the form

$$\mathrm{d}a/\mathrm{d}N = C_3\lambda^m \tag{6.34}$$

where

$$\lambda = K_{\mathrm{max}}^2 - K_{\mathrm{min}}^2 = 2K_{\mathrm{mean}}\Delta K$$

It is now possible to calculate the amount of crack growth from a_1 to a_2 over a number of cycles N by integrating the modified Paris law. For the easy case where $K_{\mathrm{min}} = 0$ (i.e. the crack is always stressed in tension, and not in compression):

$$N = (a_1^{1-m} - a_2^{1-m})/[(n-1)C_3(Y\sigma\sqrt{\pi})^{2m}] \tag{6.35}$$

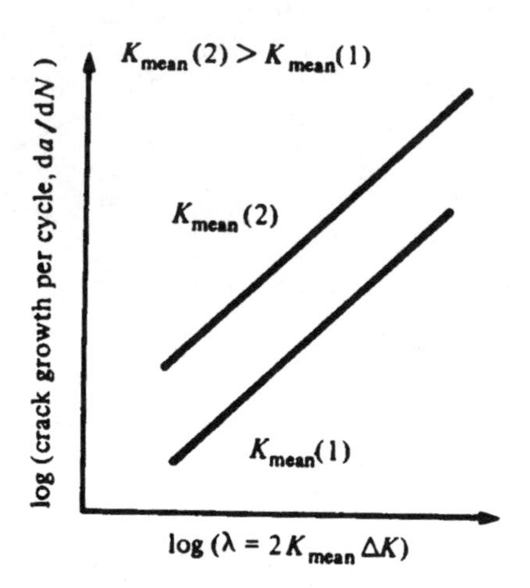

Fig. 6.34 Schematic modified Paris law (cycling loading)

Where the minimum value of applied stress is negative (compressive), the crack does not grow. This is sometimes observed on the crack surface as a set of striations parallel to the crack front. If the number of cycles to failure is small, these lines can be counted by using a low-power microscope.

6.3.5 Ductile–brittle transition revisited

We now further discuss ideas developed in sections 6.2.2(a) and 6.2.3.

(*a*) *Influence of crack depth* a

If a specimen such as that in Fig. 6.35 from a material which is able to yield is loaded, yield will occur at a load which is related to the net area under the notches:

$$P_Y = \sigma_Y \times H(D - 2a) \tag{6.36}$$

unless the crack a gives rise to earlier brittle fracture at a load

$$P_b = (K_{Ic}/Y\sqrt{(\pi a)}) \times HD \tag{6.37}$$

Apparently the material will yield at low values of a; at some higher value a_{trans} a transition to brittle failure will take place. a_{trans} apparently is small if σ_Y is high and K_{Ic} low and vice versa (Fig. 6.36). If σ_Y is very low, the specimen will always behave ductile; a low value of σ_Y is a good safeguard against brittle failure, but a bad one against high load!

The above suggests two strategies for making materials with good damage resistance. The obvious strategy is to increase K_{Ic}, the less obvious is to reduce the yield stress. Both strategies are followed in the plastics pipe industry: a modified mPVC has been developed to have a much higher K_{Ic} than uPVC (as it is able to promote local yielding near the crack tip). Medium-density polyethylene is tougher than high-density polyethylene because of its lower σ_Y.

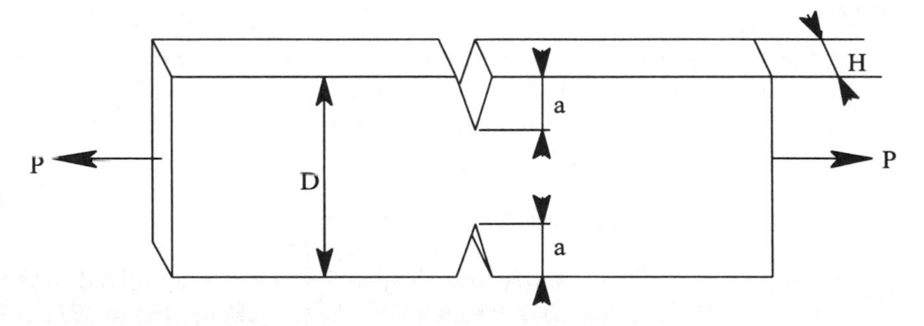

Fig. 6.35 Double-notched tensile specimen

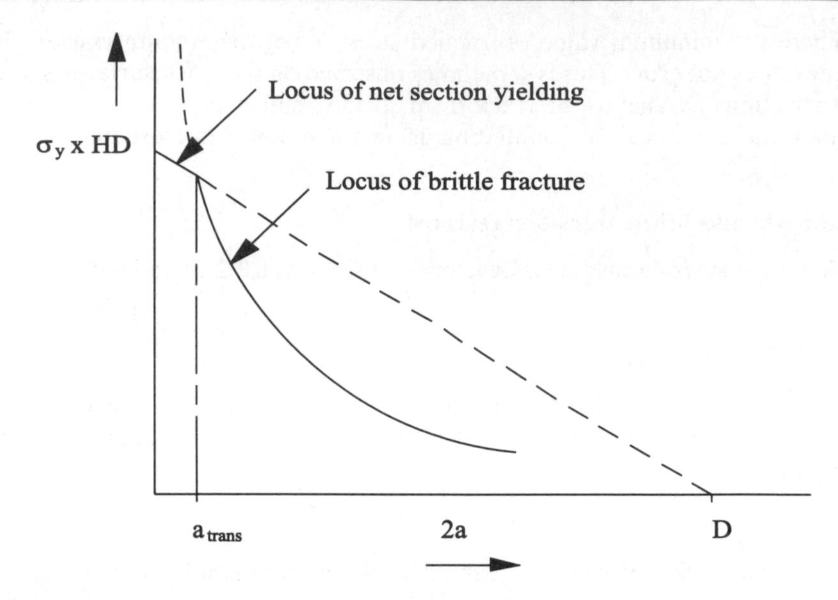

Fig. 6.36 Strength of a cracked body in relation to yield stress and fracture toughness

A small value of a_{trans} implies that small defects may render the material brittle. Note, however, that brittle behaviour only indicates the way the product fails, but that the load or energy needed for failure is determined by K_{Ic} or G_{Ic}.

The temperature or time dependence of σ_Y and of K_{Ic} are usually not the same. Thus at higher temperature or longer time under load, the magnitude of a_{trans} will not usually be the same as for standard conditions; if a_{trans} decreases, the material becomes more brittle, which is the case for standard high-density polyethylene pipe. The reader is invited to interpret Fig. 6.10 against this background, bearing in mind that small defects are always present in real materials.

(b) Local yielding near the crack tip

So far, the stress in a cracked body has only been described near the crack tip. For polymers which yield, the local stress σ_y cannot exceed the yield stress σ_Y. So putting $\sigma_y = \sigma_Y$ in equation (6.20) defining the local stress gives a rough estimate of the size of the yield zone $r = r_Y$:

$$r_Y = K_I^2/(2\pi\sigma_Y^2) = a\sigma^2/(2\sigma_Y^2) \tag{6.38}$$

Thus, the yield zone r_Y – often seen in plastics as a white voided area ahead of the crack tip in materials which are prone to yielding (Fig. 6.37) – is proportional to the crack size and to the square of the applied stress. A more

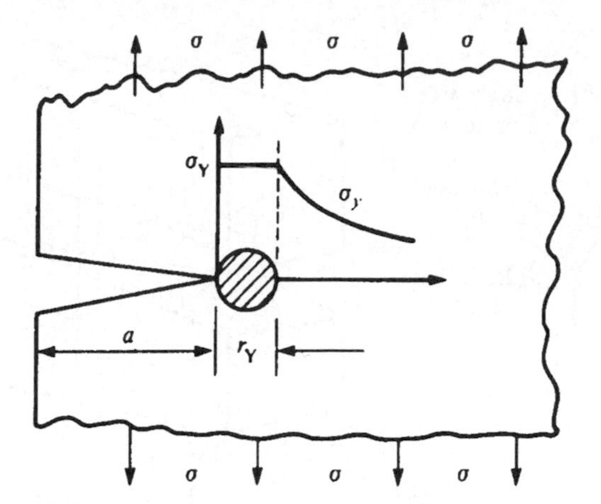

Fig. 6.37 Concept of plastic zone ahead of crack tip

precise estimate confirms that the plastic zone size is about twice the size indicated above.

Yielding normally occurs at constant volume, and the process can absorb a lot of energy. Obviously this yielding is a useful process in that it makes a polymer much more resistant to crack growth, because the energy available to grow a crack is used in causing yield. On the other hand, the yielding results in a non-linear stress–strain curve and renders the analysis of linear elastic fracture mechanics (LEFM) invalid. It is usually reckoned that the LEFM approach can be used provided that the plastic zone size is less than half the crack size, $r_Y < 0.5a$, and calculations are then based on an effective crack length $a_{eff} = a + r_Y$. If $r_Y < 0.5a$, then some form of yield fracture mechanics would be needed, perhaps the J-integral approach, as indicated in the introduction.

Once the concept of yielding is admitted, even on a small scale, so that LEFM can be used, another problem has to be examined. Why is it that, in a polymer which is able to yield, a thick section containing a crack is more likely to be brittle than a similar thin section under the same stress? Yielding is a process which is assumed to take place at constant volume; this is why a tensile specimen necks down locally under load (Fig. 6.38a). In a thin specimen containing a crack, the surfaces of the specimen are load free and hence free to be sucked inwards as the yield zone develops ahead of the crack tip. Material away from the yield zone is at a lower stress, deforms elastically and has a smaller Poisson's ratio, so it is not sucked in so much and acts as a restraint. In a thick specimen, the polymer at the load-free surface (B) can still be sucked in to form dimples on the surface at the crack tip; however, underneath the dimples (A), an element of material no longer has a stress-free surface – all the material is restrained from yielding,

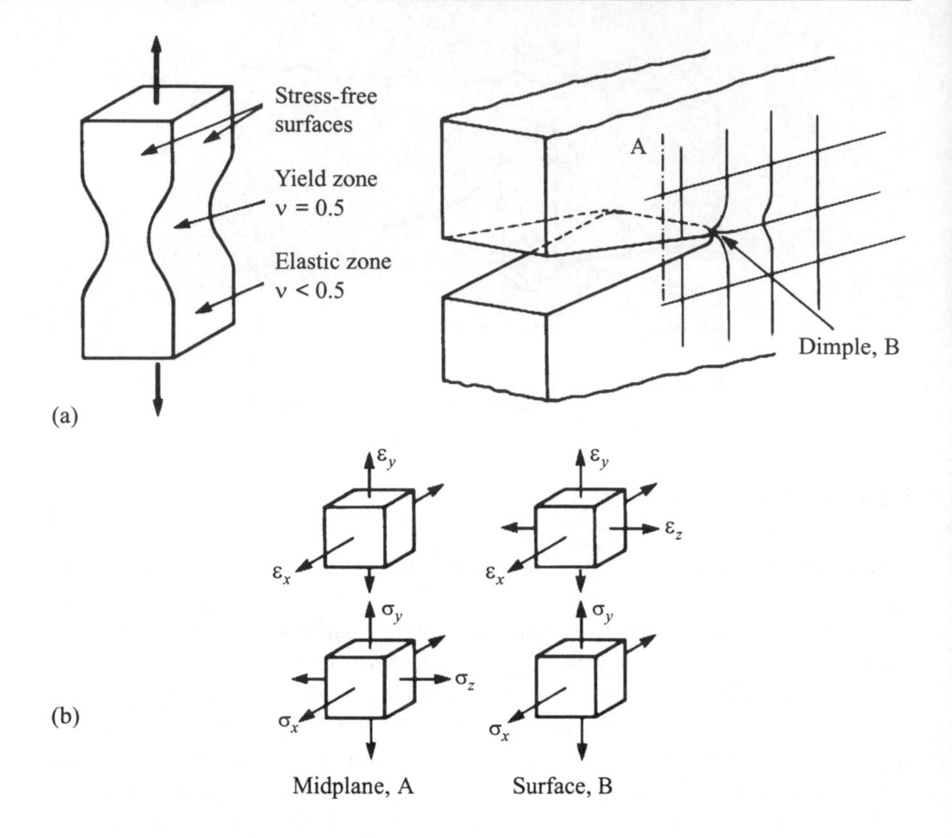

Fig. 6.38 (a) Yield in uniaxial tension. (b) State of stress and strain at crack tip

and hence becomes subject to triaxial tension which suppresses yield (Fig. 6.38b).

Thus, in a thin specimen containing a crack, the free surface (B) is important and permits yield to occur, hence absorbing a lot of energy and giving a high value of fracture toughness. In a thick specimen containing a crack, the behaviour of the inside of the specimen (A) swamps any surface effect, and fracture occurs at much lower levels of absorbed energy, i.e. a low fracture toughness. In analytical terms, the thin specimen is in a state of plane stress, and the thick one in plane strain: plane strain is the more dangerous because failure occurs at the lower value of fracture toughness. A clear consequence is that the idea of increasing the thickness in order to improve the fracture toughness could have exactly the opposite effect, if a transition from the higher plane stress G_{Ic} to the lower plane strain G_{Ic} value occurs.

How thick does a specimen have to be before a test can be regarded as plane strain? Test protocols usually suggest that, to ensure plane strain, the

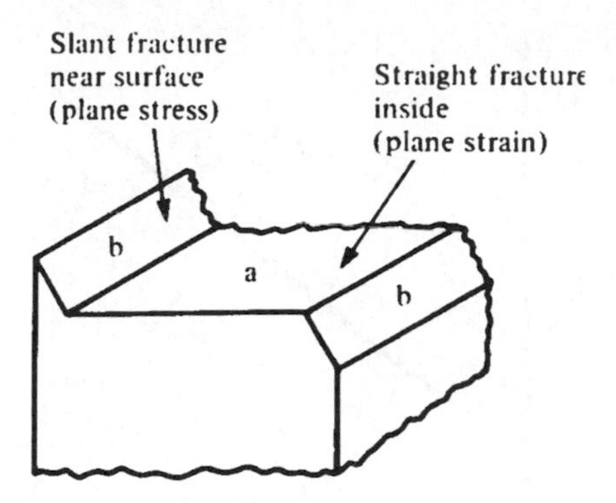

Fig. 6.39 Shape of fracture surface

specimen should have a thickness greater than four times the size of the plastic zone. In broken parts, the plane strain region is seen as a straight fracture (Fig. 6.39, a), the plane stress region at the surface as a slanting region at 45° to both the line of advance of the crack and the direction of the applied load (Fig. 6.39, b).

6.3.6 Measurement of fracture parameters

Fracture data can be obtained from many kinds of specimens, including single-edge cracked plates in tension, single and double cantilever beams, and three-point bending. (A variation on the latter is C-rings taken as rings from pipes, which are cut open at one side and notched internally at the other side in order to have a well-defined axial crack. When such a ring is loaded with a dead weight (a constant force) to open the ring, the bending moment at the crack introduces there a well-defined K_I, see results in Fig. 6.33.) The relevant stress intensity factors (for linear elastic brittle fracture) are given in the handbooks. Fully ductile fracture is still not properly understood and research on this lies outside the scope of this book. However, the linear elastic analysis can be used to give some estimate of K_{Ic} or G_{Ic} where small yield zones form ahead of the crack tip. This approach seems not too unreasonable if $r_Y < 0.5a$. To suppress yield by ensuring triaxiality, it is recommended that the width of a specimen should be at least $4r_Y$, in which case most of the fracture is then under plane strain conditions.

Power-law relationships (e.g. equations (6.29) and (6.34) are obtained by direct observation of crack length for the applied load or load range, using Y for the observed crack length at each moment of measurement.

Fracture toughness can be obtained using equations (6.24) and (6.26). Measuring the loads required to break specimens with different final crack lengths (i.e. taking any slow growth into account), a graph of $Y^2\sigma^2\pi$

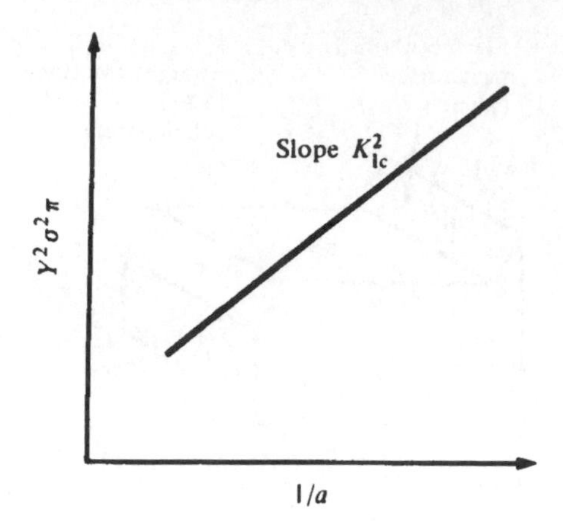

Fig. 6.40 Estimation of fracture toughness

versus $1/a$ has a slope K_{Ic}^2 (Fig. 6.40). A plot of $(Y^2\sigma^2\pi)/E$ versus $1/a$ has a slope G_{Ic}^2, where E is the modulus relevant to the loading time of the experiment.

6.3.7 Impact strength

The traditional method of assessing whether a plastic is brittle or tough is to carry out an impact test, with a fast rate of loading (often developing triaxial stress locally in notched specimens) to promote brittle failure; the test is quick and easy to carry out and can provide data for comparing different materials under the conditions of test. The main test factors are the amount of energy available for breaking the specimen, the test temperature, stress concentrations and molecular orientation. There are many additional factors; details may be found in Brown (1981). There are two kinds of traditional test: the drop test and the pendulum test.

(*a*) *Drop testing*

There are two basic procedures: either drop the product (e.g. a barrel full of liquid or a crate full of full bottles) onto a rigid surface, or drop a mass onto the plastics product (e.g. a pipe). The energy or test speed (or both) can be varied between individual drops by adjusting the mass or the drop height. Results are usually expressed as the impact energy needed to give 50% failures at a given test temperature (BS 2782). Drop testing examines the behaviour of a plastics product, including design and processing: results are therefore specific, and can involve the testing of many samples.

A further step to improve the assessment for performance of a product is to instrument the drop tests. The dart (the mass to be dropped) is then

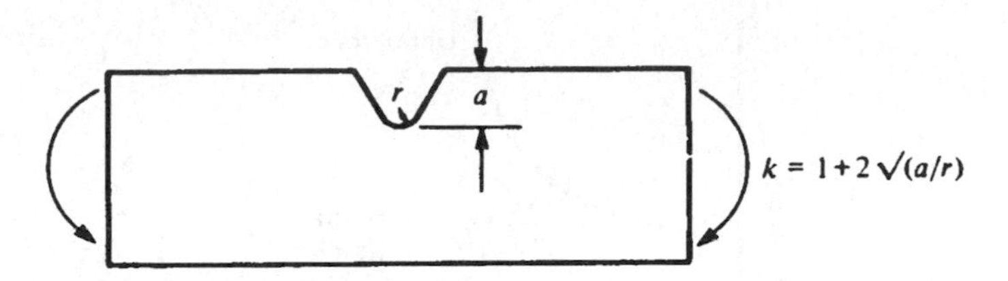

Fig. 6.41 Edge-notched bar

equipped with a high-frequency force transmitter and its signal is recorded as a function of time.

(b) Pendulum impact tests

Pendulum impact test machines are smaller versions of those used for metals, and supply a fixed amount of energy (fixed mass and length of pendulum) at a fixed speed. Charpy-type specimens are tested in three-point bending; Izod-type specimens are tested as cantilever beams. Tests are made over the service temperature range on unnotched and notched bars. Notched bars have a 45° notch with various notch-tip radii r (Fig. 6.41), and the linear elastic stress concentration factor k is given by

$$k = 1 + 2\sqrt{(a/r)} \tag{6.39}$$

where a is the notch depth. In BS 4618, tests are made in the weak direction, if the specimen is anisotropic. Results are expressed as energy to break per unit fracture surface area (Fig. 6.42). This test has been widely used to examine other variables such as the effects of changes in processing conditions, composition and environment, as reviewed in detail by Vincent (1971).

The instrumenting of impact tests has revealed that the actual impact is not a smooth, short transmission of energy from the hammer to the specimen, but consists of a series of transmissions of momentum, each of which speeds up the specimen, which is pushed by the hammer soon thereafter to provide a further momentum. Each of these steps imparts energy to the specimen, which may or may not be enough to initiate or continue crack growth. So the last step of this series may be just enough for full breakage or to transmit a surplus, which is one of the causes of spread in the impact results.

(c) Pendulum impact tests on cracked bars

Neither the pendulum test on notched bars nor the drop test give fundamental data for quantitative design. However, it is possible to obtain fracture toughness data on plastics by carrying out pendulum impact tests on

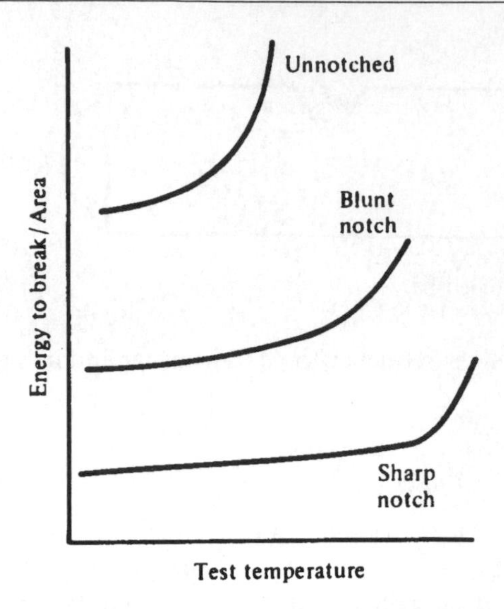

Fig. 6.42 Effect of notching on pendulum impact test results

bars with cracks in them, and this forms the basis of the latest BS 4618 on impact strength.

Consider a specimen of length L and span S, simply supported and centrally loaded (Fig. 6.43), and made from a linearly elastic material.

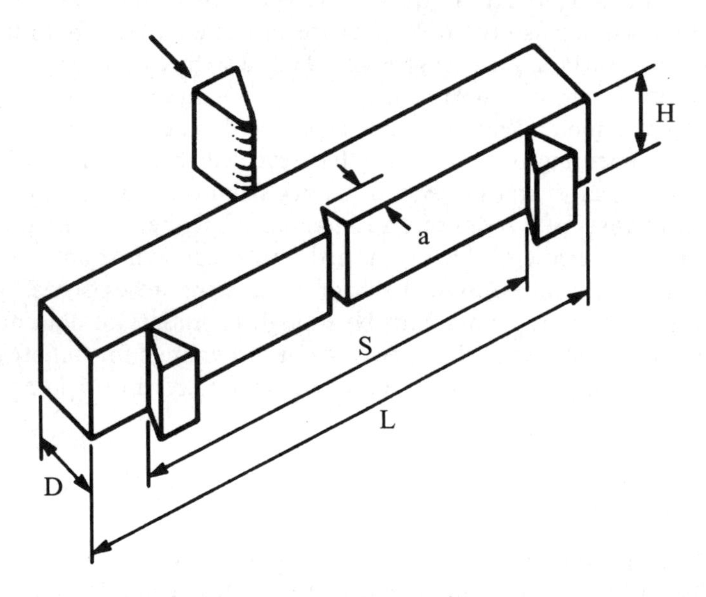

Fig. 6.43 The leading dimensions for a cracked bar for a pendulum impact test

For an uncracked bar, the maximum bending stress a and the compliance C_0 are

$$\sigma = 3PS/2HD^2 \quad \text{and} \quad C_0 = S^3/4EHD^3 \tag{6.40}$$

For a cracked bar containing a crack of depth a, the same reasoning as for the cracked plate in section 6.3.2 can be followed. The measured failure energy w can be equated to the elastic strain energy as expressed in equation (6.14), but now for a fracture load P_f at midspan displacement x_f:

$$w = \frac{1}{2}P_f x_f = \frac{1}{2}P_f^2 C(a) \tag{6.41}$$

The calibration factor $\phi(a)$ can be evaluated using equations (6.15)–(6.17):

$$D\frac{dC}{da} = \frac{dC}{d(a/D)} = \frac{C}{\phi} = \frac{G_{Ic}HDC}{w} = \frac{2G_{Ic}HD}{P_f^2} \tag{6.42}$$

The fracture load P_f is related to maximum bending stress by equation (6.40), which is related to fracture toughness by equation (6.26), hence

$$\frac{dC}{d(a/D)} = \frac{9Y^2S^2\pi(a/D)}{2EHD^2} \tag{6.43}$$

This formula describes the increase in C as a consequence of the increase in a crack a in a bar with initial compliance C_0 according to equation (6.40):

$$\int_0^{a/D} \frac{9Y^2S^2\pi(a/D)}{2EHD^2} d(a/D) = C - \frac{S^3}{4EHD^3}$$

This expression can be rewritten to give an expression for C:

$$C = \frac{9\pi S^2}{2EHD^2}\int Y^2(a/D)d(a/D) + \frac{S^3}{4EHD^3} \tag{6.44}$$

Substituting expressions for C from equation (6.44) and $dC/d(a/D)$ from equation (6.43) into equation (6.17) gives the required result:

$$\phi = \frac{\int Y^2(a/D)d(a/D)}{Y^2(a/D)} + \frac{1}{18\pi}\frac{S}{D}\frac{1}{Y^2(a/D)} \tag{6.45}$$

For $a/D < 0.2$, $Y \simeq 1$ and hence the calibration factor is

$$\phi \simeq \frac{a}{2D} + \frac{1}{18\pi}\frac{S}{D}\frac{1}{a/D} \tag{6.46}$$

If $a/D > 0.2$ the integral in equation (6.45) will have to be evaluated graphically because Y is a function of a/D.

This procedure enables the determination of G_{Ic} using equation (6.18) from the measured values of w of specimens with different cracks a.

PROBLEMS

1. (Answer each part of this question in two lines or less.)
   (a) Specimens for an impact test are 3 mm thick and 6.5 mm wide. The notch depth is 2.75 mm and the notch tip radius is either 1 mm or 0.25 mm. What are the values of the linear elastic stress concentration factors?
   (b) State the difference between 'compliance' used in fracture mechanics and 'compliance' used in deformation studies.
   (c) Write down an expression for the compliance of (i) a bar of uniform cross-section under uniaxial tensile load, (ii) a cantilever beam carrying a point load at the free end.
   (d) When a body with a small crack in it carries a constant load, the crack grows until catastrophic failure occurs. Does the failure time depend most on the initial crack or the final crack size? State why.
   (e) What are the principal stresses and maximum shear ahead of the crack tip in a linear elastic material loaded in the opening mode?
   (f) At what crack speed and stress intensity factor does the transition from 'slow' to 'intermediate' crack growth rate in cast acrylic occur at 20°C?
   (g) If in a ('quick') tensile test at 20°C in air an uncracked acrylic plate breaks at a stress of 67 MN/m^2, what is the approximate inherent flaw size?
   (h) In a plastics pipe containing internal pressure, which is more severe: a semi-elliptical crack running along the pipe or around the circumference? Which is more severe, a flaw on the outside or the inside surface of the pipe? If an extruded pipe is cooled only from the outside (by contact with a vacuum sizing die), would the cold pipe contain a residual stress distribution and, if so, would this be beneficial or detrimental to the performance of the pipe in service?
2. At what proportion of the yield pressure would you expect short-term failure in a thin-walled plastic pipe if the depth of an external surface flaw is 0.5 mm? For the plastic used, the yield stress is 60 MN/m^2 and the fracture toughness is 2 $MN/m^{3/2}$.
3. Precracked bars of a high-density polyethylene at −100°C have been broken in three-point bending in a pendulum impact test. The following data record the energy absorbed in breaking the specimens, the width, depth, crack length measured after fracture, and the appropriate calibration factor. The specimens are 90 mm long and the distance between supports is 72 mm. Estimate the fracture toughness of this polymer.

Specimen number	*Energy absorbed (mJ)*	*Specimen width, D (mm)*	*Specimen depth, H (mm)*	*Crack length, a (mm)*	ϕ
1	39.7	6.20	6.03	2.90	0.380
2	74.3	6.22	6.03	2.37	0.505
3	77.9	6.01	6.03	1.85	0.678
4	92.5	6.01	5.95	1.59	0.785
5	216.6	6.05	6.05	0.73	1.675
6	326.2	6.13	6.10	0.52	2.305
7	375.2	5.86	6.04	0.35	3.236

4. Derive an analytical expression for the functions $C(a)$ and $\phi(a)$ for a test piece as given in Fig. 6.44. It may be regarded as consisting of two cantilever beams of length a (further dimensions to be taken from the figure). A load P will give a displacement x caused only by the bending of the beams.
5. A test plate made from short-glass-fibre-reinforced polystyrene is 400 mm square and 2 mm thick. It contains an initial edge crack 3.6 mm deep. After a test run at 12 000 N uniaxial distributed tensile force at 20°C, the crack was found to have grown to 6.8 mm deep. Estimate the duration of the test. What would be the worst error in your estimate if the power-law index were in error by 5% and also the constant in the power law were in error by 10%? $K_I = 2\sigma\sqrt{a}$, where σ is the stress remote from the flaw and a is the depth of the defect. The relationship between crack growth rate $\dot{a}$ and K_I at 20°C for this material is given by:

$\dot{a}$(m/s)	7.08×10^{-7}	5.00×10^{-6}	2.24×10^{-5}	2.4×10^{-4}
K_I(MN/m$^{3/2}$)	2.0	2.5	3.0	4.0

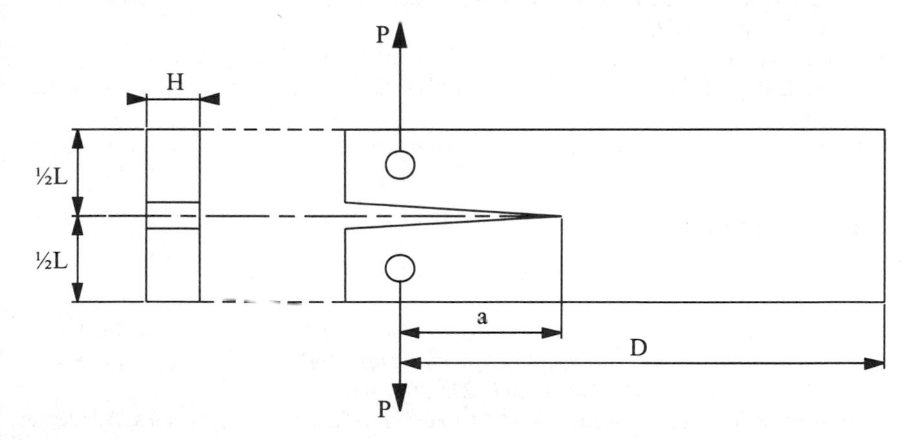

Fig. 6.44 Double cantilever beam

6. In a display mechanism operating at 20°C, a horizontal cantilever beam made from cast acrylic is 10 mm wide, 40 mm deep and 400 mm long. It carries a constant bending moment of 27.5 N m acting in the vertical plane at the free end. During your first inspection, you discover that a uniform edge crack 1 mm deep has developed across the full width of one face of the beam. The complete mechanism is required to operate safely for a further 6 months. How often would you wish to inspect the mechanism and when would you recommend that the mechanism should be switched off and the defective beam replaced? The stress intensity factor K_I is related to the crack depth a and the maximum bending stress σ by $K_I = 1.04\sigma\sqrt{(\pi a)}$.
7. A gas pipe is extruded from a medium-density polyethylene, and has a short-term fracture toughness at 20°C of 3 $MN/m^{3/2}$. To avoid creep rupture, the ratio of diameter to wall thickness is 20:1. If the installed pipe has to pass a short-term pressure test at 0.6 MN/m^2, what is the maximum allowable pipe size?

FURTHER READING

Traditional approach

Birley, A.W., Haworth, B. and Batchelor, J. (1992) *Physics of Plastics; Processing, Properties and Materials Engineering*, Hanser, Munich.
Brostow, W. and Cornelissen, R.D. (eds) (1986) *Failure of Plastics*, Hanser, Munich.
Kinloch, A.J. and Young, R.J. (1983) *Fracture Behaviour of Polymers*, Elsevier Applied Science, London.
(1995) *Fatigue and Tribological Properties of Plastics and Elastomers*, Plastics Design Library, Morris, New York.
Plastics and Rubber Institute (1976) *Design Factors for Unreinforced Thermoplastics Products with Specific Reference to Pipe.*
Sächtling, H. (1977) *Kunststoff-Taschenbuch*, Carl Hanser Verlag, Munich.
Treloar, L.R.G. (1975) *The Physics of Rubber Elasticity*, Oxford University Press, Oxford.
Williams, J.G. (1980) *Stress Analysis of Polymers*, 2nd edn, Ellis Horwood, Chichester.

Fracture mechanics approach

Anderson, G.P., Bennett, S.J. and DeVries, K.L. (1977) *Analysis of Adhesive Bonds*, Academic Press, New York.
Broek, D. (1986) *Elementary Engineering Fracture Mechanics*, Noordhof, Leiden.*
Brown, R.P. (ed.) (1981) *Handbook of Plastics Test Methods*, Godwin, Harlow.
Bucknall, C.B. (1978) *Adv. Polym. Sci.*, **27**, 121–148.
Hertzberg, R.W. and Manson, J.A. (1980) *Fatigue of Engineering Plastics*, Academic Press, New York.
Kausch, H.H. (1978) *Polymer Fracture*, Springer-Verlag, Berlin.

Rooke, D.P. and Cartwright, D.J. (1976) *Compendium of Stress Intensity Factors*, HMSO, London.
Vincent, P.I. (1971) *Impact Tests and Service Performance of Thermoplastics*, Plastics Institute, London.
Williams, J.G. (1978) *Adv. Polym. Sci.*, **27**, 67–120.
Williams, J.G. (1984) *Fracture Mechanics of Polymers*, Ellis Horword, Chichester.

* This book is not specifically about polymer fracture but provides analytical background on linear elastic fracture mechanics.

7 Fibre–Polymer Composites

This chapter introduces the properties and design methods for fibre–polymer composite structures. The discussion concentrates on fibre–plastics composites and is in five sections. The first provides an overview of the attractive qualities of composites in relation to the qualities of the fibres and the plastics, indicates two approaches to design for stiffness and strength, and indicates some 'unusual' features of behaviour of composites. The second revises the relationships between the properties of composite sheets (loaded in-plane in the principal directions) and the properties of the fibres and plastics, and introduces some new concepts. The third discusses the behaviour of a sheet or lamina containing aligned fibres, and describes how such an anisotropic lamina behaves when loaded in-the-plane at an angle to the fibre direction (the off-axis loading problem). The fourth and fifth describe the stiffness and strength behaviour of laminates under in-plane loading and include reference to the design of pipes and pressure vessels. The emphasis in this chapter is on plastics composites based on continuous fibres; the principles can be extended to indicate the likely behaviour of fibre–rubber composites and can give a semi-quantitative insight into the behaviour of short-fibre–plastics composites and also the behaviour of oriented polymers containing no fibres at all.

7.1 INTRODUCTION

What are composites? Composites consist of two or more physically distinct different materials which are combined in a controlled way to achieve a mixture having more useful properties (to defined criteria) than any of the constituents on their own.

In this book, one component will be a polymer-based compound; the other main component could be a gas, a particle or a fibre. Within the given description, a foamed plastic or rubber could be a composite, so could a rubber compound reinforced with carbon black, or polystyrene blended with rubber particles, or plasticized vinyl chloride copolymers containing inert fillers such as chalk. These are important 'composites' and some remarks were made about them in Chapter 2. The emphasis in this chapter is on fibre–polymer composites. Examples of fibre–plastics composites include the housing for the home handyman's electric drill, braided hose, frames of tennis rackets, pipes, tanks and chemical process equipment, canoes, yacht and ship hulls, sail-plane fuselage and wings, helicopter blades, and rocket motor casings. Examples of fibre–rubber composites include car and lorry tyres, pressure hose and belts for conveyors and power transmission.

In fibre–plastics composites, the most widely used fibre in load-bearing applications is glass; for special high-performance duties, carbon fibres and aromatic polyamide fibres are also used. Natural fibres such as cellulose are also employed. In fibre–rubber composites the main fibres used are textile (rayon and nylon) fibres and steel (wire). Almost all polymers can be reinforced with fibres.

Comparing the properties of the fibres with those of the polymer, the fibres are much stiffer and stronger and usually are more stable dimensionally, while the polymer has good chemical resistance (and other special qualities, such as abrasion resistance and toughness in rubber).

The fibres may well be strong and stiff along their axes (and prone to damage unless protected), but are only useful if loads can be transferred to them; in particular, fibres are at their best when carrying axial loads. The matrix has but little stiffness and strength, but has the job of holding the fibres together in the required configuration, transferring external loads to the fibres and transferring loads between fibres. The matrix also protects the fibres from damage, e.g. from development of flaws, or deterioration caused by environment.

The subject of composites is very broad. To contain the discussion within a reasonable length, the treatment in this book is limited to fibre–plastics composites, with the main emphasis on long (continuous) fibres and cross-linkable plastics. The properties of the main constituents are given in Table 7.1. In general, glass fibres are much cheaper than carbon fibres, and unsaturated polyesters are cheaper than epoxies.

7.1.1 Approaches to design

Designers may well seek to make the best possible use of the fibres: this requires them to design the material as well as the structure to carry the desired loads. The designer usually has to answer the following types of question.

- Where and how might the fibres be placed to make the best use of them? How much fibre is needed?
- How should the fibres be held in place? In this text, the plastic holds the fibre in place – how much plastic?

Table 7.1 Typical properties of fibres and crosslinkable plastics used to make load-bearing resin-bonded fibre composites

				Crosslinked	
Property	*High-modulus carbon fibres*	*High-strength carbon fibres*	*E glass*	*Epoxy*	*Unsaturated polyester*
Density (kg/m^3)	1950	1750	2560	1100–1400	1200–1500
Diameter (μm)	8	8	10	–	–
Tensile modulus (GN/m^2)	390[a]	250[a]	75	3–6	2–4.5
Poisson's ratio			0.27	0.39	0.38
Tensile strength (MN/m^2)	2200	2700	1750	35–100	40–90
Strain at break (in tension) (%)	0.5	1	2–3	1–6	2
Linear expansion coefficient (/K)	-0.5 to -1.2×10^{-6}[a]	-0.1 to 0.5×10^{-6}[a]	5×10^{-6}	6×10^{-5}	$1–2 \times 10^{-4}$
Shrinkage on curing (%)	–	–	–	1–2	4–6

[a] Carbon fibres are anisotropic; tensile moduli in the radial direction are only 10 GN/m^2 or so, and expansion coefficients in the radial direction are up to 10^{-5}/K.

- How can the fibres be put in the required places? This is a question of manufacturing technology, and strongly interacts with (i) and (ii). Also, how can manufacturing defects be allowed for in design?

These questions have led to the development of two approaches to design with fibre–plastics composites.

The first approach involves deliberately placing the fibres in those directions which result in the most efficient use of the composite material as a whole to make 'high-performance' composites. Such a composite is anisotropic, that is, the properties vary with direction. The behaviour of such a composite can be rather unusual (when judged by conventional engineering experience and the irrelevance of stiffness and strength theories for isotropic solids, which have the same properties in all directions) and can be taken into account by suitable analytical techniques. Some of these unusual features are described below, and the analysis forms the core of this chapter. This approach works best for simple structural elements such as flat plates, tubes and cylindrical or spherical shells which can be readily and precisely made using methods which confer a high volume fraction of fibres. The analytical theories indicate the very best that can be achieved in composites; in practice, some allowance has to be made for defects and difficulties incurred during manufacture – the final part may not be perfect, and it may prove impracticable to put fibres exactly where theory suggests they would be most effective. In addition, manufacturing has to be possible and transport to the location of use may impose other loads than the (use of) the application itself.

The second approach seeks to minimize the directional effects by trying to make the composite material as isotropic as possible. It is quite easy to manufacture by contact or compression moulding processes structural shells and panels which have fibres oriented randomly in the plane. This suggests that the volume fraction and effectiveness of fibres will be low – but such structures are never called 'low-performance' composites. They account for much of the usage of fibre-reinforced plastics. The advantage of such a structure is that the calculation of stiffness and strength under in-plane loads can be made using the familiar elasticity and strength theories for isotropic materials.

In both approaches to design, there is seldom any reinforcement in the through-thickness direction, and hence in laminates the strengths between individual layers depends on, but is not as high as that of the polymer matrix: this interlaminar strength must be checked in design by either method if shear or bending are likely modes of deformation.

In the older literature there is reference to another simplified design approach called 'netting analysis'. This has been used for in-plane loading of thin symmetric laminates with particular application to thin-walled pipes. The essence of this approach is that the applied loads are only taken by the fibres along their length, and that stresses transverse to the fibres or in shear are taken as zero. For pipes having a balanced helical construction, i.e. fibres wound at $\pm\theta$ to the axis of the pipe, netting analysis confirms that the strongest pipe per unit mass has $\theta = 54.75^\circ$ for internal pressure loading and 45° for applied axial torsion moments. However, the netting analysis gives no useful guidance on such important factors as in-plane stiffness and bending behaviour. Moreover the actual strength is not only determined by the strength of the fibres but also by the cohesive strength of the resin (section 7.2.3). Netting analysis will not therefore be further discussed.

7.1.2 Some interesting features of fibre–plastics composites

This section introduces features of the behaviour of flat plates made from composites which are not encountered in plates made from an isotropic material. The purpose here is to summarize qualitatively how composites can be expected to behave, based on more or less common sense reasoning. Later in the chapter we develop quantitative analysis necessary for design with composites.

(a) Unidirectional lamina

All the long fibres are parallel, regularly spaced and perfectly bonded to a plastics matrix (Fig. 7.1). The principal directions of this flat panel are along the fibres (1), across the fibres (2) and through the thickness (3). The properties of the fibres are denoted by suffix 'f' and of the plastic by suffix 'p'. The properties of the composite in the directions 1, 2 and 3 can be related to those of the constituents by micromechanics, as will be discussed in section 7.2. For fibre volume fractions V_f in the range of technical interest:

$$E_1 \simeq V_f E_f = \text{longitudinal modulus}$$
$$E_2 \simeq E_p = \text{transverse modulus}$$
$$G_{12} \simeq G_p = \text{in-plane shear modulus}$$
$$\nu_{12} = \nu_f V_f + \nu_p(1 - V_f) = \text{lateral contraction ratio}$$

Because $E_f \gg E_p$, it follows therefore that $E_1 \gg E_2$, i.e. the panel is stiff when stressed along the fibres but flexible when stressed in the transverse

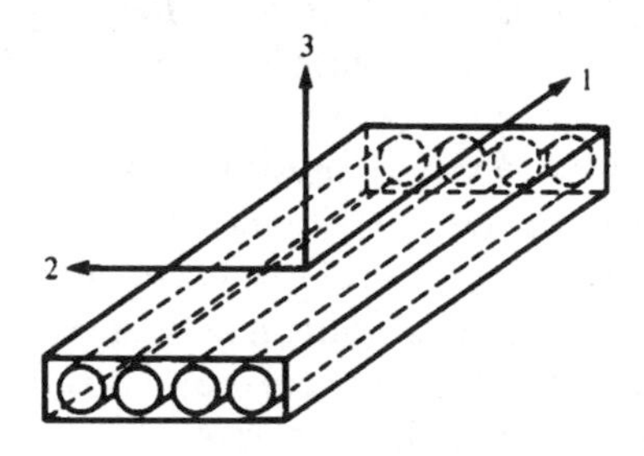

Fig. 7.1 Coordinates for an orthotropic lamina with aligned fibres

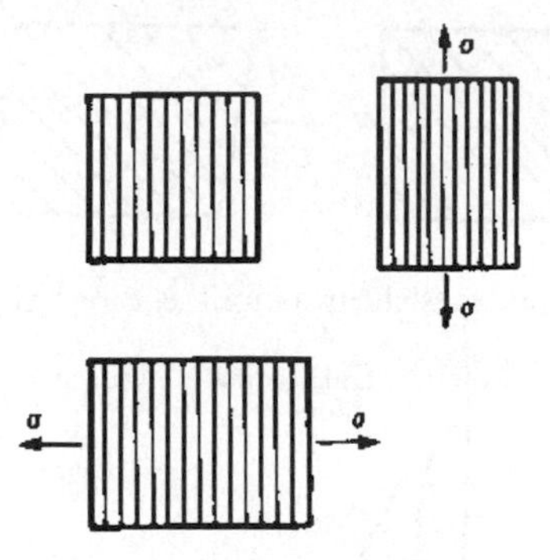

Fig. 7.2 The lamina is stiffer longitudinally than transversely

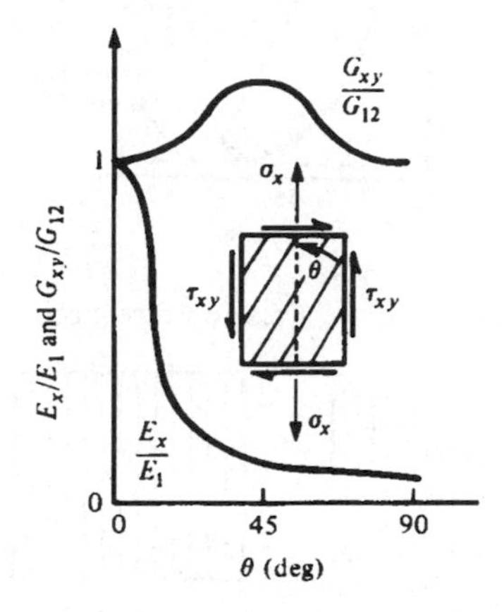

Fig. 7.3 Dependence of normalized stiffnesses on direction

direction. It also follows that $\nu_{21} < \nu_{12}$: lateral contraction under transverse stress is very small, but much larger under longitudinal stress (Fig. 7.2). Typically, ν_{12} is about 0.3, whereas ν_{21} is usually less than 0.02. The expansion coefficient of the fibres is much less than that of the plastic, so the result of a temperature change is a small value of α_1 but a large $\alpha_2 \simeq \alpha_p$.

When such a lamina is stressed in the plane in the reference x-direction at an angle θ to the fibre direction, the moduli depend markedly on θ, as shown for a given volume fraction in Fig. 7.3. Whereas E_x has a maximum value at $\theta = 0$, the in-plane shear modulus G_{xy} is biggest at 45°. Moreover, on applying a direct stress σ_x at θ to the fibre direction, not only does the sheet extend by ε_x and contract by ε_y, but the sheet distorts to a strain γ_{xy} (Fig. 7.4). This coupling between direct stress and shear strain does not occur in an isotropic sheet.

The strength of a lamina also depends on the directions in which stresses are applied. The longitudinal tensile strength $\hat{\sigma}_{1T}$ is usually high, being

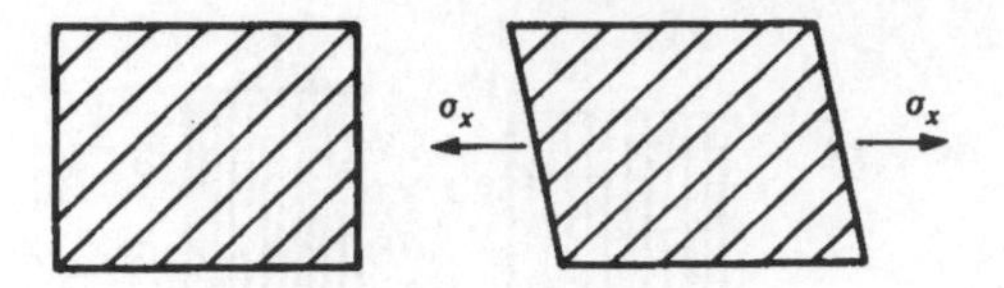

Fig. 7.4 Off-axis loading induces shear as well as direct strains

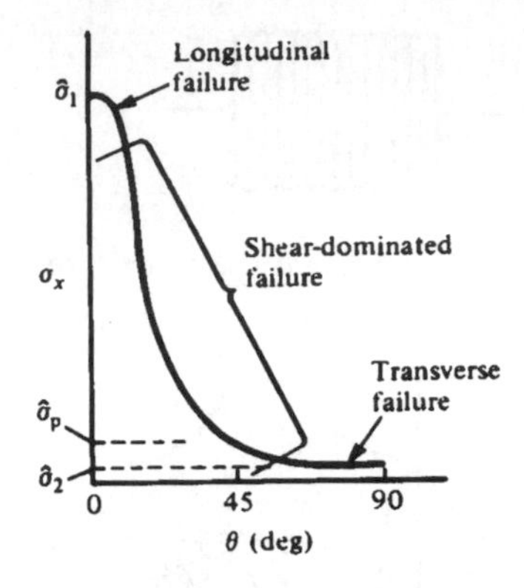

Fig. 7.5 Dependence of tensile failure stress on direction

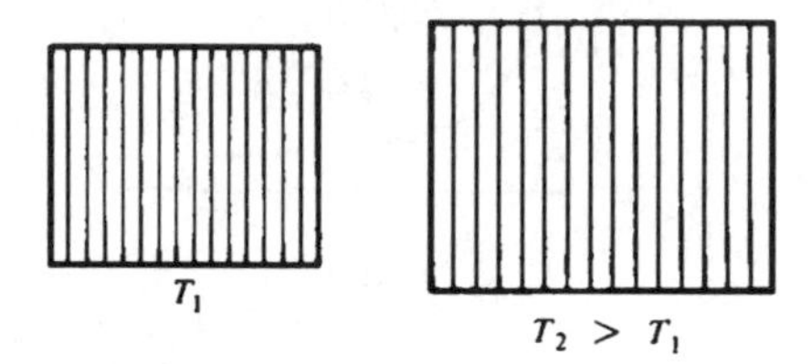

Fig. 7.6 Expansion of lamina caused by increase in temperature

dominated by the strength of the fibres, $\hat{\sigma}_f$. The transverse strength $\hat{\sigma}_2$ is lower than that of the polymer alone, $\hat{\sigma}_p$. The in-plane shear strength $\hat{\tau}_{12}$ is comparable with that of the polymer.

When the lamina is stressed in-plane by σ_x at θ to the 1-direction, the mode of fracture depends markedly on θ. If $\theta \simeq 0$, $\hat{\sigma}_x \simeq \hat{\sigma}_1$. If $\theta > 60°$, a transverse failure is likely ($\hat{\sigma}_x \simeq \hat{\sigma}_2$), and at intermediate angles failure is by shear (Fig. 7.5). Failure under multiaxial in-plane stresses can be estimated using a modified form of the von Mises yield criterion (even though the failure is seldom ductile). A rectangular lamina with two edges parallel to the fibres cured hot and cooled down will shrink more in the transverse direction than in the longitudinal direction because $\alpha_1 < \alpha_2$, but the sheet will stay flat and rectangular (Fig. 7.6). A rectangular lamina with two edges at an angle θ to the fibres will stay flat but will distort to a parallelogram (Fig. 7.7).

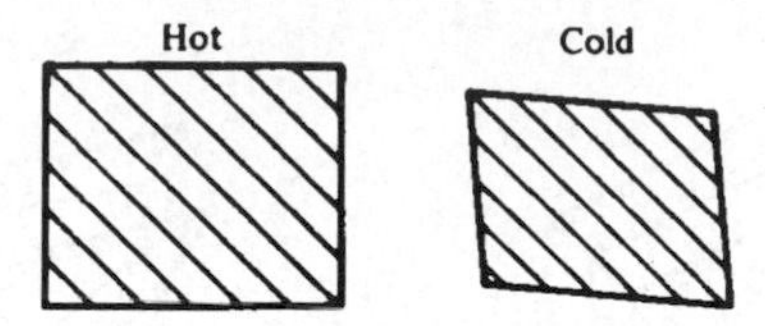

Fig. 7.7 Contraction and distortion when fibres are not parallel to sides of lamina

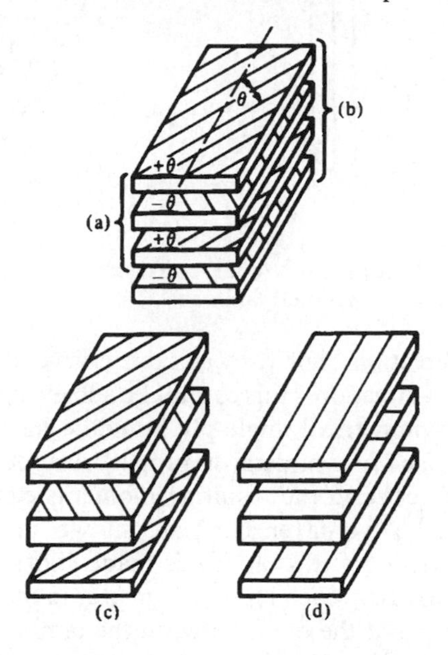

Fig. 7.8 Exploded views of different kinds of laminate construction: (a) symmetrical angle-ply, (b) non-symmetrical angle-ply, (c) symmetrical angle-ply and (d) symmetrical cross-ply

(b) Angle-ply laminate

The properties of a single unidirectional lamina are too sensitive to direction θ to be usable in design. In laminates the dependence on direction is controlled to make the best use of the fibres.

A laminate is a bonded stack of laminae, here taken for simplicity as identical laminae with unidirectional fibres. The directions of fibres in successive laminae can be different, and exploded views of a variety of laminates are shown in Fig. 7.8.

A symmetrical angle-ply laminate is one where the laminae and properties are symmetrical about the midplane. Such a laminate is well behaved under in-plane loading, and shows properties not very different in general character from those indicated in Fig. 7.3, although volume fraction will be lower. Under biaxial stress, σ_x and σ_y, it is even possible to realize no strain

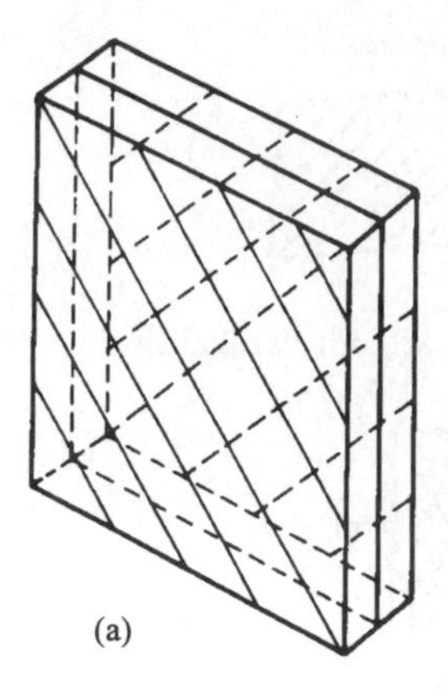

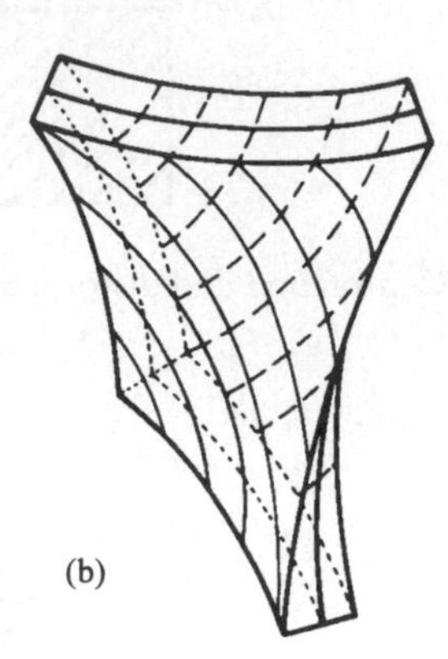

Fig. 7.9 Distortion on heating a flat 'cold'-cured unsymmetrical angle-ply laminate based on two identical unidirectional laminae

in one of the load directions x or y, by suitable choice of angle of orientation $\pm\theta$ to the reference x-direction – this cannot be achieved in an isotropic sheet.

However, an unsymmetrical angle-ply laminate has some even stranger characteristics. The lack of symmetry in properties about the midplane leads to out-of-balance forces and the resulting moments can cause unusual distortions. Consider the two-layer laminate shown in Fig. 7.9a, which is evidently not symmetric. If this lamina is made (flat) and cured hot, then on cooling, the shrinkage in each lamina is greater in the transverse direction than along the fibres; but the bond between the laminae acts as a restraint, leading to warping. On the other hand, if the two-layer flat laminate of Fig. 7.9a is made 'cold' and then heated until it reaches a uniform higher temperature, it will warp as indicated in Fig. 7.9b. The intuitive way of flattening such a warped panel is to apply a twist to correct the curvature. Another way, not so obvious at first glance, is to apply an axial load in the x-direction. This result stems from the off-axis direct loading of each lamina, affording not only in-plane shear but also twisting curvature: again, this is not encountered in two identical isotropic panels stuck together. (On the other hand, if two sheets of different isotropic metals having different expansion coefficients are bonded together, then on heating this bimetallic plate, a domed curvature (not twisted) will result, with the metal with the lower expansion having a concave surface (Fig. 7.10).) Note that, if the bimetallic plate is axially loaded, it will also bend because of the lack of symmetry about the midplane (Fig. 7.11).

It is also possible for an unsymmetric laminate in plane uniaxial stress to bend as well as twist (and even shear!), as shown in Fig. 7.12.

Usually, these out-of-plane responses to in-plane loading of unsymmetrical laminates are unwanted and can be avoided by specifying symmetrical

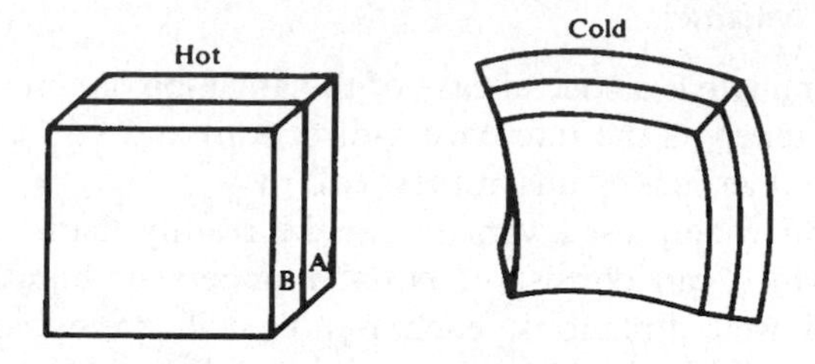

Fig. 7.10 Distortion of bimetallic plate, $\alpha_A < \alpha_B$, where A and B are isotropic

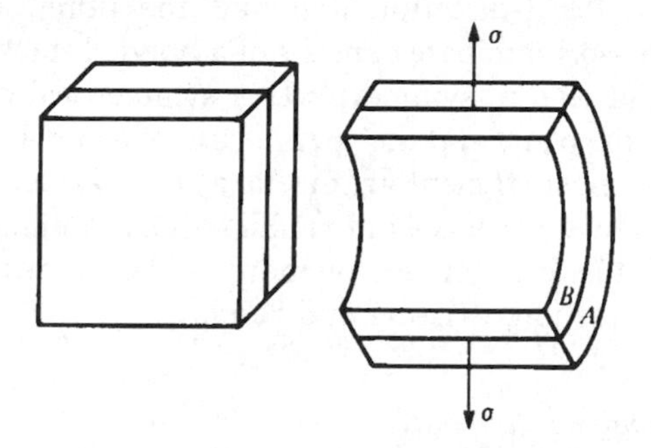

Fig. 7.11 Bending of bimetallic plate under direct stress, $E_B > E_A$

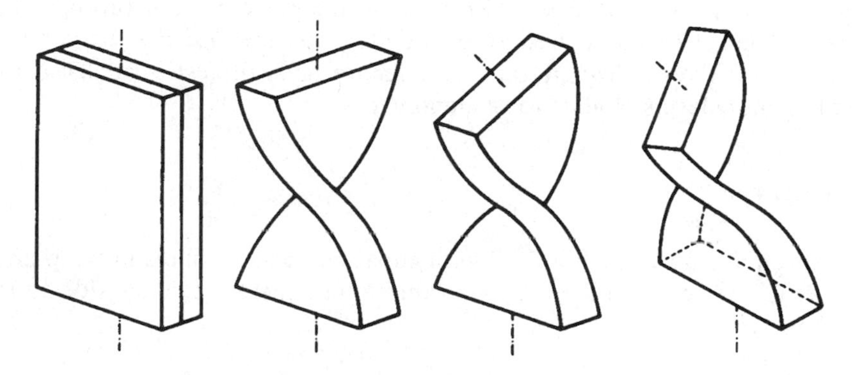

Fig. 7.12 Twisting, bending and shear of an unsymmetrical laminate based on orthotropic laminae under direct stress

laminates – this also greatly simplifies the analyses and calculations. Where specific out-of-plane responses have to be exploited, then more involved design procedures will be necessary; a fan blade is an example where the twist in the blades can be made to stay constant or even to increase as the speed of rotation increases, resulting in constant or increasing efficiency over the speed range used in service.

(*c*) *Cross-ply laminate*

A cross-ply laminate is a special case of the angle-ply laminate with laminae at 0° (usually taken as the reference x-direction) and 90°. Cross-ply laminates can be symmetrical or unsymmetrical.

A useful and widely used version can be readily made up from woven glass cloth, which can consist of equal numbers of bundles of fibres in the warp and weft directions; each bundle will consist of hundreds or thousands of individual fibres. One piece of flat cloth fully wetted out with resin and then cured will make a single lamina which has half the fibres running in the 1-direction and half the fibres in the 2-direction (Fig. 7.13). A cross-ply laminate consists of a bonded stack of these bidirectional laminae, and will be symmetrical. A symmetrical cross-ply laminate can be rotated in the plane and the special case of $\theta = \pm 45°$ (to the reference x-direction) is the most efficient arrangement for transferring pure shear.

A cross-ply laminate made from unidirectional laminae can be unsymmetrical and will then exhibit one or more of the following features under in-plane loads: distortion, twisting and bending.

(*d*) *Chopped strand mat laminate*

Bundles of glass fibres some 25–50 mm long are randomly oriented in the plane and held in place by a binder. This mat is wetted out with crosslinkable plastic and cured. This sheet is isotropic in the plane but not through the plane. The stiffness and strength in the plane are low because the volume fraction of fibres is low. Additional strength and stiffness is achieved by adding more layers of glass to the laminate.

PROBLEM

1. Given a lamina like Fig. 7.13 with equal numbers of fibres in warp and weft, it is clear that $E_1 = E_2$. Can this lamina therefore be considered to be isotropic?

Fig. 7.13 A lamina made from woven glass cloth

7.2 MICROMECHANICS

Micromechanics is concerned with ascertaining how effective the fibres are in the composite. The simplest level of discussion is based on a unidirectional lamina, with continuous or discontinuous parallel fibres, loaded in the principal directions. This depends on three factors.

First, it is necessary to have a good bond between the polymer and the fibre. This is usually achieved by coating the fibre with a material which bonds well to both fibre and polymer. (For example, in the glass fibre–polyester resin system, the coating is often a vinylsilane; in simple terms, the 'vinyl' ensures a bond to the resin and the 'silane' to the glass. This will not be discussed further here – seek advice from experts if you need it; we shall assume a good bond.)

Second, what effect does the concentration ('volume fraction') of fibres have on the properties of the composite?

Third, what effect does fibre length have on the properties of a composite? The concept of a critical fibre length will be discussed.

7.2.1 Vocabulary and assumptions of simple micromechanics

Fibre–polymer composites are inhomogeneous and anisotropic. Most of this part of the book centres on the behaviour of an orthotropic lamina. 'Orthotropic' means that properties are different in three mutually perpendicular directions at a point, with three mutually perpendicular planes of material symmetry. A lamina is a thin flat plate, as shown in Fig. 7.1.

The model orthotropic lamina discussed below consists of unidirectionally aligned, straight, equispaced fibres in a polymer matrix. It is assumed to be macroscopically homogeneous, linear elastic and initially stress-free, with no voids in the fibre or polymer or between them, with perfect bonding.

The fibres are assumed to be homogeneous and isotropic, linear elastic, much stiffer (by a factor of, say, 20 or more) than the polymer, regularly spaced and perfectly aligned, and have the same length L, diameter D and surface strain, and all break at the same strength $\hat{\sigma}_f$. The polymer matrix is assumed to be homogeneous and isotropic, and linear elastic. The interface between the fibre and the polymer is an anisotropic transition region which must provide a stable bond between the two.

7.2.2 Stiffness

The elastic constants for a lamina stressed in one of the principal directions alone are only stated here because derivations are assumed to have been described in earlier courses. The simple model used here lumps the fibres together into a ribbon the full thickness of the lamina and having a rectangular cross-section (Fig. 7.14). The full derivation from this and from more

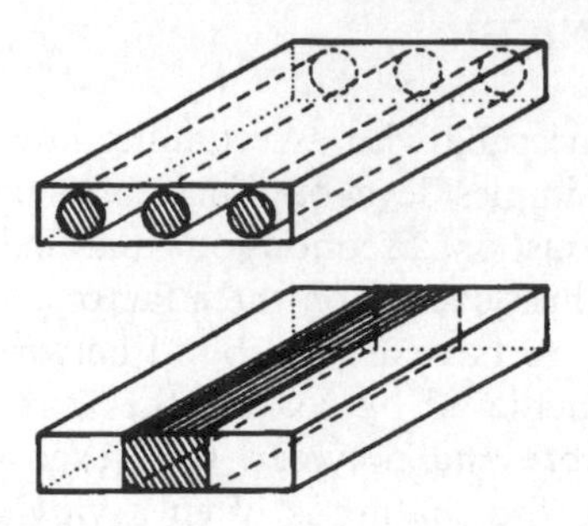

Fig. 7.14 Representation of fibres by strip of same area

realistic and complicated models will be found in any of the books cited in the further reading list.

If the composite lamina is stressed uniformly in the fibre direction, perfect bonding ensures that the strains in the lamina, the polymer and the fibres are the same:

$$\varepsilon_l = \varepsilon_p = \varepsilon_f \tag{7.1}$$

A force balance leads to the rule of mixtures for the apparent longitudinal tensile modulus, E_l:

$$E_l = E_p V_p + E_f V_f \tag{7.2}$$

This rule of mixtures works quite well in practice. If $E_p \ll E_f$, which it usually is, then E_l is dominated by the fibres unless the volume fraction is very low.

If the lamina is uniformly stressed in the transverse direction, then notionally the stress in the polymer and fibres is the same; in the simplest model

$$\sigma_2 = E_2\varepsilon_2 = E_p\varepsilon_p = E_f\varepsilon_f \tag{7.3}$$

Compatibility then leads to

$$\frac{1}{E_2} = \frac{V_f}{E_f} + \frac{V_p}{E_p} \tag{7.4}$$

In practice, this expression for the apparent transverse elastic modulus E_2 is only its lower limit. The Poisson's ratios of the fibre and the polymer are unlikely to be the same; there is a mismatch of transverse strain within individual fibres and the surrounding resin, and these, together with the stress-concentrating effect of the fibres, all mean that the stresses in the fibres and the polymer are not the same. Moreover, local differences of fibre concentration are of influence. Nonetheless, it is reasonable to infer from equation (7.4) that the presence of fibres has little effect on the transverse modulus unless the volume fraction is high.

When the lamina carries only a longitudinal stress σ_1 the major Poisson's ratio ν_{12} is defined as

$$\nu_{12} = -\varepsilon_2/\varepsilon_1 \tag{7.5}$$

The longitudinal strains in the fibre and polymer are the same, and a simple analysis shows that the major Poisson's ratio follows the rule of mixtures:

$$\nu_{12} = \nu_f V_f + \nu_p V_p \tag{7.6}$$

Values of ν_p and ν_f do not differ widely so there is no question of one of them dominating ν_{12}.

The minor Poisson's ratio ν_{21} is relevant when the unidirectional lamina is stressed in the transverse or 'minor' direction:

$$\nu_{21} = -\varepsilon_1/\varepsilon_2 \tag{7.7}$$

It is important to note that ν_{12}/ν_{21} is not defined through a combination of equations (7.5) and (7.7) because each of these equations derives from a different applied stress system. However, it can be shown that

$$\nu_{12}/E_1 = \nu_{21}/E_2 \tag{7.8}$$

Under in-plane shear, a rectangular orthotropic lamina distorts to a parallelogram. The shear stress along the fibre direction, τ_{12}, is matched by its complementary shear stress τ_{21}. In the simplest model it is assumed that the polymer and the fibres carry the same stress:

$$\tau_{12} = G_{12}\gamma_{12} = G_f\gamma_f = G_p\gamma_p \tag{7.9}$$

and the in-plane apparent shear modulus G_{12} is then

$$\frac{1}{G_{12}} = \frac{V_p}{G_p} + \frac{V_f}{G_f} \tag{7.10}$$

Thus the modulus of the polymer dominates G_{12} unless the volume fraction of fibres is very large.

7.2.3 Strength

A lamina loaded in tension in one of the principal directions can fail after the resin fails, the fibres fail or the interface fails. The main emphasis in this section is on factors dominating the longitudinal tensile strength, $\hat{\sigma}_{1T}$, the transverse tensile strength, $\hat{\sigma}_{2T}$, and the in-plane ('intralaminar') shear strength, $\hat{\tau}_{12}$. Comments are also made on failure under compressive stress, and under combined stress. It is assumed here that the fibres all have the same length and strength – more complicated models are detailed in standard texts.

(*a*) *Longitudinal tensile strength,* $\hat{\sigma}_{1T}$

From a simple force balance, the stress in the composite under a longitudinal load is

$$\sigma_1 = \sigma_f V_f + \sigma_p(1 - V_f) \tag{7.11}$$

What happens depends on the volume fraction of fibres and on whether the breaking strain of the fibres, $\hat{\varepsilon}_f$, is greater or less than that of the plastic, $\hat{\varepsilon}_p$. The data in Table 7.1 indicate which situation is likely to apply in the most common systems.

If $\hat{\varepsilon}_f > \hat{\varepsilon}_p$ (Fig. 7.15), then at large volume fractions the fibres take most of the load until

$$\hat{\sigma}_1 = \hat{\sigma}_f V_f \tag{7.12}$$

At very small volume fractions, when the matrix breaks, the load is thrown onto the fibres: complete failure of the laminate follows if there are insufficient fibres to take the load:

$$\hat{\sigma}_1 = \sigma_f^* V_f + \hat{\sigma}_p(1 - V_f) \tag{7.13}$$

The crossover point between resin-dominated failure and fibre-dominated failure occurs at V_f by equating (7.12) and (7.13):

$$V_f' = \hat{\sigma}_p/(\hat{\sigma}_f - \sigma_f^* + \hat{\sigma}_p) \tag{7.14}$$

If $\hat{\varepsilon}_p > \hat{\varepsilon}_f$ (Fig. 7.16), which is the usual case except for steel-reinforced concrete, then at large volume fractions, once the fibres break, the polymer cannot take the extra load and breaks:

$$\hat{\sigma}_1 = \hat{\sigma}_f V_f + \sigma_p^*(1 - V_f) \tag{7.15}$$

At very small fractions of fibres, when the fibres break, the polymer can still carry the extra load until

$$\hat{\sigma}_1 = \hat{\sigma}_p(1 - V_f) \tag{7.16}$$

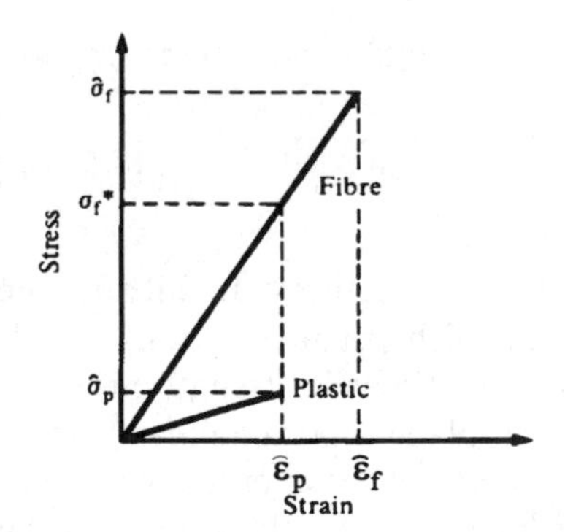

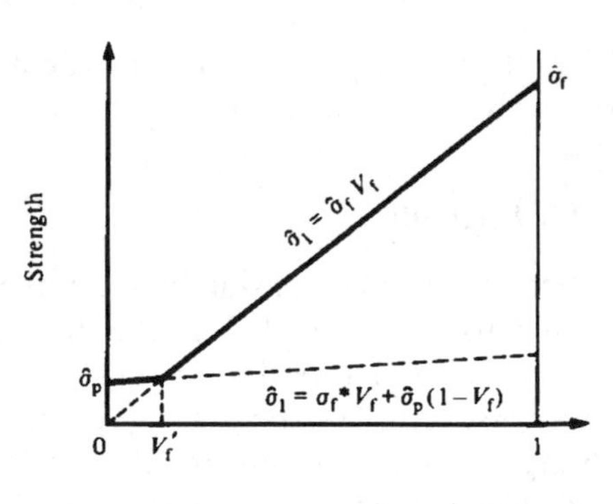

Fig. 7.15 Longitudinal tensile strength of lamina when fibre fracture strain is greater than that of the matrix

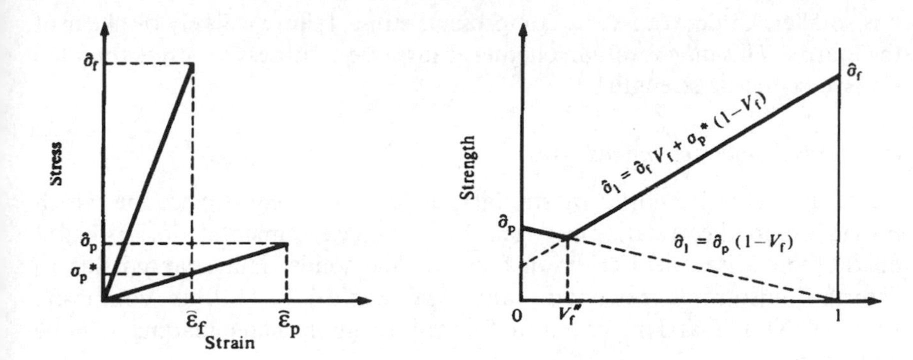

Fig. 7.16 Longitudinal tensile strength of lamina when fibre fracture strain is less than that of the matrix

The crossover point occurs at V_f'' given by:

$$V_f'' = (\hat{\sigma}_p - \sigma_p^*)/(\hat{\sigma}_f + \hat{\sigma}_p - \sigma_p^*) \tag{7.17}$$

(b) Transverse tensile strength, $\hat{\sigma}_{2T}$

Transverse strength is the major problem in lamina design and is often the cause of failure. Its usual appearance is in the form of cracks parallel to the fibres along the interface of fibres and resin or in the resin. It is evident that $\hat{\sigma}_{2T}$ will be less than the tensile strength of the matrix $\hat{\sigma}_p$, because the load is taken essentially by the matrix alone (with a cross-sectional area reduced by the presence of the fibres) and because the fibres introduce stress concentrations. The matter is further complicated by the presence of internal stress and strain within the fibres and the matrix. Much also depends on how well the polymer bonds to the fibres.

(c) In-plane shear strength, $\hat{\tau}_{12}$

The in-plane shear strength in an orthotropic lamina is dominated by the strength of the matrix $\hat{\tau}_p$ because failure of the plastic can occur by shear of the matrix without disturbing the fibres.

(d) Compressive strength, $\hat{\sigma}_{1c}$ *and* $\hat{\sigma}_{2c}$

There are several modes of failure, depending on the system used. Under longitudinal compressive stress, models have considered the fibres as struts supported along their length by a flexible matrix, but the agreement between theory and experiment is not good, perhaps because of the non-ideal arrangement of fibres in a real lamina. It is unlikely that longitudinal compressive strength is the same as the longitudinal tensile strength; usually it is smaller. Under transverse compressive stress, failure is likely by shear of the matrix. This may result in a higher transverse compressive stress than the transverse tensile strength.

(e) Strength under combined stress

There are several theories of strength under combined stress. One which provides a good correlation between theory and experimental work is based on the von Mises yield criterion for isotropic solids. Hill adapted this to describe anisotropic behaviour, and Tsai applied it to fibre composite laminae. This Tsai–Hill criterion for failure in in-plane loading can be written as

$$\left(\frac{\sigma_1}{\hat{\sigma}_1}\right)^2 - \frac{\sigma_1\sigma_2}{\hat{\sigma}_1^2} + \left(\frac{\sigma_2}{\hat{\sigma}_2}\right)^2 + \left(\frac{\tau_{12}}{\hat{\tau}_{12}}\right)^2 = 1 \tag{7.18}$$

where $\hat{\sigma}_1$ and $\hat{\sigma}_2$ are here taken as tensile failure stresses.

Equation (7.18) indicates the first failure, which often appears in the form of cracks in the resin. This does not necessarily imply total failure of the product, as different layers may each be able to sustain their relevant σ_1 according to netting theory. However, as the integrity of the layers is violated, the product may start to leak, or to deform or otherwise dysfunction. We shall say more about this in section 7.5.1.

It should be noted that in equation (7.18), σ_1 means the stress in the longitudinal direction along the fibres in the lamina. σ_1 does not have the connotation of a principal stress in any equations in this chapter and the presence of the shear stress term in equation (7.18) emphasizes this.

7.2.4 Coefficients of thermal expansion

As the coefficients of linear thermal expansion of the plastic and the fibre are markedly different, it comes as no surprise that the expansion coefficients of an orthotropic lamina depend strongly on direction. The fibres have a high modulus and a small expansion coefficient, and restrain the matrix, so it is not surprising that α_1 is small; the plastic has a low modulus and a high expansion coefficient, and the fibres offer virtually no transverse restraint, so α_2 is usually fairly large. The coefficients can be related to those of the constituents using an analysis by Schapery based on energy considerations:

$$\begin{aligned} \alpha_1 &= (\alpha_f E_f V_f + \alpha_p E_p V_p)/E_1 \\ \alpha_2 &= (1+\nu_f)\alpha_f V_f + (1+\nu_p)\alpha_p V_p - \alpha_1 \nu_{12} \end{aligned} \tag{7.19}$$

where ν_{12} and E_1 relate to the same volume fraction V_f.

If a lamina is cured hot and then cooled down, it is evident that if the stiffnesses of both fibres and matrix are assumed independent of temperature (which is not quite true for the polymer), then in the absence of externally applied forces the internal forces in the fibres must balance the internal forces in the matrix. In glass fibre–polymer laminae cured hot and then cooled, the fibres have an internal compressive stress, and the matrix is in tension, often carrying a longitudinal stress which is a significant part of the tensile strength of the polymer. In viscoelastic polymers, some of this stress can relax away, but most crosslinked resins are glassy so this relaxation is not very substantial.

7.2.5 Laminae based on other arrangements of fibres

Reference has already been made (in sections 7.1.2(c) and 7.1.2(d)) to the common practice of using glass fibres in the form of woven cloth or chopped strand mat: these are much easier to handle than parallel sets of unconnected fibres.

An approximate estimate of some of the behaviour of a single layer of balanced woven glass cloth (equal warp and weft), impregnated with crosslinkable plastic and then cured, can be made using the procedures of micromechanics. The principal directions along each axis of the weave are evidently axes of symmetry. So $E_1 = E_2$ and as half the glass is in each direction, equation (7.2) can be adapted to give

$$E_1 \simeq \tfrac{1}{2} E_f V_f + E_p V_p \qquad (7.20)$$

Transverse strength $\hat{\sigma}_{2T}$ will be the same as longitudinal strength $\hat{\sigma}_{1T}$ at roughly half the value for a unidirectional lamina of the same fibre volume fraction; and shear strength will be much better than that of a unidirectional lamina because any in-plane shear must disturb fibres. The weaving of bundles of fibres does reduce the volume fraction of fibres, and there will be localized pockets rich in resin in the interstices of the cloth. Typical volume fractions of fibres in this system range from 25 to 45%.

A single layer of chopped strand mat impregnated with plastic has fibres oriented randomly in the plane and is therefore isotropic in the plane. The volume fraction will be low and the tensile modulus in the plane is given approximately by

$$E \simeq \tfrac{3}{8} E_f V_f + \tfrac{5}{8} E_p V_p \qquad (7.21)$$

Typical volume fractions for chopped strand mat range from 15 to 25% fibres.

Other structures can be made from short-fibre–polymer composites. One example is the glass and filler systems based on unsaturated polyester used as dough or sheet moulding compounds and having fibres from 3 to 12 mm long. This system is usually compression moulded and cured hot. Another example is the family of glass-fibre-reinforced thermoplastics, and most of these compounds are injection moulded. In either example, the fibres tend to align themselves in the directions of flow, which is not easy to predict in advance: this is a real problem and still the subject of much current research.

7.2.6 Prediction of performance

It is apparent from the previous sections 7.2.1–7.2.4 that it is not always easy to predict quantitatively the behaviour of a single orthotropic lamina containing unidirectionally aligned fibres from the properties of the constituents. There is reasonable agreement on E_1, ν_{12}, $\hat{\sigma}_1$ and α_1, but it is less easy to be certain of predicting 'the others'. There is always the doubt about the influence of manufacture on 'the others'. For this reason many designers prefer to make a lamina by the same process and under the same conditions as will be used in production, and measure the various physical properties directly. The measured properties are then used in design.

Unidirectional laminae, usually made by special mechanized processes, are hardly ever used as such because of the tremendous variations in stiffness and strength between the different principal directions. But it is possible to estimate how such a structure will behave provided that stresses and strains are only applied in the plane and in the principal directions. The response of flat panels can then be calculated using Hooke's law (in two dimensions); this approach, which is only a small part of the first approach to design mentioned in section 7.1.2, will be developed in section 7.3. In addition beams under lateral load can be designed, always being careful to check that failure does not occur by shear as well as by direct stress or excessive deflection.

The performance under in-plane stress of the cloth and mat laminae can also be predicted from their properties, following the advice given in the previous paragraph. These laminae are often made by contact moulding, and the designer should be alert to the more obvious sources of variation in behaviour. A given cloth has a tolerance of perhaps 10% on volume of fibres within a given area, and this can be increased by handling, especially to form a complicated shape. The operator is responsible for making up the resin system correctly, for wetting out the fibres properly, and also controls how much resin is used, and where, in an open-mould process.

7.2.7 Fibre length

In fibre–polymer composites, the aim is to exploit the load-bearing capabilities of the fibres. The matrix has the job of transferring the external load to the fibre. The end face of the fibre alone could not transfer much load – the interface is not strong in tension, and the fibre creates a stress concentration in an already weak polymer. Although the shear strength of the interface is low – typically 20 MN/m^2 – there is sufficient area around the outside of the fibre to transfer the load to the fibre, if the fibre is long enough. How long is

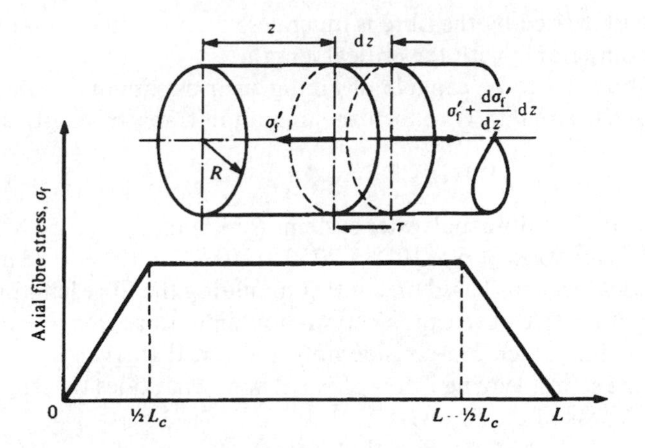

Fig. 7.17 Stress transfer from polymer to fibre and critical fibre length, L_c

'long enough'? Continuous fibres pose no problems, but the load transfer is dominated by end effects in short fibres.

Consider a fibre of radius R. At the end, the stress in the fibre is zero. If the shear stress applied by the matrix to the fibre is τ and assumed constant near each end of the fibre, then a force balance on an element of the fibre (Fig. 7.17) gives

$$\sigma_f' \pi R^2 + \tau 2\pi R \mathrm{d}z = \left(\sigma_f' + \frac{\mathrm{d}\sigma_f'}{\mathrm{d}z}\mathrm{d}z\right)\pi R^2 \tag{7.22}$$

i.e.

$$\frac{\mathrm{d}\sigma_f'}{\mathrm{d}z} = \frac{2\tau}{R} \tag{7.23}$$

By integration, the stress σ_f' increases along the fibre as

$$\sigma_f' = 2\tau z / R \tag{7.24}$$

The maximum stress in the fibre is related to the stress in the composite (which has the same strain) by

$$\varepsilon_1 = \sigma_f E_f = \sigma_1 E_1 \tag{7.25}$$

The length of fibre over which the stress σ_f' builds up to σ_f is called the transfer length or critical length, L_c, i.e. $\frac{1}{2}L_c$ at each end. Thus the critical length L_c depends on the interface stress τ, radius R and applied load:

$$L_c/2 = R\sigma_f/2\tau \quad \text{or} \quad L_c = R\sigma_f/\tau \tag{7.26}$$

If the fibre length L is much greater than L_c, then almost all the load in the composite is carried by the fibre – this minimizes end effects. However, the average load carried by the fibre is much less than the maximum if the fibre length is comparable with the critical length.

If the fibres are to be capable of taking their maximum stress $\hat{\sigma}_f$ and the interface shear strength is τ_i, then the minimum transfer length is

$$L_c = R\hat{\sigma}_f/\tau_i \qquad (7.27)$$

For a typical glass fibre–polyester system, $R \simeq 5\mu m$, $\hat{\sigma}_f \simeq 1750\,MN/m^2$ and $\tau_i \simeq 25\,MN/m^2$ so $L_c = 5 \times 10^{-6} \times 1750 \times 10^6/25 \times 10^6 = 0.35\,mm$, and in most practical systems based on contact moulding the fibre lengths are much larger than this. (However, in injection moulding, fibres may be quite short, e.g. 2 or 3 mm, which are considerably shortened further by the shearing action of the screw, leaving fibres of 0.5–2 mm: what does this suggest about the effectiveness of the fibres?)

7.2.8 Select database for continuous fibre-reinforced composites

Table 7.2 gives representative properties which will be helpful for simple feasibility studies. The volume fractions can be varied by using the principles of micromechanics.

Table 7.2 Representative properties of continuous fibre-reinforced composites

Property	*Unit*	*gfrp*[a]	*cfrp*[a]	*SMC*[a]	*CSM*	*EG*	*HM-CARB*	*RR*	*PULT*
E_1	GPa	38.6	181	13	8	45.6	180	1.74	17.2
E_2	GPa	8.27	10.3	13	8	10.73	8	0.014	5.52
G_{12}	GPa	4.14	7.17	3	2.75	5.14	5	0.0025	2.93
ν_{12}	–	0.26	0.28	0.3	0.32	0.274	0.25	0.547	0.33
α_1	10^6/K	8.6	0.02	20–30	20–30	7	6	10	9.36+
α_2	10^6/K	22.1	22.5	20–30	20–30	30	35	200	+
$\hat{\sigma}_{1T}$	MPa	1062	1500	85	108	1000			207
$\hat{\sigma}_{2T}$	MPa	31	40	85	108	100			48.3
$\hat{\sigma}_{1C}$	MPa	610	1000	120(?)	148	800			207
$\hat{\sigma}_{2C}$	MPa	118	246	120(?)	148	270			103
$\hat{\tau}_{12}$	MPa	72	68	70(?)	85	80			20.7
V_f	–	0.45	0.7	0.2	0.19	0.7			0.5
ρ	kg/m^3	1800	1600	1700	1450	2180			1700

[a] Recommended for design data; gfrp is unidirectional glass fibres/epoxy prepreg; cfrp is unidirectional carbon fibres/epoxy prepreg; SMC is random glass fibres in polyester (sheet moulding compound); CSM is a representative random glass-fibre chopped strand mat with polyester epoxy and polyester as thermosetting resins; EG is unidirectional glass reinforced epoxy for use in text problems only; HM-CARB is high-modulus carbon fibre epoxy for use in text problems only; RR is a Rayon cord-reinforced rubber; PULT is a pultruded mat material (from commercial literature); V_f is the volume fraction of fibres; ρ is density; for PULT the thermal expansion coefficient is α_x; α_y is not quoted but is probably about 25×10^{-6}/K.

PROBLEMS

1. Estimate the maximum volume fractions of fibre having a circular section in the following model composites:
   (a) All fibres straight, close-packed and aligned in one direction.
   (b) All fibres straight in layers one fibre thick, alternating at $\pm\theta^\circ$ to a reference direction.
   (c) All fibres straight and adjacent fibres mutually perpendicular to give three principal directions of reinforcement.
2. At what ratio of critical length to overall fibre length does a fibre have an average strength which is 95% of that for a continuous fibre?
3. A panel is to be made having the following proportions of constituents by weight: 20% glass fibres, 55% polyester resin (density $1200\,kg/m^3$) and 25% calcium carbonate filler (density $2700\,kg/m^3$). What is the volume fraction of fibres in the panel?
4. A long slender beam is made from a resin-bonded unidirectional fibre composite of given volume fraction. When simply supported and carrying a lateral load at midspan, what span-to-depth ratio ensures that the beam is equally likely to fail by shear or by direct stress? What is the practical significance of this for isotropic materials, and for the orthotropic materials described in section 7.2? Is the problem more severe at high or low volume fractions of fibres?
5. At 20°C, an orthotropic lamina is 200 mm long, 100 mm wide and 2 mm thick and consists of parallel high-modulus carbon fibres 200 mm long bonded by epoxy resin. The elastic constants for this lamina are $E_1 = 208\,GN/m^2$, $E_2 = 7.6\,GN/m^2$, $\nu_{12} = 0.3$, and $G_{12} = 4.8\,GN/m^2$. The tensile strengths of the lamina are $\hat{\sigma}_{1T} = 828\,MN/m^2$, $\hat{\sigma}_{2T} = 41\,MN/m^2$ and $\hat{\tau}_{12} = 55\,MN/m^2$. All data refer to 20°C. Calculate:
   (a) The width of the sheet under load when the uniformly distributed applied longitudinal force is 80 kN.
   (b) The length of the sheet when a uniformly distributed 80 kN is applied longitudinally and a uniformly distributed 5 kN is applied transversely.
   (c) The transverse stress causing failure when the uniformly distributed shear load is 10 kN in the plane along the long edges of the sheet.
   (d) The weight fraction of carbon fibres in the lamina.
6. A long slender beam is to be made from a unidirectional array of glass fibres bonded together with a thermosetting resin. Estimate the optimum volume fraction of fibres necessary to achieve the best bending stiffness per unit weight of composite. Outline the assumptions you make in your analysis.
7. A mild steel rod is 10 mm in diameter and 500 mm long. It is proposed to replace it by a pultruded rod 500 mm long in which the fibres are axially aligned. Using either the material gfrp or cfrp (see database, section 7.2.8) as a data source, calculate the composite rod diameter and weight saving

(if any) for the same: (a) axial tensile stiffness, (b) axial compressive buckling load, (c) torsional stiffness, (d) bending stiffness, (e) axial tensile strength, (f) torsional strength, (g) bending strength. Assume the properties of this mild steel are $E = 208$ GPa, $G = 81$ GPa, tensile strength = compressive strength = 200 MPa, shear strength = 100 MPa, density $\rho = 7800\,\mathrm{kg/m^3}$.

8. A thin symmetric unidirectional laminate is to be made using two types of ply, both with fibres parallel to the direction of load. One ply, 0.25 mm thick, is based on carbon fibres and epoxy resin, and has $E_c = 208$ GPa. The other is 0.15 mm thick and is based on glass fibres and epoxy with $E_g = 40$ GPa. Calculate the modulus of the laminate in the fibre direction.
9. An experimental closed-coil spring has been made using parallel bundles of wet-resin-impregnated fibres wound by hand into the helical grooves of a large (greased) bolt (this is a messy operation!). After the resin has crosslinked, the bolt is withdrawn. What factors would affect the stiffness of the spring when loaded (a) in axial tension and (b) in torsion? How would you expect the spring to fail in torsion? How could you easily and simply improve the strength of the bolt by slightly adjusting the method of manufacture (still by hand)?

   If you had a metal spring of the same coil diameter and axial stiffness, would you expect the metal spring to have the same ratio of axial stiffness/torsion stiffness as the reinforced plastics spring?

7.3 MACROMECHANICS OF A LAMINA

Section 7.2 properly regarded an element of an orthotropic lamina as inhomogeneous on the scale of a few micrometres, and showed how the properties of a lamina in the principal directions depend on the volume fraction of fibres present. We also looked at some simple calculations of behaviour of structural elements. The dependence of properties on fibre content pervades all that follows, but will not be elaborated upon – simply taken for granted. It is now convenient to change the scale on which inhomogeneity is recognized, from micrometres to millimetres. We shall from now on regard the lamina as homogeneous from an engineering point of view, but having different stiffnesses and strengths in different directions. This approach is known as macromechanics.

We now outline the mechanics needed to predict the behaviour of rods, pipes, beams and plates to be made from one unidirectional ply in which the fibres are aligned with or at an angle to a reference direction in the structure. It is worth noting the following points before going into the detail.

- The analysis closely resembles conventional treatments for isotropic materials: the basic framework of understanding must be the same, only the relationships between stresses and strains are new. Apart from the

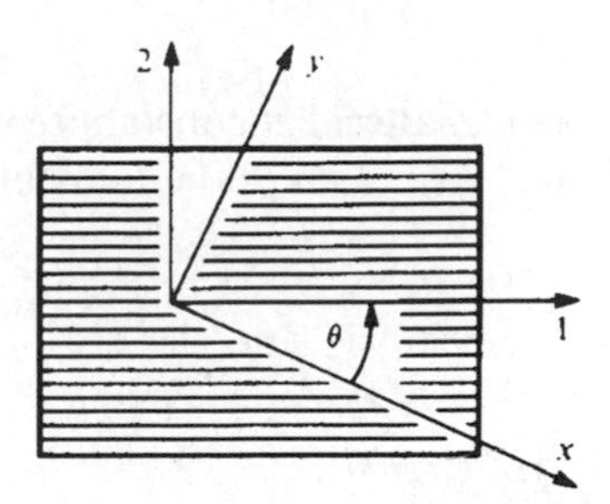

Fig. 7.18 Coordinate system (x,y) related to principal directions (1,2)

substantial differences in stiffness and strength possible between longitudinal and transverse directions, we find it quite normal that the values of Poisson's ratios ν_{xy} for laminates can be between 0 and +2, depending on the composition of the composite and the orientation of the fibres.

- The convention in composites is to use (x,y,z) for global coordinates of the laminate in problem-solving. Coordinates $(1,2,z)$ relate to directions along and across the fibres of a lamina, and are called the principal directions. The coordinates $(1,2,z)$ have nothing to do with the concepts of principal stresses used in analysis of isotropic materials. Transformation of axes for stresses and for strains is simple revision of conventional mechanics of isotropic materials, but transformation of axes for 'modulus' is new.

It is evident from section 7.1 that the properties of a lamina consisting of axially aligned fibres bonded with a crosslinkable resin vary markedly with the direction of in-plane loading in which they are measured. This section explores the stiffness and strength behaviour of an orthotropic panel stressed in the plane at some angle θ to the principal directions (Fig. 7.18), in relation to the properties measured in the principal directions (E_1, E_2, G_{12}, ν_{12}, $\hat{\sigma}_1$, $\hat{\sigma}_2$, $\hat{\tau}_{12}$). When only one of the stresses σ_x, σ_y or τ_{xy} is applied, an expression can be obtained for the apparent modulus in the direction of stress. But the general approach also permits calculation of all the in-plane strains resulting from one or more stresses. Similarly, it is possible to derive expressions for the strength of a lamina when stressed off-axis.

What are now needed are the following: a formalized relationship between in-plane stresses and strains in the principal directions, rules for the transformation of stresses or strains when the coordinate system is rotated in the plane, and relationships between in-plane stresses and strains where the reference directions (x, y) are at an angle θ to the principal directions (1, 2).

7.3.1 Hooke's law for principal directions

Before discussing the behaviour of an orthotropic lamina stressed in the plane in the principal directions, it is worth revising Hooke's law for an isotropic material.

(*a*) *Isotropic material*

For an isotropic linear elastic material at constant temperature, Hooke's law in the Cartesian coordinates (1,2,3) takes the familiar form:

$$\begin{aligned}\varepsilon_1 &= \sigma_1/E - \nu(\sigma_2+\sigma_3)/E\\ \varepsilon_2 &= \sigma_2/E - \nu(\sigma_1+\sigma_3)/E\\ \varepsilon_3 &= \sigma_3/E - \nu(\sigma_1+\sigma_2)/E\\ \gamma_{23} &= \tau_{23}/G\\ \gamma_{31} &= \tau_{31}/G\\ \gamma_{12} &= \tau_{12}/G\end{aligned} \tag{7.28}$$

As matrix algebra is a convenient way of handling relationships between stresses and strains, and will be used in the rest of this chapter, the equations (7.28) can be written in the equivalent form:

$$\begin{bmatrix}\varepsilon_1\\ \varepsilon_2\\ \varepsilon_3\\ \gamma_{23}\\ \gamma_{31}\\ \gamma_{12}\end{bmatrix} = \begin{bmatrix}1/E & -\nu/E & -\nu/E & 0 & 0 & 0\\ -\nu/E & 1/E & -\nu/E & 0 & 0 & 0\\ -\nu/E & -\nu/E & 1/E & 0 & 0 & 0\\ 0 & 0 & 0 & 1/G & 0 & 0\\ 0 & 0 & 0 & 0 & 1/G & 0\\ 0 & 0 & 0 & 0 & 0 & 1/G\end{bmatrix}\begin{bmatrix}\sigma_1\\ \sigma_2\\ \sigma_3\\ \tau_{23}\\ \tau_{31}\\ \tau_{12}\end{bmatrix} \tag{7.29}$$

Where only in-plane loads need be considered, equation (7.29) for a thin sheet reduces to:

$$\begin{bmatrix}\varepsilon_1\\ \varepsilon_2\\ \gamma_{12}\end{bmatrix} = \begin{bmatrix}1/E & -\nu/E & 0\\ -\nu/E & 1/E & 0\\ 0 & 0 & 1/G\end{bmatrix}\begin{bmatrix}\sigma_1\\ \sigma_2\\ \tau_{12}\end{bmatrix} = [S]\begin{bmatrix}\sigma_1\\ \sigma_2\\ \tau_{12}\end{bmatrix} \tag{7.30}$$

$[S]$ is the **compliance matrix**, deliberately not written out in full just yet, relating the in-plane strain vector $[\varepsilon]$ to the applied stresses $[\sigma]$. Individual elements of the compliance matrix have the units m^2/N and the connotations first introduced in section 5.2.1.

If in-plane strains are applied to the thin isotropic sheet, the induced stresses are found by inverting relation (7.30) to give

$$\begin{bmatrix}\sigma_1\\ \sigma_2\\ \tau_{12}\end{bmatrix} = \begin{bmatrix}E/(1-\nu^2) & \nu E/(1-\nu^2) & 0\\ \nu E/(1-\nu^2) & E/(1-\nu^2) & 0\\ 0 & 0 & G\end{bmatrix}\begin{bmatrix}\varepsilon_1\\ \varepsilon_2\\ \gamma_{12}\end{bmatrix} = [Q]\begin{bmatrix}\varepsilon_1\\ \varepsilon_2\\ \gamma_{12}\end{bmatrix} \tag{7.31}$$

where $[Q]$ is called the **reduced stiffness matrix** such that

$$[Q] = [S]^{-1} \quad \text{or} \quad [S] = [Q]^{-1} \tag{7.32}$$

Any of the equations (7.28)–(7.31) confirm the experience that applying a direct stress does not induce a shear strain.

(*b*) *Anisotropic behaviour*

In an anisotropic material there *is* coupling between all stresses and all strains. In order to describe such behaviour in a manageable way, it is necessary to review the nomenclature. Direct stresses and strains have two subscripts, e.g. σ_{11}, ε_{22}, γ_{12}, and moduli should therefore have two subscripts as well: E_{11}, E_{22} and G_{12}. By convention, engineers use a contracted form of notation, where possible, so that repeated subscripts are reduced to just one: σ_{11} becomes σ_1, E_{11} becomes E_1, but G_{12} stays as it is. This book follows this convention. An inconvenience arises in the full subscript system when trying to relate stresses and strains, because the coefficients of the compliance or reduced stiffness matrices should then have four suffices. This is cumbersome and unwieldly. So the convention of contracted subscripts has been extended. Stresses have one subscript as follows:

$$\begin{aligned} &\text{Standard notation}: \sigma_{11} \quad \sigma_{22} \quad \sigma_{33} \quad \tau_{31} \quad \tau_{32} \quad \tau_{12} \\ &\text{Contracted notation}: \sigma_1 \quad \sigma_2 \quad \sigma_3 \quad \sigma_4 \quad \sigma_5 \quad \sigma_6 \end{aligned} \tag{7.33}$$

Strains are labelled similarly, e.g. $\varepsilon_1, \ldots, \varepsilon_6$.

The coupling between stresses and strains (i.e. generalized Hooke's law) in an anisotropic material can now be expressed in the rather daunting form:

$$\begin{bmatrix} \varepsilon_1 \\ \varepsilon_2 \\ \varepsilon_3 \\ \varepsilon_4 \\ \varepsilon_5 \\ \varepsilon_6 \end{bmatrix} = \begin{bmatrix} S_{11} & S_{12} & S_{13} & S_{14} & S_{15} & S_{16} \\ S_{21} & S_{22} & S_{23} & S_{24} & S_{25} & S_{26} \\ S_{31} & S_{32} & S_{33} & S_{34} & S_{35} & S_{36} \\ S_{41} & S_{42} & S_{43} & S_{44} & S_{45} & S_{46} \\ S_{51} & S_{52} & S_{53} & S_{54} & S_{55} & S_{56} \\ S_{61} & S_{62} & S_{63} & S_{64} & S_{65} & S_{66} \end{bmatrix} \begin{bmatrix} \sigma_1 \\ \sigma_2 \\ \sigma_3 \\ \sigma_4 \\ \sigma_5 \\ \sigma_6 \end{bmatrix} \tag{7.34}$$

where $[S]$ is a symmetric matrix. The reason for including this equation is to indicate the source of the unusual nomenclature. For an anisotropic sheet under in-plane loads, equation (7.34) reduces to:

$$\begin{bmatrix} \varepsilon_1 \\ \varepsilon_2 \\ \varepsilon_6 \end{bmatrix} = \begin{bmatrix} \varepsilon_1 \\ \varepsilon_2 \\ \gamma_{12} \end{bmatrix} = \begin{bmatrix} S_{11} & S_{12} & S_{16} \\ S_{12} & S_{22} & S_{26} \\ S_{16} & S_{26} & S_{66} \end{bmatrix} \begin{bmatrix} \sigma_1 \\ \sigma_2 \\ \tau_{12} \end{bmatrix} \tag{7.35}$$

in which the standard notation is retained for shear stress and strain. Note carefully now that S_{66} represents shear, but S_{12} involves coupling between ε_1 and σ_2 or ε_2 and σ_1 – and not shear.

(*c*) *Orthotropic behaviour*

For an orthotropic lamina under in-plane stresses in the principal directions, equation (7.35) reduces to

$$\begin{bmatrix} \varepsilon_1 \\ \varepsilon_2 \\ \gamma_{12} \end{bmatrix} = \begin{bmatrix} S_{11} & S_{12} & 0 \\ S_{12} & S_{22} & 0 \\ 0 & 0 & S_{66} \end{bmatrix} \begin{bmatrix} \sigma_1 \\ \sigma_2 \\ \tau_{12} \end{bmatrix} \tag{7.36}$$

with no coupling between direct stresses and shear strain.

But what significance do the terms in the compliance matrix have? It would be helpful to relate them to the moduli E_1, E_2 and ν_{12} which are measured in conventional uniaxial tensile tests, and to the in-plane shear modulus G_{12}. This can be readily done by applying the single stress to the general expression (7.36).

Putting $\sigma_2 = \tau_{12} = 0$ in (7.36) gives $\varepsilon_1 = S_{11}\sigma_1$, so

$$S_{11} = 1/E_1 \tag{7.37}$$

Similarly, with $\sigma_1 = \tau_{12} = 0$,

$$S_{22} = 1/E_2 \tag{7.38}$$

and $\sigma_1 = \sigma_2 = 0$ gives

$$S_{66} = 1/G_{12} \tag{7.39}$$

The meaning of S_{12} is found by recalling the definition of the major Poisson's ratio (equation (7.5)):

$$\nu_{12} = -\varepsilon_2/\varepsilon_1$$

which relates to a panel under direct stress σ_1 with $\sigma_2 = \tau_{12} = 0$. However, under this circumstance, $\varepsilon_1 = \sigma_1/E_1$ from expression (7.36) and hence

$$\nu_{12} = -\varepsilon_2/\varepsilon_1 = -\varepsilon_2 E_1/\sigma_1$$

and from (7.36) $\varepsilon_2 = S_{12}\sigma_1$:

$$S_{12} = -\nu_{12}/E_1 \tag{7.40}$$

Putting $\sigma_1 = \tau_{12} = 0$ gives $S_{21} = \varepsilon_1/\sigma_2 = -\nu_{21}/E_2$ and as $[S]$ is a symmetric matrix, $S_{ij} = S_{ji}$, so

$$\nu_{12}/E_1 = \nu_{21}/E_2 \tag{7.41}$$

In summary, therefore, the compliance matrix for an orthotropic lamina can be expressed as

$$[S] = \begin{bmatrix} 1/E_1 & -\nu_{12}/E_1 & 0 \\ -\nu_{12}/E_1 & 1/E_2 & 0 \\ 0 & 0 & 1/G_{12} \end{bmatrix} \tag{7.42}$$

which obviously reduces to equation (7.30) for an isotropic material by putting $E_1 = E_2 = E, \nu_{12} = \nu$, and $G_{12} = G$. For the orthotropic lamina, equation (7.42) can be inverted to give the stress response to plane strain:

$$\begin{bmatrix} \sigma_1 \\ \sigma_2 \\ \tau_{12} \end{bmatrix} = [S]^{-1} \begin{bmatrix} \varepsilon_1 \\ \varepsilon_2 \\ \gamma_{12} \end{bmatrix} = \begin{bmatrix} Q_{11} & Q_{12} & 0 \\ Q_{12} & Q_{22} & 0 \\ 0 & 0 & Q_{66} \end{bmatrix} \begin{bmatrix} \varepsilon_1 \\ \varepsilon_2 \\ \gamma_{12} \end{bmatrix} \tag{7.43}$$

The terms in the reduced stiffness matrix [Q] can be readily obtained and [Q] reduces to equation (7.31) for an isotropic sheet.

In developing the understanding of the mechanics, the concept of 'compliance' is most useful. In designing with composites we find that the concept of reduced stiffness is especially useful for calculation of stresses (and thus strength) in a given (unidirectional) ply.

7.3.2 Transformation of axes for stress or strain

Hitherto, stresses or strains have been applied in the principal directions of the lamina. This is of direct interest because properties are most often measured in the principal directions. However, because of the highly directional nature of the properties, laminae are used in laminates and the stresses or strains are applied in the plane in directions (x,y) at an angle θ to the principal directions (1,2) as shown in Fig. 7.18. The strains in the (x,y) directions are related to the stresses in the (x,y) directions using the measured properties in the (1,2) directions and the transformation matrix $[T]$.

The T matrix will have been derived in an earlier course on stress analysis for isotropic materials (though probably not in matrix form), and still applies for the orthotropic lamina. Remember that σ_1, σ_2 and τ_{12} refer to stresses in the principal directions (and not principal stresses). On rotation of the coordinate system (1,2) to an angle θ to the (x,y) system (Fig. 7.18), the stresses transform according to

$$\begin{bmatrix} \sigma_1 \\ \sigma_2 \\ \tau_{12} \end{bmatrix} = \begin{bmatrix} \cos^2\theta & \sin^2\theta & 2\sin\theta\cos\theta \\ \sin^2\theta & \cos^2\theta & -2\sin\theta\cos\theta \\ -\sin\theta\cos\theta & \sin\theta\cos\theta & \cos^2\theta - \sin^2\theta \end{bmatrix} \begin{bmatrix} \sigma_x \\ \sigma_y \\ \tau_{xy} \end{bmatrix} = [T] \begin{bmatrix} \sigma_x \\ \sigma_y \\ \tau_{xy} \end{bmatrix} \qquad (7.44)$$

This equation can be readily inverted:

$$\begin{bmatrix} \sigma_x \\ \sigma_y \\ \tau_{xy} \end{bmatrix} = \begin{bmatrix} \cos^2\theta & \sin^2\theta & -2\sin\theta\cos\theta \\ \sin^2\theta & \cos^2\theta & +2\sin\theta\cos\theta \\ \sin\theta\cos\theta & -\sin\theta\cos\theta & \cos^2\theta - \sin^2\theta \end{bmatrix} \begin{bmatrix} \sigma_1 \\ \sigma_2 \\ \tau_{12} \end{bmatrix} = [T]^{-1} \begin{bmatrix} \sigma_1 \\ \sigma_2 \\ \tau_{12} \end{bmatrix} \qquad (7.45)$$

The strains transform according to

$$\begin{bmatrix} \varepsilon_1 \\ \varepsilon_2 \\ \frac{1}{2}\gamma_{12} \end{bmatrix} = [T] \begin{bmatrix} \varepsilon_x \\ \varepsilon_y \\ \frac{1}{2}\gamma_{xy} \end{bmatrix} \quad \text{and} \quad \begin{bmatrix} \varepsilon_x \\ \varepsilon_y \\ \frac{1}{2}\gamma_{xy} \end{bmatrix} = [T]^{-1} \begin{bmatrix} \varepsilon_1 \\ \varepsilon_2 \\ \frac{1}{2}\gamma_{12} \end{bmatrix} \qquad (7.46)$$

Note carefully that the $[T]$ matrix can transform stresses having the same format as that used in the previous section, whereas for strains the $[T]$ matrix only applies when the forms $[\varepsilon_1, \varepsilon_2, \frac{1}{2}\gamma_{12}]$ and $[\varepsilon_x, \varepsilon_y, \frac{1}{2}\gamma_{xy}]$ are used.

Analysis involving transformation of axes often needs to relate the two types of strain; this is most readily achieved using the Reuter's matrix $[R]$ defined as

$$[R] = \begin{bmatrix} 1 & 0 & 0 \\ 0 & 1 & 0 \\ 0 & 0 & 2 \end{bmatrix}$$

and hence

$$[R]^{-1} = \begin{bmatrix} 1 & 0 & 0 \\ 0 & 1 & 0 \\ 0 & 0 & \frac{1}{2} \end{bmatrix} \tag{7.47}$$

Using the $[R]$ matrix, the relationship between tensor strains and engineering strains is evidently

$$\begin{bmatrix} \varepsilon_1 \\ \varepsilon_2 \\ \gamma_{12} \end{bmatrix} = [R] \begin{bmatrix} \varepsilon_1 \\ \varepsilon_2 \\ \frac{1}{2}\gamma_{12} \end{bmatrix} \quad \text{and} \quad \begin{bmatrix} \varepsilon_1 \\ \varepsilon_2 \\ \frac{1}{2}\gamma_{12} \end{bmatrix} = [R]^{-1} \begin{bmatrix} \varepsilon_1 \\ \varepsilon_2 \\ \gamma_{12} \end{bmatrix} \tag{7.48}$$

The $[R]$ matrix applies for any coordinate system including the (x, y) set at θ to (1,2). For example:

$$\begin{bmatrix} \varepsilon_x \\ \varepsilon_y \\ \gamma_{xy} \end{bmatrix} = [R] \begin{bmatrix} \varepsilon_x \\ \varepsilon_y \\ \frac{1}{2}\gamma_{xy} \end{bmatrix} \tag{7.49}$$

Having now transformed the stresses or the strains, the next step in the argument is to transform the stiffnesses and strengths. For stiffness, this leads to transformed compliances $[\bar{S}]$. When a panel carries only one of the in-plane off-axis stresses σ_x, σ_y or τ_{xy}, some of the transformed compliances can be directly related to the 'apparent moduli' E_x, E_y and G_{xy}, and also to ν_{xy}. The more general case of stiffness under combined stress can also be tackled using $[\bar{S}]$. Strength under a single stress or a combination of stresses will also be discussed.

7.3.3 Stress–strain relationships under off-axis loading

Hooke's law is defined in terms of elastic constants in the principal directions by equation (7.42):

$$\begin{bmatrix} \varepsilon_1 \\ \varepsilon_2 \\ \gamma_{12} \end{bmatrix} = \begin{bmatrix} 1/E_1 & -\nu_{12}/E_1 & 0 \\ -\nu_{12}/E_1 & 1/E_2 & 0 \\ 0 & 0 & 1/G_{12} \end{bmatrix} \begin{bmatrix} \sigma_1 \\ \sigma_2 \\ \tau_{12} \end{bmatrix} = [S] \begin{bmatrix} \sigma_1 \\ \sigma_2 \\ \tau_{12} \end{bmatrix} \tag{7.50}$$

We seek the apparent elastic constants E_x, E_y, ν_{xy} and G_{xy} which are defined when only one in-plane stress is applied to the lamina:

$$\begin{aligned} E_x &= \sigma_x/\varepsilon_x \quad & G_{xy} &= \tau_{xy}/\gamma_{xy} \\ E_y &= \sigma_y/\varepsilon_y \quad & \nu_{xy} &= -\varepsilon_y/\varepsilon_x = -E_x\varepsilon_y/\sigma_x \end{aligned} \tag{7.51}$$

The stresses or strains in (x,y) can be related to (1,2) using the $[T]$ matrix (together with Reuter's matrix for strains), and the stresses and strains in the principal directions are related by Hooke's law. This leads to:

$$\begin{bmatrix} \varepsilon_x \\ \varepsilon_y \\ \gamma_{xy} \end{bmatrix} = \bar{S} \begin{bmatrix} \sigma_x \\ \sigma_y \\ \tau_{xy} \end{bmatrix} \tag{7.52}$$

$[\bar{S}]$ is the transformed compliance matrix which relates the stress and strain vectors in an orthotropic lamina under in-plane off-axis load. It can be evaluated simply (but tediously) using

$$[\bar{S}] = \begin{bmatrix} \bar{S}_{11} & \bar{S}_{12} & \bar{S}_{16} \\ \bar{S}_{12} & \bar{S}_{22} & \bar{S}_{26} \\ \bar{S}_{16} & \bar{S}_{26} & \bar{S}_{66} \end{bmatrix} = [R][T]^{-1}[R]^{-1}[S][T] \tag{7.53}$$

Because $[S]$ only involves four independent constants (E_1, E_2, G_{12} and either ν_{12} or ν_{21}), $[\bar{S}]$ only involves the same number of constants (together with θ), even though it is a fully populated matrix.

The objective was to find E_x, E_y, G_{xy} and ν_{xy}. If the lamina only carries a stress σ_x, then from equation (7.52):

$$\varepsilon_x = \bar{S}_{11}\sigma_x \quad \text{so} \quad 1/E_x = \bar{S}_{11} \tag{7.54}$$

and

$$\varepsilon_y = \bar{S}_{12}\sigma_x \quad \text{so} \quad \nu_{xy} = -E_x\bar{S}_{12} \tag{7.55}$$

If the lamina only carries an in-plane shear stress τ_{xy}, then

$$\gamma_{xy} = \bar{S}_{66}\tau_{xy} \quad \text{so} \quad 1/G_{xy} = \bar{S}_{66} \tag{7.56}$$

Equations (7.54)–(7.56) are the off-axis equivalents of (7.37), (7.40) and (7.39).

From equation (7.53) it is easy to show that the individual components of the transformed compliance matrix $[\bar{S}]$ are

$$\begin{aligned}
\bar{S}_{11} &= S_{11}\cos^4\theta + S_{22}\sin^4\theta + (2S_{12} + S_{66})\cos^2\theta\sin^2\theta \\
\bar{S}_{12} &= (S_{11} + S_{22} - S_{66})\cos^2\theta\sin^2\theta + S_{12}(\cos^4\theta + \sin^4\theta) \\
\bar{S}_{22} &= S_{11}\sin^4\theta + S_{22}\cos^4\theta + (2S_{12} + S_{66})\cos^2\theta\sin^2\theta \\
\bar{S}_{66} &= 4(S_{11} - 2S_{12} + S_{22})\cos^2\theta\sin^2\theta + S_{66}(\cos^2\theta - \sin^2\theta)^2 \\
\bar{S}_{16} &= (2S_{11} - 2S_{12} - S_{66})\cos^3\theta\sin\theta - (2S_{22} - 2S_{12} - S_{66})\cos\theta\sin^3\theta \\
\bar{S}_{26} &= (2S_{11} - 2S_{12} - S_{66})\cos\theta\sin^3\theta - (2S_{22} - 2S_{12} - S_{66})\cos^3\theta\sin\theta
\end{aligned} \tag{7.57}$$

Expressing each of the compliances in terms of E_1, E_2, ν_{12} and G_{12} now gives the required results:

$$\bar{S}_{11} = \frac{1}{E_x} = \frac{\cos^4\theta}{E_1} + \frac{\sin^4\theta}{E_2} + \left(\frac{1}{G_{12}} - \frac{2\nu_{12}}{E_1}\right)\cos^2\theta \sin^2\theta \quad (7.58)$$

$$\bar{S}_{22} = \frac{1}{E_y} = \frac{\sin^4\theta}{E_1} + \frac{\cos^4\theta}{E_2} + \left(\frac{1}{G_{12}} - \frac{2\nu_{12}}{E_1}\right)\cos^2\theta \sin^2\theta \quad (7.59)$$

$$\bar{S}_{66} = \frac{1}{G_{xy}} = 4\left(\frac{1}{E_1} + \frac{2\nu_{12}}{E_1} + \frac{1}{E_2}\right)\cos^2\theta \sin^2\theta + \frac{1}{G_{12}}(\cos^2\theta - \sin^2\theta)^2 \quad (7.60)$$

$$-E_x\bar{S}_{12} = \nu_{xy} = E_x\left[\frac{\nu_{12}}{E_1}(\sin^4\theta + \cos^4\theta) - \left(\frac{1}{E_1} + \frac{1}{E_2} - \frac{1}{G_{12}}\right)\cos^2\theta \sin^2\theta\right] \quad (7.61)$$

For laminae which have been properly made and carefully tested, there is close agreement between theoretical predictions and experimental results. It is now possible to calculate the dependence of stiffness on angle for an orthotropic lamina having given properties in the principal directions; some authors plot modulus versus θ, some plot normalized moduli, i.e. E_x/E_1, E_y/E_1 and G_{xy}/G_{12}. Representative plots are shown in Figs 7.19 and 7.20.

Attention has already been drawn to the coupling between shear strain and direct stress when an orthotropic lamina is loaded in-plane but off-axis (see Fig. 7.4). It is now apparent that the transformed compliances $\bar{S}_{16}$ and $\bar{S}_{26}$ provide a quantitative basis for this coupling.

The argument detailed above has focused on the strain response to applied stresses. A similar argument can be developed in order to find the stresses in a lamina resulting from the application of in-plane strains, leading to

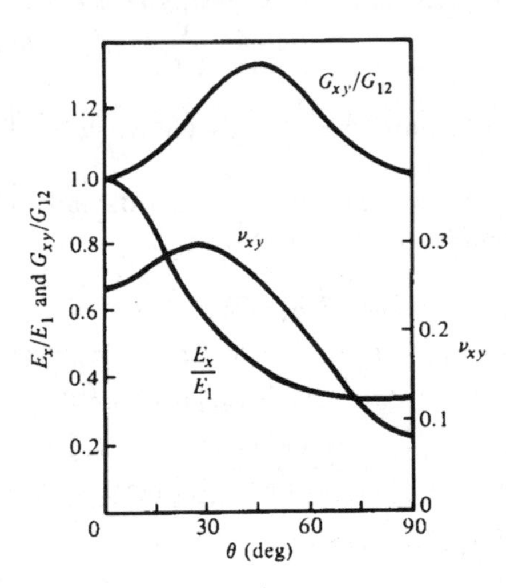

Fig. 7.19 Elastic constants for a glass fibre–epoxy lamina: $E_1 = 42\,\text{GN/m}^2$, $E_2 = 14\,\text{GN/m}^2$, $G_{12} = 7\,\text{GN/m}^2$ and $\nu_{12} = 0.25$

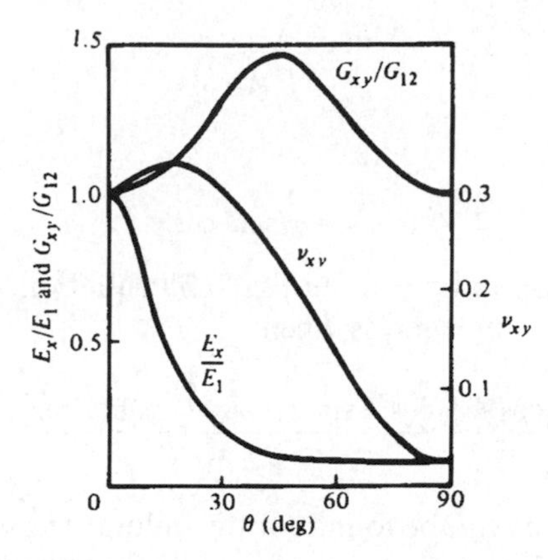

Fig. 7.20 Elastic constants for a graphite–epoxy lamina: $E_1 = 168\,\text{GN/m}^2$, $E_2 = 14\,\text{GN/m}^2$, $G_{12} = 8.4\,\text{GN/m}^2$ and $\nu_{12} = 0.3$

$$\begin{bmatrix} \sigma_x \\ \sigma_y \\ \tau_{xy} \end{bmatrix} = \begin{bmatrix} \bar{Q}_{11} & \bar{Q}_{12} & \bar{Q}_{16} \\ \bar{Q}_{12} & \bar{Q}_{22} & \bar{Q}_{26} \\ \bar{Q}_{16} & \bar{Q}_{26} & \bar{Q}_{66} \end{bmatrix} \begin{bmatrix} \varepsilon_x \\ \varepsilon_y \\ \gamma_{xy} \end{bmatrix} \tag{7.62}$$

$[\bar{Q}]$ is known as the **transformed reduced stiffness matrix**. The individual elements $\bar{Q}_{ij}$ are related to the reduced stiffnesses Q_{ij} by

$$\begin{aligned}
\bar{Q}_{11} &= Q_{11}\cos^4\theta + Q_{22}\sin^4\theta + (2Q_{12} + 4Q_{66})\cos^2\theta\,\sin^2\theta \\
\bar{Q}_{12} &= Q_{12}(\cos^4\theta + \sin^4\theta) + (Q_{11} + Q_{22} - 4Q_{66})\cos^2\theta\,\sin^2\theta \\
\bar{Q}_{22} &= Q_{11}\sin^4\theta + Q_{22}\cos^4\theta + (2Q_{12} + 4Q_{66})\cos^2\theta\,\sin^2\theta \\
\bar{Q}_{66} &= (Q_{11} + Q_{22} - 2Q_{12} - 2Q_{66})\cos^2\theta\,\sin^2\theta + Q_{66}(\cos^4\theta + \sin^4\theta) \\
\bar{Q}_{16} &= (Q_{11} - Q_{12} - 2Q_{66})\cos^3\theta\,\sin\theta - (Q_{22} - Q_{12} - 2Q_{66})\cos\theta\,\sin^3\theta \\
\bar{Q}_{26} &= (Q_{11} - Q_{12} - 2Q_{66})\cos\theta\,\sin^3\theta - (Q_{22} - Q_{12} - 2Q_{66})\cos^3\theta\,\sin\theta
\end{aligned} \tag{7.63}$$

7.3.4 Failure under off-axis loading

When an orthotropic lamina is subjected to in-plane stresses in the principal directions, failure will occur in accordance with the Tsai–Hill criterion (7.19) when

$$\left(\frac{\sigma_1}{\hat{\sigma}_1}\right)^2 - \frac{\sigma_1\sigma_2}{\hat{\sigma}_1^2} + \left(\frac{\sigma_2}{\hat{\sigma}_2}\right)^2 + \left(\frac{\tau_{12}}{\hat{\tau}_{12}}\right)^2 = 1 \tag{7.64}$$

If the lamina is stressed in-plane but off-axis, then the applied stresses must be transformed into those in the principal directions (using equation (7.44)) so that the Tsai–Hill criterion can be used.

Suppose the lamina carries only a direct stress σ_x at θ to the 1-direction. Putting $\sigma_y = \tau_{xy} = 0$ in equation (7.44) gives

$$\sigma_1 = \sigma_x \cos^2\theta$$
$$\sigma_2 = \sigma_x \sin^2\theta$$

and

$$\tau_{12} = -\sigma_x \sin\theta \cos\theta$$

Substituting for σ_1, σ_2 and τ_{12} in the Tsai–Hill equation, the lamina will fail under this direct and single stress when:

$$\hat{\sigma}_x = \left(\frac{\cos^4\theta}{\hat{\sigma}_1^2} - \frac{\cos^2\theta \sin^2\theta}{\hat{\sigma}_1^2} + \frac{\sin^4\theta}{\hat{\sigma}_2^2} + \frac{\sin^2\theta \cos^2\theta}{\hat{\tau}_{12}^2} \right)^{-1/2} \quad (7.65)$$

Analogous expressions can be found for the failure stresses $\hat{\sigma}_y$ and $\hat{\tau}_{xy}$. For a given ratio of applied stresses $(\sigma_x, \sigma_y, \tau_{xy})$, the corresponding $(\sigma_1, \sigma_2, \tau_{12})$ can be found using equation (7.44). Applying Tsai–Hill, if the left-hand side of equation (7.64) is greater than unity, then failure is expected. If the left-hand side of equation (7.64) is less than unity, failure is not expected and the margin of safety F is indicated by the left-hand side of equation (7.64) $= 1/F^2$.

7.3.5 Compatibility and equilibrium in a lamina

Composites are most often used in the form of thin plate-like elements. It follows therefore that the language and terminology of plates are used to describe the basic mechanics of laminated plates. We introduce here this terminology for the unidirectional ply. This may seem unnecessarily complicated at this stage, but it has the added advantage that the concepts used in the main software packages will also be understood.

It is helpful to revise three basis concepts of lamina behaviour, which are commonly used in plate and shell theory for elastic isotropic material, before applying them to the behaviour of laminates. The three concepts are strain–displacement relationships, force resultants and moment resultants. It is also helpful to introduce the new concepts of lamina stiffnesses and compliances.

(a) Strain–displacement relationships

It is customary to describe the displacements in a plate in terms of the midplane strains (ε^0) and curvatures (κ) and a through-thickness coordinate (z). (The midplane in a shell or plate is the equivalent of the neutral axis in a rectangular beam.) Consider the straight line element AB in the x, z plane of the plate (Fig. 7.21) normal to the midplane and to the x-direction. Let the point O on the midplane be displaced by u_0, v_0 and w_0 in the x, y and z directions. It is assumed that after deformation and displacement A′B′ remains normal to the plane so that there is no shear in the z-direction:

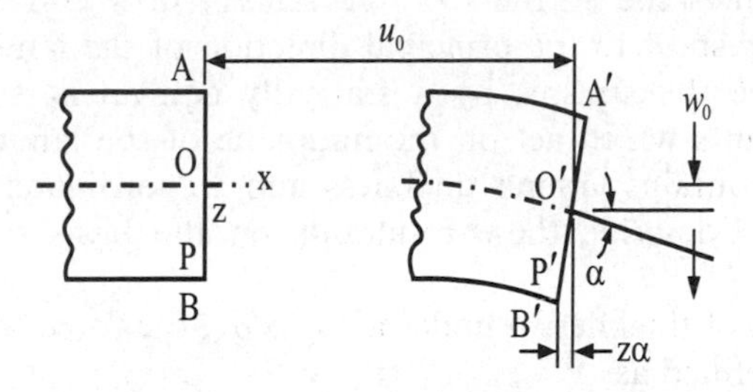

Fig. 7.21 Displacements of a thin plate

$\gamma_{xz} = \gamma_{yz} = 0$. The midplane at O′ is at a slope $\partial w_0/\partial x$ to its original direction. The displacement u of any point P is related to the midplane displacement by

$$u = u_0 - z\frac{\partial w_0}{\partial x} \tag{7.66}$$

The plate is thin and hence it can be assumed that the change in length of any part of the normal AB is insignificant compared with the displacement w_0 of AB, i.e. $\varepsilon_z = 0$. On this basis, compatibility requires that the direct strain in the x-direction at P′ must be:

$$\varepsilon_x = \frac{\partial u}{\partial x} = \frac{\partial u_0}{\partial x} - z\frac{\partial^2 w_0}{\partial x^2} \equiv \varepsilon_x^0 + z\kappa_x \tag{7.67}$$

where ε_x^0 is the midplane strain and κ_x is the plate curvature at the midplane. By a similar argument:

$$\varepsilon_y = \varepsilon_y^0 + z\kappa_y \tag{7.68}$$

Compatibility requires that the shear strain γ_{xy} is given by

$$\gamma_{xy} = \frac{\partial u}{\partial y} + \frac{\partial v}{\partial x} = \frac{\partial u_0}{\partial y} + \frac{\partial v_0}{\partial x} - 2z\frac{\partial^2 w_0}{\partial x \partial y} \equiv \gamma_{xy}^0 + z\kappa_{xy} \tag{7.69}$$

where κ_{xy} has the characteristics of a twisting curvature.

Combining these expressions, the strain distribution vector $[\varepsilon]$ across the deformed and displaced element A′B′ has the linear form

$$[\varepsilon] = [\varepsilon^0] + z[\kappa] \tag{7.70}$$

(b) Stress and moment resultants

In a deformed plate, the variation in the in-plane stresses through the thickness can be found by combining equations (7.62) and (7.70):

$$[\sigma] = [\bar{Q}][\varepsilon^0] + z[\bar{Q}][\kappa] \tag{7.71}$$

where $\bar{Q}$ describes the stiffness of the lamina in a convenient reference direction with respect to the principal direction of the lamina. It is convenient to replace the stresses by a statically equivalent set of force and moment resultants which act on the midplane of the lamina, because this set no longer contains lamina thickness and z-coordinates explicitly. It is conventional to describe these resultants on the basis of unit width of laminate.

For a lamina of thickness h under a stress σ_x, the force resultant per unit width, N_x, is defined as

$$N_x = \int_{-h/2}^{+h/2} \sigma_x \mathrm{d}z \tag{7.72}$$

The sign convention for N is the same as that for stress. The three force resultants per unit width (N_x, N_y and N_{xy}) acting on a lamina in plane stress may be written as

$$[N] = \int_{-h/2}^{+h/2} [\sigma] \mathrm{d}z \tag{7.73}$$

Similarly, the moment resultants per unit width $[M]$ in a lamina are defined by

$$[M] = \int_{-h/2}^{+h/2} [\sigma] z \mathrm{d}z \tag{7.74}$$

The directions of positive moment resultants are indicated in Fig. 7.22.

(c) Lamina stiffness and compliance

For a single ply of unidirectional material we find it convenient to define the ply in-plane stiffnesses $[A]$ as

$$[A] = [Q]h \quad \text{or} \quad [A] = [\bar{Q}]h \tag{7.75}$$

and the ply bending stiffnesses as

$$[D] = [Q]h^3/12 \quad \text{or} \quad [D] = [\bar{Q}]h^3/12 \tag{7.76}$$

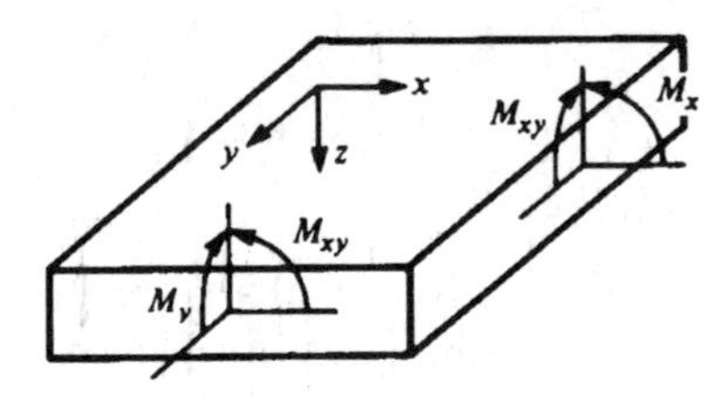

Fig. 7.22 Directions of positive moment resultants

where h is the ply thickness, and the quantity $[\bar{Q}]$ (the transformed reduced stiffness of the material) is used when fibres are aligned at an angle to the reference direction x.

We can now relate $[N]$, $[M]$, $[A]$, and $[D]$ to the midplane strains and curvatures in global coordinates $[\varepsilon^0]$ and $[\kappa]$. Thus for a thin plate under plane stress we have

$$[N] = [A][\varepsilon^0] \quad \text{and} \quad [M] = [D][\kappa] \tag{7.77}$$

and (also under plane stress)

$$[\varepsilon^0] = [A]^{-1}[N] = [a][N] \tag{7.78}$$

and

$$[\kappa] = [D]^{-1}[M] = [d][M] \tag{7.79}$$

where $[a]$ and $[d]$ are the in-plane and bending compliances of the ply.

PROBLEMS

Design in composites often relies on calculations involving the small differences between large quantities; it is therefore good practice to work with four significant figures, and only round off at the end of a chain of calculations.

Although the simple problems at the end of this section and elsewhere in Chapters 7 and 10 can be solved by hand (and we recommend doing this for a few problems as a learning aid in itself), it is more convenient in general to use one of the many readily available software packages, especially if the reader wishes to explore changes in values of variables (such as the angle θ) or to solve more open-ended problems. The Laminate Analysis Programme (LAP) is particularly useful and user-friendly, and is available from The Centre for Composite Materials, Imperial College, London SW7 2BY, UK.

1. (a) Write down the compliance and reduced stiffness matrices for steel and for aluminium.
   (b) Write down the compliance matrix for cast acrylic sheet at 20°C, using $\nu = 0.35$. How would you expect this to change if the load were maintained constant for 1 year?
   (c) Write down the reduced stiffness matrix of a gum natural rubber of hardness 35 IRHD.
   (d) For a given unidirectional glass-fibre-reinforced epoxy lamina, $E_1 = 35\,\text{GN/m}^2, E_2 = 8\,\text{GN/m}^2, G_{12} = 3.5\,\text{GN/m}^2$ and $\nu_{12} = 0.26$. What are the reduced stiffness matrix and the compliance matrix for this lamina? What is the value of ν_{21}? What is the volume fraction of glass fibres in this lamina?
2. Show that the in-plane shear modulus G_{xy} has a maximum value when shear stress is applied at 45°C to the principal directions of a unidirectional lamina.

3. The flat, uniform base of a seed tray injection moulded from 20 wt% short-glass-fibre-reinforced polypropylene is found to be orthotropic when measurements are made of the apparent tensile moduli of test bars cut at different angles to the radial flow direction. The moduli in the plane of the base along, across and at 45° to the radial flow direction are 4.4, 2.75 and 3.4 GN/m^2 at 20°C. What values of modulus would you expect to obtain on specimens cut and uniaxially stressed at 30° and 60° to the radial flow direction?
4. Derive an equation relating the transformed reduced stiffness matrix to the reduced stiffness matrix, and hence or otherwise confirm any one of the six expressions given in equation (7.63).
5. An orthotropic elastic lamina consists of unidirectional epoxy-resin-bonded straight carbon fibres and has the properties defined in problem 5 of section 7.2. The lamina is subjected to plane strains in directions x and y which are at 45° to the principal directions. If the applied strains are $\varepsilon_x = 100 \times 10^{-6}$, $\varepsilon_y = 150 \times 10^{-6}$ and $\gamma_{xy} = 200 \times 10^{-6}$, calculate the resulting shear stress τ_{12} in the lamina. What is the safety factor by which the lamina avoids failure?
6. A panel 300 mm square and 3 mm thick carries uniformly distributed in-plane tensile loads of 9 kN normal to all edges. Calculate the force resultants in the panel.
7. Strain gauges have been applied to a horizontal, thin, flat, square sheet of unidirectional material. Under certain conditions the strains on the top and bottom surfaces of the sheet parallel to the sides have been measured as follows:

Surface	ε_x%	ε_y%	γ_{xy}%
Top	0.3046	−0.2748	−0.3266
Bottom	0.3121	−0.2335	−0.1239

Calculate the curvatures $[\kappa]$ from this information. Sketch the shape of the deformed sheet. What minimum amount of information would you need in order to determine the ratio of applied in-plane stresses, σ_1/σ_2?

8. A long, weightless, thin-walled pipe, sealed at each end, is to be wound with polymer-impregnated fibres at +30° to the hoop direction. The unidirectional composite may be assumed linear elastic with the following values of the extensional compliance matrix:

$$[a] = \begin{bmatrix} 5.66 & -3.92 & -7.47 \\ -3.92 & 9.09 & 1.53 \\ -7.47 & 1.53 & 12.2 \end{bmatrix} \text{(mm/kN)}$$

The pipe is mounted by clamping vertically the upper end; the lower end is free to move in response to any applied load.

Explain, with annotated sketches and reference to components of the extensional stiffness matrix, the patterns of deformation you would expect to see in the pipe wall well away from the ends, and comment briefly on the displacements at the lower end, for each of the following loading conditions:
(a) axial tension applied at the free end;
(b) internal pressure;
(c) opposing torques at each end.

9. A square plate is made from a unidirectional material in which the fibres are laid in the plane at 45° to the reference direction x parallel to two edges. The bending compliance coefficients are:

$$[d] = \begin{bmatrix} 7144 & -2195 & -3421 \\ -2195 & 7144 & -3421 \\ -3421 & -3421 & 12210 \end{bmatrix} (\text{/MN mm})$$

Calculate (a) the midplane curvatures under an applied moment resultant $M_x = 1\,\text{N}$ and (b) the internal moment resultants and deformations when a curvature κ_y is applied to the plate.

10. A 1 mm thick sheet of prepreg cfrp is 1 m square with fibres arranged at 20° to two parallel edges of the sheet. $\alpha_1 = 0.02 \times 10^{-6}/\text{K}$, $\alpha_2 = 22.5 \times 10^{-6}/\text{K}$. Calculate the thermal expansion coefficients in global coordinates and hence the deformations in the sheet when it is heated uniformly by +40 K.

7.4 STIFFNESS OF LAMINATES

It is evident that the properties of an orthotropic lamina can depend dramatically (even by an order of magnitude or more) on the directions in which loads or deformations are applied. Unfortunately, the combination of the properties of an individual lamina and the state of in-plane stresses or strains seldom give the stiffnesses, compliances or strengths required in the final product. To achieve this in a structurally effective way, it is necessary to bond together a stack of a suitable number of suitable laminae, with the principal axes of each lamina aligned at such an angle to a reference direction in the laminate that the laminate as a whole has the required characteristics. The choice of types and numbers of laminae, and the most suitable directions, relies on a small number of straightforward concepts.

We can suggest the following outline, before going into detail.

1. For the chosen laminate, identify the stacking sequence, fibre directions, ply thicknesses and ply properties. A proper way to indicate the structure is to identify the kind of unidirectional lamina (ply) to be used (e.g. glass fibre-reinforced epoxy, thickness 0.5 mm), usually identical for all laminae, and the stacking sequence as orientation angle and number of plies, e.g.

$(45°/-45°_2/45°)$, even $(45°/-45°_2/45°)_T$, where T indicates the total number of plies. In this book we emphasize the use of symmetric laminates, for the same laminate indicated as $(45°/-45°)\hat{s}$, the s denoting 'symmetric'.
2. Calculate for the chosen laminate the in-plane and bending stiffnesses, and invert them to obtain the compliances. Use these quantities under plane stress conditions to relate force and moment resultants to midplane strains and curvatures for the laminate as a whole in global coordinates.
3. Calculate the global stress profiles through the thickness of the laminate, based on the strains in the ply and the reduced stiffnesses of the ply $[Q]$ or $[\bar{Q}]$.
4. Calculate the stress profiles in the principal directions and hence prepare for the use of the Tsai–Hill failure criterion (section 7.5).

After reading this section you will be able to explore the behaviour of thin-walled laminated plates under in-plane loads and under uniform bending moments applied at the edges. The in-plane loading of thin laminates lends itself especially to the simple design of thin-walled pipes. The bending examples prepare the way for the analysis of beam elements which will be discussed further in section 10.3.5. It is essential to have a good working knowledge of the likely stresses inside the laminate, in order to prepare for the estimates of laminate strength in section 7.5.

We shall see that the interaction between all the various parameters in a thin composite panel even under plane stress is predictable but extremely involved. It is seldom easy and much more seldom is it reliable to 'guestimate' the internal stresses. It is thoroughly recommended that you check the expected performance with detailed calculations (using relevant software) and check the results for 'reasonableness'.

7.4.1 Compatibility and equilibrium in a laminate

A laminate consists of a bonded stack of laminae. The laminate has its own midplane on which midplane strains and curvatures can be described.

It is necessary to describe in a consistent way the positions and thicknesses of each lamina throughout the laminate. The laminae are numbered starting from the top of the stack, as shown in Fig. 7.23. The total number of laminae is F. The lower surface of the fth lamina is assigned the coordinates h_f, so that the thickness of the fth lamina $h(f)$ is:

$$h(f) = h_f - h_{f-1} \tag{7.80}$$

In this coordinate system, numerical values of h_f above the midplane are negative. The total thickness of the laminate h_L is

$$h_L = \sum_{f=0}^{F} h(f) = h_0 + h_F \tag{7.81}$$

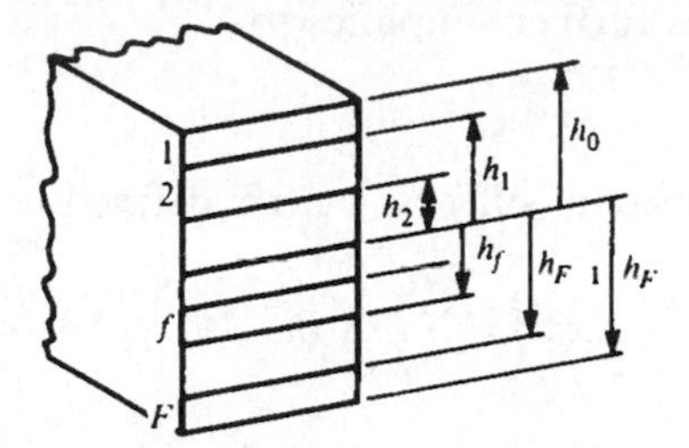

Fig. 7.23 Thickness coordinates in a laminate

For the laminate, the force resultant in the fth lamina is $[N]_f$ and hence the total force resultant in the laminate, N_L, is given by

$$[N]_L = \sum_{f=1}^{F} [N]_f \tag{7.82}$$

The moment resultant about the laminate midplane $[M]_L$ is given by the sum of the moment resultants $[M]_f$ in each lamina about the laminate midplane:

$$[M]_L = \sum_{f=1}^{F} [M]_f \tag{7.83}$$

7.4.2 Laminate stiffness

(a) General stiffness relationship

The final stage in deriving an expression for the stiffness of a laminate is to relate the force and moment resultants to the laminate midplane strains and curvatures. This can be done by combining equations (7.71), (7.73) and (7.82), and understanding that $[\varepsilon^0]$ and $[\kappa]$ are now replaced by the laminate quantities $[\varepsilon^0]_L$ and $[\kappa]_L$:

$$\begin{aligned}[N]_L &= \sum_{f=1}^{F} \int_{h_{f-1}}^{h_f} [\sigma]_f \mathrm{d}z \\ &= \sum_{f=1}^{F} \left(\int_{h_{f-1}}^{h_f} [\bar{Q}]_f [\varepsilon^0] \mathrm{d}z + \int_{h_{f-1}}^{h_f} [\bar{Q}]_f [\kappa] z \mathrm{d}z \right)\end{aligned} \tag{7.84}$$

For a thin laminate, $[\varepsilon^0]_L$ and $[\kappa]_L$ relate to the midplane only and do not depend on z, and within a given lamina $[\bar{Q}]_f$ is constant and independent of z, although $[\bar{Q}]_f$ varies from lamina to lamina. It is now possible to simplify equation (7.84) to

$$[N]_L = \sum_{f=1}^{F} \left([\bar{Q}]_f [\varepsilon^0]_L \int_{h_{f-1}}^{h_f} \mathrm{d}z + [\bar{Q}]_f [\kappa]_L \int_{h_{f-1}}^{h_f} z\mathrm{d}z \right) \tag{7.85}$$

After integration, this further simplifies to

$$[N]_{\mathrm{L}} = [A][\varepsilon^0]_{\mathrm{L}} + [B][\kappa]_{\mathrm{L}} \tag{7.86}$$

where $[A]$ is the extensional stiffness matrix defined by

$$[A] = \sum_{f=1}^{F} [\bar{Q}]_f (h_f - h_{f-1}) \tag{7.87}$$

and $[B]$ is the coupling stiffness matrix defined by

$$[B] = \frac{1}{2} \sum_{f=1}^{F} [\bar{Q}]_f (h_f^2 - h_{f-1}^2) \tag{7.88}$$

A similar argument can be used to express the moment resultants in terms of midplane strains and curvatures:

$$[M]_{\mathrm{L}} = \sum_{f=1}^{F} \int_{h_{f-1}}^{h_f} [\sigma]_f z \mathrm{d}z = [B][\varepsilon^0]_{\mathrm{L}} + [D][\kappa]_{\mathrm{L}} \tag{7.89}$$

where $[D]$ is the bending stiffness matrix defined as

$$[D] = \frac{1}{3} \sum_{f=1}^{F} [\bar{Q}]_f (h_f^3 - h_{f-1}^3) \tag{7.90}$$

The complete set of relationships between the force and moment resultants and the midplane strains and curvatures may now be written in the concise but abstract form:

$$\begin{bmatrix} N \\ M \end{bmatrix}_{\mathrm{L}} = \begin{bmatrix} A & B \\ B & D \end{bmatrix} \begin{bmatrix} \varepsilon^0 \\ \kappa \end{bmatrix}_{\mathrm{L}} \tag{7.91}$$

or in the expanded form:

$$\begin{bmatrix} N_{\mathrm{L}x} \\ N_{\mathrm{L}y} \\ N_{\mathrm{L}xy} \\ M_{\mathrm{L}x} \\ M_{\mathrm{L}y} \\ M_{\mathrm{L}xy} \end{bmatrix} = \begin{bmatrix} A_{11} & A_{12} & A_{16} & B_{11} & B_{12} & B_{16} \\ A_{12} & A_{22} & A_{26} & B_{12} & B_{22} & B_{26} \\ A_{16} & A_{26} & A_{66} & B_{16} & B_{26} & B_{66} \\ B_{11} & B_{12} & B_{16} & D_{11} & D_{12} & D_{16} \\ B_{12} & B_{22} & B_{26} & D_{12} & D_{22} & D_{26} \\ B_{16} & B_{26} & B_{66} & D_{16} & D_{26} & D_{66} \end{bmatrix} \begin{bmatrix} \varepsilon^0_{\mathrm{L}x} \\ \varepsilon^0_{\mathrm{L}y} \\ \gamma^0_{\mathrm{L}xy} \\ \kappa_{\mathrm{L}x} \\ \kappa_{\mathrm{L}y} \\ \kappa_{\mathrm{L}xy} \end{bmatrix} \tag{7.92}$$

Equations (7.91) or (7.92) state Hooke's law in two dimensions for a laminate (compare with equation (7.62) for a lamina). In the general case for an anisotropic laminate, all components of the force and moment resultants are related to all components of the midplane strains and curvatures. For example, applying just a direct strain $\varepsilon^0_{\mathrm{L}x}$ will induce bending moments as well as direct and shear forces. In addition, by inversion of equation (7.91), applying a direct stress (i.e. $N_{\mathrm{L}x}$) will introduce bending, twisting and distortion as well as the usual direct strains, as illustrated in Fig. 7.12. Equations (7.91) and (7.92) provide a quantitative basis for many of the strange features described in sections 7.1.2(b) and 7.1.2(c) (but do not describe the effects of change in temperature).

(*b*) *Special cases*

The stiffness of a generally anisotropic laminate may seem awesomely complicated and certainly lies outside the scope of this book. Fortunately, many practical applications allow the designer to choose some simplified forms of the stiffness matrices. Only a few of these simplifications are discussed here. Further details are given in Powell (1994), Jones (1975) or in Agarwal and Broutman (1980). The rationale behind the simplifications stems from the individual features making up the stiffness matrices. The variables are the nature of the laminate, the orientation of them to a reference direction in a laminate, and the stacking sequences.

The most obvious simplification is to devise a symmetrical laminate, and we shall concentrate on this practical approach in this chapter. By definition (equation (7.88)), the B matrix is identically zero, so there is no coupling between force resultants and curvatures, and no coupling between moment resultants and midplane strains. If, in addition, the laminae are all orthotropic with the principal direction at 0° or 90° to the reference direction in the laminate, then A_{16}, A_{26}, D_{16} and D_{26} are zero too because $Q_{16} = Q_{26} = 0$. This represents the symmetric cross-ply laminate which might be made from balanced woven cloth, i.e. cloth with equal proportions of warp and weft. This additional simplification does not occur if the cross-plies are aligned at some angle $0° < \theta < 90°$ to the stressed reference direction, x.

A different kind of simplification results from requiring the individual laminae to be isotropic. The terms A_{16}, A_{26}, D_{16} and D_{26} disappear. If such a laminate is symmetric, then equation (7.91) reduces to conventional theory for an elastic isotropic plate. The familiar example of this is the laminate made up from laminae based on chopped strand mat (i.e. randomly oriented fibres in the plane). However, checks must be made that the interfacial shear strength is adequate. It should be noted that the bimetallic strip is not a comparable example.

(*c*) *Calculation of laminate stiffness matrices*

The individual elements of the A, B and D matrices may be calculated using equations (7.87), (7.88) and (7.90). An awareness of the simplifications mentioned above saves some unnecessary effort.

Consider an angle-ply laminate of thickness h consisting of three identical orthotropic laminae stacked at angles in the sequence $+\alpha/-\alpha/+\alpha$ to the reference direction. By inspection, this is a regular symmetrical angle-ply

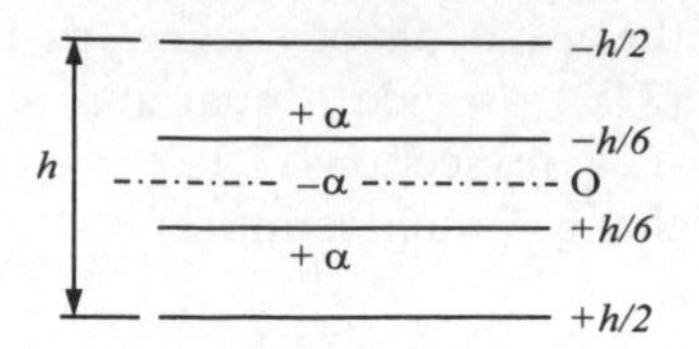

Fig. 7.24 A regular symmetrical angle-ply laminate made from three identical orthotropic laminae

laminate. The coordinates of each surface of the individual laminae are shown in Fig. 7.24.

Substituting terms in equation (7.87) gives

$$A_{11} = \bar{Q}_{11}(+\alpha)[-h/6 - (-h/2)] + \bar{Q}_{11}(-\alpha)[+h/6 - (-h/6)] + \bar{Q}_{11}(+\alpha)[h/2 - h/6]$$

Noting (from equation (7.63)) that $\bar{Q}_{11}(+\alpha) = \bar{Q}_{11}(-\alpha), A_{11} = h\bar{Q}_{11}(+\alpha)$. Similar expressions result for A_{12}, A_{22} and A_{66}. But as $\bar{Q}_{16}(+\alpha) = -\bar{Q}_{16}(-\alpha)$, so $A_{16} = (h/3)\bar{Q}_{16}(+\alpha)$. By symmetry, all B_{ij} are zero.

Hence, the force resultants resulting from the application of midplane strains are

$$\begin{bmatrix} N_{Lx} \\ N_{Ly} \\ N_{Lxy} \end{bmatrix} = h_L \begin{bmatrix} \bar{Q}_{11} & \bar{Q}_{12} & \bar{Q}_{16}/3 \\ \bar{Q}_{12} & \bar{Q}_{22} & \bar{Q}_{26}/3 \\ \bar{Q}_{16}/3 & \bar{Q}_{26}/3 & \bar{Q}_{66} \end{bmatrix} \begin{bmatrix} \varepsilon^0_{Lx} \\ \varepsilon^0_{Ly} \\ \gamma^0_{Lxy} \end{bmatrix}$$

The individual values of $\bar{Q}_{ij}$ are related to the elastic constants E_1, E_2, G_{12} and ν_{12} for each lamina using equations (7.43) and (7.63).

The D matrix is found from equation (7.90):

$$D_{11} = \frac{1}{3}h^3\bar{Q}_{11}(+\alpha)\left\{\left[\left(-\frac{1}{6}\right)^3 - \left(-\frac{1}{2}\right)^3\right] + \left[\left(\frac{1}{6}\right)^3 - \left(-\frac{1}{6}\right)^3\right] + \left[\left(\frac{1}{2}\right)^3 - \left(\frac{1}{6}\right)^3\right]\right\}$$
$$= (h^3/12)\bar{Q}_{11}(+\alpha)$$

and

$$D_{16} = (25h^3/12 \times 27)\bar{Q}_{16}(+\alpha)$$

so that the moment resultants resulting from the application of midplane curvatures are

$$\begin{bmatrix} M_{Lx} \\ M_{Ly} \\ M_{Lxy} \end{bmatrix} = \frac{h_L^3}{12} \begin{bmatrix} \bar{Q}_{11} & \bar{Q}_{12} & (25/27)\bar{Q}_{16} \\ \bar{Q}_{12} & \bar{Q}_{22} & (25/27)\bar{Q}_{26} \\ (25/27)\bar{Q}_{16} & (25/27)\bar{Q}_{26} & \bar{Q}_{66} \end{bmatrix} \begin{bmatrix} \kappa_{Lx} \\ \kappa_{Ly} \\ \kappa_{Lxy} \end{bmatrix}$$

7.4.3 Laminate compliance

It often happens that forces and moments are applied to a laminate and it is necessary to calculate the resulting midplane strains and curvatures. The compliance matrix a, b, h, d can be obtained by formal inversion of equation (7.91):

$$\begin{bmatrix} \varepsilon^0 \\ \kappa \end{bmatrix}_L = \begin{bmatrix} a & b \\ h & d \end{bmatrix} \begin{bmatrix} N \\ M \end{bmatrix}_L \tag{7.93}$$

In general, the compliance matrix contains four separate terms, although normally $h = b$ because B is usually a symmetrical matrix.

For a symmetric laminate, $B = 0$ and hence $b = h = 0$, so that the complete midplane strain response to a set of in-plane force resultants is described by

$$[\varepsilon^0] = [a][N]_L = [A]^{-1}[N]_L \tag{7.94}$$

If the symmetric laminate is subjected to a single stress in only one of the reference directions, then it is possible to calculate the apparent modulus of the laminate in that direction (by analogy with the procedure described for a lamina in section 7.3.3). For example, applying $\sigma_x = N_{Lx}/h_L$ alone, the apparent longitudinal modulus E_{Lx} is found from equation (7.94) to be

$$\varepsilon^0_{Lx} = N_{Lx}a_{11} = (N_{Lx}/h_L)/E_{Lx} \tag{7.95}$$

i.e.

$$E_{Lx} = (A_{11}A_{22} - A_{12}^2)/(h_L A_{22}) \tag{7.96}$$

Comparable expressions can be derived for the transverse modulus, the shear modulus and the major Poisson's ratio of the symmetric laminate.

7.4.4 Stresses in laminates

Application of external loads or moments will induce laminate midplane strains and curvatures, usually expressed in global coordinates. These strains and curvatures vary linearly through the thickness of the laminate. Because different plies have different properties or orientations, the stresses within the laminate will be discontinuous at interfaces between different types of ply. We have already seen how to calculate the stresses in a given ply in terms of ply properties $[\bar{Q}]_f$ and thickness coordinate z using equation (7.71). We shall see in section 7.5 that it is also necessary to calculate the stresses in each ply in the principal directions: this can be done quite simply using the stress transformation matrix $[T]$.

In an isotropic material, applying unidirectional tension induces a tensile strain and a lateral contraction strain, but no shear nor any indication of lateral stress. In a multidirectional laminate, applying unidirectional stress will induce lateral stresses through the thickness: the net lateral force within the laminate is zero in order to conserve equilibrium, of course. These lateral stresses are important for strength calculations, as we shall see in section 7.5. Can you confirm the development of these lateral stresses using classical laminate analysis on a $(0°/90°)_s$ cross-ply laminate? Can you explain why they arise, perhaps by consideration of the behaviour of a plate stressed in the x-direction in which plies are not bonded together?

7.4.5 Choosing a laminate

Given a laminate, classical lamination theory (CLT) enables you to predict performance.

It is however much less straightforward to specify a combination of loads and deformations, and then choose 'the best' laminate. We have already seen that unless you buy a ready-made laminate or a standard product, you cannot normally specify design properties such as stiffness and strength (because you do not know about the fibres and their alignment). It is possible to suggest some bases for limiting the design choice problem, but the optimization still has to be carried out.

Some constraints are relatively straightforward:

- Choose a resin which meets needs such as corrosion resistance, flame retardancy and thermal conductivity.
- Identify likely manufacturing methods for the product, which will provide information on possible lay-up angles or fibre forms and directions, and likely volume fractions. For example, contact moulding may use random mat or woven 0°/90° cloths. Compression moulding may use more or less random in-plane 'short fibre' compositions. Basic pultrusion makes a unidirectional cross-section. Filament winding of long slender sections imposes winding angles between about 20° and 70° relative to the axis of the structure.
- Work with a fibre–polymer system for which properties are known, or estimate them by using micromechanics.

It is then possible to suggest some simplifying principles for simple designs, including the following:

- use symmetric laminates;
- use plies of the same thickness and properties throughout;
- use balanced angle-ply laminates in the form $(\theta/-\theta)_s$ or $(\theta/0/-\theta)_s$

You are still left with a vast range of choices to meet the criteria of your design problem.

PROBLEMS

1. All the following laminates of total thickness h consist of stacks of bonded laminae, each lamina having the same Q_{ij}. Describe each laminate and calculate values of the stiffness matrices A, B and D. Take the reference direction in each laminate as the 0° direction. (Sketch an exploded view of each laminate if you find it helpful.)

(a) $h/4$ at $+\alpha, h/2$ at $-\alpha, h/4$ at $+\alpha$.
(b) $h/3$ at $0°, h/3$ at $90°, h/3$ at $0°$.
(c) $h/2$ at $+\alpha, h/2$ at $-\alpha$.
(d) $h/2$ at $0°, h/2$ at $90°$.

2. A laminate consists of an even number of thin identical laminae laid up alternately at $+\theta$ and $-\theta$ to a reference direction. Show that if n is large, this laminate can be sensibly regarded as symmetric.
3. A laminate of thickness h is to be assembled from orthotropic laminae, all having the same properties $E_1 = 207\,\mathrm{GN/m^2}, E_2 = 20.7\,\mathrm{GN/m^2}$, $\nu_{12} = 0.38$ and $G_{12} = 7.6\,\mathrm{GN/m^2}$. The stacking sequence relative to the reference x-direction and thickness of each lamina is $h/6$ at $+30°, h/6$ at $-30°, h/3$ at $0°, h/6$ at $-30°, h/6$ at $+30°$. Calculate the elastic constants for the laminate under in-plane loading in the orthogonal x- and y-directions. Calculate the strain response of this laminate when subjected to a biaxial stress of $60\,\mathrm{MN/m^2}$ and a shear stress of $35\,\mathrm{MN/m^2}$, all aligned in the x- and y-directions.
4. The propellant for a rocket motor is a brittle solid having the form of a cylinder with a small-diameter hole along its axis. An essential feature of the case which fits snugly round the propellant is that its diameter should not increase when the propellant is ignited and develops the required thrust. The case is to be wound from four layers of a carbon fibre–epoxy lamina having the reduced stiffnesses $Q_{11} = 207.6\,\mathrm{GN/m^2}, Q_{12} = 2.28\,\mathrm{GN/m^2}, Q_{22} = 7.6\,\mathrm{GN/m^2}$ and $Q_{66} = 4.8\,\mathrm{GN/m^2}$. If the laminate is regarded as symmetrical with two layers at $+\theta$ and two layers at $-\theta$ to the axial direction, can a suitable casing be wound to meet this design criterion? What constraints are there on materials properties for a problem like this?
5. By modelling a chopped strand mat laminate as symmetrical with a very large number of identical infinitesimally thin microlaminae oriented uniformly at all possible directions to a reference direction in the plane, show that the in-plane tensile modulus of this randomly aligned laminate E_r is approximately given by $E_r \simeq 3E_1/8 + 5E_2/8$, and that the in-plane shear modulus G_r is given by $G_r \simeq E_r/3$. E_1 and E_2 are the moduli of the microlamina in its principal directions.
6. A symmetrical, regular, cross-ply carbon–epoxy laminate $(0°/90°)_s$ is based on 0.125 mm plies of HM-CARB. The reduced stiffnesses of one ply and the compliances of the laminate are as follows:

$$[Q_{11}(0°), Q_{12}(0°), Q_{22}(0°), Q_{66}(0°)] = [180.5, 2.006, 8.022, 5]\mathrm{GPa}$$

$$[a]\ (\mathrm{mm/MN}) \qquad [d]\ (/\mathrm{MNmm})$$

$$\begin{bmatrix} 21.23 & -0.452 & 0 \\ -0.452 & 21.23 & 0 \\ 0 & 0 & 400 \end{bmatrix} \quad \begin{bmatrix} 604.5 & -41 & 0 \\ -41 & 3248 & 0 \\ 0 & 0 & 19200 \end{bmatrix}$$

(a) What is the Poisson's ratio for this laminate?
(b) Calculate the stress profiles $\sigma_x(z)$ and $\sigma_y(z)$ under a load $N_x = 100\,\text{N/mm}$.
(c) Calculate the curvatures κ_y and κ_{xy} resulting from an applied moment $M_{xy} = 0.1\,\text{N}$.

7. Rayon cord-reinforced rubber is available in unidirectional plies 1 mm thick. The plies are made into a laminate designated $(30°/-30°)_s$. The transformed reduced stiffnesses of one ply and the compliances of the laminate are as follows:

$$[Q^*(30°)] = \begin{bmatrix} 0.9867 & 0.3326 & 0.5622 \\ 0.3326 & 0.1217 & 0.187 \\ 0.5622 & 0.187 & 0.3274 \end{bmatrix} \text{(GPa)}$$

$$[a]\ (\text{mm/MN}) \qquad\qquad [d]\ (\text{/MNmm})$$

$$\begin{bmatrix} 3225 & -8817 & 0 \\ -8817 & 26160 & 0 \\ 0 & 0 & 763.6 \end{bmatrix} \qquad \begin{bmatrix} 2729 & -6736 & -629.4 \\ -6736 & 19670 & 249.1 \\ -629.4 & 249.1 & 1277 \end{bmatrix}$$

(a) Calculate the major Poisson's ratio for this laminate in terms of the relevant compliance coefficients.
(b) For an applied load $N_x = 1\,\text{N/mm}$ sketch and explain the shape of the shear stress profile in global coordinates.
(c) For an applied bending moment $M_x = 0.1\,\text{N}$, sketch the profile of the bending stress σ_x and explain its shape.

7.5 STRENGTH OF WIDE LAMINATES

Avoiding failure in composite structures is a major preoccupation of the designer. The main points discussed in this section are as follows:

1. A laminate is considered 'wide' if the ratio of breadth to laminate thickness, b/h, is greater than 10. The strength of a wide laminate at any position is assessed by applying the Tsai–Hill criterion (or another valid criterion) to the stresses in the ply principal directions at that point.
2. In bending, the largest strains are at the surfaces of a laminate, but failure may occur not at the surface but inside the laminate if the effective stiffness there is larger than at the surface, or if the internal thermal stresses there (in the principal directions) are substantial.
3. The edges of laminates are more vulnerable than the rest of a laminate: through-thickness stresses may be large enough to start delamination. Edge effects occur over a width about the same as the thickness of a laminate; narrow laminates are not as strong as wide ones. Be careful

about cut edges, and remember that the edge of a hole for bolted joints is not inherently as strong as the rest of the laminate, unless the bolt applies some compression, at least a finger-tight joint.

4. **It is most important** to take into account the internal thermal stresses in any laminate induced by hot processing and cooling down to the temperature used in service. Calculations suggest that these internal thermal stresses can often reduce the strength of the ply by between one-third and one-half. Hot-cure systems are almost always used in mass production to minimize production times.

Thermal stresses can also be induced by change of temperature during use: think of the temperature range over which a motor vehicle has to be usable. We shall discuss the effect of temperature change and the resulting internal stresses in Chapter 10. In the rest of this section we shall look only at laminates made at 'room' temperature which will be mechanically loaded at 'room' temperature.

7.5.1 First-ply failure

We have already shown in section 7.4.3 how to calculate the strains and hence the stresses in any ply within a laminate under externally applied loads or moments. The simplest prediction of laminate strength is based on calculation of stresses in a particular ply in its principal directions and use of the Tsai–Hill criterion to see which ply fails first. Thus we identify the ply which has the lowest safety factor F load failure and is most at risk. If $F < 1$ this is the ply where 'first-ply failure' will occur.

Under the applied loads, this ply fails first when the loads are all increased by the factor F. At that load, the remaining plies will not have failed if they have a larger load factor. So the structure as a whole may not fail, even though one ply has failed. Increasing the external load will cause second-ply failure when the next highest load factor is reached.

Strictly, after first-ply failure, we should disregard the stiffness of the failed plies, and recalculate the [*ABBD*] and [*abhd*] matrices before increasing the loads. Often, however, it is acceptable in global assessments to use the original *ABBD* matrix in later failure events. Obviously, the more plies that fail, the less accurate this approach becomes. Design beyond second-ply failure may be unwise.

The direction of failure or direction of loading responsible for failure is not indicated by use of the Tsai–Hill criterion: only the proportions of the total applied load in each principal direction are identified. We can form some impression of the likely cause of failure by looking at the stress/strength ratios: if one of the values $\sigma_i/\sigma_{i\max}$ is much larger than the other two, this suggests the most likely form of failure.

7.5.2 Strength in bending

For plates under a uniform bending moment or bending moments per unit width, we can use plate theory to calculate deformations, and hence obtain the stress profiles in the principal directions and then obtain the Tsai–Hill load factors. When applying a bending moment, compatibility requires that for linear elastic plies the strain(s) must be continuous through the thickness of the laminate, and that the maximum values must be at the surfaces. However, the modulus of a ply varies with direction depending on the orientation of the fibres. It is therefore quite possible to have (in a given global direction or in a given principal direction) plies which are stiffer inside the laminate than at the exposed surface. The stresses in the stiffer ply may well be larger than those in the flexible outer ply and, depending on the strength of the different plies, it is quite possible under bending loads to achieve failure inside the laminate and not at the surface. This needs to be checked very carefully: the internal damage may not be visible to the eye.

For structural elements such as thin-walled tubes or I-sections, a slightly more elaborate procedure is needed in order to determine bending strength. From the usual structural idealization, we can calculate easily where the stresses are maximum; for example, at the top and bottom of a horizontal beam under vertical loading, the shear stress is in the neutral plane of the complete cross-section for a symmetrically shaped structure during bending. We can convert these stresses into forces per unit width. We can then take an element of the chosen laminate in global coordinates and, assuming that the stress through the thickness is uniform, we can use the compliance matrix of the laminate to calculate the laminate local strains, and hence via $[\bar{Q}]$ the stresses in each ply, then via $[T]$ transform to stresses in the principal directions, and test the Tsai–Hill criterion. (We shall discuss this further in section 10.3.5.)

7.5.3 Edge effects

Laminate theory deals with wide plates. At some edges of plates, the external in-plane loads may be taken as zero. Inside the plate there are internal stresses. In a region about one plate thickness wide the internal stresses have to decrease from their proper value to zero at the free edge. To achieve this and conserve equilibrium, stresses are developed through the thickness which can affect the strength of the plate. Remember that laminate theory assumes there are no through-thickness stresses. In wide plates these edge effects are small. In tubes the axial ‘edges’ are joined together so the edge effect is not present (except at the ends). Beams are usually narrow, and we should be careful about the strength of beams which have longitudinally cut edges.

PROBLEMS

1. Why does the Tsai–Hill criterion contain a τ_{12} term? Should it not be possible to transform the stress tensor (σ_1, σ_2, τ_{12}) to the major stress tensor (σ_x, σ_y, 0) and to apply a criterion just having permissible values of $\hat{\sigma}_x$ and $\hat{\sigma}_y$?
2. Four identical laminae have the following properties: $Q_{11} = 56\,\mathrm{GN/m^2}$, $Q_{12} = 4.6\,\mathrm{GN/m^2}$, $Q_{22} = 18.7\,\mathrm{GN/m^2}$, $Q_{66} = 8.9\,\mathrm{GN/m^2}$, $\hat{\sigma}_1 = 1\,\mathrm{GN/m^2}$, $\hat{\sigma}_2 = 30\,\mathrm{MN/m^2}$ and $\hat{\tau}_{12} = 45\,\mathrm{MN/m^2}$. These laminae are bonded to form a cross-ply laminate in the sequence 0°, 90°, 90°, 0° to the reference x-direction. What is the safety factor by which failure is avoided when the following in-plane strains are applied:
   (a) $\varepsilon_x^0 = 0.1\%, \varepsilon_y^0 = 0.1\%, \gamma_{xy}^0 = 0.1\%$.
   (b) $\varepsilon_x^0 = 0$, $\varepsilon_y^0 = 0$, $\gamma_{xy}^0 = 0.3\%$.
   (c) $\varepsilon_x^0 = 0.1\%, \varepsilon_y^0 = 0.05\%$, $\gamma_{xy}^0 = 0.1\%$.
3. A long pipe with closed ends is 100 mm in diameter with a total wall thickness of 0.5 mm. The pipe wall is made as a symmetrical laminate from four identical plies of EG (section 7.2.8), two arranged at +45° and two at −45° to the axis of the pipe. The laminate is cured at room temperature and has the following coefficients for in-plane stiffness and in-plane compliance

   $$[A]\ (\mathrm{kN/mm}) \quad \begin{bmatrix} 10.49 & 5.346 & 0 \\ 5.346 & 10.49 & 0 \\ 0 & 0 & 6.42 \end{bmatrix} \qquad [a]\ (\mathrm{mm/MN}) \quad \begin{bmatrix} 128.9 & -65.7 & 0 \\ -65.7 & 128.9 & 0 \\ 0 & 0 & 155.8 \end{bmatrix}$$

   $$[Q^*(+45°)] = \begin{bmatrix} 20.97 & 10.69 & 8.874 \\ 10.69 & 20.97 & 8.874 \\ 8.874 & 8.874 & 12.84 \end{bmatrix} \quad (\mathrm{GPa})$$

   Three tests are to be carried out using (separately) three identical pipes. You are invited to predict the results of each test using the Tsai–Hill criterion, and to suggest the likely cause of failure.
   (a) What value of failure load would you expect when an increasing, uniformly distributed axial load is applied to the pipe?
   (b) What value of internal pressure will cause failure of the pipe?
   (c) The third test involves the combination of a constant uniform axial load of 4 kN together with an increasing internal pressure. At what pressure would you expect failure to occur?
4. A symmetrical $(0°, 90°)_s$ cross-ply laminate is made from a special epoxy–glass fibre material EG cured at room temperature. Each ply of EG is 0.5 mm thick and has the following properties: $Q_{11} = 46.42\,\mathrm{GPa}$, $Q_{12} = 2.993\,\mathrm{GPa}$, $Q_{22} = 10.92\,\mathrm{GPa}$, $Q_{66} = 5.14\,\mathrm{GPa}$, $\hat{\sigma}_{1T} = 1000\,\mathrm{MPa}$, $\hat{\sigma}_{2T} = 100\,\mathrm{MPa}$, $\hat{\sigma}_{1C} = 800\,\mathrm{MPa}$, $\hat{\sigma}_{2C} = 270\,\mathrm{MPa}$, $\hat{\tau}_{12} = 80\,\mathrm{MPa}$.

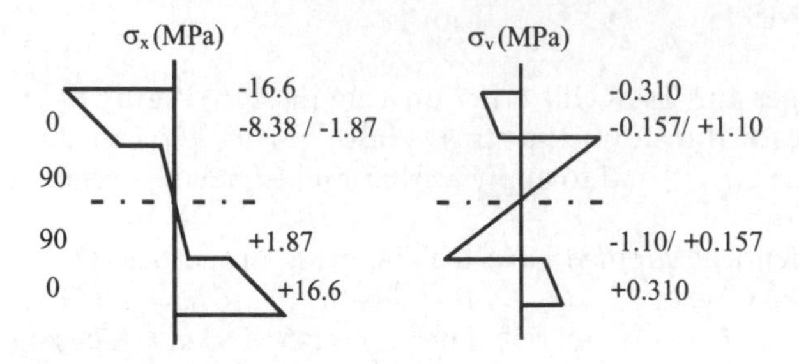

Fig. 7.25 Stress profiles.

The laminate compliance coefficients are

$$[a]\ (\text{mm/MN}) \qquad [d]\ (\text{/MNmm})$$

$$\begin{bmatrix} 17.63 & -1.84 & 0 \\ -1.84 & 17.63 & 0 \\ 0 & 0 & 97.28 \end{bmatrix} \quad \begin{bmatrix} 36.23 & -7.06 & 0 \\ -7.06 & 99.03 & 0 \\ 0 & 0 & 291.8 \end{bmatrix}$$

What are the values of Poisson's ratio for this laminate? Where is first-ply failure predicted under a uniform moment resultant per unit width of $M_x = 10\,\text{N}$? For reference, the stress profiles $\sigma_x(z)$ and $\sigma_y(z)$ are shown in Fig. 7.25.

FURTHER READING

Agarwal, B.D. and Broutman, L.J. (1980) *Analysis and Performance of Fibre Composites*, Wiley, New York.

Daniel, I.M. and Ishai, O. (1994) *Engineering Mechanics of Composite Materials*, Oxford University Press, Oxford.

Datoo, M.H. (1991) *Mechanics of Fibrous Composites*, Elsevier, Amsterdam.

Eckold, G. (1994) *Design and Manufacture of Composite Structures*, Woodhead, Cambridge, UK.

Gibson, R.F. (1994) *Principles of Composite Material Mechanics*, McGraw-Hill, New York.

Hancox, N.L. and Mayer, R.M. (1994) *Design Data for Reinforced Plastics, a Guide for Engineers and Designers*, Chapman & Hall, London.

Jones, R.M. (1975) *Mechanics of Composite Materials*, McGraw-Hill, New York.

Matthews, F.L. (ed.) (1987) *Joining Fibre Reinforced Plastics*, Elsevier, Amsterdam.

Matthews, F.L. and Rawlings, R.D. (1994) *Composite Materials: Engineering and Science*, Chapman & Hall, London.

Middleton, D.H. (ed.) (1990) *Composite Materials in Aircraft Structures*, Longman, Harlow.

Niu, M.C. *Composite Airframe Structures, Practical Design Information and Data*, Conmillt, Hong Kong.

Powell, P.C. (1994) *Engineering with Fibre/Polymer Laminates*, Chapman & Hall, London.

Tsai, S.W. (1988) *Composites Design*, 4th edn., Think Composites, Dayton, Ohio.

Vasiliev, V.V. (1993) *Mechanics of Composite Structures*, Taylor & Francis, Washington, DC.

Vinson, J.R. (1993) *Behaviour of Shells Composed of Isotropic and Composite Materials*, Kluwer, Dordrecht.

8 Fluid Flow and Heat Transfer in Melt Processing

8.1 INTRODUCTION

Chapter 4 gave a simplified description of the main methods for making articles from polymer compounds. Almost all polymers are shaped in the fluid state, and many processes involve heating or cooling before, during or after a shaping operation.

This chapter introduces the principles of fluid flow and heat transfer, together with properties which are relevant to the melt processing of (thermoplastic) polymers. Not aiming to be comprehensive, this does, however, provide examples which are amenable to a simplified quantitative approach, leading to estimates of the sizes of equipment, the heat duties and pressures or forces needed to make articles from polymers, and this indicates some of the common sources of faults in products which melt processing can introduce. These principles also have implications for process control, although these will not be explored here. The next chapter provides examples of the interactions between processing conditions, properties and design.

The basic principles of fluid mechanics and heat transfer will be familiar to the reader. However, some attention is paid to the analysis here, because the values of and the nature of some of the properties of polymers are likely to be unfamiliar, and these lead to simplifying assumptions and effects which are rather different from those in much of traditional engineering analysis. The main examples of these features are the low density (about $1000\,kg/m^3$), the low thermal conductivity (about 0.1 W/m K) and the high shear viscosity (typically 10^2–$10^4\,N\,s/m^2$). The first two give rise to diffusivities of the order of $10^{-7}\,m^2/s$, and result in long heating and cooling times and a preoccupation with non-steady-state conduction. The first and third are responsible for polymer melt flows always being comfortably laminar, with Reynolds numbers usually much less than unity. Most polymer melts are substantially non-Newtonian over the range of rates of deformation encountered in commercial equipment.

To illustrate this point further, it is instructive to consider two processes which have some similarities, at least superficially: pressure die casting of zinc at, say, 450°C, and injection moulding of a thermoplastic at, say, 250°C. Zinc has a density of 6500 kg/m^3, a melt viscosity of the order of 0.01 N s/m^2 at 450°C and a thermal conductivity greater than 100 W/m K. The energy needed to heat up the zinc would be about 1000 MJ/m^3, but only about 300–500 MJ/m^3 for the thermoplastic. In trying to fill a similar 'difficult' cavity with a large ratio of flow length to thickness, the zinc would require very little pressure but a major worry would be freeze-off near the injection point leading to a short moulding. Filling the cavity with thermoplastic would require much larger pressures, typically 100 MN/m^2, in order to fill the cavity within an acceptable time, and cooling times could be long if the part thickness were substantial.

Reverting to more general matters, it is now appropriate to describe the various kinds of flows which are commonly encountered in polymer melt processing. Some shear flows between fixed boundaries are caused by applied pressures, e.g. one-dimensional flows in circular or thin slot channels of constant cross-section in extrusion dies, or in runners to moulds. Other shear flows result from dragging the fluid between moving boundaries, e.g. between screw root and barrel in an extruder or in an internal mixer. Some flows involve pressures but no boundaries, and induce tensile deformations, e.g. the inflation of parisons in extrusion blow moulding and the inflation of the bubble in the tubular film blowing process. Yet other flows produce combinations of shear and tensile deformations, as found in spreading disc flows between parallel plates (e.g. compression moulding of rubber and sheet or dough moulding compounds, and injection moulding) or in flows within tapered channels (e.g. in the nip region of a two-roll mill). There are many other examples.

Heat transfer during processing may occur in a stationary or in a moving polymer by one or more of the following mechanisms. If more than one mechanism occurs, one may dominate one process but not another because of the different physical conditions. Conduction occurs in polymers in moulds, for example in the cooling of injected thermoplastics or in the heating of compressed or injected crosslinkable polymers. Convection is important within melt flows, in cooling of thermoplastics extrudates in air or in a water bath, and in heating of rubber extrudates as they pass through a vulcanization chamber. Radiation can account for a substantial amount of heat loss as thin films cool during the manufacture of tubular blown film. Because polymer melts have high viscosities, viscous shear heating may occur near flow boundaries when large velocity gradients are present, especially in the runners in injection machines, the lands in wire-covering dies, and in the melting zones of a plasticating screw extruder. Crosslinking chemical reactions can be exothermic, thus causing some rise in temperature (and hence an increase in cure rate) during the curing of a rubber or cross-linkable plastics compound. A change of state or phase may occur which can have a significant effect: the melting or freezing of a partially crystallizable thermoplastic delays the achievement of a desired temperature (profile) compared with the heating or cooling of an amorphous polymer compound, and the vaporization of a blowing agent to produce a cellular structure contributes to the cooling process.

The general approach to designing for flows within the elements of polymer processing equipment relies on the standard procedures of continuum mechanics, namely solving simultaneously the equations of continuity, equilibrium and energy (together with the particular properties of the fluid (or solid) such as equations of state and relationships between stress and rate of deformation. Completely general models for polymer flow are dauntingly difficult, even taking advantage of powerful computers and numerical methods. Design for specific flows relies on the reasonableness of making many simplifying assumptions. For example, almost all steady flows are laminar, and hence inertia forces can be ignored compared with viscous forces. Body forces can often be ignored, too. If pressure drops are modest, the fluid can be taken as incompressible, and yet further simplification results if the flow can be assumed isothermal: the equation of energy can now be disregarded, and many physical properties such as density, conductivity and specific heat become constants for a given temperature. The flow of polymer melts can usefully be described by a power law relating shear stress and rate of shear deformation. Flows in many items of equipment can be taken as only two-dimensional, and even some of these can be simplified further by the use of the lubrication approximation to give useful insight into the flow behaviour in the region of interest.

For reasons of space, the discussion of flow phenomena in this book is restricted to some simple problems which have obvious analytical solutions, and to the limitations on the range of applicability of this approach. Many standard specialist books have developed this analytical approach in great detail. Further advances in solving practically important flow problems have also been made by the use of numerical techniques.

The rest of this chapter covers the following points. It revises the analysis of particular one-directional isothermal Newtonian flows caused by pressure or drag, reviews the viscous shear behaviour of melts, and reworks the analyses for power-law fluids. After a discussion of aspects of the equation of energy, particularly heat transfer by conduction and convection, and heat generation by viscous dissipation, there follows an overview of real flows in injection moulds. The chapter finishes with a more qualitative account, and the practical significance, of elasticity and tensile deformations in melt flows.

8.2 UNIDIRECTIONAL ISOTHERMAL NEWTONIAN FLOW

This section revises aspects of the mechanics of pressure and drag flow covered in the first 2 years of an engineering degree course and applies them to simplified models of situations encountered in melt processing. The assumptions made in this account are that the flow is steady, isothermal and laminar, with a fully developed velocity profile in the downstream direction and no slip at the flow boundary, and that the fluid is Newtonian (constant shear viscosity), incompressible and time independent. Flow channels are very much longer than the minimum transverse dimension, so that end effects are ignored.

8.2.1 Simple pressure and drag flows

The procedure should be familiar: consider a representative elemental volume, derive an expression for the momentum flux, and transform this into a differential equation for the fluid velocity. Then integrate this with appropriate boundary conditions to give expressions for velocity distributions v_z, volumetric flow rate Q, pressure drop Δp, shear rates $\dot{\gamma}$, shear stress distributions τ and forces on flow boundaries.

Using the sign convention that quantities increase in the positive directions of the coordinates shown in Fig. 8.1, the law of conservation of momentum reduces to a simple force balance. For elements in cylindrical polar and Cartesian coordinates, this leads to

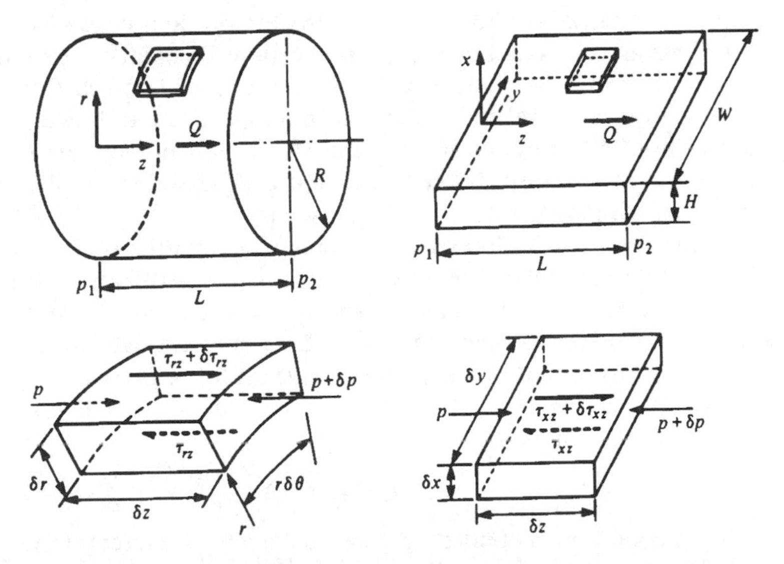

Fig. 8.1 Unidirectional flow channels and representative elemental volumes

$$\frac{1}{r}\frac{\partial}{\partial r}(r\tau_{rz}) = \frac{\partial p_z}{\partial_z}$$
$$\frac{\partial \tau_{xz}}{\partial x} = \frac{\partial p_z}{\partial z} \tag{8.1}$$

(Equation (8.1) could also have been obtained directly from the general equations of change by deletion of unwanted terms.) For a Newtonian fluid of shear viscosity η, the relationship between shear stress and shear rate is

$$\tau_{rz} = \eta\frac{\partial v_z}{\partial r} = \eta\dot{\gamma}_{rz}$$
$$\tau_{xz} = \eta\frac{\partial v_z}{\partial x} = \eta\dot{\gamma}_{xz} \tag{8.2}$$

Combining equations (8.1) and (8.2) leads to

$$\frac{\eta}{r}\frac{\partial}{\partial r}\left(r\frac{\partial v_z}{\partial r}\right) = \frac{\partial p_z}{\partial z}$$
$$\eta\frac{\partial^2 v_z}{\partial x^2} = \frac{\partial p_z}{\partial z} \tag{8.3}$$

These equations can now be readily integrated. The boundary conditions and resulting equations are summarized for pressure flows at constant flow rate and drag flow at constant boundary speeds in Table 8.1. Readers are most strongly encouraged to confirm for themselves the development of argument in the equations given in Table 8.1, in preparation for the new analysis presented later in this chapter.

Pressure flows cause parabolic velocity distributions and hence the shear rate and stress have maximum values at the flow boundary. Once these maximum values have been identified, the relationship between flow rate and pressure drop can be related to the maximum shear rate and shear stress, and the two are related by the shear viscosity. (This may seem excessively long-winded when viscosity is a constant, but is a helpful approach when viscosity varies with shear rate, as will be evident later.)

In the 'drag flow' caused by one plate being fixed and the other moving, the velocity is not parabolic. The maximum shear rate is still at one boundary but, in general, the zero value of shear rate is not at the midplane. Associated with the drag flow there may be a pressure which increases along the flow direction; this is sometimes called a 'back-pressure' because it also acts against the flow. The volume flow rate Q (Table 8.1, (O)) therefore has a drag component q_{d} offset by a back-pressure component q_{p}:

$$Q = q_{\mathrm{d}} + q_{\mathrm{p}} \tag{8.4}$$

Two limiting conditions bound drag flows of practical interest. Pure drag flow occurs when the system operates in open discharge so that no pressure

Table 8.1 Isothermal Newtonian flow in channels of constant cross-section (assumptions are given at the beginning of section 8.2)

(1) Pressure flow in circular cross-section, radius R, length $L \gg R$

Boundary conditions:

$$v_z = 0 \quad \text{at} \quad r = R \quad \partial v_z/\partial r = 0 \quad \text{at} \quad r = 0 \tag{A}$$

Velocity distribution:

$$v_z = \frac{1}{4\eta}\left(-\frac{\partial p}{\partial z}\right)(R^2 - r^2) \tag{B}$$

Volume flow rate:

$$Q = \frac{\pi R^4}{8\eta}\left(-\frac{\partial p}{\partial z}\right) \tag{C}$$

Pressure drop:

$$\Delta P = (p_2 - p_1) = 8\eta QL/\pi R^4 \tag{D}$$

Shear rate at wall:

$$(\dot{\gamma}_{rz})_R = \partial v_z/\partial r = -4Q/\pi R^3 \tag{E}$$

Shear stress at wall:

$$(\tau_{rz})_R = \eta(\dot{\gamma}_{rz})_R = R\Delta P/2L \tag{F}$$

(2) Pressure flow in slot cross-section, thickness H, width $W \gg H$, length $L \gg H$

Boundary conditions:

$$v_z = 0 \text{ at } x = \pm H/2 \tag{G}$$

Velocity distribution:

$$v_z = \frac{1}{2\eta}\left(-\frac{\partial p}{\partial z}\right)\left(\frac{H^2}{4} - x^2\right) \tag{H}$$

Volume flow rate:

$$Q = \frac{WH^3}{12\eta}\left(-\frac{\partial p}{\partial z}\right) \tag{I}$$

Pressure drop:

$$\Delta P = (p_2 - p_1) = -12\eta QL/WH^3 \tag{J}$$

Shear rate at wall:

$$(\dot{\gamma}_{xz})_{H/2} = -6Q/WH^2 \tag{K}$$

Shear stress at wall:

$$(\tau_{xz})_{H/2} = H\Delta P/2L \tag{L}$$

(3) Drag flow in slot cross-section

Boundary conditions:

$$v_z = 0 \text{ at } x = -H/2 \qquad v_z = v_1 \text{ at } x = +H/2 \tag{M}$$

Velocity distribution:

$$v_z = v_1\left(\frac{x}{H} + \frac{1}{2}\right) + \frac{1}{2\eta}\left(-\frac{\partial p}{\partial z}\right)\left(\frac{H^2}{4} - x^2\right) \tag{N}$$

Volume flow rate:

$$Q = \frac{W}{2}\left[v_1 H + \frac{H^3}{6\eta}\left(-\frac{\partial p}{\partial z}\right)\right] \tag{O}$$

is developed, i.e. $q_p = 0$. The other limit is that the system develops maximum pressure when the outlet is closed: if $Q = 0$, then $q_p/q_d = -1$.

8.2.2 Simplified analysis of flow in metering zone of an extruder

In a single-screw extruder, the relationship between output rate Q, pressure P and screw speed N is usually determined by the flow in the metering zone. In the simplest possible model of the metering zone, which gives an interesting if somewhat naive qualitative insight into the basic operating characteristics, drag between the surfaces of the barrel and the root of the screw dictates the maximum output rate.

The screw rotates within a closely fitting barrel of internal diameter D. The leading dimensions of the metering zone of the simple screw shown in Fig. 8.2 are the axial length L, the depth of the channel H, the width of the channel W normal to the flight, the width of the flight e, the radial clearance c between the flight tip and the barrel, and the lead L_s which is the axial length of one complete turn of the screw. The helix angle θ of the screw is defined by $\tan\theta = L_s/\pi D$.

The simple analysis makes all the assumptions listed at the beginning of section 8.2. It is further assumed that the screw dimensions are uniform, that there is no leakage over the flights, and that the depth of flight is small compared with the diameter and the helical length of the metering zone. Whilst in a real extruder the screw rotates within a stationary barrel, it is analytically more convenient to model flow in the equivalent notional system where the barrel rotates around a stationary screw. This embodies the assumptions that the centrifugal forces are negligible and that the gravity forces on the fluid are negligible compared with the viscous shear and pressure forces. The final simplification is to unwrap or develop the flow channel, to give a shallow wide rectangular stationary slot full of a Newtonian fluid and bounded by a plate moving with velocity v_p as shown in Fig. 8.3. Drag flow in such a system is described by the equations in Table 8.1, (M) to (O). What remains to be done is to relate model dimensions to barrel and screw dimensions.

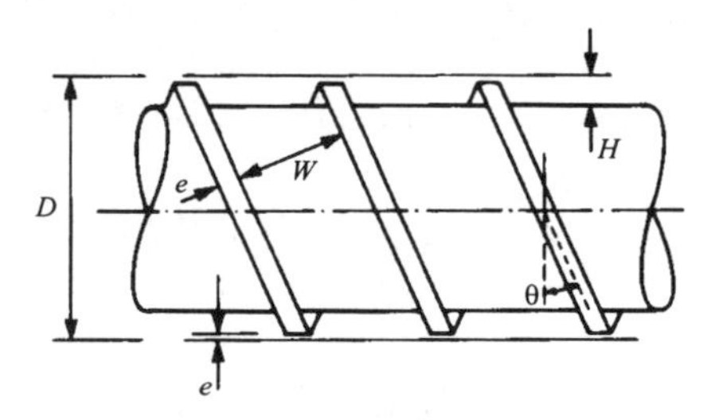

Fig. 8.2 Metering zone in extruder

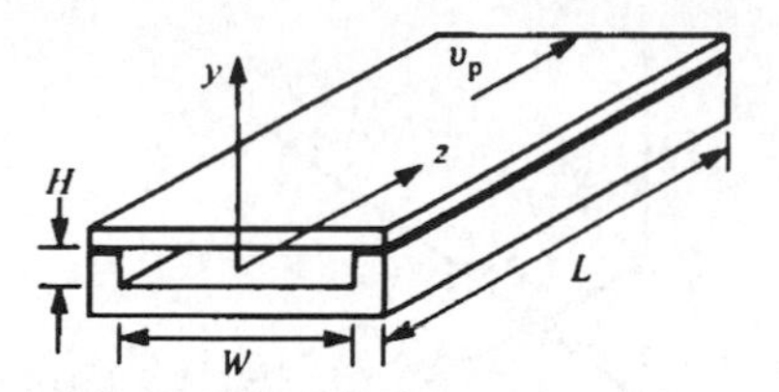

Fig. 8.3 Developed form of screw flight

For a single-start screw, the channel width is $W + e \simeq W$. The distance down the channel, z, is related to the axial length along the screw l by $z = l/\sin\theta$, with a total channel length $L_z = L/\sin\theta$. If the screw rotates at an angular speed N, the linear circumferential velocity of the barrel is $v_l = \pi ND$, and hence the velocity of the plate along the developed channel is $v_p = v_l \cos\theta$. Making the reasonable assumption of a linear pressure gradient, $dp/dz = \Delta P/L_z$, the volume flow rate in Table 8.1, (O) now becomes

$$Q = \frac{1}{2}\pi WHND\cos\theta - WH^3\Delta P\sin\theta/12\eta L \tag{8.5}$$

The first term is the drag-induced flow, and the second term is the reduction to the drag flow caused by the 'back-pressure' ΔP. The ratio of pressure flow to drag flow is

$$\frac{q_p}{q_d} = -\frac{H^2\Delta P\tan\theta}{6\pi\eta NDL} \tag{8.6}$$

For the metering zone of a given screw, W, H, D, L and θ are constant, so the output rate can be expressed in the form

$$Q = C_1N - C_2\Delta P \tag{8.7}$$

known as the screw characteristic (for the metering zone). This shows that the greater the screw speed, the greater the output. In operation, the pressure drop is controlled by such features as the barrel exit dimensions and any filters, and by the die fitted to the end of the barrel to shape the extrudate.

For a simple capillary or slot die, the output rate increases linearly with pressure drop (Table 8.1, (D) or (J)) but only for a Newtonian fluid. Combining the die and screw characteristics defines the operating point for the system, as shown in Fig. 8.4.

With an eye on commercial practice, it should be borne in mind that this simple model of flow in the metering section of the extruder screw needs to be refined in several ways to be of any quantitative use in extrusion of polymer melts. These refinements are mentioned now but not discussed in detail. First, no mention has been made of transverse circulatory flows within the flights – these interact with the main longitudinal flow Q to

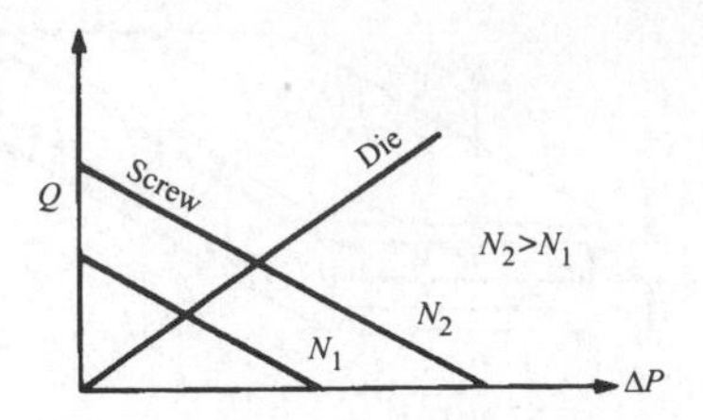

Fig. 8.4 Operating characteristics of single-screw and die system for Newtonian Fluids

promote a mixing action so that the screw delivers a more homogeneous melt. Second, leakage over the flights has been ignored. Third, heat transfer from the barrel walls and heat generation by viscous dissipation have not been taken into account. In addition, the fluid has been assumed to have a constant viscosity – this, taken with leakage over the flights, introduces a major error in calculation of the power needed to deliver a given output using a non-Newtonian fluid.

8.2.3 Lubrication approximation

The flow channels considered so far (for example those having circular or wide thin rectangular cross-sections) have had a constant cross-section over their entire length. However, many flows involve channels having a varying section from inlet to outlet. In general, such flows are (at least) two-dimensional and therefore more difficult to analyse than their one-dimensional counterparts. But if the angle of the tapering channel is small, less than 10°, then calculations of shear rates, velocities and pressure drops can be made by invoking the lubrication approximation. On this basis, the local steady flows between smooth surfaces can be regarded as if they were steady, fully developed flows between plane parallel surfaces. If the relationship between flow boundary separation and length is known, then the pressure drop can be found by integrating the appropriately modified expression for volumetric flow rate.

The remainder of this discussion of isothermal Newtonian flows describes the application of the lubrication approximation to three situations commonly encountered in polymer processing. Pressure flows within linearly tapered slit or circular channels are found in extrusion dies. Drag flows within the wide narrow gap between two large counter-rotating rolls dictate the operation of the two-roll mill and the calender. Spreading disc flows result from central injection into a cavity defined by parallel plates, and these occur widely in injection moulding.

8.2.4 Flow in tapered channels

Consider a conical flow channel or a linearly tapered narrow slot channel of constant width, having the leading dimensions shown in Fig. 8.5. At any

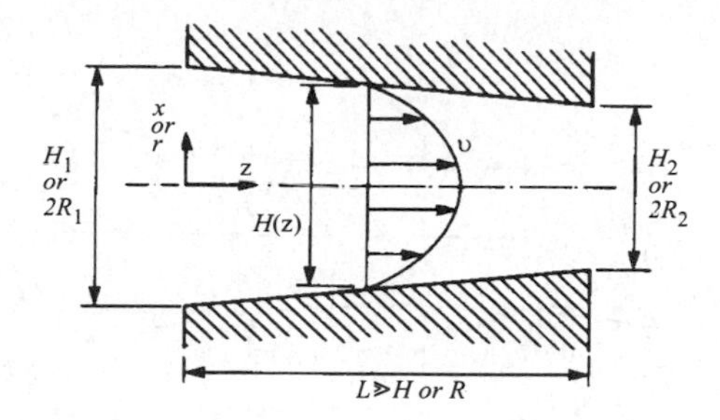

Fig. 8.5 Linearly tapered channel

intermediate distance z, the transverse dimension has the form

$$H = H_1(1 - Kz) \quad \text{with} \quad K = (H_1 - H_2)/H_1\,L \tag{8.8}$$

In a conical channel of inlet radius R_1, outlet radius R_2 and length L, the pressure drop caused by shear flow is obtained by integrating equation (D) from Table 8.1:

$$\begin{aligned} p_2 - p_1 &= -\int_0^L \frac{8\eta Q}{\pi R^4}\,dz \\ &= -\frac{8\eta Q}{\pi R_1^4}\int \frac{dz}{(1-Kz)^4} \\ &= \frac{8\eta Q}{3\pi R_1^3 \tan\theta}\left[\left(\frac{R_1}{R_2}\right)^3 - 1\right] \end{aligned} \tag{8.9}$$

The strain rate at the wall at the inlet is $\dot{\gamma}_1 = 4Q/\pi R_1^3$, and the shear stress at the wall at the inlet is $\tau_1 = \eta\dot{\gamma}_1$. Hence, equation (8.9) can be expressed as

$$\Delta P = \frac{2\tau_1}{3\tan\theta}\left[\left(\frac{R_1}{R_2}\right)^3 - 1\right] \tag{8.10}$$

A comparable expression can be readily derived for flow in a tapered slot.

It should be remembered that these equations are only valid when θ is less than 10°; they ignore the small accelerations which induce extensional flows which can account for a pressure drop of up to 10% additional to that calculated just for shear flow.

8.2.5 Simplified melt flow in a two-roll mill

The two-roll mill is widely used to make compounds from rubber and, in the form of a calender (several rolls in series), to make rubber sheet and plastics foil. The simplest model involves an analysis of isothermal drag flow of a Newtonian fluid. The rolls are each of radius R and axial length W and rotate at a surface speed v_1. The gap H_0 between the rolls is very small

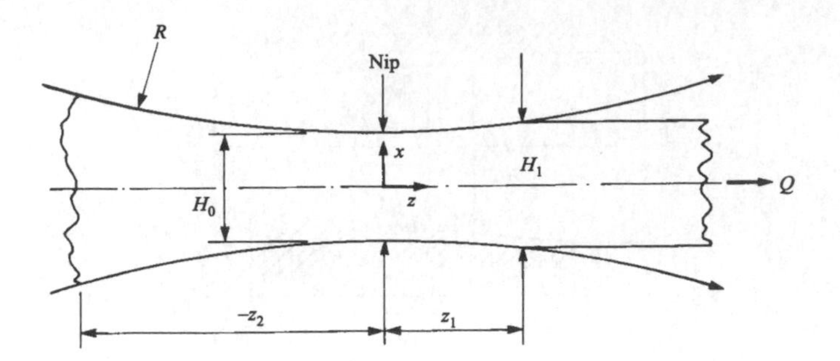

Fig. 8.6 The nip region in a two-roll mill

compared with the roll radius. Using the Cartesian axes shown in Fig. 8.6, the equation of motion (8.3) can be integrated with the boundary conditions $v = v_1$ at $x = \pm H/2$ to give the velocity distribution:

$$v_z = v_1 + \frac{1}{2\eta}\left(-\frac{\partial p}{\partial z}\right)\left(\frac{H^2}{4} - x^2\right) \tag{8.11}$$

and hence the output rate is

$$Q = 2W\int_0^{H/2} v_z \, \mathrm{d}x = WHv_1 + \frac{WH^3}{12\eta}\left(-\frac{\partial p}{\partial z}\right) \tag{8.12}$$

At the point of separation from the rolls, $H = H_1$, the melt pressure is the same as ambient pressure and the pressure gradient is zero. By making the simple assumption that the final sheet thickness is the same as the roll gap at the point of separation, i.e. no swelling occurs, the output rate becomes

$$Q = v_1 WH_1 \tag{8.13}$$

To calculate the pressure gradient and the pressure along the midplane of the flow, it is now appropriate to invoke the lubrication approximation. The relationship between H and x is

$$(H - H_0) = 2[R - \surd(R^2 - z^2)] \tag{8.14}$$

but in the region of the nip where $H \ll R$, second-order terms can be neglected so that

$$H \simeq H_0 + z^2/R \tag{8.15}$$

Combining equations (8.12), (8.13) and (8.15), the pressure drop becomes

$$\frac{\mathrm{d}p}{\mathrm{d}z} = \frac{12\eta v_1(z^2 - z_1)^2)}{R(H_0 + z^2/R)^3} \tag{8.16}$$

which clearly indicates that the maximum pressure occurs well upstream from the nip at $z = -z_1$.

The pressure distribution along the midplane of the melt may be found by integrating equation (8.16). It is marginally less tedious to use the dimensionless distance $\alpha = z/\sqrt{(RH_0)}$, so that

$$\frac{dp}{d\alpha} = \frac{12\eta v_1\sqrt{(RH_0)}}{H_0^2}\frac{(\alpha^2 - \alpha_1^2)}{(1+\alpha^2)^3} \tag{8.17}$$

i.e.

$$p = \frac{3\eta v_1\sqrt{(RH_0)}}{2H_0^2}\left(\frac{\alpha[\alpha^2(1-3\alpha_1^2) - (1+5\alpha_1^2)]}{(1+\alpha^2)^2} + (1-3\alpha_1^2)\tan^{-1}\alpha + B\right) \tag{8.18}$$

The constant of integration may be evaluated by recalling that $p = 0$ where the sheet leaves the rolls at $\alpha = \alpha_1$, and hence

$$B = [\alpha_1(1+3\alpha_1^2) - (1+\alpha_1^2)(1-3\alpha_1^2)\tan^{-1}\alpha_1]/(1+\alpha_1^2) \tag{8.19}$$

B can be expressed as an infinite series; ignoring terms in α_1 with powers higher than 3 gives $B \simeq 16\alpha_1^3/3$. The maximum pressure therefore occurs at $\alpha = -\alpha_1$ and has the value

$$\hat{p} = 16\eta v_1 z_1^3/RH_0^3 \tag{8.20}$$

It is evident that the maximum pressure is extremely sensitive to the nip setting, H_0.

Before calculating the force exerted by the melt on the rolls, it is necessary to ascertain where upstream the feed engages with the rolls. This occurs at $z = -z_2$, where the pressure between the rolls reaches ambient pressure. This can be found by putting $p = 0$ in equation (8.18), and solving for α numerically as a function of the downstream separation point α_1. Typically, $\alpha_2 = -k\alpha_1$ with k having a value in the range 2–4.

The total force F_x acting on each roll is now found from

$$F_x = \int_{-z_2}^{+z_1} Wp\,dz = W\sqrt{(RH_0)}\int_{-\alpha_2}^{\alpha_1} p\,d\alpha \tag{8.21}$$

Although this model provides a useful appreciation of the nature of flow in the calender, it needs to be refined to take into account the following factors. First, polymer melts are non-Newtonian – section 8.4 will show that it is relatively simple to adapt equations (8.11)–(8.13) to take account of power-law behaviour, but then the pressure distributions and forces on the rolls can only be evaluated numerically. Second, in practice, the thickness of the discharged sheet is different from that of the roll gap at separation because of swelling caused by elastic behaviour (see section 8.6) and because of drawdown caused by haul-off conditions: the analysis described above is based on roll gap at separation and not the final measured thickness of the sheet. Third, the analysis has only examined shear deformation, whereas the converging and diverging flows between the rolls induces some extensional flow which develops some additional pressure.

8.2.6 Spreading disc flow at constant volumetric flow rate

In unidirectional pressure flow between parallel plates, the velocity profile was found to be

$$v_z = \frac{1}{2\eta}\left(-\frac{\partial p}{\partial z}\right)\left(\frac{H^2}{4} - x^2\right) \tag{8.22}$$

Providing the radius is not too small, the lubrication approximation permits the use of this expression to estimate the radial flow between parallel plates. For the circumferential element shown in Fig. 8.7, equation (8.22) transforms to

$$v_r = \frac{1}{2\eta}\left(-\frac{\partial p}{\partial r}\right)\left(\frac{H^2}{4} - x^2\right) \tag{8.23}$$

The average velocity $\langle v_r \rangle$ is then

$$\langle v_r \rangle = \int_{-H/2}^{+H/2} v_r \mathrm{d}x \bigg/ \int_{-H/2}^{+H/2} \mathrm{d}x = \frac{H^2}{12\eta}\left(-\frac{\partial p}{\partial r}\right) \tag{8.24}$$

The volumetric flow rate is the product of the cross-sectional area and the average velocity, and hence the expression for Q:

$$Q = 2\pi r H \frac{H^2}{12\eta}\left(-\frac{\partial p}{\partial r}\right) \tag{8.25}$$

can be integrated to find the pressure difference needed to fill the cavity:

$$p_1 - p_2 = \frac{6Q\eta}{\pi H^3}\ln\left(\frac{R_2}{R_1}\right) \tag{8.26}$$

This analysis gives a fairly good description of isothermal Newtonian flow, except in the region near the injection point, $r \simeq R_1$. In real injection

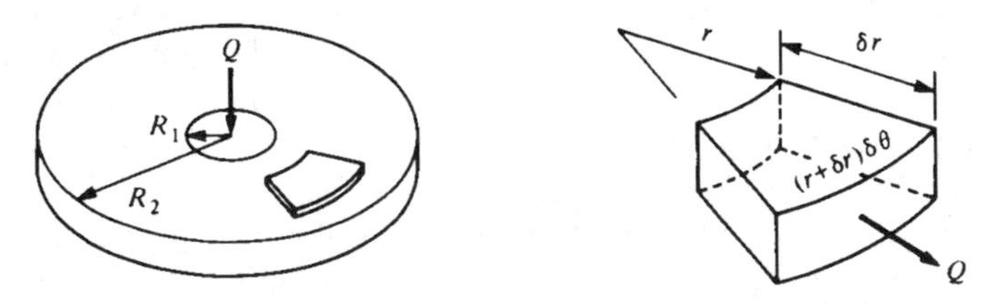

Fig. 8.7 Spreading disc flow by central injection between parallel plates

moulding, some additional features need to be incorporated to make the model quantitatively reliable. The high velocities near the gate may lead to some shear heating (this also applies within the gate itself) which may cause premature localized crosslinking in some polymers, and the large range of shear rates makes it essential to consider non-Newtonian behaviour. The flow is never isothermal in practice – either the moulds are much hotter than the melt (to promote crosslinking) or much colder (to effect cooling prior to ejection), and in each case conduction leads to a progressive solidification of polymer on the mould surface during injection. This is particularly important when it is desired to fill a cavity with a long flow path between narrowly separated plates. High flow rates with fluids of high viscosity generate high pressures and therefore some account should be taken of the compressibility of the fluid (see section 8.8.2). The flow is assumed to be steady – this is manifestly untrue at the flow front. Moulds are assumed properly vented so that the incoming fluid can readily displace the air in the cavity.

PROBLEMS

In all these problems, you should use the assumptions listed at the beginning of section 8.2 and, where relevant, those in section 8.2.4.

1. (a) Sketch the velocity profiles for the following flows between parallel plates:
   (i) pressure flow between stationary plates;
   (ii) drag flow between one fixed and one moving plate for open and closed discharge.
   (b) Sketch the velocity profiles in the two-roll mill at the following places: $z = z_1, 0, -z_1$ and $-z_2$. Optional: make a rough sketch of the pressure distribution along the midplane of the two-roll mill (ignore roll curvature).
2. For flow by central injection into a uniform circular disc cavity, answer briefly (e.g. not more than two lines) each of the following:
   (a) Sketch the pressure profile along a radial flow path and the velocity and shear stress profiles near the inlet and near the end of the flow for the conditions where the mould is almost completely full.
   (b) What is the ratio of maximum to minimum shear rate in the radial direction at the parallel plate boundaries?
   (c) What is the time needed to fill a circular cavity of uniform thickness at constant volumetric flow rate?
   (d) What minimum force is needed to keep the mould shut if the minimum injection pressure to just fill the cavity is maintained after the mould is filled?
3. Show that, in the simplest model of isothermal flow of a Newtonian melt in the metering section of a given single-start screw extruder coupled to a

capillary die, the optimum output rate is achieved at a helix angle of 30° when only the helix angle and channel depth are allowed to vary.

4. A hollow cylinder 2 mm thick, 200 mm long and with an internal radius of 100 mm is sealed at one end by a disc of the same thickness. This pot is to be made by injecting an incompressible fluid of viscosity $1000\,\mathrm{N\,s/m^2}$ at a constant flow rate of $2 \times 10^{-5}\,\mathrm{m^3/s}$ through a circular runner 8 mm in diameter and 300 mm long into the centre of the disc. What is the minimum pressure needed at the feed end of the runner to ensure that the cavity can be filled? What is the minimum clamping force needed to ensure that the two-part mould stays shut during injection?
5. In the metering zone of a square-pitch single-screw extruder, the axial length is 200 mm, the channel depth is 2.0 mm and the barrel diameter is 25 mm. The screw rotates at 60 rev/min and delivers a polymer of melt density $800\,\mathrm{kg/m^3}$ into a capillary die of uniform radius 5 mm and length 0.2 m. Estimate the output rate of the rod in kg/h. What assumptions have you made? What is the ratio of pressure flow to drag flow along the screw channel? (The lead in a single-start screw has the same value as the diameter of the tip of the flight.)
6. A laboratory-scale two-roll mill has rolls 100 mm in diameter and 400 mm long with a nip of 1 mm. The distance between the rolls is 1.16 mm at the point of separation of melt from the mill at a speed of 60 rev/min. The melt viscosity is $10^4\,\mathrm{N\,s/m^2}$. What is the maximum pressure in the melt? Where does this occur? Where does the feed bite on the rollers? (are your assumptions valid?) *Optional*: what is the roll separation force?
7. A long slot die has inlet dimensions $2H_1$ and W_1, and outlet dimensions $2H_2$ and W_2, with linear tapers of half-angle θ for thickness and φ for width. Derive an expression for the pressure drop caused by steady isothermal developed shear flow of a Newtonian fluid through this die.

8.3 SHEAR VISCOSITY OF POLYMER MELTS

Under small velocity gradients in unidirectional flows, e.g. $\dot{\gamma} < 1/s$, polymer melts are approximately Newtonian and, at the low pressure drops involved, virtually incompressible. However, most melt processing machinery operates at substantially higher shear rates where the polymer has a much lower viscosity at a given temperature, and in some processes pressures may be sufficiently high, e.g. $100\,\mathrm{MN/m^2}$, that compressibility does need to be taken into account.

This section outlines some methods for measuring shear viscosity, identifies some of the main factors on which shear viscosity depends, and comments on how to describe this behaviour. Much more detailed discussion will be found in the books listed in the further reading list.

8.3.1 Measurement of shear viscosity

The objective in rheometry work is to measure the relationship between force and flow rate for a given polymer melt at the required temperature and perhaps as a function of pressure, too. These measurements should provide data for the design of flow channels, and they may describe a fundamental property of the polymer which may relate to its molecular structure.

The two main classes of rheometer devolve on two modes of deformation: drag flow between a moving and a fixed plate, and pressure flow in extrusion through a capillary. Rheometers based on drag flow can operate up to a maximum shear rate of about 1/s, but those based on pressure flows can offer a range from about 0.1 up to about 10^5/s. The drag flow rheometers operate at constant or cyclic shear rates or stress, where the fluid is usually Newtonian; analysis of steady flows essentially follows the methods used in section 8.2, subject to making any necessary corrections (e.g. for end effects). Interpretation of cyclic loading data is not discussed here but would require extension of the analysis to incorporate features of linear viscoelasticity comparable with those discussed in section 5.2.2 for mechanical properties.

(*a*) *Drag flow rheometer*

The **cone-and-plate** (Weissenberg) rheometer, shown schematically in Fig. 8.8, has a flat, horizontal heated baseplate onto which the polymer is placed and allowed to reach the required test temperature. The cone, having a vertical axis, is lowered into the centre of the polymer melt until its tip just contacts the metal plate. The cone (which is shielded to prevent too much heat loss) is then made to rotate either at fixed torque or fixed speed of rotation. The angle ϕ that the cone makes with the plate is usually less than 5°, and hence the simple shear is uniform throughout the flow. If the rate of rotation of the cone is $d\theta/dt$, the shear rate $\dot{\gamma}$ is

$$\dot{\gamma} = (d\theta/dt)/\phi$$

If the measured torque is T and the maximum radius of the cone is R, then the shear stress τ is given by

$$\tau = 3T/2\pi R^3$$

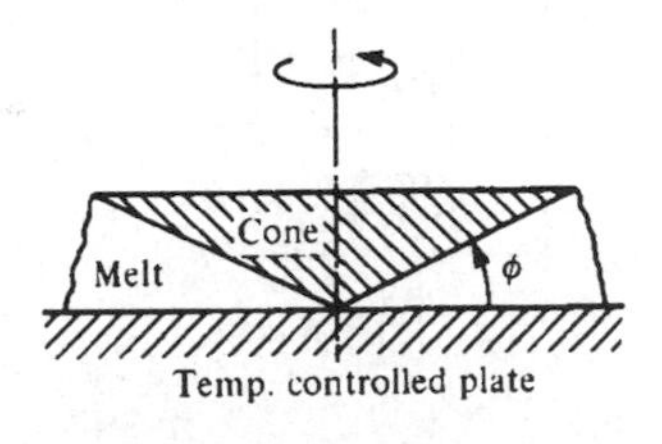

Fig. 8.8 Cone-and-plate rheometer (schematic)

The main disadvantage of this rheometer is the tendency for slip to occur at the boundaries at high stress.

The **concentric cylinder** (Couette) viscometer involves two coaxial cylinders, the outer one being fixed, and the inner one being rotated at constant rotational speed, or at constant torque. Torque and rotational speed can be related to shear stress and shear rate using the standard approach of section 8.2.

(*b*) *Pressure flow rheometer*

The **capillary** rheometer shown schematically in Fig. 8.9 consists of a heated barrel, typically some 250 mm long and 20 mm in diameter, with a capillary die at one end having a radius in the range 0.25–2 mm. The rheometer may operate at constant load or pressure, or at constant volumetric flow rate. For a Newtonian fluid, the relationships between pressure drop and shear stress at the wall, and between volumetric flow rate and shear rate at the wall, are given in Table 8.1, (E) and (F).

It is, however, necessary to make corrections to the analysis in order to obtain the correct value of viscosity for the given test conditions. First, the pressure in the barrel should be measured as near to the entry to the capillary die as possible, in order to eliminate changing pressure drop along the length of the barrel as material is extruded. Second, to allow for pressure drop in the die entry region and for the full development of a velocity profile within the capillary, it is good practice to carry out two tests at the same flow rate using a long die L_2 and a short die L_1. Arguing that the viscous drag in the entry region and the development of a parabolic profile should be the same in both dies, the true pressure drop in a die of length $L_2 - L_1$ should be $\Delta P_2 - \Delta P_1$. However, the normal procedure is to correct the measurements for one die by plotting for a number of dies (of constant diameter and different ratios of length to diameter) the pressure drop ΔP_i versus L_i/D (Fig. 8.10). Extrapolation to zero length gives the pressure drop ΔP_0 associated with a

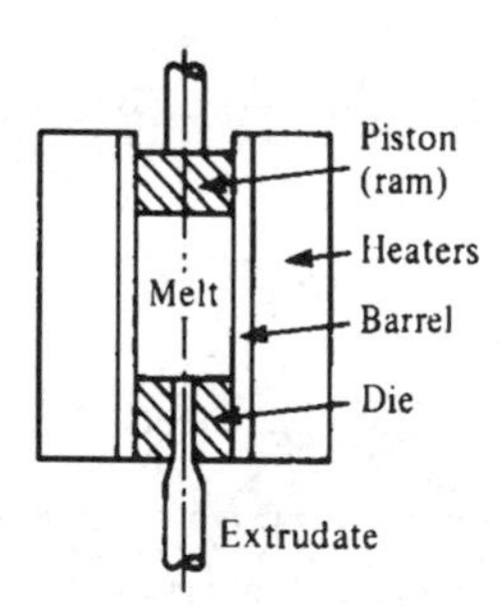

Fig. 8.9 Capillary rheometer (schematic)

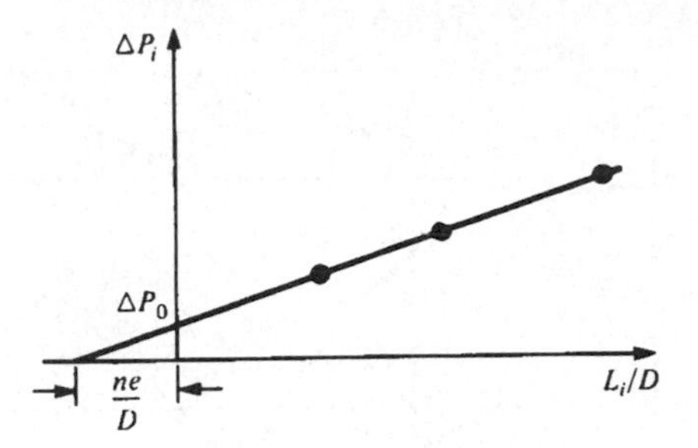

Fig. 8.10 Bagley plot for end-correction of capillary flow data

notional die of zero length, and extrapolation to zero pressure gives the effective extra length e of the capillary which can be five or more diameters. The true pressure drop for any given die i at a given flow rate is then

$$\Delta P_{\text{true}} = \frac{\Delta P_i - \Delta P_0}{L_i} = \frac{\Delta P_i}{L_i + e} \tag{8.27}$$

The true stress at the wall is now calculated from ΔP_{true} using Table 8.1, (F). Both ΔP_0 and e increase with increase in flow rate, which can be evaluated experimentally.

8.3.2 Presentation of shear viscosity data

The analysis of section 8.2 assumed that the fluid was Newtonian, i.e. viscosity independent of shear rate. Analysis of flows in the rheometers made the same assumption. To draw attention to this, rheologists use the term **apparent shear rate** $\dot{\gamma}_a (= 4Q/\pi R^3$ for a capillary and calculated as if the fluid were Newtonian). For any given apparent shear rate, the corresponding **apparent shear viscosity** η_a is defined by

$$\eta_a \equiv \tau/\dot{\gamma}_a \tag{8.28}$$

where τ is in fact a true stress calculated from corrected data as described in section 8.3.1. These quantities are now related by flow curves presented on double logarithmic scales of apparent viscosity versus shear stress, as shown schematically in Fig. 8.11 (the diagonal constructions for constant apparent shear rate follow from equation (8.28)). Some workers prefer curves of shear stress versus apparent shear rate (Fig. 8.12). Both presentations are often called ‘flow curves’.

The flow curves show that, at low shear rates, the behaviour is Newtonian. However, at the higher shear rates relevant to many processing operations, the apparent viscosity decreases markedly with increase in shear rate: the melt exhibits **shear thinning** and is called **pseudoplastic**. The shear rate at which shear thinning becomes apparent, and the degree, depend on the test conditions, on the type of polymer, on molecular details such as molecular

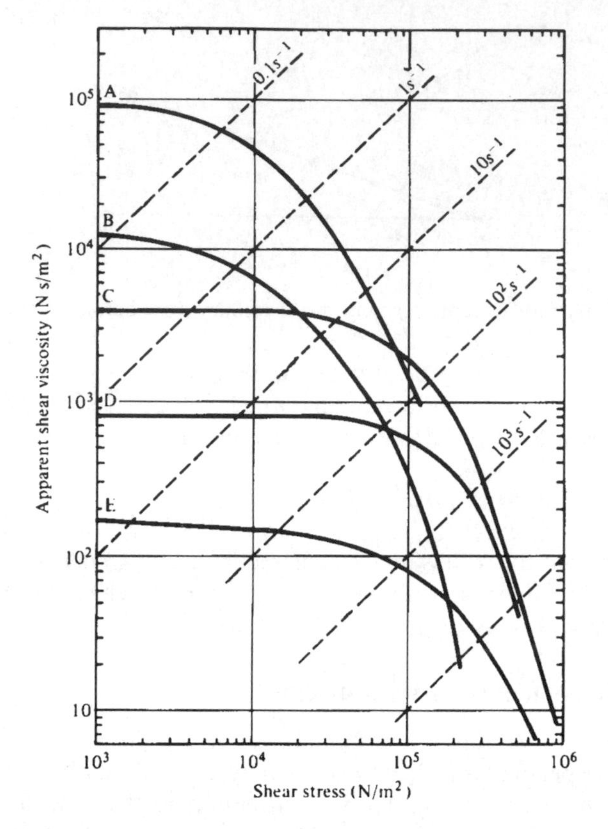

Fig. 8.11 Typical flow curves for polymer melts: A, extrusion-grade low-density polyethylene, 170°C; B, extrusion-grade propylene–ethylene copolymer, 230°C; C, moulding-grade acrylic, 230°C; D, moulding-grade acetal copolymer, 200°C; E, moulding-grade nylon-6,6 285°C (data courtesy ICI)

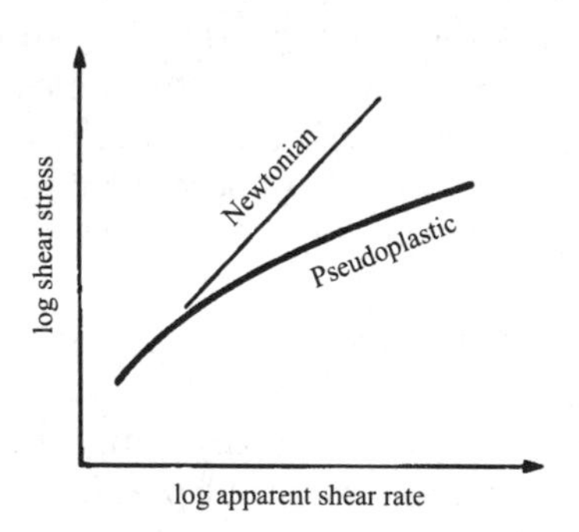

Fig. 8.12 One presentation of shear flow data

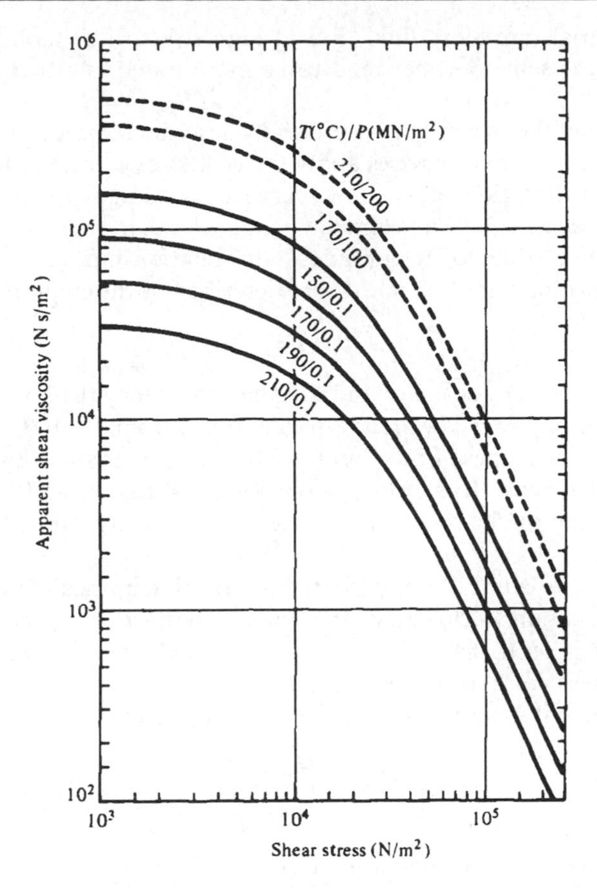

Fig. 8.13 Typical flow curves for low-density polyethylene showing effect of temperature and pressure (data courtesy ICI)

weight and molecular weight distribution, and on any other ingredients such as lubricants or plasticizers present in the compound.

Shear viscosity varies with temperature – the higher the temperature, the more readily can adjacent molecules disentangle and slip past each other. Results can be expressed as families of flow curves (log apparent shear viscosity versus log shear stress) which are parallel to each other but displaced vertically. Shear viscosity also varies with hydrostatic pressure – the higher the pressure, the more difficult it is for adjacent molecules to pass each other because they are squashed into more intimate contact. The effects of temperature and pressure are indicated in Fig. 8.13.

In a melt, the molecules of a polymer having a low molecular weight could disentangle more readily than in one having a high molecular weight. At low stress (e.g. where flow is Newtonian), the shear viscosity is directly proportional to molecular weight raised to a power of about 3.5, so low molecular weight materials are more fluid. But at high stress, high molecular weight versions of the same polymer tend to be more pseudoplastic (i.e. a steeper flow curve).

The extent of the shear thinning is usually described by the slope of the flow curve. Simple algebraic analyses apply the well-known technique of approximating a small region of the flow curve to a straight line – here, using logarithmic scales. Over a restricted range of shear rates – typically not more than one order of magnitude – the relationship between apparent viscosity and apparent shear rate can be described by the empirical power law

$$\eta_a = C\dot{\gamma}_a^{n-1} \tag{8.29}$$

where n is the power-law index, and C is the consistency (with rather unusual units). Values of power-law index typically range from 0.2 to 0.5 at the shear rates found in high-speed flows, but from 0.75 to 1.0 for slow flows. C has no physical significance. It is evident from Fig. 8.14 that C and n for a given material depend on the magnitude of the range of shear rates of interest in a given process.

There are two sources of awkwardness in the empirical power law embodied in equation (8.29). First, the units of both η_a and $\dot{\gamma}_a$ depend on n, which varies. Second, the analysis of flows often yields negative values of shear rate. Both these points can be eliminated, as well as the value of C, by defining C as the apparent shear viscosity η_{a0} at reference shear rate $\dot{\gamma}_{a0} = 1/\text{s}$, so that

$$\eta_a = \eta_{a0}|\dot{\gamma}_a/\dot{\gamma}_{a0}|^{n-1} \tag{8.30}$$

8.3.3 True shear viscosity and true shear rate

The discussion of viscosity has so far centred on apparent shear viscosity and apparent shear rate. The true rate $\dot{\gamma}_T$ is important because this is the velocity gradient (e.g. $\partial v_z/\partial x$ or $\partial v_z/\partial r$) used in the analysis of non-Newtonian flows. This may be obtained by applying the Rabinowitsch correction to the

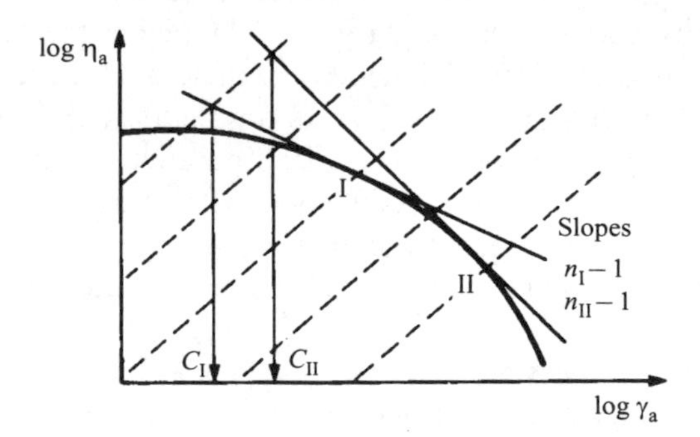

Fig. 8.14 Construction of power-law approximations to apparent shear flow curve

apparent rate. The correction, derived in section 8.4, depends on the geometry of flow:

$$\begin{aligned} \dot{\gamma}_T &= \dot{\gamma}_a(3n+1)/4n \quad \text{for circular channels} \\ \dot{\gamma}_T &= \dot{\gamma}_a(2n+1)/3n \quad \text{for slot channels} \end{aligned} \tag{8.31}$$

The values of n are derived from apparent flow curves. For a Newtonian fluid, no correction is needed, but at $n = 0.2$ the true rate is about twice the apparent rate.

A true viscosity η_T can be defined for any given true shear rate by the equivalent of equation (8.28):

$$\eta_T \equiv \tau/\dot{\gamma} \tag{8.32}$$

and hence it is possible to construct 'true' flow curves of, for example, log η_T versus log τ, which have a very similar general appearance to the apparent flow curves.

The power-law representation of flow over a decade of true shear rate can be expressed in the form

$$\eta_T = C_T \dot{\gamma}_T^{n'} \tag{8.33}$$

and fortunately a generalized analysis (details are given in the standard tests) shows that n' from the true curves is the same as n from the apparent curves. The true consistency C_T can be eliminated from equation (8.33) using the approach explored above, giving

$$\eta_T = \eta_{T0} |\dot{\gamma}_T / \dot{\gamma}_{T0}|^{n-1} \tag{8.34}$$

Using the definition from equation (8.32), the shear stress can be expressed in the form

$$\tau = \eta_T \dot{\gamma}_T = \eta_{T0} |\dot{\gamma}_T / \dot{\gamma}_{T0}|^{n-1} \dot{\gamma}_T \tag{8.35}$$

Where true shear rates have negative values (which occurs in the positive coordinates of the channel geometries (below), it is helpful to recast (8.35) in terms of a positive quantity, $-\dot{\gamma}_T$. Recalling that $-\dot{\gamma}_{T0} = +1$, this leads to the form

$$\tau = -(\eta_{T0}/\dot{\gamma}_{T0}^{n-1})(-\dot{\gamma}_T)^n \tag{8.36}$$

which will be used in further analysis.

8.3.4 Melt flow index (MFI)

This term is widely used in the plastics industry to describe the fluidity of a polymer melt at a standardized value of melt temperature, load, die length and die diameter. The essence of the test system is a single small short capillary ($L/D \simeq 4$) through which polymer is forced by applying a constant load to a piston in the reservoir. A detailed critique is given by Brown (1981). The following remarks comment on the value of the MFI.

- The test is easy to carry out. A high MFI corresponds to a low value of viscosity at very low shear rate, and gives an impression of molecular mass. In practice there is often a compromise between choice of high MFI for easy processing and low MFI for good quality products.
- MFI cannot be related directly to any form of shear viscosity useful in design calculations because only one short die is used in the test, and this precludes a sensible estimate of effective die length.
- No provision is made in the MFI test to show how fluidity varies with shear rate, temperature or pressure.

PROBLEMS

1. Below which value of shear rate can the following polymer melts be regarded as sensibly Newtonian: extrusion-grade low-density polyethylene at 170°C and at 210°C, and a moulding grade of acetal copolymer at 200°C?
2. Calculate the power-law index for a moulding grade of nylon at 285°C for the following range of shear rates: 10 to 100/s and 10^4 to 10^5/s. Note how poor the power law is at representing the experimental behaviour over more than one decade of shear rate.
3. Comment briefly on the likely effect of an increase in (uniform) melt temperature on the output pressure drop characteristic of the metering zone of a single-screw extruder coupled to a die.
4. How does the magnitude of the Rabinowitsch correction for true shear rate in a wide narrow slit change over the range of apparent shear rates quoted for acrylic at 230°C? What is the difference between shear stress based on an apparent shear rate of 100/s and the shear stress based on the true shear rate (at the same volumetric flow rate)?
5. The effect of different uniform melt temperature on the shear viscosity of polymer melts can be expressed as families of flow curves parallel to each other but displaced vertically (at constant stress) in proportion to the temperature change. At constant shear rate:

$$\eta_a(T) = \eta_a^* \exp[-b_T(T - T^*)]$$

   where T^* denotes the reference temperature. Calculate b_T from the data given in Fig. 8.13 for a low-density polyethylene. How would you relate b_T to that in an expression for viscosity $\eta_a(T)$ relating to shift at constant stress?

8.4 UNIDIRECTIONAL POWER-LAW FLOWS

Section 8.2 revised the analysis of isothermal laminar flows of Newtonian fluids, and clearly stated the assumptions made. Section 8.3 showed that polymer melts are non-Newtonian and that, over a small range of shear rates (about a decade), the flow curve can be approximately described by a power law of the form

$$\eta_{\mathrm{T}} = C\dot{\gamma}_{\mathrm{T}}^{n-1} \tag{8.37}$$

where C and n are constants for the given material and test conditions. This section introduces the analysis of unidirectional power-law flows.

It would seem reasonable to replace the constitutive equation for Newtonian behaviour (8.2) by one for power-law behaviour (8.36) and rework all the analyses given in section 8.2 for output rates and pressure drops. Setting up the analysis is straightforward, although algebraically more cumbersome. This analytical approach can be readily applied to those pressure flows which display an obvious symmetry. However, there are minor difficulties with drag flows and unsymmetrical pressure flows where the location of the zero-stress surface is not readily identifiable, and which dictate the use of numerical methods.

8.4.1 Pressure flows through channels of constant cross-section

The main emphasis here is on flow in a channel having either a circular or a wide thin slot cross-section. In the circular cross-section, steady flow in the positive z-direction results in a negative velocity gradient or shear rate at any finite radius from the centreline, hence the preoccupation with negative signs in equations (8.35) and (8.36). In the slot channel the velocity gradient is negative only in the positive x- and z-directions (regardless of position in the width direction), and the flow is symmetrical about the yz plane by inspection. Our analysis will be confined to the region where shear rate is negative and relies on symmetry.

The procedure is given in full for steady flow in a channel of constant rectangular slot cross-section. All the assumptions made at the beginning of section 8.2 apply here, except that the fluid behaviour is described by the power law. Combining equation (8.36) with the elemental momentum balance in equation (8.2) now gives

$$\tau_{xz} = x\left(\frac{-\partial p}{\partial z}\right) = \frac{\eta_{\mathrm{T0}}}{\dot{\gamma}_{\mathrm{T0}}^{n-1}}\left(\frac{-\partial v_z}{\partial x}\right)^n \tag{8.38}$$

There is no difficulty in taking the nth root of this equation because $-\partial p/\partial z$ is positive, and hence

$$\frac{\partial v_z}{\partial x} = -\left[\frac{\dot{\gamma}_{\mathrm{T0}}^{n-1}}{\eta_{\mathrm{T0}}}\left(-\frac{\partial p}{\partial z}\right)\right]^{1/n} x^{1/n} = -A_1 x^{1/n} \tag{8.39}$$

where A_1 represents the constant terms. This expression may now be integrated and the boundary condition is that $v_z = 0$ at $x = +H/2$, giving

$$v_z = -\frac{A_1 n}{1+n}\left[x^{1+1/n} - \left(\frac{H}{2}\right)^{1+1/n}\right] \tag{8.40}$$

When $n = 1$, this reduces to Table 8.1, (H), but for pseudoplastic behaviour, the smaller the value of n, the greater the shear rate near the wall and the flatter the profile in the central region about the midplane.

The volumetric flow rate, Q, is

$$Q = 2\int_0^{H/2} W v_z \mathrm{d}x = \frac{2Wn}{2n+1}\left[\frac{\dot{\gamma}_{\mathrm{T0}}^{n-1}}{\eta_{\mathrm{T0}}}\left(-\frac{\partial p}{\partial z}\right)\right]^{1/n}\left(\frac{H}{2}\right)^{2+1/n} \tag{8.41}$$

where $W(\gg H)$ is the width of the channel, and hence the pressure gradient is given by

$$-\frac{\partial p}{\partial z} = \frac{2}{H}\left(\frac{2n+1}{3n}\frac{6Q}{WH^2}\right)^n \frac{\eta_{\mathrm{T0}}}{\dot{\gamma}_{\mathrm{T0}}^{n-1}} \tag{8.42}$$

By integration, the pressure drop along a slot of length $L \gg H$ is

$$p_1 - p_2 = \frac{2L}{H}\left(\frac{2n+1}{3n}\frac{6Q}{WH^2}\right)^n \frac{\eta_{\mathrm{T0}}}{\dot{\gamma}_{\mathrm{T0}}^{n-1}} \tag{8.43}$$

The momentum balance $\tau_{xz} = x(-\partial p/\partial z)$ can be used to show that the shear stress at the wall is

$$(\tau_{xz})_{H/2} = (H/2)(-\partial p/\partial z) \tag{8.44}$$

and hence by combining equations (8.42)–(8.44):

$$(p_1 - p_2) = (2L/H)(\tau_{xz})_{H/2} \tag{8.45}$$

In unidirectional flow of a Newtonian fluid (Table 8.1, (J) and (L)), the shear stress at the wall was related by viscosity to the shear rate at the wall and hence to flow rate Q. For a power-law fluid, the same approach can be used, but it is necessary to distinguish whether the flow data relate to apparent viscosity and shear rate, or to true viscosity and true shear rate. The data in this book relate to apparent viscosity, so the design procedure will involve calculating the apparent shear rate at the boundary, $\dot{\gamma}_a = 6Q/WH^2$. The value of τ_{xz} at the boundary is taken from the apparent flow curve (e.g. Fig. 8.11). The pressure drop is then calculated using equation (8.45).

Finally, by eliminating A_1 between equations (8.39) and (8.42), the true shear rate is

$$\dot{\gamma}_{xz} = \left(\frac{2n+1}{3n}\frac{6Q}{WH^2}\right)\left(\frac{2x}{H}\right)^{1/n} \tag{8.46}$$

The true shear rate at the wall ($x = H/2$) is now seen to be related to the apparent shear rate at the wall by the Rabinowitsch correction first quoted in equation (8.31):

$$(\dot{\gamma}_{xz})_{H/2} = \frac{2n+1}{3n}(\dot{\gamma}_{\mathrm{a}})_{H/2} \tag{8.47}$$

(It is formally correct to quote data in terms of true rather than apparent quantities. Where true data are given, the design procedure is to calculate the true shear rate (e.g. equation (8.47)) and read the shear stress from the true data. The pressure drop is then calculated from equation (8.45) as before. The pressure drops calculated by either procedure will be the same within about 2% for the same polymer melt and dimensions. What must not be done is to mix the basis of the calculations: using true shear rate with apparent data is wrong and can lead to values of pressure drop which are wrong by up to 100%.)

Analyses similar to that for flow between parallel plates can be readily made for other symmetrical unidirectional flows such as those within circular channels of constant cross-section and within thin annular gaps of constant cross-section (treated as a slot having a width $W = 2\pi R_{\mathrm{mean}}$).

8.4.2 Pressure flows through channels of gradually changing cross-section

In symmetrical flows, the lubrication approximation can be applied to power-law fluids in a cross-section which changes gradually along the main flow direction. This method can be applied to estimate the relationship between pressure drop and flow rate in, for example, tapered channels of circular or slot cross-section, or for spreading disc flow in a centre-gated disc bounded by parallel plates (or even slightly converging plates). The principles involved are quite straightforward, but the algebra becomes tedious and messy.

The procedure is illustrated for the flow of a power-law fluid in a gradually tapering capillary. All the other assumptions made at the beginning of section 8.2 apply here, and the dimensions of the channel are given in Fig. 8.5. Combining the constitutive relationship of equation (8.36) with the momentum balance of equation (8.1) gives

$$(-\dot{\gamma})^n = \frac{\dot{\gamma}_{\mathrm{T0}}^{n-1}}{\eta_{\mathrm{T0}}}\frac{r}{2}\left(-\frac{\partial p}{\partial z}\right) \tag{8.48}$$

i.e.

$$\frac{\partial v_z}{\partial r} = -\left[\frac{\dot{\gamma}_{\mathrm{T0}}^{n-1}}{2\eta_{\mathrm{T0}}}\left(-\frac{\partial p}{\partial z}\right)\right]^{1/n} r^{1/n} = A_2 r^{1/n} \tag{8.49}$$

Integration, with the boundary condition $v_z = 0$ at $r = R$, gives the velocity distribution:

$$v_z = \frac{A_2 n}{1+n}\left(r^{1+1/n} - R^{1+1/n}\right) \tag{8.50}$$

The volume flow rate Q can now be found as

$$Q = 2\pi \int_0^R r v_z \mathrm{d}r = \frac{\pi n R^{3+1/n}}{3n+1} \left[\frac{\dot{\gamma}_{\mathrm{T0}}^{n-1}}{2\eta_{\mathrm{T0}}} \left(-\frac{\partial p}{\partial z} \right) \right]^{1/n} \tag{8.51}$$

Noting that the true shear rate at the boundary $r = R$ can be found from equation (8.49), and that the apparent shear rate at the wall is $\dot{\gamma}_a = 4Q/\pi R^3$ from Table 8.1, (E), the Rabinowitsch correction for flow in a circular channel derives from equation (8.51):

$$(\dot{\gamma}_{rz})_R = \frac{3n+1}{4n} (\dot{\gamma}_a)_R \tag{8.52}$$

Hitherto, the analysis applies to a circular channel of constant or tapering cross-section. Invoking the lubrication approximation and substituting

$$R = R_1(1 - kz) \quad \text{with} \quad k = (R_1 - R_2)/R_1 L$$

into (8.51) gives

$$-\int_{p_1}^{p_2} \mathrm{d}p = \left(\frac{(3n+1)Q}{\pi n} \right)^n \left(\frac{2\eta_{\mathrm{T0}}}{\dot{\gamma}_{\mathrm{T0}}^{n-1}} \right) \frac{1}{R_1^{3n+1}} \int_0^L \frac{\mathrm{d}z}{(1-kz)^{3n+1}} \tag{8.53}$$

After integration and some rearrangement, the result is

$$p_1 - p_2 = \left(\frac{3n+1}{4n} \frac{4Q}{\pi R_1^3} \right)^n \left(\frac{2\eta_{\mathrm{T0}}}{\dot{\gamma}_{\mathrm{T0}}^{n-1}} \right) \frac{1}{3n \tan\theta} \left[\left(\frac{R_1}{R_2} \right)^{3n} - 1 \right] \tag{8.54}$$

in which θ is the half-angle of the taper. The term between the first parentheses is the true shear rate and by comparison with (8.48) or (8.33), the first two groups of terms in equation (8.54) represent twice the shear stress at the wall at the inlet of radius R_1. On this basis, equation (8.54) simplifies to:

$$p_2 - p_1 = -\frac{2\tau_1}{3n \tan\theta} \left[\left(\frac{R_1}{R_2} \right)^{3n} - 1 \right] \tag{8.55}$$

(*a*) *Example*

A part of a runner system in an injection mould consists of a channel of circular cross-section 100 mm long, of 12 mm diameter at the inlet and 6 mm diameter at the outlet. Acrylic plastic at 230°C is to be pumped through this tube at a flow rate of $2 \times 10^{-5}\ \mathrm{m^3/s}$. Assuming the flow to be isothermal, calculate the pressure drop along this part of the runner system, using the data in Fig. 8.11.

Figure 8.11 provides apparent flow data, so it is necessary to calculate the apparent shear rates at the wall at each end of the tube in order to specify the range of shear rates on which the power-law index can be calculated:

$$\text{at the inlet}: \qquad \dot{\gamma}_{a1} = 4Q/\pi R_1^3 = 118/\text{s}$$
$$\text{at the outlet}: \qquad \dot{\gamma}_{a2} = 4Q/\pi R_2^3 = 943/\text{s}$$

For the acrylic moulding compound at 230°C, Fig. 8.11 gives the corresponding apparent shear viscosities: $\eta_{a1} = 1250\,\text{N s/m}^2$ and $\eta_{a2} = 320\,\text{N s/m}^2$. The power-law index can now be found from equation (8.29):

$$\begin{aligned} n - 1 &= \frac{\log(\eta_1/\eta_2)}{\log(\dot{\gamma}_{a1}/\dot{\gamma}_{a2})} \\ &= \frac{\log(1250/320)}{\log(118/943)} \\ &= 0.5918/(-0.9027) = -0.656 \\ n &= 0.344 \end{aligned}$$

To find the pressure drop from equation (8.55), the maximum shear stress at the inlet must correspond to the maximum apparent shear rate at the wall at the inlet, $\dot{\gamma}_{a1}$: Figure 8.11 shows $\tau_1 = 1.5 \times 10^5\,\text{N/m}^2$. The half-angle of the taper is given by $\tan\theta = (R_1 - R_2)/L = (6 - 3)/100 = 0.03$. The pressure difference is thus

$$\Delta p = -\frac{2 \times (1.5 \times 10^5)}{3 \times 0.344 \times 0.03}[(2)^{1.032} - 1] = -10.18 \times 10^6\,\text{N/m}^2$$

8.4.3 Drag flow of power-law fluids

In the previous two sections on pressure flows, the location of the neutral stress plane at the midplane was found by inspection; analysis of the power-law flow relied on symmetry about the midplane. However, in drag flows the location of the neutral surface is not known in advance and, in general, this precludes a full analytical solution for relationships such as pressure difference versus flow rate.

There is a special case where the pressure drop can be calculated analytically as a function of flow rate, namely drag flow in a two-roll mill having rolls of the same size rotating at the same speeds. The general form of the pressure profile for a power-law fluid obtained by numerical integration is much the same as that for a Newtonian fluid, although the feed bites rather earlier upstream of the nip.

Further details of the approach to the analysis of drag flows lie outside the scope of this introductory text but the ideas are developed in the standard texts noted at the end of this chapter.

PROBLEMS

1. (a) Show that for a Newtonian fluid, equations (8.41) and (8.43) reduce to the forms in Table 8.1, (I) and (L).

(b) Distinguish the concepts embodied in equations (8.28) and (8.29).
(c) For a flow in a given capillary at the same volumetric rate, compare the shear rate at the wall for a Newtonian fluid with that for a pseudoplastic fluid having a power-law index of 0.3.

2. Calculate the pressure drop in the tapered runner example in section 8.4.2 if the flow rate is increased to $2 \times 10^{-4}\,m^3/s$.
3. The extrusion grade of low-density polyethylene, for which data are given (Fig. 8.13), is to be injection moulded at 210°C. The feed channel to the mould is 5 mm in diameter and 300 mm long, and maintained at 210°C. If the polymer were injected at $10^{-5}\,m^3/s$, estimate the pressure drop resulting from flow through this channel, using the flow curve labelled $210°C/0.1\,MN/m^2$. Suggest (in one sentence) how you might take the effect of pressure into account in arriving at a more accurate assessment of the pressure drop. Would this correction be more, or less, important if the flow rate were $10^{-6}\,m^3/s$ or less?
4. A runner 6 mm in diameter feeds into the centre of a disc-shaped cavity 700 mm in diameter with a constant channel thickness of 2 mm. If acrylic at 230°C is injected into the cavity at a flow rate of $2 \times 10^{-4}\,m^3/s$, what pressure is required to just fill the cavity under the artificial condition that the mould is at the same temperature as the injected melt? If the mould were maintained at 20°C, would you expect to need a larger or a smaller pressure to just fill the cavity at the same rate? (Detailed calculations for filling the cold cavity are not required.)
5. A long circular cross-section die has inlet radius R_1, outlet radius R_2, length L and a linear taper. Under constant flow rate, where is the largest pressure gradient? Calculate the pressure gradient along the die for (a) a Newtonian fluid, and (b) a power-law fluid with $n = \frac{1}{3}$.
6. For a pressure-driven flow within a circular cross-section die of radius R and length L, derive expressions for (a) the minimum residence time, (b) the average residence time, and (c) the residence time for a fluid particle at radius $r = 0.99R$. Compare your results for (i) a Newtonian fluid and (ii) a power-law fluid with $n = 0.25$.

8.5 MIXING AND BLENDING OF POLYMER MELTS

The mixing of additives changes a polymer into a plastic. Additives tailor a specific plastic for a specific application. The additives can be powders, usually consisting of agglomerate parts, or fluids, such as oils or molten material, e.g. rubber. The purpose of the mixing is to disperse added material into small particles and to distribute these as evenly as possible in the melt. Dispersive mixing means breaking up the added agglomerates into the smallest possible parts, or to emulgate (blend) the added fluid into small droplets.

The main characteristic of mixing, as dealt with here, is that it involves fluid flow. With dispersive mixing this flow has to transmit the forces necessary to break particles or drops. The high viscosity of a polymer melt is quite suitable to transmit quite high forces to break up agglomerates and drops.

Distributive mixing, however, only requires movement and thus flow, so here only the forces to realize the required flow pattern are necessary. For distributive mixing, turbulence would be ideal. However, the high viscosity of polymer melts makes this unattainable. The only way here is to prescribe a flow pattern which simulates turbulence. Another route is to follow the baker's routine of stretching and doubling (folding) again and again.

In most cases, for both types of mixing, the same device is used, the choice of the device should then be determined by the combination of mixing requirements. In some cases a dispersive and a distributive mixer are used in series.

An additional problem after emulgation is the tendency for drops to coalesce when they touch. It is thus necessary to distribute the drops and to freeze the melt as soon as possible.

Mixing is performed in the domain of laminar flow. The application of laminar flow theory to mixing and blending problems is therefore required.

8.5.1 Dispersive mixing

To break up an agglomerate, a certain minimum force F is needed. Consider the agglomerate to consist of two cohering balls with radii r_1 and r_2. The maximum force that can be delivered in shear flow is

$$F_{\max} = 3\pi\eta\dot{\gamma}r_1r_2$$

and in extensional flow

$$F_{\max} = 6\pi\eta\dot{\varepsilon}r_1r_2$$

This will be effective if the balls are in a proper position. As in shear flow, however they tumble in space with rotational speed $\omega = \frac{1}{2}\dot{\gamma}$; the time $t = 2\pi/\dot{\gamma}$ is necessary before they experience $F_{\max}$. This determines the minimum length of a shear zone. Whether the force is sufficient depends on the coherence force F_{coh} in the agglomerate and the values of $\eta, \dot{\gamma}, r_1$ and r_2.

For emulgation also, a certain residence time is needed in order to deform a drop such that it will break up. This means that a certain flow length in shear or extensional flow is required. The question as to whether and how far the drops will break up is determined by the capillary number $\tau R/\sigma$ and the viscosity ratio $\eta_{\mathrm{D}}/\eta_{\mathrm{C}}$ (Fig. 8.15). The capillary number $Ca = \tau R/\sigma$ is the ratio of viscous (τ_{C}) to interfacial tension (σ/R) forces. η is the viscosity, D denotes the disperse phase and C the continuous phase (the polymer in

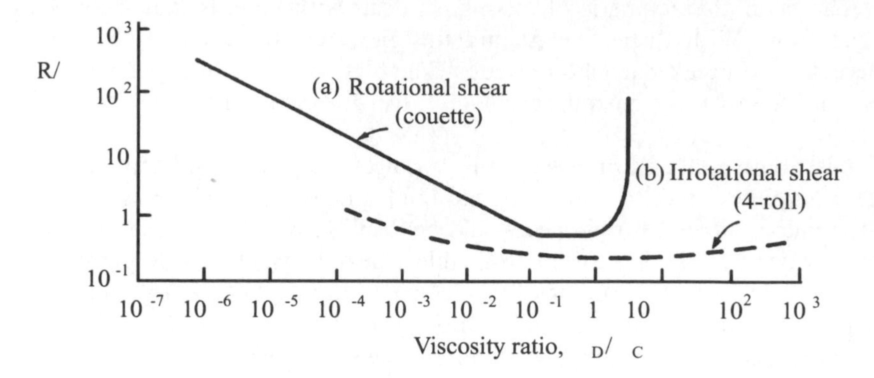

Fig. 8.15 Critical capillary number of the break-up of drops in (a) shear flow and (b) extensional flow (from H.P. Grace, *Chem. Eng. Commun.*, **14**, 225–77, 1982)

which the added melt is to be emulgated). The lines in Fig. 8.15 indicate the lower boundaries to which the radius R of the drops can be brought down. Note that the boundary for extensional flow (such as that generated in a four-roll device) is lower than that for shear flow (such as that generated in a Couette device).

The force needed to break up drops needs to be maintained for some time in order to allow for the deformation of the drops, which precedes the breaking up. The apparatus therefore has to provide the necessary length of the shear or extension zones to the fluid.

8.5.2 Distributive mixing

In laminar flow a certain domain, e.g. that containing the colourant master-batch, may deform but it will stay coherent as one domain. Deformation backwards yields the domain in the same shape again. The most effective way to obtain a good distribution is to deform the domain such that it forms a thin layer next to the equally deformed layer of the main mass. If we call the total thickness of these two the striation thickness t_s, then t_s is a measure of the quality of the distribution. Following the baker's routine makes multiple use of deformation and doubling, as shown in Fig. 8.16. This process is very effective since the striation thickness decreases exponentially. It can be finished when, for example, the layer thickness is too small to be visible or when the diffusion of particles generates the required distribution.

8.5.3 Axial mixing

Melt processing machines always have a certain residence time for transport and heat exchange. This in itself is not a problem; it only means a delay between entrance to and exit from the machine.

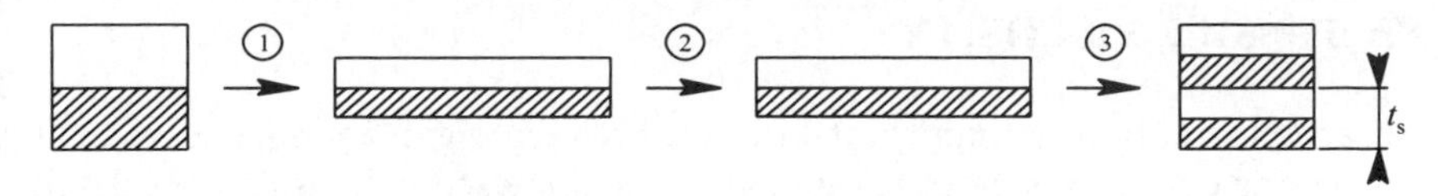

Fig. 8.16 Stretching and doubling n consecutive times results in 2^n layers

More relevant for product quality is the residence time distribution. This can be desirable if the composition of the feed changes over time. However, it can also be unwanted, for example with a changeover in colour (products with mixed colours are rejected) or with risk of thermal degradation (in dead corners of the process flow).

8.5.4 Mixing devices

A multitude of mixing devices are in use for a large number of mixing tasks. The accent usually falls on one of the mixing mechanisms, but the others function well or badly at the same time. For dispersive mixing, elongational flow is the most effective. It can be shown to be present in two-roll mills and locally in some other devices. Usually the device is based on drag flow, driven by a rotating extruder screw. Pressure flow is also used, but is less efficient as pressure has to be built up with other means, e.g. an extruder screw.

Primary distributive mixers may be static elements that split the flow and either rotate or deform it so that, after removing the split (at the end of an element), the flow is reunited but in a different order. It should now have a greater contact surface area between the domains and therefore a smaller t_s. Repetition of this sequence with new elements should reduce t_s to the required level. This type of mixer is based on pressure flow, which is economic if relatively wide channels are used.

Some other distributive mixers are extruder driven and prescribe a flow which circulates in separate cavities and therefore shows some resemblance with turbulence, but only because of its flow pattern. The flow pattern is such that it is in a way equivalent with that of the baker's cut-and-fold routine. The system is highly effective as even elongational flow can be present in the transfer between cavities.

PROBLEM

1. A warm hardening resin has to leave a heated pipe with a certain degree of reaction. The resin–hardener system was mixed cold just before entering the pipe. Describe nature and causes of the distribution of the degree of reaction at the exit cross-section. Indicate a measure which might be taken to improve the distribution.

8.6 TENSILE VISCOSITY

Shear deformation is important in flows between rigid boundaries in melt processing operations. Shear flows have been analysed in considerable detail both for the unidirectional situations and for more complicated situations. However, there are many processes in which substantial tensile deformations occur – some were mentioned in section 8.2.1. These have only recently received attention and the literature is not as yet well documented on the phenomena and analysis. Tensile deformations occur most obviously in flows where there are no constraining surfaces, but can also occur in flows within bounded channels.

The simple phenomenon of tensile deformation is easy to envisage. Applying a tensile force to a rod of uniform cross-section will induce viscous extension (accompanied by contraction in the transverse direction) for as long as the force is applied. Gripping the ends of the rod can be tricky, but is possible with care at low rates of deformation. One procedure is to vary the applied tension so that the gauge length of the specimen is under constant tensile stress σ. In another experiment devised to overcome the problem of gripping the ends, the specimen in sheet form is extended by two sets of two-roll mills a fixed distance apart. The speed of the rolls controls the extension rate in the sheet. Both experiments are conducted in a fluid medium at constant temperature. The tensile viscosity λ is related to the tensile strain rate $\dot{\varepsilon}$ by

$$\lambda = \sigma/\dot{\varepsilon} \tag{8.56}$$

The quantity λ is sometimes called the elongational or extensional viscosity. For fluids which show Newtonian behaviour, Trouton explained and derived the fact that the tensile viscosity is exactly three times bigger than the shear viscosity. This pattern of behaviour also applies to polymer melts at low stress where the shear behaviour is Newtonian.

Obtaining data at high rates of extension has proved difficult by direct experiment, but some useful information has been obtained by inference from pressure flows in converging channels of circular cross-section. An attempt has been made to separate out the shear and the tensile components of flow. The results of these studies show that at high stresses (at which shear behaviour is pseudoplastic), the tensile viscosity is a function of stress (Fig. 8.17). The main point is that the tensile viscosity is always much higher than the shear viscosity: the ratio of λ/η can range from a minimum of 3 to a few hundred. So even if tensile deformation rates are low compared with shear rates, the associated tensile stresses may still be large.

For some polymer melts, the tensile viscosity is independent of tensile stress even at high stress – linear polymers such as polyacetal and polyamide (nylon) show this behaviour. For other polymer melts, still linear but having a wide distribution of molecular weights, such as polypropylene

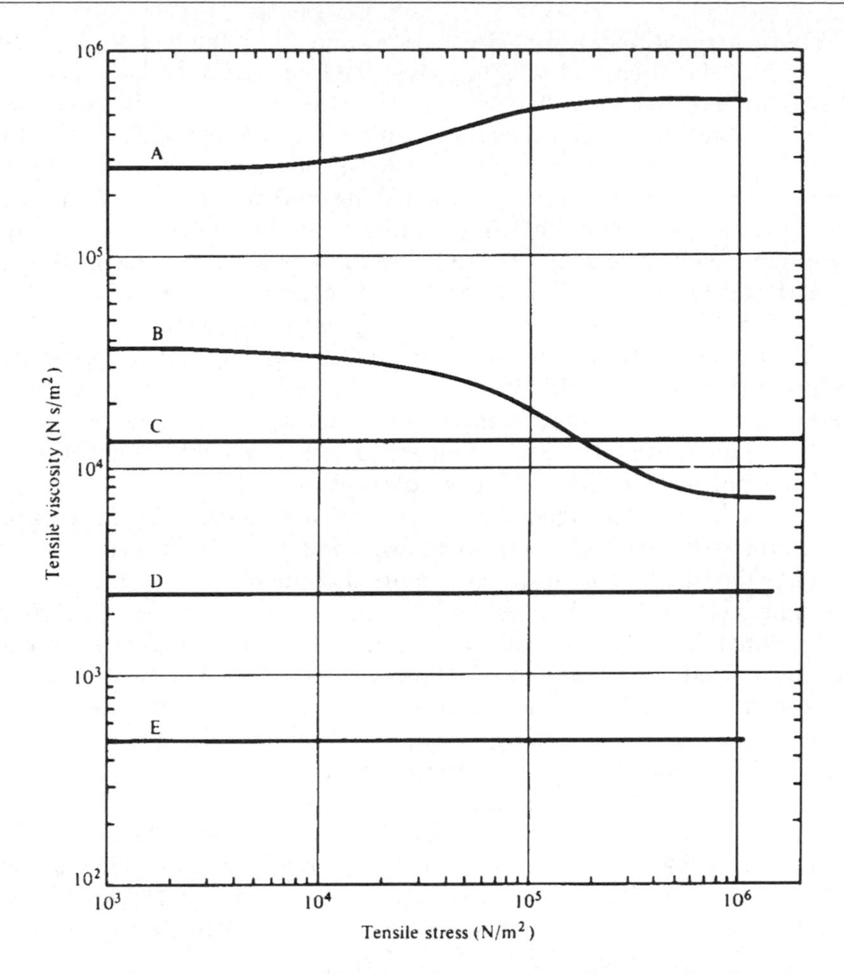

Fig. 8.17 Typical curves of tensile viscosity versus tensile stress for five thermoplastics at atmospheric pressure: A, extrusion-grade low-density polyethylene, 170°C; B, extrusion-grade propylene–ethylene copolymer, 230°C; C, moulding-grade acrylic, 230°C; D, moulding-grade acetal copolymer, 200°C; E, moulding-grade nylon-6,6, 285°C (data courtesy ICI)

and high-density polyethylene, the tensile viscosity drops to a plateau with increasing stress, and the behaviour may be termed 'tension thinning'. In highly branched polymers such as low-density polyethylene and *cis*-polyisoprene, high extension rates inhibit disentanglement and the viscosity rises to a plateau, and these melts show 'tension stiffening' behaviour.

There is a dearth of data on tensile viscosity under conditions other than uniaxial tension. Under biaxial tension it is thought that the viscosity would be twice that under uniaxial tension at the same stress, but this has yet to be investigated in detail.

Fluids are not very strong in tension, and most fluids have a tensile strength of the order of 1 MN/m^2, and therefore readily cavitate. Polymer melts are no exception, having tensile strengths usually in the range 0.5–5 MN/m^2, the value depending mainly on the type of polymer and to a smaller extent on the test temperature. Many flow defects are caused by the tensile stresses somewhere in a process having exceeded the maximum which the melt can withstand, so it is essential to design processes to operate below this limit, unless defects are to be deliberately exploited. In flow in dies, the tensile deformation can occur in a deliberately tapered channel, or just before the entrance to an abrupt change in cross-section.

The further significance of viscous tensile deformations can be appreciated by discussing two examples. Packaging film is usually made by extruding a very thin-walled tube which is inflated by internal air pressure and axially drawn before it solidifies. Each of these processes introduces uniaxial stress, and the polymer extends. A feature of vital interest in production is the behaviour of the tube when it contains a local thin spot. With a tension-stiffening melt, the high local stress will tend to stabilize the thin spot because the surrounding material is more deformable. But with a tension-thinning melt, the local thin spot becomes thinner still. Thus, all other things being equal, it is easier to make close-tolerance thin-gauge film from low-density polyethylene than from high-density polyethylene. Tests involving shear flows are unlikely to reveal this significant difference between the two types of polyethylene. (Cooling by radiation and convection from the tube surface also occurs during the inflation stage, and this also helps to stabilize the process particularly in tension-thinning polymers.)

The second example concerns the flow in the two-roll mill. At any position along the midplane of flow, the volume flow rate is constant. It therefore follows that the melt accelerates towards the nip and becomes extended. This extension is accompanied by tensile stress which therefore absorbs power in addition to that needed to induce shear flow.

Detailed calculations lie outside the scope of this introductory text. Further information will be found in the further reading list.

PROBLEMS

1. Your job is to make domestic dustbins from polymer X. Two commercial grades of X are available, one having a high molecular weight and the other a low one. If the dustbins could be made by either injection moulding or extrusion blow moulding, would both grades be equally suitable for each process? Which grade would you prefer for which process?
2. A parison 1 m long is extruded from low-density polyethylene at 170°C over a period of 5 s and is left hanging for a further 3 s. Estimate roughly the viscous sag in the parison over the 3 s period. Suggest a method for making a tubular parison of uniform wall thickness immediately prior to blowing.

3. Within a process, two circular flow channels of constant cross-section have radii R_1 and R_2 with $R_1 > R_2$. They are joined by a linearly tapered circular channel of length L. Basing your analysis on the average velocity at any cross-section in the tapered channel, show that the average rate of tensile deformation $\dot{\varepsilon}$ is related to the half-angle of convergence θ and the shear rate at the wall γ_ω by $\dot{\varepsilon} = (\dot{\gamma}_\omega/2)\tan\theta$. On what basis would you now decide whether the tensile stresses in the converging flow were dangerously high?

8.7 ELASTICITY OF POLYMER MELTS

8.7.1 Elasticity phenomena

Chapter 5 showed that the stiffness of solid polymers is in general time dependent, although there were limiting states where the behaviour was sensibly elastic, e.g. polymers well below T_g or lightly crosslinked polymers well above T_g. For non-crosslinked polymers, creep under constant load was shown to have a 'time-dependent elastic' character provided the solid polymer had not yielded. This provided the basis for a superposition theorem valid in the linear viscoelastic region.

Polymer melts are also viscoelastic: under some circumstances, the response to stress is essentially viscous and has been discussed earlier in this chapter, and under other circumstances the response is essentially elastic. At short timescales of loading, the polymer melt stores strain energy which can be partially recovered on removing the stress. The following examples illustrate elastic behaviour.

First, at the end of a test run using a capillary rheometer at constant rate of displacement after the drive has been switched off, if the thin rod of polymer melt hanging from the die is stretched axially by pulling with a (gloved) hand and then let go, the rod recovers back to its unstretched state. Second, if the rheometer drive is suddenly switched off during a high-speed test, the leading edge of the melt will bounce – this is often seen in extrusion blow moulding where a ram accumulator is used to deliver a large volume of melt quickly through an annular die, and the parison bounces when the feed is stopped. Third, the melt emerging from a die swells and has a greater cross-sectional area than that of the die, often by a factor between 1.5 and 2. Fourth, if some molten polymer is moulded in a gloved hand to a spherical shape, it will bounce when dropped on a hard surface.

8.7.2 Spring and dashpot model

Useful but incomplete insight into polymer melt behaviour derives from representing melt behaviour by a simple Maxwell model of a spring and a dashpot in series. The two elements can work together either both in shear or both in tension. Under tensile stress σ the strain rate $\dot{\varepsilon}$ is

$$\dot{\varepsilon} = \frac{1}{E}\frac{d\sigma}{dt} + \frac{\sigma}{\lambda} \tag{8.57}$$

At a constant stress σ_0 the resulting strain is

$$\varepsilon = \sigma_0\left(\frac{1}{E} + \frac{t}{\lambda}\right) \tag{8.58}$$

It is apparent that, at very short times under stress, the strain is dominated by the elastic response of the spring, and at long times under stress the strain response is essentially viscous. The transition between elastic and viscous behaviour occurs when the two responses are comparable, i.e. the relaxation time, or the natural time, of the material under tensile stress is given by

$$t_T = \lambda/E \tag{8.59}$$

where the tensile viscosity λ and the tensile elastic modulus E relate to the same stress conditions. In a similar approach, the natural time for a shear deformation t_s would be given by

$$t_s = \eta/G \tag{8.60}$$

Because the shear and tensile stresses in a given process are likely to be different, the natural times are likely to be different too.

Each melt processing operation or part of it can be assigned a characteristic timescale t_p, e.g. $t_p = L/\langle v \rangle$. Given the necessary data, it is possible now to ascertain whether the processing operation is likely to be dominated by elastic or viscous behaviour. If $t_p < t_T$ or t_s, then elasticity dominates (e.g. thermoforming) but if $t_p > t_T$ or t_s, then viscous effects are important (e.g. parison sag in blow moulding). The ratio of process time to the natural time of the polymer is sometimes called the Deborah number, *De*.

8.7.3 Elasticity

When a tensile or shear stress is applied to a polymer melt, the molecules uncoil somewhat and become preferentially aligned in the direction of the stress. On removing the stress, the oriented molecules seek to regain their randomly coiled-up configuration. This recovery leads to a contraction in the stress direction. The amount of elastic recovery observed, ε_R in tension or γ_R in shear, can be related to the applied stress σ in tension, or τ in shear, thus defining the spring constant or elastic modulus in the Maxwell model:

$$E = \frac{\sigma}{\varepsilon_R} \quad \text{or} \quad G = \frac{\tau}{\gamma_R} \tag{8.61}$$

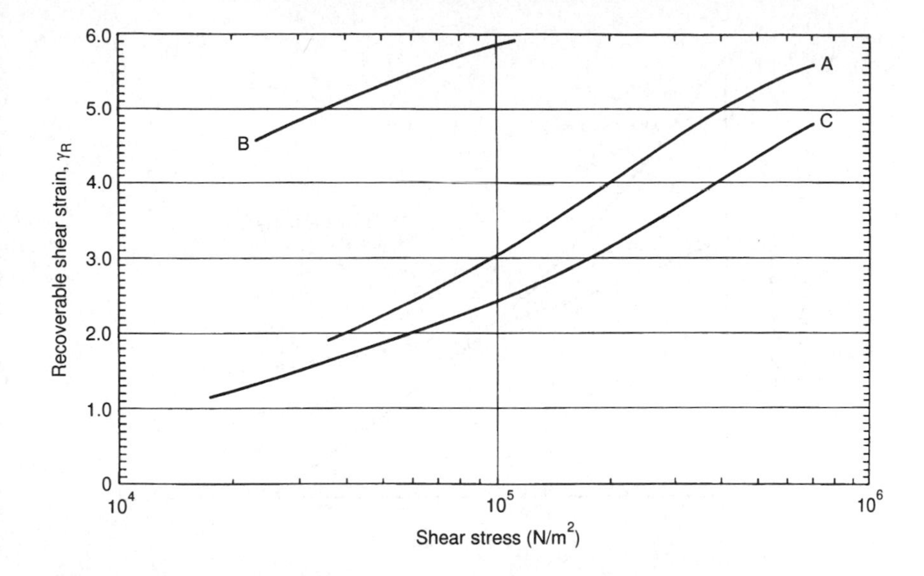

Fig. 8.18 Elastic recovery behaviour of three thermoplastics melts: low-density polyethylenes A and B at 170°C and a propylene–ethylene copolymer C at 230°C (data courtesy ICI)

It is understood that the elastic modulus depends not only on the polymer but also on the processing conditions, particularly the applied stress and the temperature. This suggests that the spring and dashpot system should more properly be described as a 'Maxwell model with variable coefficients'.

Figure 8.18 shows some typical data relating shear stress at the wall in preceding shear flow in a capillary to the recovered shear strain; most of these data relate to the same polymers for which shear viscosity data are given in Fig. 8.11. Experiments suggest that the maximum amount of strain which can be recovered is only a few units of shear. The tensile modulus has a value three times that of the shear modulus at the same stress. The elastic shear modulus of most polymer melts is of the order of 10^5–10^6 N/m^2. Shear modulus does vary with temperature but not nearly as much as does the shear viscosity, and this is illustrated in Fig. 8.19.

8.7.4 Post-extrusion swelling

The term 'die swell' is widely and incorrectly used in informal conversation: the die is of course made of solid metal and does not swell at all, and the context makes it clearly understood that it is the emerging polymer which swells. The shorthand 'polymer swell' is alas not in common usage.

As already mentioned in section 8.7.1 swelling occurs when the melt leaves the extrusion die. To understand why this happens, consider the flow of melt

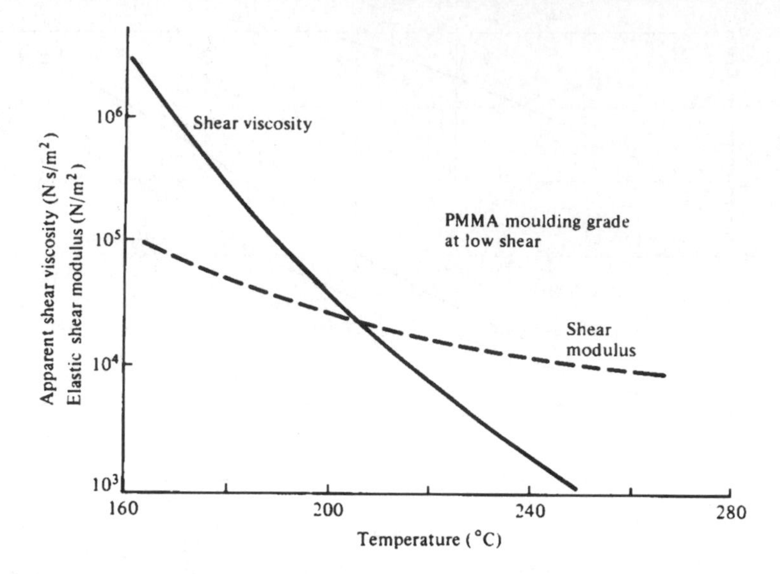

Fig. 8.19 Comparison of temperature dependence of shear viscosity and modulus

within a long die of constant circular cross-section. At modest rates of extrusion, the fully developed velocity profile is approximately parabolic; the polymer near the die wall is highly sheared and hence highly oriented, whereas the melt near the centreline is hardly sheared at all. On emerging from the die, the highly oriented molecules in the material near the surface of the rod want to curl up to a random kinked state and hence wish to shrink in the extrusion direction and swell in the radial direction to conserve constant volume. In contrast, the material in the centre of the rod is not oriented and does not need to curl up further on emerging from the die. However, as the two regions have a common boundary, the outside contracts but is constrained by the core, and the core elongates in response to the outside, until equilibrium is reached. The faster the rate of extrusion, the more pronounced is the orientation profile within the die and hence the greater the swelling. If the rod is free to contract axially, the swelling ratio B_R is defined as

$$B_R = \frac{\text{diameter of extrudate}}{\text{diameter of die}}$$

where both diameters are measured at the melt temperature.

Cogswell (1981) has presented a relationship between B_R and γ_R for a rod formed from a die having a length/diameter ratio greater than 16, shown graphically in Fig. 8.20. Such a relationship (which is not universal) facilitates the design of dies: there are two common problems. First, the simpler one: given a die, what will be the rod size R_d for a given flow rate Q? The

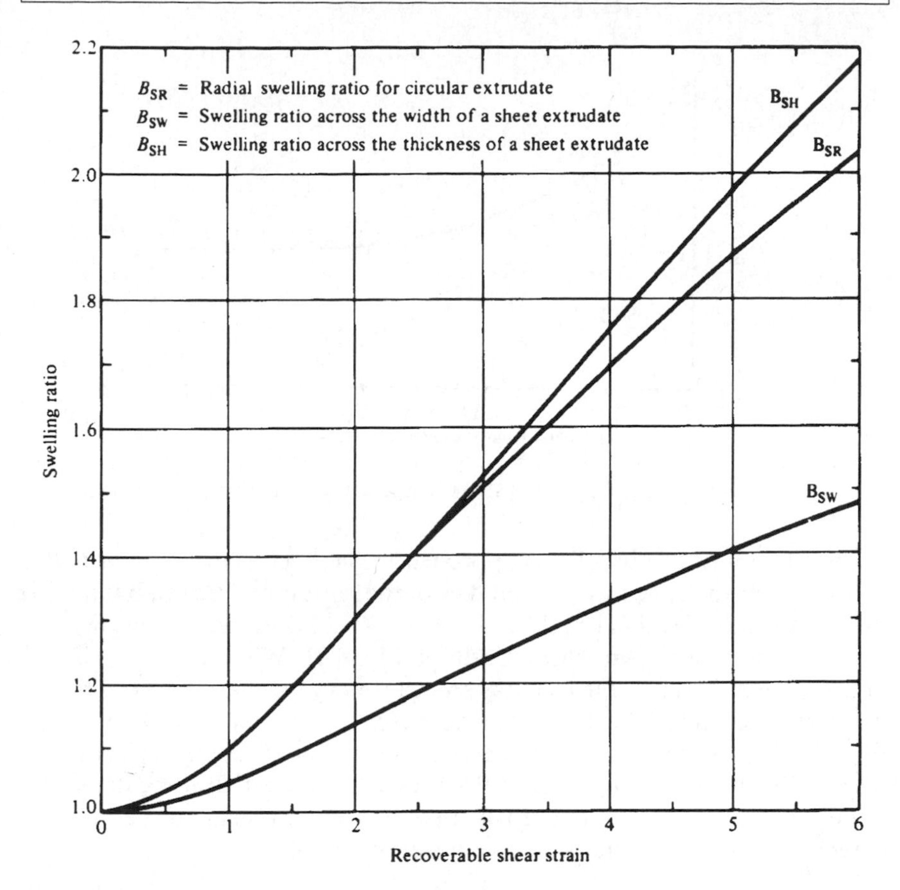

Fig. 8.20 Dependence of swelling ratio on recoverable shear strain for long capillary and slot dies (data courtesy ICI)

steps are to calculate the apparent shear rate at the wall $\dot{\gamma}_w = 4Q/\pi R_d^3$, from which the shear stress at the wall can be read from the flow curve (e.g. Fig. 8.11); elasticity data for that stress (Fig. 8.18) give the recoverable strain γ_R and hence (Fig. 8.20) the swelling ratio B_R. Second, the more common problem is to calculate the die size needed to make a product of a given size at a given flow rate. This is a more tedious problem to solve because a value of R_d has to be found (by iteration) which satisfies all dimensions and properties; fortunately, convergence is rapid.

A similar approach can be followed where a tube or a wide thin sheet is extruded through a die, the exit region of which has a constant ratio of length to die gap with a ratio greater than 16. The die swell for the thickness of the extrudate B_H is greater than that for the width or circumference B_W, as indicated in Fig. 8.20; both relate to the same shear rate at the wall, $\dot{\gamma}_W = 6Q/WH^2$. Cogswell suggests that $B_H = B_W^2$.

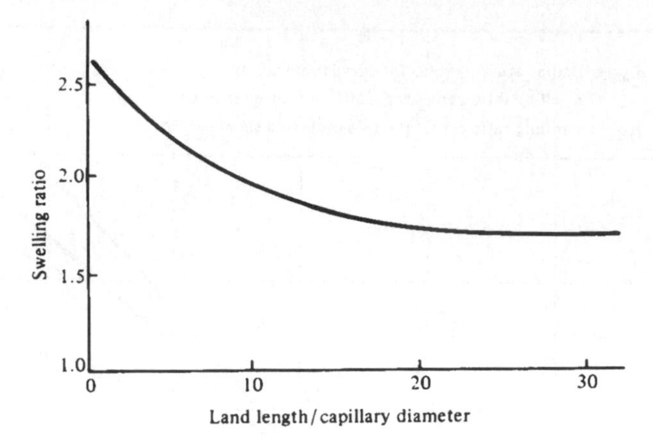

Fig. 8.21 Die swell versus *L*/*D* at constant temperature and constant flow rate

The lack of generality of this approach to die swell is evident when B is measured for the same flow rate on dies of the same cross-section but having different ratios of (flow) length to diameter. As L/D decreases, the swelling ratio increases, as shown schematically in Fig. 8.21. Why should this be so? One explanation is that the melt has some memory of what happened before it entered the land of the die, and as discussed in section 8.6, flow converges into the land region so that orientation results from tensile deformation (especially about the centreline of flow) as well as shear. In long die lands, the orientation from previous tensile flow has time to relax out before melt emerges from the die, and hence the lower die swell.

PROBLEMS

1. Answer the following in not more than three or four lines:
   (a) Why is the die swell across the width of a sheet less than across the thickness?
   (b) Sketch the likely shape of an extrudate taken from a wide rectangular slot die.
   (c) Sketch the likely shape of extrudate from a die of square cross-section.
   (d) Sketch the shape of die needed to ensure that a sheet of rectangular cross-section can be made.
2. The annular gap in a pipe die has a mean radius of 60 mm and a gap of 2 mm, with a land length of 40 mm. Low-density polyethylene at 190°C is extruded through this die at a flow rate of $5 \times 10^{-6}\,m^3/s$. Estimate the dimensions of the extruded pipe and the haul-off speed required to ensure no drawdown of the pipe. The melt density of this polyethylene is $750\,kg/m^3$ at 190°C.

3. Derive an expression for the shear strain profile for flow of a Newtonian fluid in a long capillary. Would you expect a power-law fluid to swell more or less than a Newtonian fluid under otherwise identical conditions?
4. A rod of 8 mm diameter is to be extruded from a propylene–ethylene copolymer at 230°C (for which data are given in this chapter). The haul-off gear applies no tension to the rod and is set to run at 15 mm/s. Recommend suitable die exit dimensions. (Details of the body of the die are not required but you may assume that the ratio of land length to diameter is at least 16.)

8.8 HEAT TRANSFER IN POLYMER PROCESSING

8.8.1 General comments on heat transfer

It is evident that heat transfer plays an important role in polymer processing. Major preoccupations in melt processing include melting, solidification, control of temperatures during heating, melt transport and cooling, and (where relevant) chemical reactions such as foaming or crosslinking. These set sizes on conventional services such as electric motors, heaters, water baths and refrigerators, and affect capital costs, running costs, production rates and product quality.

Steady-state heat transfer in solid polymers is more likely to be encountered in products in service rather than during processing. However, in some (assumed steady) flows of polymer melts between boundaries at the same temperature, viscous dissipation can induce a modest increase in average temperature and a high local temperature rise.

Of much greater relevance to polymer processing is non-steady-state heat transfer. This sometimes involves heating static slabs of solid polymer and often involves cooling static masses of polymer with or without a phase change. The heat transfer may result from imposed temperature changes at the surface (or from a liquid in contact with the surface), or may result from internal heating sources such as chemical reaction. Non-steady heat transfer may also occur in a moving polymer melt; the melt may move within channels which are at a lower or higher temperature and transverse temperature profiles may dramatically affect the cross-section available to melt flow. In many processes, heating or cooling is accompanied by a change of phase from liquid to solid or vice versa.

A solution to these illustrative and other problems derives from applying the principles of conservation of energy, momentum and mass. Some of these solutions are standard ones encountered in earlier engineering courses, but the orders of magnitude of numbers may be rather unfamiliar. Others solutions will not be familiar in detail. This section introduces the main ideas relevant to polymer engineering but there is not sufficient space to develop them in detail.

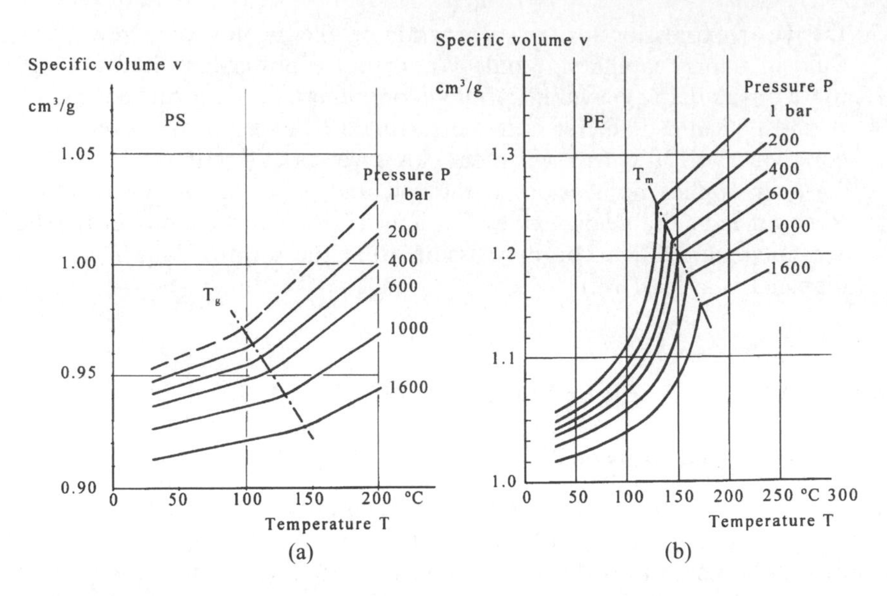

Fig. 8.22 *pvT* diagrams of (a) (amorphous) polystyrene and (b) (semi-crystalline) high-density polyethylene measured at mean cooling rates of 0.04 and 0.03 K/s, respectively (courtesy VDMA, 1979)

8.8.2 Thermal properties of polymers

The thermal properties of polymers can be understood with reference to the remarks on molecular mobility in section 2.6. A rise in temperature in an unrestrained polymer is accompanied by an increase in volume and a decrease in density. Linear expansion coefficients increase somewhat with increase in temperature. For an amorphous polymer, the relationship between specific volume and temperature shows a change in slope by a factor of about 3 at the glass transition temperature (see the 1 bar curves of the *pvT* diagram, Fig. 8.22), as does enthalpy (Fig. 8.23).

In the solid state it is customary to use the term density, whereas in analysis of melt processing the term specific volume is often used: the one is the reciprocal of the other. We have already seen in the 1 bar curves of Fig. 8.22 that the density decreases with increase in temperature. The compressibility of a polymer as a function of hydrostatic pressure p and temperature T is presented as a *pvT* diagram in the manner shown in the full Fig. 8.22. There are two general forms, that for amorphous polymers and that for partially crystallizable polymers, the main difference between the two being the step involving melting or crystallization at about the melting point.

The most obvious conclusion from the *pvT* diagram is that polymers are indeed somewhat compressible in the melt: the simple approach to process analysis we have used in this chapter has ignored this. For low-pressure

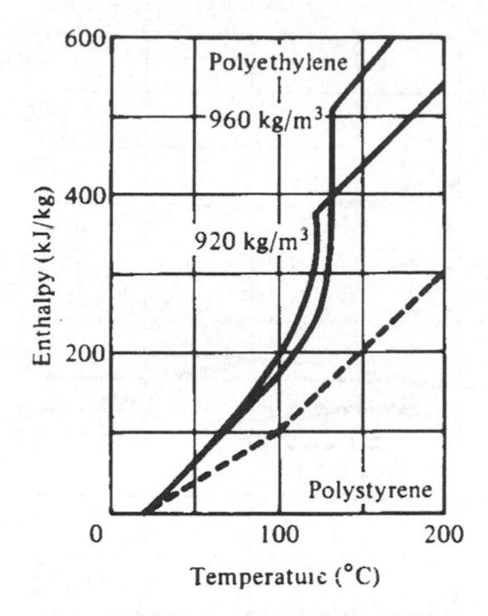

Fig. 8.23 Enthalpy versus temperature (data courtesy Hoechst)

processes this is not a problem, but for high-pressure processes, such as injection moulding, the compressibility should not be ignored, and this is further discussed in the books in the section on further reading. In particular, the *pvT* diagram is useful in helping to explain the role of 'after pressure' applied after the cavity of an injection mould has been filled but before solidification has fully taken place, and we comment further on this in section 9.2.1(a).

The phase change from solid crystallite to amorphous melt occurs at nominally constant temperature and there is therefore a step in the enthalpy–temperature diagram (Fig. 8.23); correspondingly, the specific heat shows a sharp peak at the melting point. Above the melting point, any further increase in temperature merely increases enthalpy and specific volume. The main differences between amorphous and partially crystallizable polymers when heated from room temperature to their appropriate melt processing temperature range lie in the much bigger volume changes (roughly double) and the latent heat of fusion for the crystalline polymers.

The thermal conductivity of amorphous polymers changes little over a wide range. However, for the partially crystallizable polymers there is a significant decrease in conductivity with increase in temperature until the melting point is reached (Fig. 8.24).

An important question for the processor is the choice of processing temperature for making a thermoplastics product. It is not easy to give fixed rules: normally processors have advice from the raw materials suppliers and draw on their own experience. Certainly there is seldom a fixed temperature, rather we should aim for a window; too low a temperature and the viscosity is too low with risk of premature freeze-off or in extreme cases the process cannot be done, and too high a temperature leads to too low a viscosity and loss of control together with risk of thermal degradation. Some

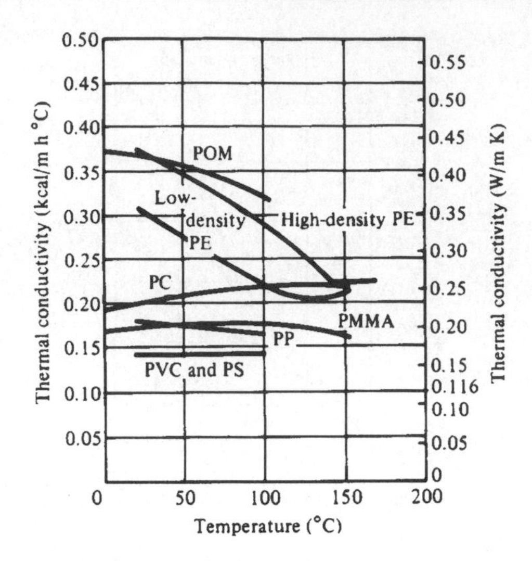

Fig. 8.24 Thermal conductivity versus temperature (data courtesy Hoechst)

control over viscosity is available by choosing a suitable molecular mass. We suggest some ground rules here for processes which occur at low and at high deformation rates.

As a rough guide for the newcomer to plastics, we might suggest that for thick-walled extrudates many crystalline thermoplastics can be shaped at 50–70°C above their melting point T_m and many amorphous thermoplastics at about 80–100°C above their glass transition temperature T_g. Thus unplasticized PVC has $T_g = 80$°C and is extruded at about 170°C (within a narrow temperature range dictated by thermal stability). Polypropylene has a melting point of about 165°C and is typically extruded at about 230°C (with a wider range than for PVC). Thin-walled sections such as tubular film may be extruded at higher melt temperatures to give acceptable properties at acceptable throughputs and pressure drops.

For injection moulding the processing melt temperatures will normally be rather higher than those given above in order to achieve an acceptable pressure drop during injection, whilst retaining sufficiently high molecular mass to confer acceptable strength and toughness.

For moulding of rubber and hot crosslinking of unsaturated polyester or epoxide resins, typical process temperatures are in the range 140–170°C.

There are exceptions to these broad generalizations, and the reader is strongly advised to consult the relevant trade literature.

8.8.3 Steady-state heat transfer in static polymers

The approach is the energy equivalent of that outlined for momentum transfer in section 8.2, helped by the fact that energy is a scalar quantity.

It is well documented in standard texts and will have been amply covered in elementary courses. It is assumed that the polymer mass is uniform and isotropic with constant properties over the temperature range of interest and that the rate of heat transfer is constant over the timespan of interest. Making an energy balance over a representative element of fluid leads to Fourier's law of heat conduction, which relates the heat flux q in the positive coordinate direction, the temperature gradient ∇T and the thermal conductivity k:

$$q = -k\nabla T \tag{8.62}$$

Equation (8.62) may be integrated to give the required temperature distribution in the body. The constants of integration are evaluated from appropriate boundary conditions such as specified surface temperatures, or rates of heat transfer across interfaces, sometimes in terms of a heat transfer coefficient h defining the rate of heat transfer across a boundary between a solid polymer surface at T_s and a convecting fluid having a bulk temperature T_f:

$$q = h(T_s - T_f) \tag{8.63}$$

The thermal conductivity of polymers is in the range 0.15–0.45 W/m K (compared with > 100 W/m K for metals) so that temperature gradients can be large even though heat fluxes are modest.

Heat transfer into a convecting fluid is of interest in cooling free surfaces after shaping. Typical values of heat transfer coefficients encountered in polymer processing are given in Table 8.2.

It is helpful to ascertain whether heat transfer is likely to be dominated by internal conduction or external convection, before delving into detailed calculations. This can be determined by calculating the ratio of thermal conductance by convection and conduction, expressed as the Biot number

$$Bi = hx/k \tag{8.64}$$

Here x is a characteristic thickness dimension: the radius of a solid cylinder, the total thickness of a slab heated or cooled from one face only or the half-thickness if heated or cooled from both sides. If $Bi < 1$, convection

Table 8.2 Typical values of heat transfer coefficient

Medium	h(W/m^2 K)
Polymer to still air	10–20
Polymer to air moving at 25 km/h	30–40
Solid polymer to metal surface	250–500
Polymer to oil bath	400–600
Polymer to stirred water	1500–3000
Polymer melt at high pressure to metal surface (assume perfect thermal contact)	> 10 000

dominates; if $Bi > 1$, conduction dominates. Most processing is dominated by conduction; convection is only important in thin polymers (e.g. films) surrounded by air.

8.8.4 Non-steady-state heat transfer

Because of the low diffusivities of molten polymers, most processing involves temperatures which change with time as well as position. Returning to the equation of energy for a polymer mass having no internal velocity gradients, the energy balance across a representative element must now include the rate of change of temperature with time. For an isotropic polymer of constant density, the governing equation now becomes

$$\frac{\partial T}{\partial t} = \alpha \nabla^2 T \tag{8.65}$$

where $\alpha = k/\rho C_p$ is the thermal diffusivity [m^2/s] and C_p is the specific heat. For convenience, the ensuing discussion is limited to bodies which initially have a uniform temperature T_0. For heat flow in one direction only, the situations of major interest are the slab of uniform thickness $x = 2H$ and very large cross-sectional area (normal to flow), for which equation (8.65) reduces to

$$\frac{\partial T}{\partial t} = \alpha \frac{\partial^2 T}{\partial x^2} \tag{8.66}$$

and the long cylinder of uniform radius R, for which equation (8.65) reduces to

$$\frac{\partial T}{\partial t} = \alpha \left(\frac{\partial^2 T}{\partial r^2} + \frac{1}{r} \frac{\partial T}{\partial r} \right) \tag{8.67}$$

These equations must now be solved with appropriate boundary conditions. The two most common ones encountered in standard texts are a sudden change in surface temperature T_s at $t = 0$, and sudden exposure to a convecting fluid at bulk temperature T_f at $t = 0$. The analytical solutions are rather cumbersome, and it is useful to express the results graphically in terms of dimensionless quantities. Dimensionless time is described by the Fourier number

$$Fo = \alpha t/H^2 \quad \text{or} \quad \alpha t/R^2 \tag{8.68}$$

The dimensionless distance ξ from the midplane or axis is

$$\xi = x/H \quad \text{or} \quad r/R \tag{8.69}$$

It is convenient to use different expressions for dimensionless temperature Y depending on the boundary conditions. For a change in surface temperature:

$$Y_1 = \frac{T(x,t) - T_s}{T_0 - T_s} \quad \text{or} \quad \frac{T(r,t) - T_s}{T_0 - T_s} \tag{8.70}$$

and for immersion in a fluid at bulk temperature T_f:

$$Y_2 = \frac{T(x,t) - T_f}{T_0 - T_f} \quad \text{or} \quad \frac{T(r,t) - T_f}{T_0 - T_f} \tag{8.71}$$

For a sudden change in surface temperature of a solid, information relating Fo, Y_1 and ξ are indicated for a slab in Fig. 8.25, and for a cylinder in Fig. 8.26. For sudden immersion in a convecting fluid, information relates Fo, Y_2, ξ and Bi, as indicated for a slab at the midplane and the cylinder at the axis in Figs 8.27 and 8.28, respectively. More detailed information is given in the standard texts.

It is worth adding some comments on the implications of Figs 8.25–8.28, all of which apply to both heating and cooling situations. Let us consider Fig. 8.25. At small values of Fourier number, say $Fo < 0.01$ in the region A, for example for short times after a surface temperature change, there is only a very small penetration of temperature change, and the bulk of the sheet is unaffected. At $Fo \approx 0.2$ in the region B the midplane has just started to change temperature, and there is a substantial temperature gradient through the plate. At $Fo \geqslant 1$ most of the temperature change has taken place and the temperature gradient through the plate is small. Similar remarks can be made about Figs 8.26–8.28.

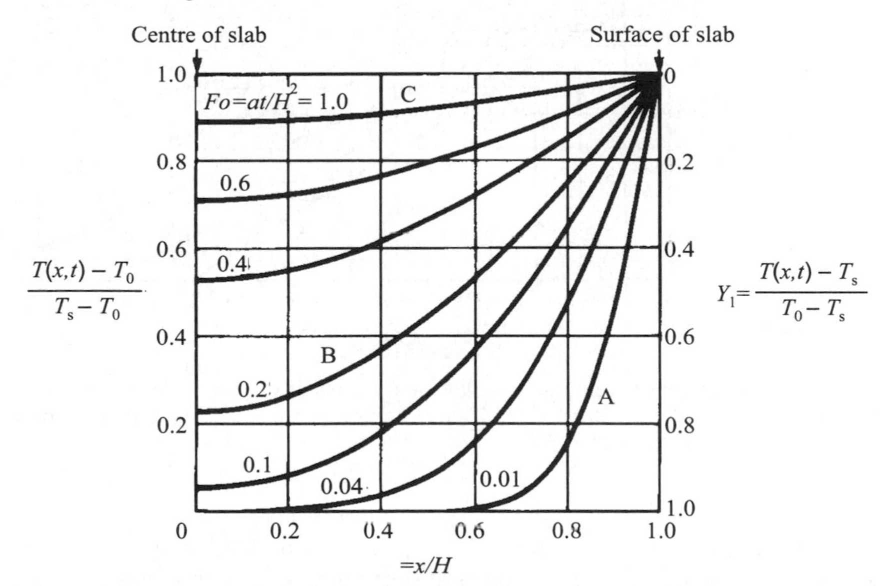

Fig. 8.25 Temperature profiles versus position in space for unsteady–steady heat conduction. Initial temperature T_0 (uniform), surface temperature T_s for $t > 0$. Slab of thickness $2H$ (from Carslaw and Jaeger, 1959)

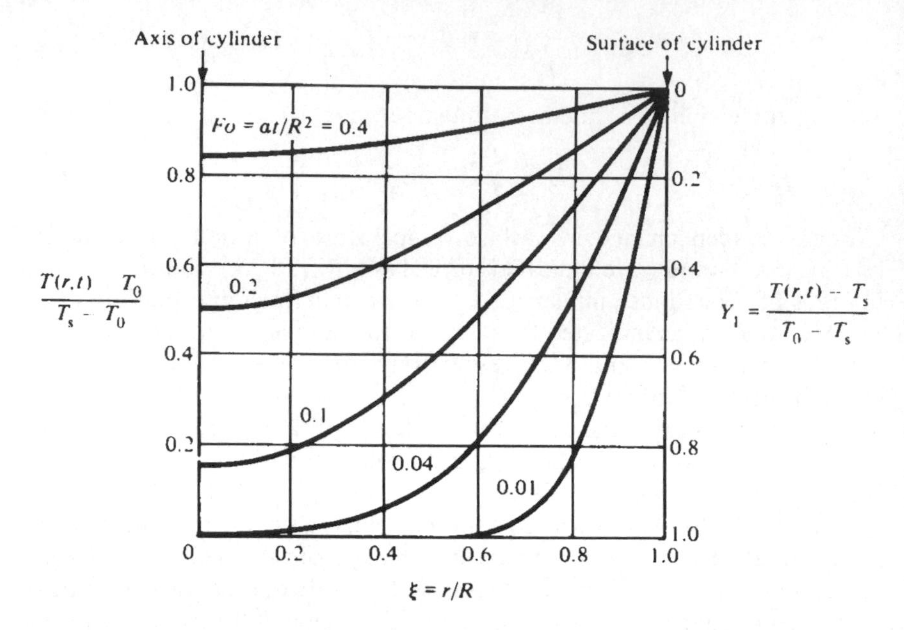

Fig. 8.26 As Fig. 8.25, cylinder of radius R

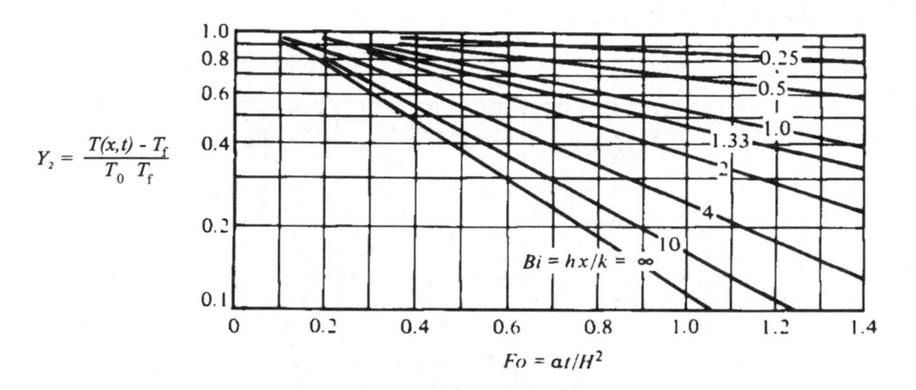

Fig. 8.27 Temperature development with time for unsteady–steady heat conduction. Initial temperature T_0 (uniform), fluid temperature T_f for $t > 0$. Slab of thickness $2H$. $\xi = 0$ (from Middleman, 1977)

The main conclusion from this account of non-steady-state heat transfer is that if a fast cooling time is needed in a process, thin walls are particularly attractive with the added advantage that the volume of material used is small.

This approach can be adapted to estimate temperature profiles in bodies of finite size in which conduction occurs in more than one direction. For a parallelepiped initially at T_0 and at $t = 0$ suddenly cooled to T_1 at the

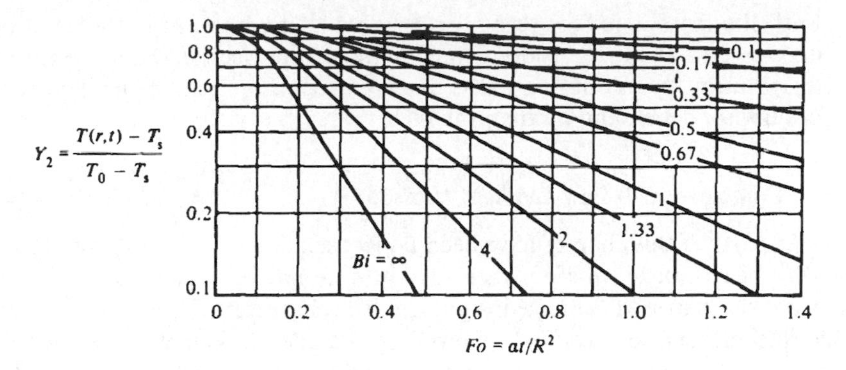

Fig. 8.28 As Fig. 8.27, cylinder of radius $R, \xi = 0$

surfaces, the individual dimensionless temperatures in the x-, y- and z-directions would be Y_{1x}, Y_{2x} and Y_{3x}. By superposition, the temperature $T(x,y,z,t)$ is then given by

$$Y_1 = Y_{1x} Y_{1y} Y_{1z} \tag{8.72}$$

It is of course tempting to apply these data to any situation where the surface temperature of a polymer mass is changed, and melt processing would seem an obvious candidate. This classical theory of non-steady-state conduction does not apply exactly – and may be quite misleading – under the following circumstances (where it is better to consult more advanced texts).

- The polymer mass is not in perfect thermal contact with a solid heat sink (possibly caused by polymer shrinkage).
- The specific volume or other properties change substantially over the temperature interval of interest; this may be troublesome if there is a change of phase.
- Crystallization occurs during cooling or melting during heating. This will lead to a dwell time at nearly constant temperature on a plane or shell moving through the body. It can be allowed for by splitting up the process into solid and liquid states with the solid–liquid boundary advancing with time.
- Heat is supplied or generated within the body during the timescale of interest. This can take the form of a crosslinking exotherm (itself temperature dependent), or viscous dissipation in a moving melt, or a flow of melt between boundarics at a different temperature from the melt.
- The initial temperature of the mass may not be uniform. This can occur in thick sections which are heated externally to effect crosslinking but removed from the vulcanizing chamber before the centre is fully up to temperature, and then allowed to cool slowly in air. In thick-walled blown containers, the hollow article is cooled in the mould from the outside only

until the mass is judged to be form-stable as a whole. In both cases, on exposure to air, the outside surface temperature rises as the temperature distribution evens out internally. This redistribution may be beneficial or otherwise, depending on the process.

8.8.5 Combined melt flow and heat transfer

Earlier parts of this chapter have been concerned either with isothermal flow or with heat transfer in a polymer with no internal velocity gradient. These approaches provided valuable insight into the character of flows where these assumptions can be justified. Examples include slow melt flows within extrusion dies maintained at the melt temperature, cooling of extruded sections in water baths in the absence of haul-off tensions, and heating of rubber or crosslinkable plastics in hot compression moulds: either the melt flow or the heat flow can be disregarded.

But there are situations of great practical importance where flows are not isothermal. For example, the melt can either generate heat during flow between channels held at some temperature roughly that of the inlet melt, or the melt surface can be deliberately maintained at a temperature which is quite different from that of the inflowing melt, or both. Practical examples include viscous dissipation in high-speed flows within heated dies and mould runners, flow in heated or cooled mould cavities, and melt flow in air with superposed external forces, as seen in tubular film or tape-making processes. In these representative situations the distribution of temperatures takes time (because of the low diffusivity) and may well not become fully developed over the timescale of interest – this is in contrast to velocity profiles in steady isothermal flows which usually are fully developed.

Rather than delve into a more detailed approach, it is again helpful to identify the relative importance of these various effects, using several dimensionless groups. The factors of major interest are characteristic values of (average) shear stress τ, (average) shear rate $\dot{\gamma}$, channel separation $2R$ or H, (average) fluid velocity v, and the temperature coefficient of viscosity b defined by

$$\eta/\eta_0 = \exp[-b(T - T_0)]$$

The Griffiths number

$$G = b\tau\dot{\gamma}H^2/k$$

indicates the relative importance of viscous dissipation and thermal conduction. The Brinkman number

$$Br = \tau\dot{\gamma}H^2/k(T_b - T_0)$$

compares viscous dissipation with the effect of an imposed boundary temperature T_b relative to an initial melt temperature T_0. If G and G/Br are both much less than unity, then the flow may be regarded as isothermal, using the kind of approach described in sections 8.2 and 8.4.

The Graetz number

$$Gr = \rho C_p v H^2 / kL$$

compares the relative importance of thermal convection and conduction over a flow length L. If $Gr \ll 1$, heat transfer in the flowing melt is dominated by thermal conduction from the boundaries and the temperature profiles do not change in the direction of flow. Often Gr is large, indicating that temperature profiles develop very slowly in the direction of flow.

Where fast flows of melt occur within boundaries held at the inlet melt temperature, i.e. $G \gg 1$, there is a temptation to model the flows as simply adiabatic in the first instance. The energy balance on an element of unit volume equates the rate of increase of thermal energy to the rate at which work is done by the prevailing pressure:

$$Q\rho C_p \mathrm{d}T = Q\mathrm{d}p \tag{8.73}$$

where dT is an average temperature. The melt is regarded as incompressible and Newtonian with a viscosity η related to the inlet (η_0, T_0) by

$$\eta = \eta_0 \exp[-b(T - T_0)]$$

In isothermal flow in a capillary of radius R and length L, the pressure gradient is described by the Hagen–Poiseuille relationship (Table 8.1, (C)). Combining this with equation (8.73) leads to the average temperature rise:

$$T - T_0 = b^{-1} \ln[1 + (b\Delta P_0 / \rho C_p)] \tag{8.74}$$

where ΔP_0 is the pressure drop in the channel if the conditions were completely isothermal at $T = T_0$.

The form of equation (8.73) (or its non-Newtonian equivalent) confirms the expectation that the faster the flow, the higher ΔP_0 and hence the higher the average temperature rise (and the actual pressure drop would be less than ΔP_0, of course). In flows in high-speed processes such as injection moulding, the average temperature rises calculated by this method are seldom large. The important word here is average. A 10 or 20°C rise in average temperature is unlikely to be worrying in itself, but the peak temperature rise (near the flow boundary) could be at least 3–5 times higher than this, and the residence time could be significant for a layer near the channel wall. This could be sufficient to cause some local premature crosslinking or degradation if the inlet melt temperature T_0 is already high, and this may be detrimental to the quality of the end-product. It is possible to extend this argument a stage further: if the flow boundary is maintained at a temperature above that of the inflowing melt, the temperature rise (and its peak) will be greater than that calculated above – a danger in crosslinking systems –

but if the boundary temperature is lower, this will offset the viscous dissipation and may be a benefit in thermoplastics processing.

In order to estimate the temperature profile in a channel containing flowing melt, it is necessary to make a proper energy balance over a representative element of fluid. Conservation of energy requires that the rate of increase of thermal energy plus the rate of increase in energy by conduction plus the rate of increase in energy by viscous dissipation is zero. The heat-generation term is the product of stress and velocity gradients. Ignoring axial conduction of heat and assuming a constant thermal conductivity, the energy equation for this capillary flow problem takes the form

$$\rho C_p v_z \frac{\partial T}{\partial z} = k \frac{1}{r} \frac{\partial}{\partial r}\left(r \frac{\partial T}{\partial r}\right) + \frac{\eta_{T0}}{\dot{\gamma}_{T0}^{n-1}}\left(-\frac{\partial v_z}{\partial r}\right)^{n-1}\left(\frac{\partial v_z}{\partial r}\right)^2$$

The equation of motion for isothermal flow takes the form

$$-\frac{\partial p}{\partial z} = \frac{1}{r} \frac{\partial}{\partial r}\left[r \frac{\eta_{T0}}{\dot{\gamma}_{T0}^{n-1}}\left(-\frac{\partial v_z}{\partial r}\right)^{n-1} \frac{\partial v_z}{\partial r}\right]$$

where the viscosity index $\eta_{T0}/\dot{\gamma}_{T0}^{n-1}$ will be temperature dependent. These two equations will have to be solved numerically; the flow boundary conditions are no slip at the wall and no velocity gradient at the axis; the thermal boundary conditions are uniform inlet temperature T_0, wall temperature T_R and no temperature gradient at the axis.

Valuable analyses of a wide variety of combinations of heat and momentum transfer found in a range of polymer processes have been developed, but these lie outside the scope of this introductory book. For further information the reader is advised to consult more specialized texts such as Middleman (1977), Agassant (1991) or Tadmor and Gogos (1979).

In restricted circumstances, it is helpful to examine simple descriptions of flow situations which have been devised partly by theory and partly by careful experimentation. For example, Barrie studied the cooling process during spreading disc flow of molten thermoplastic in a cold cavity with a cavity thickness in the range 2–5 mm. He found that the amount of material ΔH frozen onto the mould surface during the cavity filling operation was related to the filling time by

$$\Delta H \simeq 2 Y_1 \alpha^{1/2} t^{1/3}$$

The exponent of α is that required by non-steady-state conduction. Y_1 is defined by equation (8.70) with $T(x, t)$ taken as the apparent freeze-off temperature for which some values are given in Table 8.3. The exponent of t is based on experimental investigation and allows for the competing effects of mould cooling, viscous dissipation and phase change. The value of t at any distance x from the gate is taken as zero until the melt front passes that point. This study provides a simple means of estimating whether a cavity can

Table 8.3 Typical values of apparent freeze-off temperature occurring in injection moulding (after Barrie, 1978)

Polymer	*Apparent freeze-off temperature* (°C)
Low-density polyethylene	90
Polypropylene	135
General-purpose polystyrene	130
Polycarbonate	200
Nylon-6,6	240
PVC	140
PMMA (acrylic)	160
Polyacetal	135

be completely filled before the gate region freezes-off and stops the flow. Perhaps surprisingly, this approach does not distinguish between the behaviour of amorphous and partially crystalline polymers, although it must be remembered that this may be hidden in the values of apparent freeze-off temperatures. Amorphous polymers will be form-stable only when approaching T_g after cooling, i.e. well below melt processing temperatures, whereas partially crystallizable polymers will be form-stable only a few tens of degrees below their melting point because of the reinforcing effects of the crystallites.

PROBLEMS

1. A large flat polymer sheet 10 mm thick has a thermal diffusivity of 10^{-7} m^2/s and a thermal conductivity of 0.2 W/m K. It is initially at a uniform temperature of 120°C. How long will it take to cool the midplane of the sheet to 80°C under the following conditions: (a) perfect thermal contact with two flat metal mould surfaces kept at 20°C, (b) immersion in stirred water at a bulk temperature of 20°C, (c) exposure to still air at a bulk temperature of 20°C.
2. A cylindrical bucket-shaped product of maximum wall thickness 3 mm is to be moulded from a polymer having a thermal diffusivity of 10^{-7} m^2/s and a thermal conductivity of 0.2 W/m K. The melt, at 220°C, is shaped in a mould cavity held at 20°C. It is recommended that the product can be removed from the mould when the melt reaches 60°C. Estimate and compare the cooling time if the bucket is made by (a) injection moulding, and (b) extrusion blow moulding.
3. If the thermal properties of a polymer melt are taken as $\rho C_p = 2 \times 10^6$ J/m^3, estimate the likely average temperature rise in melt flows in (a) extrusion, (b) injection moulding, assuming adiabatic conditions, and comment briefly on the results.
4. A low-density polyethylene rod 15 mm in diameter at 20°C is to be extruded at a mass flow rate of 23.5 kg/h. After extrusion the rod is to

be cooled in a water cooling bath. Recommend the length of cooling bath, and discuss the strategy you would use to dimension the extruder die and to estimate the size of single-screw extruder needed. Assume $\alpha = 10^{-7}\,m^2/s$.

FURTHER READING

Unidirectional isothermal Newtonian flow

Bird, R.B., Armstrong, R.C. and Hassager, O. (1977) *Dynamics of Polymeric Liquids*, Vol. 1, Wiley, New York.
Bird, R.B., Stewart, W.E. and Lightfoot, E.N. (1960) *Transport Phenomena*, Wiley, New York.
Massey, B.S. (1979) *Mechanics of Fluids*, 4th edn, Van Nostrand Reinhold, New York.
Middleman, S. (1977) *Fundamentals of Polymer Processing*, McGraw-Hill, New York.
Rauwendaal, C. (1990) *Polymer Extrusion*, Hanser, Munich.

Shear viscosity of polymer melts

Brown, R.P. (ed.) (1981) *Handbook of Plastics Test Methods*, Godwin, Harlow.
Brydson, J.A. (1981) *Flow Properties of Polymer Melts*, Godwin, Harlow.
Cogswell, F.N. (1981) *Polymer Melt Rheology*, Godwin, Harlow.
Whorlow, R.W. (1980) *Rheological Techniques*, Ellis Horwood, Chichester.

Unidirectional power-law flows

Agassant, J.-F., Avenas, P., Sergent, J.-Ph. and Carreau, P.J. (1991) *Polymer Processing: Principles and Modelling*, Hanser, Munich.
Fenner, R.T. (1979) *Principles of Polymer Processing*, Macmillan, London.
Hensen, F. (ed.) (1988) *Plastics Extrusion Technology*, Hanser, Munich.
Michaeli, W. (1992) *Extrusion Dies for Plastics and Rubber: Design and Engineering Computations*, Hanser, Munich.
Middleman, S. (1977) *Fundamentals of Polymer Processing*, McGraw-Hill, New York.
Tadmor, Z. and Gogos, C.G. (1979) *Principles of Polymer Processing*, Wiley, New York.
Tanner, R.I. (1985) *Engineering Rheology*, Oxford University Press, Oxford.

Tensile viscosity

Brydson, J.A. (1981) *Flow Properties of Polymer Melts*, Godwin, Harlow.
Cogswell, F.N. (1978) in *Polymer Rheology* (ed. R.S. Lenk), Applied Science, London.
Cogswell, F.N. (1981) *Polymer Melt Rheology*, Godwin, Harlow.
Ferguson, J. and Kemblowski, Z. (1991) *Applied Fluid Rheology*, Elsevier Applied Science, London.

Petrie, C.J.S. (1979) *Elongational Flows*, Pitman, London.
Tadmor, Z. and Gogos, C.G. (1979) *Principles of Polymer Processing*, Wiley, New York.

Mixing and blending of polymer melts

Ottino, J.M. (1989) *The Kinematics of Mixing: Stretching, Chaos and Transport*, Cambridge University Press, Cambridge.
Rauwendaal, C. (ed.) (1991) *Mixing in Polymer Processing*, Marcel Dekker, New York.

Elasticity of polymer melts

Brydson, J.A. (1981) *Flow Properties of Polymer Melts*, Godwin, Harlow.
Cogswell, F.N. (1981) *Polymer Melt Rheology*, Godwin, Harlow.
Middleman, S. (1977) *Fundamentals of Polymer Processing*, McGraw-Hill, New York.
Tadmor, Z. and Gogos, C.G. (1979) *Principles of Polymer Processing*, Wiley, New York.

Heat transfer in polymer processing

Barrie, I.T. (1978) The rheology of injection moulding, in *Polymer Rheology*, (ed. R.S. Lenk), Applied Science, London, ch. 13.
Bird, R.B., Stewart, W.E. and Lightfoot, E.N. (1960) *Transport Phenomena*, Wiley, New York.
Carslaw, H.S. and Jaeger, J.C. (1959) *Conduction of Heat in Solids*, Oxford University Press, Oxford.
Middleman, S.P. (1977) *Fundamentals of Polymer Processing*, McGraw-Hill, New York.
Tadmor, Z. and Gogos, C.G. (1979) *Principles of Polymer Processing*, Wiley, New York.
VDMA (Verein Deutscher Maschinenbau-Anstalten e.V.) (ed.) (1979) *Kenndaten für die Verarbeitung thermoplastischer Kunststoffe, Teil 1, Thermodynamik*, Hanser, Munich.

9 Some Interactions between Processing and Properties

9.1 INTRODUCTION

Simple design methods usually assume that the product is homogeneous and isotropic before being put into service. In products made from fibre-reinforced composites, this is so obviously untrue that special design procedures have been developed. These were discussed in Chapter 7. However, in other systems, practical designs largely ignore inhomogeneity, anisotropy, and internal stress and dimensional instability as well. This chapter introduces the causes of these factors and their effects, and introduces other influences of (changes of) processing conditions on the attributes of finished parts, such as stability in its dimensions. The discussion illustrates a range of effects, mainly for thermoplastics, in qualitative terms.

The main causes of these (usually) unwanted effects are heat transfer, flow of polymer melts, or a combination of the two. The effects range from visual imperfections, through parts with out-of-tolerance dimensions or distortions, to impaired mechanical performance in service. The scope of the book does not allow us to go into details; for those we refer to the further reading list at the end of this chapter and to more specialized texts.

9.2 THERMAL EFFECTS

When a polymer cools down from its melt processing temperature to ambient temperature, it shrinks. Amorphous polymers contract typically by about 6% *by volume*, and partially crystallizable polymers by at least twice as much. This temperature change is responsible for the following effects, each of which may be latent or strident in a given product, depending on the process used and the design of the product: free shrinkage, residual stress, restrained shrinkage, and inhomogeneity. Initially the effects of flow on thermal phenomena are disregarded; combinations of thermal and flow effects are outlined in section 9.4.

When a melt cools down, it will reach a temperature below which it will behave as a solid. We will call this the solidification temperature T_s. For amorphous polymers the glass transition temperature T_g acts as such, and for semi-crystalline polymers it is the crystallization temperature T_{cr}. In modelling a cooled melt flow there is a somewhat higher 'no-flow temperature' below which melt will not flow and it is caught in the solid layer at the wall.

9.2.1 Residual stress after free shrinkage

If a polymer melt shaped as a part is free to shrink in all directions during the cooling process, the linear dimensions of the product will be at least 2% smaller than that of the molten product. This must be allowed for when dimensioning the flow channels such as moulds and dies. Unfortunately, parts made from polymers do not respond exactly to such a simple approach.

(*a*) *Principles*

Consider a cavity completely filled with an isotropic mass of polymer melt, and disregard how the melt got there. If the mould were suddenly cooled at $t = 0$ to ambient temperature, and assuming perfect thermal contact, the skin of the polymer, in contact with the mould, would instantaneously freeze solid and there would exist in the polymer mass a temperature gradient. At some later time $t > 0$, there would be a solidified outer layer of polymer encasing a molten core. This outer layer is mainly at about ambient temperature and has shrunk ('freely') as much as it needed to. The core, still very much molten, ideally wants to shrink by the same amount as it cools. However, the skin is much stiffer than the core and prevents this core shrinkage. The net result is that the skin becomes compressed by the contraction of the core, and the core becomes extended by the restraint of the skin until stress equilibrium is reached. Upon further cooling another thin layer solidifies and is supposed to be stress-free immediately after its transition to the solid phase, but its further cooling will increase the compression stress in the outer skin. This so-called free quench cooling induces a parabolic residual stress distribution in the moulding with a compressive stress in the plane on the surface and hydrostatic tension inside, with a resultant stress of zero to realize equilibrium.

A typical example is the cooling phase during pipe extrusion. The molten pipe is cooled only from the outside. Here the outer skin of the wall is free to shrink in circumferential and axial directions, whereas the internal skin or the wall is also free to move, but can be regarded as an adiabatic surface. The gradually deposited new solidified layers against the outer skin will cool and

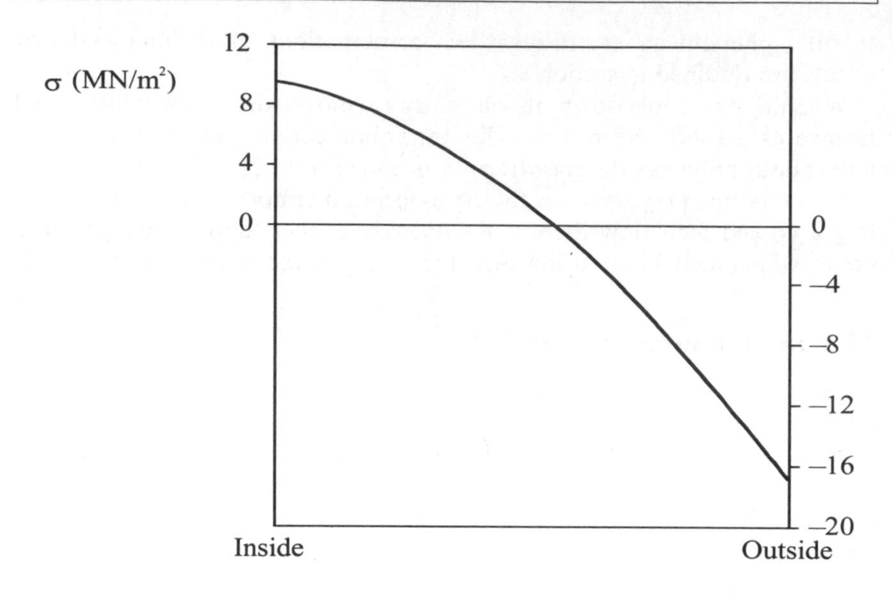

Fig. 9.1 Residual stress in the wall of an extruded externally cooled polyethylene pipe (D = 80 mm, wall thickness H = 18 mm)

shrink. Shrinking, however, is restrained by the outer skin, which therefore will come under compression stress. Continuation of this process results in a stress distribution as in Fig. 9.1. This stress is biaxial by nature, due to the biaxial shrinkage parallel to the surface of the pipe. Tensile stresses at the internal surface of the pipe wall, which (for HDPE pipe) can be (depending on the wall thickness) as high as 10 MN/m^2 (a large figure if compared with the long-term strength). If such a pipe is cut open by a longitudinal slit, the circumferential stress distribution in the wall gives it a tendency to over-close and thus overlap if the parts are allowed to do so.

One is inclined to add these residual stresses to the stresses due to internal pressure of the pipe, but this is seldom done in practice. The usual consideration is that their joint effect is presented in graphs and tables of pipe strength, such as Fig. 6.10. This, however, implies a neglect of different issues, such as the effect of pipe wall thickness on inside tensile stress, variation of crystallinity and changed restraint, so that thin pipes are tougher than thick ones. Attempts have been made to increase the pipe strength by changing the profile of internal stress by cooling from the inside as well as from the outside. This, however, is hampered by major technological difficulties and lack of acceptance of standardization. Several implications follow for moulded products which have been cooled on both surfaces.

First, if the polymer is flexible when cold, the thermal contraction of the core will pull the skin in to cause a sink mark (Fig. 9.2a); the thicker the polymer mass, the greater the tendency to sink. Partially crystallizable polymers

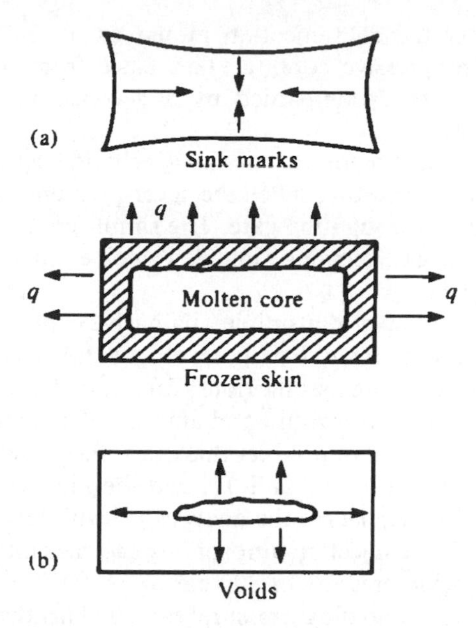

Fig. 9.2 Formation of sinks and voids

will shrink more than amorphous ones. Polymers which are stiff at room temperature, including ball- and cylinder-shaped parts, will resist sinking, thus developing higher tensile stresses in the middle; if this hydrostatic tension becomes too great before the core has frozen, the melt will rupture to form 'voids' (Fig. 9.2b), which are particularly evident in transparent polymers. Locally thick sections will be more likely to cause sinks or voids (or both), so it is good practice to aim for uniform wall thickness where possible.

Second, the cold moulding is in static equilibrium. If a surface layer of polymer is machined away from a flat sheet, the remaining polymer is no longer in equilibrium and will distort until it is. Machining of thin plastic parts after melt processing is likely to cause distortion. The size of the distortion is a measure of the internal stress that was present before machining. Machining of successive layers and measuring the distortion is the most reliable way to determine the magnitude and profile of the internal stresses.

Third, such a distribution of internal stress is advantageous in parts destined to carry bending loads – the residual compressive stress on the surface helps to resist damage if defects are present. This argument is reversed if in-plane tensile loads are applied. Residual thermal stress is as real as stress from an external load!

Fourth, a potential way of reducing these residual stresses is to apply a pressure to the cooling core of polymer until it has frozen-off. This is done by the routine application of holding pressure during injection moulding to some extent (section 9.2.2), but not in most other processes.

During and after the cold-injection mould cavity has been filled with molten polymer, progressive cooling takes place from the mould surface inwards. This must be accompanied by a decrease in polymer volume, leading to shrinkage.

To compensate (partly) for this shrinkage, it is customary to apply a secondary hydrostatic pressure, called the 'after-pressure', 'packing pressure' or 'holding pressure', through the gate. The simple effect is that for as long as the polymer in the gate remains molten, a small extra amount of melt can be packed into the cavity.

We can offer an example of a simple qualitative explanation of the effect of holding pressure using the *pvT* diagram (Fig. 8.22). To do this we have to make a range of simplifying assumptions; for example, we ignore temperature gradients, variation of pressure and amount of solidification with position in the cavity, variation of product thickness, the presence of anisotropy, the effect of cooling rate on T_m and T_g, and the kinetics of crystallization and physical aging. Nevertheless the analysis is instructive.

Suppose (Fig. 9.3) we inject an amorphous polymer melt into a flat plate cavity with an injection pressure of 800 bar.

If we apply too low a holding pressure, say 200 bar, then cooling follows path 1. After the gate has frozen, the cavity pressure reaches atmospheric

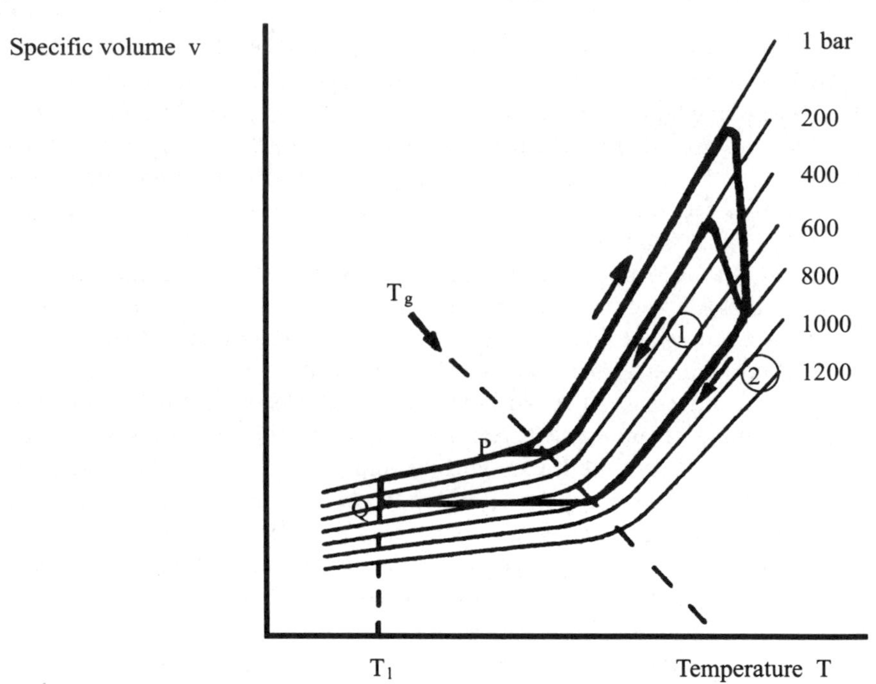

Fig. 9.3 Schematic *pvT* diagram for an amorphous polymer. Effect of using a holding pressure which is (1) too low and (2) too high (courtesy D. J. Van Dijk)

pressure (point P) at a temperature well above room temperature. Further cooling from P to room temperature allows the moulding to shrink, so that it stands loose in the cavity. Thus the product is smaller than the cavity, and the efficiency of cooling is greatly impaired because of the poor thermal contact with the mould wall.

If we apply too high a holding pressure, say 800 bar, the effect is that the cooling follows path 2. Thus the moulding solidifies under hydrostatic pressure, has good contact with the mould and cools efficiently. When the moulding is ejected from the mould (point Q), the remaining pressure relieves itself by a modest expansion.

All other things being equal, we would seek (experimentally) a value of after-pressure such that the polymer in the moulding reaches atmospheric pressure just at the time of ejection. This explanation can be further refined, including the application to partially crystallizable polymers, but it is not discussed here.

The effect of holding pressure on internal stresses is discussed in sections 9.2.2. and 9.4.1

(*b*) *Restraints on free shrinking*

The above has shown how a flat plate of polymer contains a residual stress distribution even though it was free to shrink. However, the polymer in a product may not be free to shrink as it pleases, because of restraint by a much more rigid material. One example occurs during extrusion wire covering; the plastic cannot shrink freely round the wire in the circumferential direction (and at the wire surface, it is not able to contract to give a fully developed die swell). A similar situation occurs in the hot curing of fibre laminate structures, and unsymmetrical laminates will inevitably warp as a result.

In injection or compression moulding, there may be a metal core pin, forming the inside of a cup, box or bucket (Fig. 9.4). Not only is a thermal

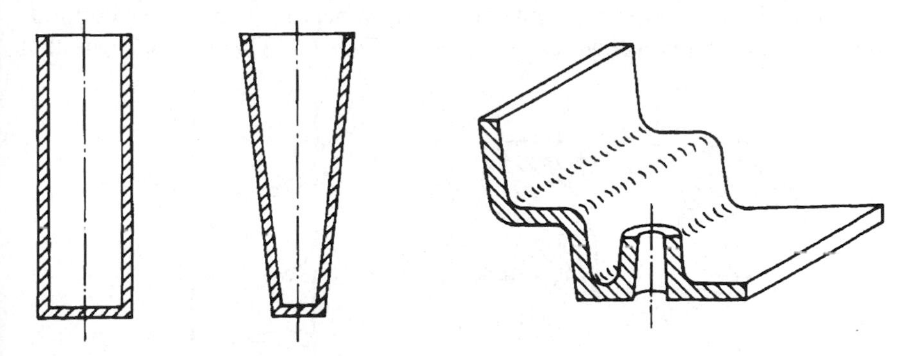

Fig. 9.4 Taper aids extraction from mould

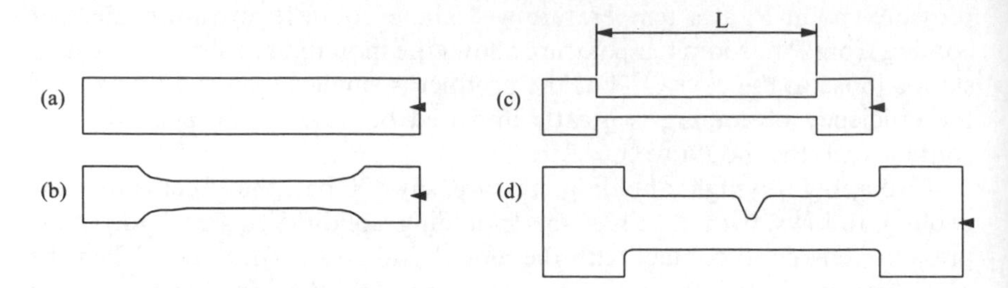

Fig. 9.5 Areas not free to shrink may develop internal stress

stress imparted to the structure, but there is the added difficulty that the moulding shrinks tightly onto the metal core, making extraction difficult. The remedy is to allow generous tapers in the line of draw, typically 0.5–2° per side, and any surface texture design should be in the same direction as the taper.

Restriction of shrinkage can also occur where a shoulder at the end of a narrow section or the rim of a jar lid prevents free longitudinal or radial contraction during cooling. Figure 9.5 shows a sequence of designs which build up the problem. In Fig. 9.5a free shrinkage can occur. The design in Fig. 9.5b is not quite so good, although the generous radius on the shoulders may allow some axial contraction as the shoulders shrink in width. In Fig. 9.5c the shoulders are rigidly held in the mould, and in a brittle material there is a risk of cracks developing at the sharp internal corners of the moulding. The design in Fig. 9.5d will develop a high hydrostatic tension at the notch root, and in polymers used well below T_g this tension will much more than offset the skin compression and could well lead to the development of crack-like defects in the polymer before or during demoulding, thus increasing the chance of premature failure in service.

Inserts may restrict shrinkage of the surrounding polymer as it cools. Around the circumference, tensile stress develops in the circumferential direction. These stresses may turn out to be even worse if the insert contains sharp thread forms, shown exaggerated in Fig. 9.6. There is also an axial

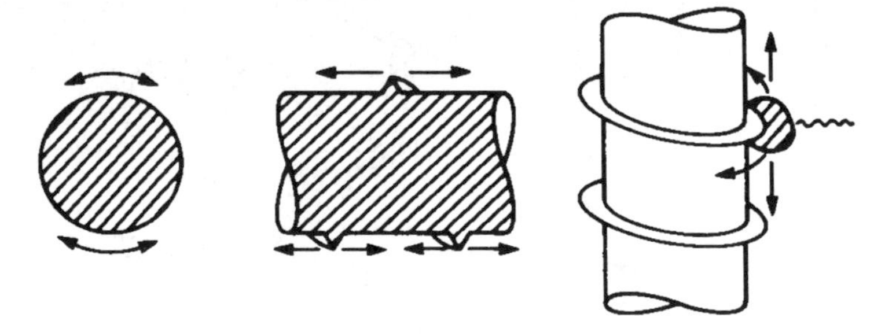

Fig. 9.6 Sources of stress from threaded metal insert

stress between thread tips. In brittle polymers these features are likely to induce crack-like defects which would radiate from the tips of the thread.

9.2.2 Residual stress due to the injection moulding process

In injection moulding it is common practice to apply holding pressure on the melt after the mould has been filled for as long as the gate of the mould is not frozen in (9.2.1(a), fourth implication). This results in a situation where the melt at the mould wall is forced to stick to the wall and cannot shrink. After solidification at T_s it will cool further to the mould wall temperature T_w, but because the corresponding shrinkage $\alpha(T_s - T_w)$ is restrained, a tensile stress of $\alpha(T_s - T_w)E/(1 - \nu)$ will develop. The $(1 - \nu)$ stems from longitudinal and transverse directions, whereas the material is free to shrink perpendicular to the wall, fresh melt being supplied by the holding pressure. Upon further cooling another thin layer will solidify and will experience the same restraint to shrinking. If the core were to stay accessible to melt until all of the melt has solidified, the product would contain a homogeneous tensile stress of the indicated value. If the product was then demoulded, it would shrink and be stress-free afterwards. (Note that constant α and E have been assumed, together with an absence of relaxation.)

However, as the gate of the mould freezes during cooling, the holding pressure is not maintained inside the product thereafter. During the remainder of the cooling path the internal pressure will vanish and the restraint changes from external (the mould wall) to internal (the product wall). The final cooling may result in a product containing tensile stress at its surface (up to $\simeq 10\,\mathrm{MN/m^2}$), compressive stress underneath and tensile stress again in the core (Fig. 9.7). We come back to the origin of this stress distribution in section 9.4.1.

The surface tensile stress can in some cases lead to cracking if the product comes into contact with environments that promote stress cracking (section 6.2.3(e)) or is affected by degrading radiation.

The magnitude of the surface stress can be diminished, or even changed into a compression stress, by lowering the holding pressure or by increasing the mould wall temperature. Either one or both of these changes, however, will also affect other process and product aspects such as cycle time, shrinkage and warpage.

9.2.3 Dimensional effects and tolerances

Injection-moulded products show a linear shrinkage compared with the mould dimensions of 0.4–0.7% for amorphous and 1–3% for crystalline polymers. Shape, gating and moulding conditions (in particular holding pressure, but also mould temperature, melt temperature and injection velocity) of the

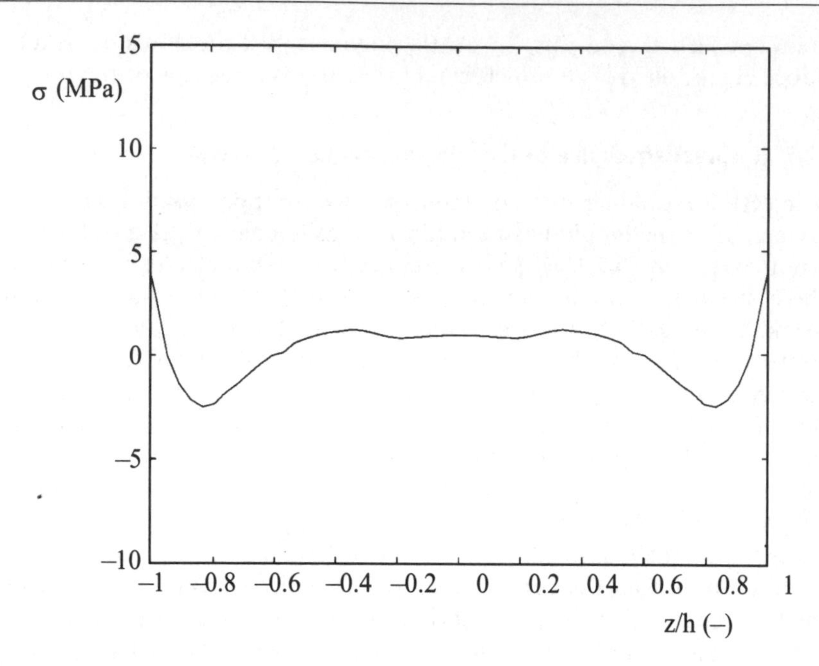

Fig. 9.7 Measured in-plate stress profile as a function of dimensionless thickness in an injection-moulded polystyrene plate (courtesy of W. F. Zoetelief)

product have a strong influence on the magnitude of these quantities; a compound filled with short glass fibres reduces them considerably.

The dimensions are further subject to 'after-shrinkage'; this is the shrinkage which occurs after the product has been ejected from the mould cavity. It is caused by the fact that release from the mould implies release of the relaxed tensile stress that held the product in its place in the mould. This stress was relaxed as it arose during cooling in the mould, at a relatively high temperature, so it could pass through its full time–temperature path. The release can be expressed as superposition of a compression stress, similar to the indicated tensile stress but of opposite sign. This compression stress results in compression creep (= shrinkage), which slows down in time in a way similar to normal creep.

In normal practice shrinkage is determined 24 h after moulding; it therefore already contains a considerable fraction of the 'after-shrinkage'. The remaining 'after-shrinkage' is higher for crystalline than for amorphous plastics. Therefore for high-tolerance products, amorphous plastics are preferred.

'After-shrinkage' can be accelerated in a cycle of change of temperature. The maximum temperature should, however, stay well below T_g in view of eventual rubber elastic stresses that might otherwise deform the product.

Tolerances have their origin in mould size accuracy and in variations in processing conditions. A major factor in the mould lies in the accuracy with which the two mould halves (and possible inserts) fit together. Regarding the process, it should be realized that a great number of variables have to be controlled and that control of a variable nearly always implies setting a point and allowing a certain deviation from it. The greater the shrinkage, the greater the tolerances have to be. It is difficult to reach a tolerance level of the order of 10^{-4}. Usually, in view of the costs involved in high tolerances, one is content with 2×10^{-3} or even larger.

Warpage is caused by an unsymmetrical stress distribution, resulting from local differences in shrinkage, due to differences in temperature and pressure during cooling and by anisotropy in the material. These in turn are strongly dependent on shape and gating of the mould cavity.

Warpage can occur, for example, in products that contain thick and thin elements. As thin parts cool faster than thick parts, they will have shrunk while the thick elements still are fluid or still have to shrink after solidification. Elements farther from the gate will be better packed (and therefore show less shrinkage) than those near the gate because the gate freezes at some point. Centrally gated parts may show warpage due to molecular orientation; in the direction of orientation, shrinkage is smaller than perpendicular to it.

The same parameters of the injection moulding cycle which affect shrinkage influence warpage. Warpage may be diminished by increasing the melt temperature, the mould temperature, the injection velocity or the holding pressure. The first and second clearly have a negative influence on cycle time.

9.2.4 Inhomogeneity

Inhomogeneity resulting from heat transfer is exemplified by a crosslink density distribution in rubber or thermoset plastic, by a foam density distribution in cellular thermoplastics and by a variation in texture in a partially crystallizable polymer. In the first and third examples, the inhomogeneity is tolerated, but in foamed plastics the density distribution may be welcome for structural reasons.

If a thick, shaped, rubber part is heated up, either under pressure in a mould or in a hot-gas vulcanization chamber, the outside heats up first and therefore has longer to crosslink than the centre of the rubber. Too short a heating time will mean that the centre is undercured, too long that the outside may degrade. Much can be done to strike the right balance by a suitable choice of compounding ingredients, and by removing the moulding before the centre has fully cured and allowing the moulding to cool slowly in air; this, with the curing exotherm, will enable the temperature to continue to rise in the centre and complete the cure.

As explained in section 5.1, foamed plastics are widely used to achieve high bending stiffness per unit weight of panel. In a one-shot moulding process, the foam is achieved by injecting an intimate mixture of gas and molten polymer into the mould. Foam bubbles collapse as the surface of the moulding is formed because the mould is cold. This gives a poor surface finish with a characteristic swirl pattern, and the surface is essentially a solid polymer encasing a low-density foam core. The pressure of the blowing agent is sufficient to prevent sinks and voids forming, and presumably the levels of residual stress would be small too, in spite of the chunky sections.

9.2.5 Crystallization

In partially crystallizable thermoplastics (see sections 2.5 and 2.6.2), the amount of crystalline material and its texture depends on the rate of cooling: quenching results in fine-grained structure and low crystallinity, and annealing in coarse structure and high crystallinity.

The background of this phenomenon is as follows. The formation of crystalline structure is controlled by two mechanisms, nucleus formation and crystal growth. Each has its own dynamics, dependent on temperature. Both are promoted by cooling below the melting point T_m. The interval just below T_m is a metastable zone in which nuclei hardly form but disappear as well, but where crystals can grow because of supercooling, although at a very low speed. As the temperature T is lowered, both the growth of crystals and rate of appearance of new nuclei grow exponentially with $T_m - T$ to a certain temperature. As the substance is cooled further, the viscosity increases such that crystal growth and the formation of nuclei are inhibited until the formation of new crystalline structure ceases in the neighbourhood of T_g (Fig. 9.8).

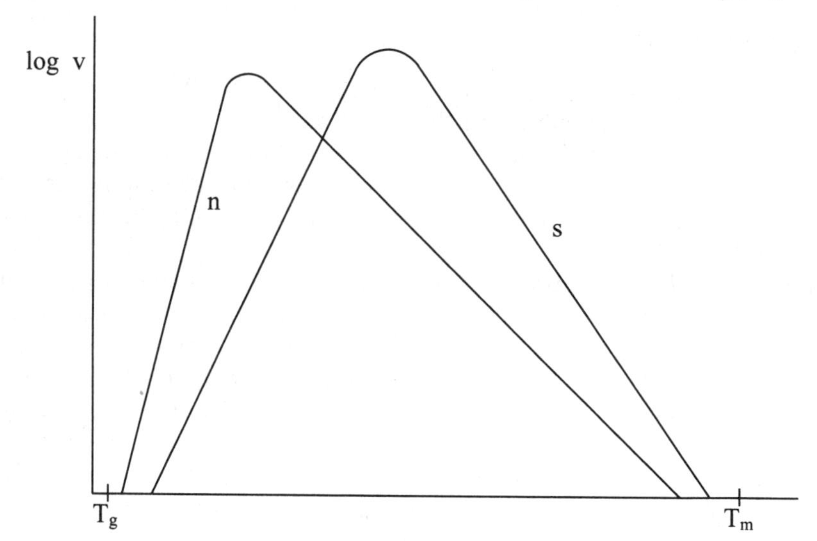

Fig. 9.8 Temperature dependence of rate of growth *s* of spherulites and of rate of appearance *n* of new nuclei (based on Van Krevelen)

Crystalline structure is the result of a process controlled by temperature and residence time at that temperature. These variables lead to a distribution of morphology through the thickness rather than a uniform texture.

Quenching (rapid cooling) promotes development of nuclei at a high rate, possibly combined with slow crystal growth, leading to an open amorphous structure with many small spherulites (left-hand part of Fig. 9.8). Even more rapid cooling might suppress all spherulite growth. Annealing below T_m of such an open structure allows these numerous spherulites to grow until they have absorbed as much amorphous material as possible, leading to a more dense structure.

Slow cooling of the melt allows the few nuclei, which are formed just below T_m, to grow to a coarse structure of large spherulites.

At a cooled surface the cooling rate is high. Well below the surface the cooling rate is much lower because of the low coefficient of heat conduction of the polymer mass. This explains why, at the surface and in thin-walled structures (which are especially 'all-surface'), the texture is likely to be essentially uniform or even rather amorphous, except when the wall acts as a surface covered with nuclei. In the centre of thick-walled products the slow rate of cooling will lead to a fully developed coarse spherulitic structure. As yet the effect of these morphological features is not well documented in quantitative detail.

In order to control crystallinity, nucleus-forming agents (e.g. chalk) may be added to the polymer. This permits the crystal structures to grow as soon as the temperature has fallen below T_m, independently of the nucleus-forming mechanism. It will contribute to reaching a homogeneous dense packing of small spherulites.

Where crystallinity is locally different, the properties are also different. Compared with the amorphous phase, the crystalline phase has a tighter packing and therefore a higher density and may tend to be more brittle. This implies a higher shrinkage (again problems with thin and thick elements in one product), and also a much higher modulus and a lower toughness. It has also been established that the resistance against wear is better. A coarse structure is more likely to have more contaminants on the boundaries of the spherulites than a dense structure of small spherulites; this results in poorer impact properties. Local differences in properties increase the need for control of moulding parameters such as mould and melt temperature, holding pressure and injection speed.

PROBLEMS

1. A thin-walled product has a 90° L-shaped cross-section. When moulded, will the product have the same shape as the 90° mould? Tip: note that the cooling of an outer edge of a mould is less efficient than the cooling of an inner edge.

2. A flat injection-moulded plate has an extension along its length of half the plate thickness. Indicate whether this product is subject to any dimensional effect on cooling.
3. In what way will the holding pressure influence shrinkage and why?
4. A product, shaped as in Fig. 9.5c, is injection moulded and released only after it has been cooled fully to the wall temperature of 20°C. Describe the course of shrinkage of the active length L after release from the mould. Tip: the average stress σ_l in the cross-section before release can be considered to be fully relaxed and thus constant, because it was still warm during its build-up.

9.3 FLOW EFFECTS

Chapter 8 showed that tensile and shear deformations occur during the transport of polymer and shaping of the end-product. The flow or deformation history is responsible for a number of phenomena. Some can be accepted and allowed for in design – extrusion die swell is an example. Others cause defects which need to be eliminated or controlled in order to make a saleable product. Phenomena in this category include molecular orientation, orientation of short fibres, the formation of weld-lines in many processes, 'melt fracture' and 'sharkskin'.

9.3.1 Molecular orientation

Different shaping processes introduce different flows. In shear flow and, more effectively, in extensional flow of polymer melts, chain molecules will be oriented. When the flow stops, disorientation caused by the natural thermal movement of chain segments will take place for as long as the polymer is in the molten state. The rate of disorientation is temperature dependent and slows down to zero just below the solidification temperature T_s. A kinetic equilibrium of orientation and disorientation is frozen in during solidification of melt.

In compression moulding, flows are small, and usually of a spreading nature. So the degree of orientation will be low as well and not of interest if the product is to be cured.

Extension in the melt phase during blow moulding or tubular film blowing results in unidirectional or bidirectional molecular orientation, which is usually uniform through the thickness. Extension in the rubbery phase occurs during thermoforming and causes more effective orientation, as disorientation is impeded. In all these processes disorientation is counteracted by cooling the deformed product. The extension itself is elastic as long as the oriented state lasts. If, however, the product remains too long above T_g or close to T_m or is heated up again without external constraints, the extension is recovered, leaving a collapsed tube or flat sheet.

When a rubber band is extended, it is in its rubbery phase and molecules will align in the extension direction and disorientation is prevented by the crosslinks. Natural rubber will even crystallize under sufficient extension. These crystallites scatter light so the rubber goes pale under the strain. On removing the load, the crosslinks ensure complete recovery back to the amorphous state and the change of colour disappears.

An example where uniaxial extension is applied is in the manufacture of fibres and tape by extruding an appropriate cross-section and hauling-off to achieve high extension ratios or draw ratios. The resulting product is stiff and strong in the draw direction but usually very flimsy in the transverse direction, as is evident in polypropylene string. In polymers which crystallize, the high degree of molecular alignment at high draw ratios is just right for encouraging extremely rapid crystallization in the orientation direction. The crystalline structure under these conditions is linear rather than spherulitic. Control of this process is then quite tricky. At extremely high stress, the melt can even crystallize above the nominal melting point it has in its unstressed condition.

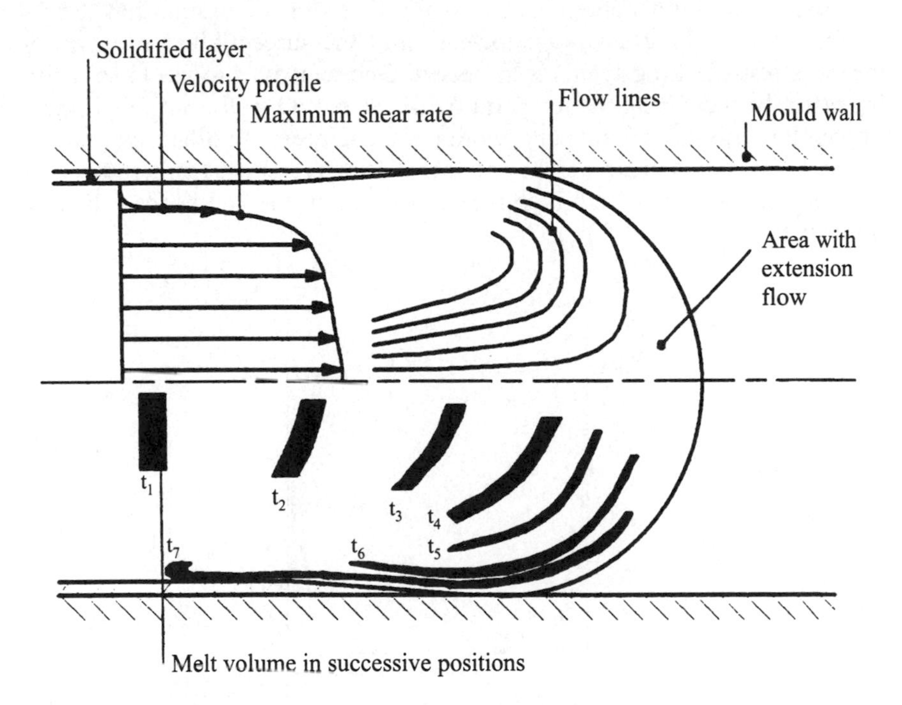

Fig. 9.9 Fountain flow in the mould gap and solidifying layer at the mould wall. Note the flat velocity profile near the centre, which is characteristic of a power-law fluid. Note that this region is usually studied with a fixed free surface and relative motion of the wall, i.e. with moving coordinates. In the bottom half successive positions of a melt volume in time are represented ($t_1 < t_2 < t_3 < t_4 < t_5 < t_6 < t_7$)

During injection moulding both orientation in extension and in shear will be present in the melt when it flows in the mould cavity; extension occurs at the free front (where it is called the fountain flow; Fig. 9.9), and the material there will be stretched, placed along the wall and frozen in because the mould wall is cold (this frozen-in extended layer is very thin). Shear occurs due to the pressure flow in the channel, with a maximum at the outer layer between fluid and solidified polymer. As oriented extended polymer shows birefringence, the degree of orientation including the maxima stemming from both mechanisms (a stretched fountain front frozen in along the wall and shear maximum near the solidified layer) can be made visible (Fig. 9.10). This figure also illustrates the low level of orientation in the middle: shear forces are low and the material stays longer in the molten phase, which allows time for disorientation.

The fast cooling rates with injection moulding prevent disorientation of the surface of the polymer, so that the high residual orientation is maintained. This resultant orientation, and the connected anisotropy (section 2.9), provides some tendency to split along planes of weakness and consequently a decrease of strength perpendicular to the direction of orientation. Stress crazing and cracking levels are lowered and the susceptibility to environmental stress-cracking agents is increased. One remedy is to use lower cooling rates (by increasing the temperature of the mould wall), but this is most unpopular with accountants and production engineers. Another approach is to use a lower molecular weight polymer – at a given melt temperature this has a lower viscosity than a high molecular weight one and hence will relax faster before freezing off.

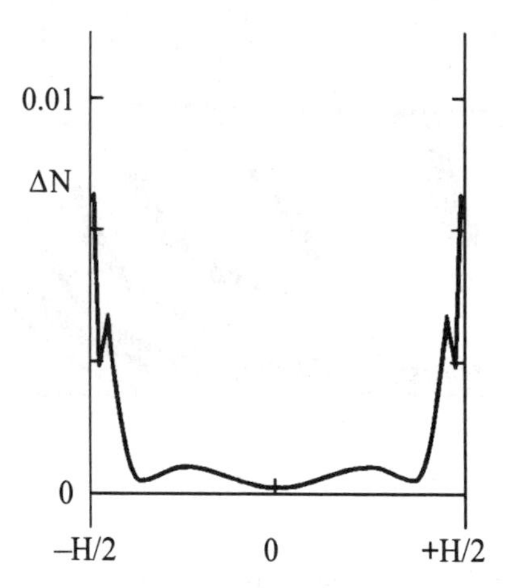

Fig. 9.10 Variation of birefringence (as a measure of orientation) as a function of thickness in an injection-moulded polystyrene not far from the gate (courtesy G. Nijman)

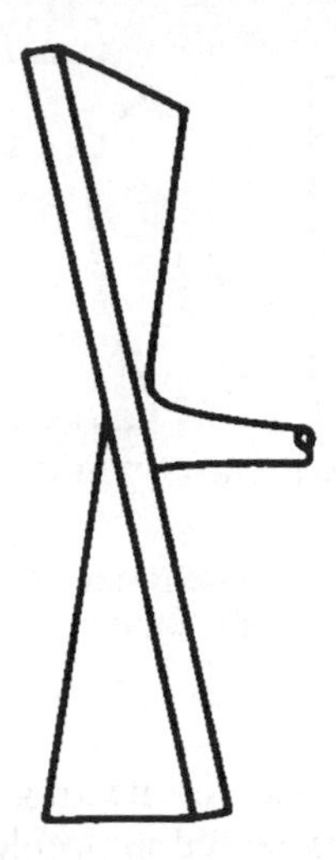

Fig. 9.11 Warping of a thin 'flat' centre-gated panel

In centre-gated mouldings the radial flow induces radial molecular orientation. This leads to anisotropy in thermal expansion coefficient and in strength. The first will result in tangential residual tensile stress, the second in lower tangential strength. In serious cases the combination of these becomes evident as a radial crack in, for example, the flat bottom of a tub.

In less serious cases the difference in shrinkage along and across the radial flow direction leads to buckling out-of-plane to a dome, or to a saddle shape (Fig. 9.11). This defect is well known, e.g. in thin lids to domestic items, and not easy to correct in a given cavity. It is preferable to hide the deformation of the part by deliberate curvature or to increase the stiffness by ribbing to improve the buckling resistance, but better still to use multiple gates or edge flash gates to promote essentially unidirectional flow.

Heat treatment is not always useful and may even cause harm, as it may disturb the sensitive equilibrium between frozen-in residual stress and frozen-in molecular orientation. Heating of a moulded product first results in accelerated relaxation of residual stresses. In this phase the shape of the product is generally maintained unless the stresses (or the anisotropy caused by molecular orientation) were not evenly distributed through the thickness.

Upon further heating, when T_s is approached, disorientation of chain molecules becomes possible, resulting in shrinkage in the flow direction, followed by loss of shape of the product; the aligned molecules obtain freedom to shrink back to their random coiled configuration. Obviously shrinkage along the flow direction now will be greater than that across it. A square plaque fed uniformly from one edge by a properly designed strip or flash gate will only be obtained from a slightly rectangular cavity. The preferential

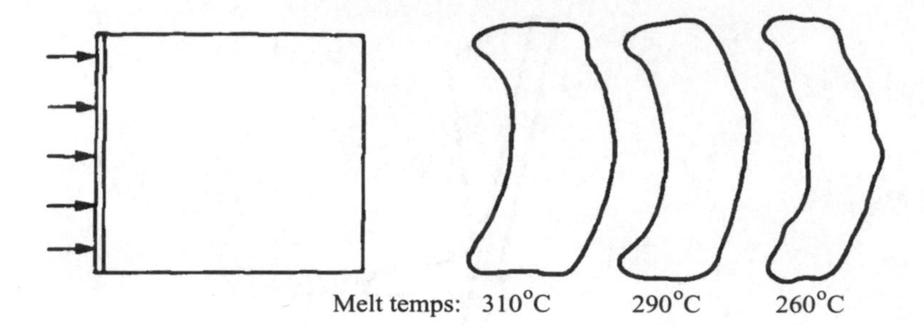

Fig. 9.12 Film-gated high-density polyethylene plaques before and after heating to 170°C (curvature indicates slight radial component of flow rather than a linear flow front)

direction of alignment can be seen by heating this kind of moulding in talc above the melting point. The oriented molecules in the skin of the moulding are now free to retract in the flow direction, pulling the unoriented core with them, as shown in Fig. 9.12. Note that the retraction in the transverse direction is negligible, as expected. The slight curvature of the retracted mass suggests that the original flow into the mould was not quite unidirectional. Figure 9.12 shows that, at higher melt temperatures, the amount of retraction lessens: this is consistent with the notion that, since the rate of freezing off was slower, disorientation has had more chance so that the resulting oriented skin was thinner.

9.3.2 Short-fibre orientation

The flow of fibre-filled polymers influences fibre length and results in a certain orientation of the fibres in the moulded product. The distribution of orientation of the fibres is, however, not the same as the orientation of molecules, since the fibres are relatively stiff, compared with the surrounding fluid, and as each fibre will, in most cases, lie on more than one flow line. Therefore fibres show a tendency to tumble in the flow. As a general rule it nevertheless can be said that both extension flow and shear flow will orient the fibres in the direction of their main action. In a moulded product this may result in fibres oriented along the flow direction in the surface or 'skin' and transversely oriented fibres in the core. It seems that the core orientation is greatly influenced by the flow pattern in and just before the gate. Fibre orientations in both skin and core of the moulding are also influenced by features which induce strongly convergent or divergent flows.

In view of the above, it is not sensible to regard the resulting product as being isotropic. It can be better regarded as a layered structure (Fig. 9.13), with properties depending mainly on the amount and the angle of the fibre

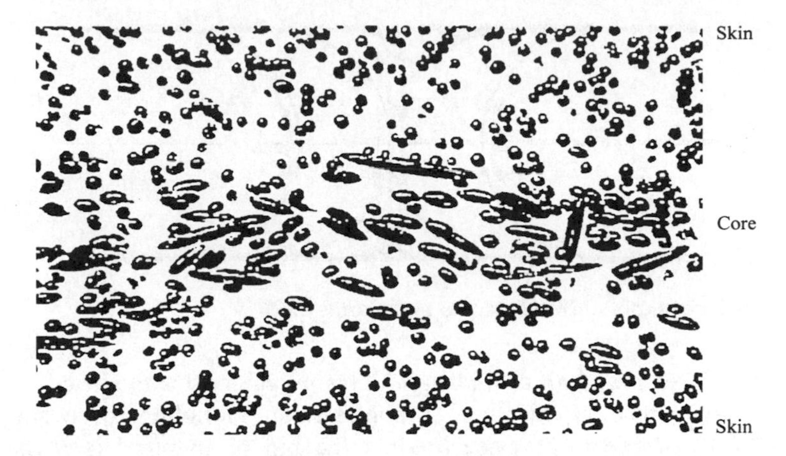

Fig. 9.13 Cross-section perpendicular to flow direction of an injection-moulded product, showing the circular cross-sections of the glass fibres. Their ellipsoid axes indicate their orientation with respect to flow direction (courtesy M. Freriksen)

orientation of each layer (Chapter 7). Some progress is being made in the prediction of orientation of fibres in mouldings, but current results are qualitative and not yet comprehensively quantitative. The effects of pressure distribution through the thickness and the slight increase of fibre content through the length of the flow path are much less pronounced than for unreinforced plastics and do not need to be taken into consideration.

If flow conditions are not symmetrical in the plane, the fibre orientation distribution will not be symmetrical either, including the distribution of anisotropic stiffness and thermal expansion. Thermal residual effects will therefore affect warpage.

Flow also causes fracture of the fibres. For the injection moulding of fibre-reinforced thermoplastics, granules containing an average fibre length of 2–3 mm are usually used. Relative movement against each other, especially in the screw, is the main cause of fibre fracture, resulting in an average fibre length of about 0.2–1 mm in the moulded product. If the fibres are present in bundles in the granules, they may stay together in the flow and protect each other against fracture. A longer average fibre length and a worse distribution in space will be the result.

9.3.3 Weld-lines

Weld-lines form where two melt fronts meet and join together, usually with no shear or mixing of the two flows. The join is usually not as strong as that of the homogeneous polymer, but the strength strongly depends on the processing conditions.

Weld-lines form in one or more of the following situations:

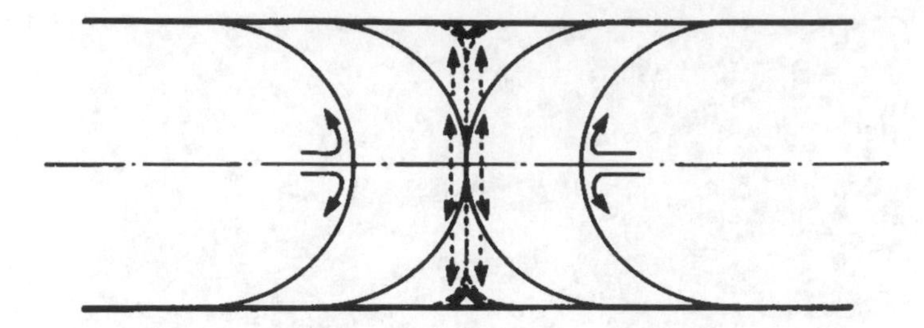

Fig. 9.14 Extensional flow where two melt fronts meet

1. a melt flows round an obstacle and joins up again: this might be a pillar defining a hole in injection or compression moulding, or a support ('spider') or centring device holding the pin or mandrel in a die for extruding tube or covering wire;
2. the melt enters an injection mould cavity through two or more gates, or two or more separate charges are placed in a compression mould;
3. in a single-gated mould cavity there is preferential flow in a thick section and restricted flow in a narrow channel;
4. two free melt surfaces may be compressed to seal a tube, as happens in extrusion blow moulding or in making a welded joint such as a butt weld in a pipeline;
5. jetting may occur during filling of certain types of injection mould.

Why is a weld-line a source of weakness? Within a single flow, polymer molecules are oriented along the direction(s) of flow at the flow boundary. So when two melt fronts meet – either head-on or at a convergent angle – there is little incentive for molecules to bridge the boundary, which may be slightly cooled and therefore stiffer, even when pressure is applied. The weakness of weld-lines will be more pronounced in mineral- or short-fibre-reinforced polymer compounds because of the absence of chain molecules or fibres bridging the joint.

Local circumstances dictate the details of the formation of weld-lines. Where two separate melt fronts approach each other head-on, as may occur in an injection mould, the midplane joins first (Fig. 9.14). As the join extends through the thickness of the part to reach the mould faces, transverse elongation occurs which impedes diffusion between the fronts and introduces local stresses caused by differential shrinking. There may be some risk of entrapment of air and a real risk of a slight score or surface blemish. This blemish could result in a latent crack-like defect. In extreme circumstances, air entrapped between the advancing melt fronts could become so hot that the polymer actually burns. Local venting of the mould could help avoid this.

A weld formed at the end of a flow path is called a cold weld. An example is shown in a detail of the wall of an open-mesh basket (Fig. 9.15a, point c). It

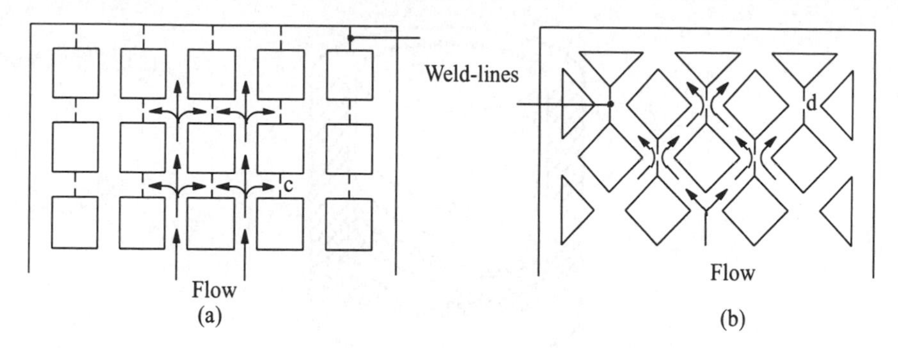

Fig. 9.15 Injection-moulded baskets containing (a) cold and (b) hot welds

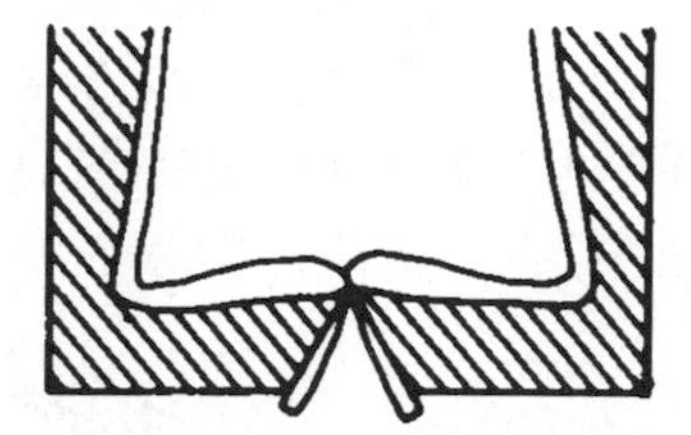

Fig. 9.16 Weld-line at pinch-off in an extrusion blow moulding

can only cool down (partly under pressure). If the weld-line is positioned such that after first contact fresh and hot melt is supplied at a convergent angle near the front (or any other source of heat), the added heat will help to keep the temperature high for some time and promote thermally induced diffusion of chain molecules between the fronts (hot weld; Fig. 9.15b, point d). If it is hot enough for long enough and if there is enough pressure between surfaces, the chain molecules will bridge the fronts of the hot weld and reinforce the joint up to the strength of the surrounding material. Long-chain molecules need more time to penetrate through the front than short ones. This build-up of strength is called healing. Every weld-line needs healing in order to reach the required level of strength.

Extension flow in the fountain at the flow front of fibre-reinforced thermoplastics leads to weld-lines containing fibres parallel to the weld surface. This means always a weak spot in the product, as fibres are not involved in the healing process like macromolecules are.

In blow moulding, the main surface blemish will occur on the inside of the product (Fig. 9.16), unless the mould has been poorly made or maintained.

Longitudinal weld-lines would be particularly disastrous in the manufacture of very thin-walled tubes destined for subsequent inflation into film. Every effort is made to smear the flow so that the effect of the weld-lines is minimized, and Fig. 9.17 shows schematically one way of achieving this. (The spiral mandrel die eliminates weld-lines through the film thickness:

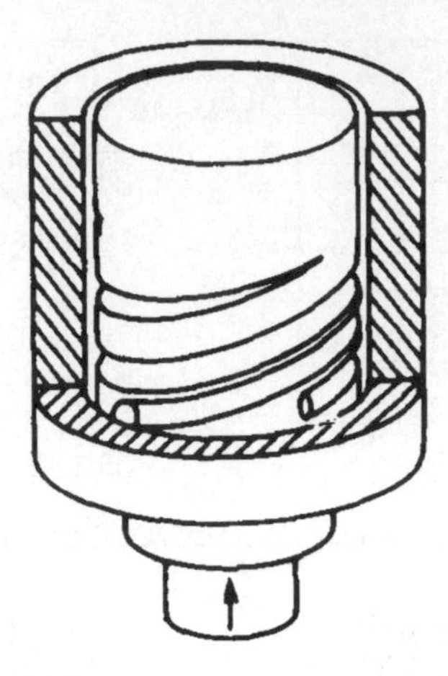

Fig. 9.17 Spiral mandrel film die; spiral distributes melt evenly around the circumference

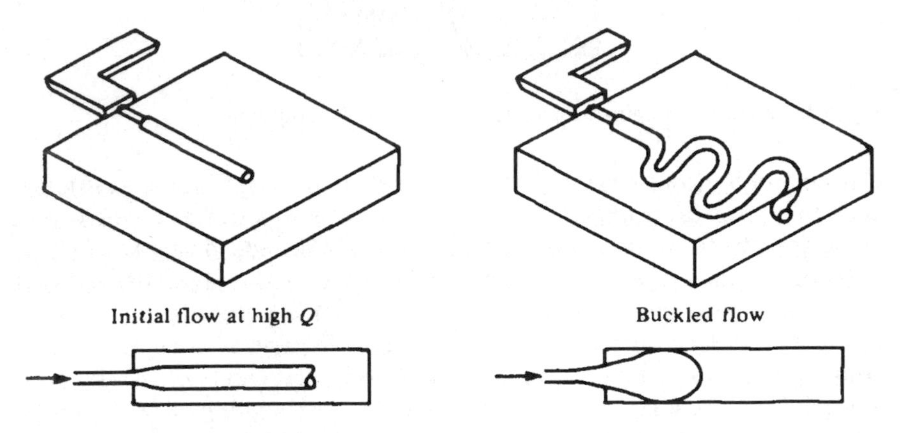

Fig. 9.18 Jetting phenomena

there are now weld-planes in the thickness of the film, with plenty of surface area for good load transfer through the thickness in, for example, adhesive joints.) This elegant approach is not adopted for thick-walled tube, although many pipe dies have a small reservoir downstream from the spider to encourage the weld-line to knit together.

Weld-lines can sometimes occur in injection moulds where a small gate feeds a cavity in the same plane; if the wall opposite the gate is remote, then jetting may result (Fig. 9.18). One possible remedy is to increase the gate size to reduce the momentum of the jet or to redirect the flow so that on leaving the gate the melt almost immediately meets a nearby wall of the mould.

9.3.4 Melt fracture and sharkskin

These two defects can occur in extrudates. In appearance, they range from a slight (and unintended) loss of surface gloss, through gross distortion and a knobbly surface, even to spitting out of separate lumps from the die under extreme conditions. The appearance of both types of defect can be similar and the cause can be attributed to excessive tensile stress; how the tensile stress causes the defect differs and therefore the remedy is different.

Melt fracture occurs in converging flow channels at high flow rates. The melt accelerates and if the rate of extension is too great, the melt simply breaks, i.e. fractures. Under the prevailing hydrostatic pressure, the melt fuses together again of course, but now different elements of the melt have had different experiences – some have fractured, some have not. So when the melt leaves the die, the swelling is uneven (Fig. 9.19), leading to loss of gloss or even break-up of the integrity of the product. Assuming that a high output rate is essential, the remedy is to reduce the extension rate by using a shallower taper within the die; the higher the output rate, the shallower the taper needed.

Sharkskin looks similar to melt fracture but its cause and effect are different. Consider an element of fluid in the die just upstream from the exit. At the wall, its velocity is (notionally) zero. As the element leaves the die, it undergoes an extremely rapid rate of extension (Fig. 9.20) until it reaches the uniform velocity of the fully swollen extrudate. If the rate of extension is too high, then the surface will fracture, although the core will

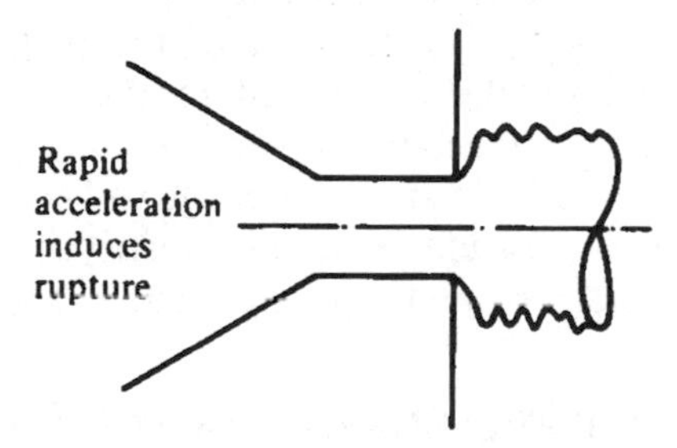

Fig. 9.19 A cause of 'melt fracture'

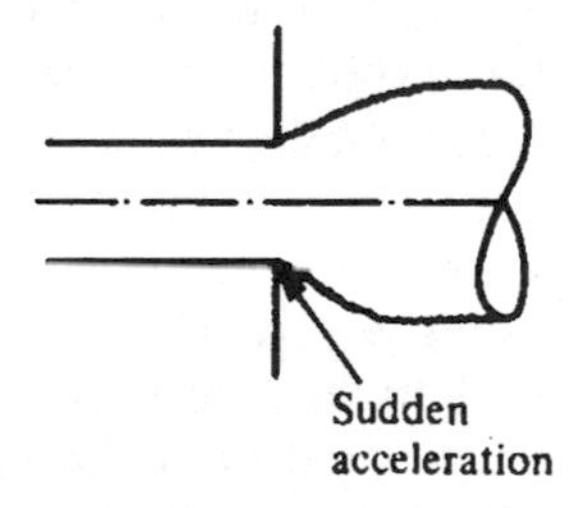

Fig. 9.20 A cause of 'sharkskin'

maintain its integrity as it is in axial compression. The remedy is to make the melt surface more deformable, a popular method being to heat the surface of the extrudate locally just as it emerges from the die.

Other causes can also lead to instability in polymer products. These range from variations in output rate, temperature and pressure ('surging') caused by break-up of the solid bed in the extruder screw, to (visco)elastic instabilities involving either tensile deformations and tension-thinning polymers or resonances related to surface tension effects. These can cause periodic fluctuations in product dimensions. A useful introduction to these phenomena is given by Middleman (1977).

PROBLEMS

1. Describe why disorientation is prevented (or not) in the following processes: (a) extension of a rubber band, (b) thermoforming, (c) tubular film blowing and (d) blow moulding.
2. Describe the change in length of a sheet of uniaxial (lengthwise) molecular oriented material, when heated from a temperature well below T_g.
3. Identify the various process parameters that promote healing of a weld-line.
4. A rectangular plate was injection moulded in short-fibre-reinforced thermoplastic through a film gate at one end of the mould. The mould and the flow inside the mould cavity were symmetrical in thickness about the midplane and the plate was perfectly flat. Lying on a flat cooled surface, which permits a change of dimensions of the plate, it was heated by infrared radiation until the upper half of the thickness was molten. Then the radiation was taken away and the plate was allowed to cool down freely. Determine in what way warpage will show up when the plate is fully cold. Discuss possible orientation distribution, internal residual stress, modulus anisotropy and expansion anisotropy.

9.4 SOME COMBINED PRESSURE, FLOW AND THERMAL EFFECTS

In section 9.3 residual stress and orientation were treated as if they were caused by cooling or flow alone. This, however, is an oversimplification. In most processes pressure, flow, cooling and reaction kinetic effects occur simultaneously. Only as a result of their interaction can frozen-in stresses, molecular orientation and crystalline structure be fully understood.

In this section we will look somewhat further into injection moulding (with unwelcome complications, e.g. residual stresses in mouldings, or anisotropy of stiffness and strength) and into processes, where orientation is deliberately sought (e.g. in tapes and fibres, in films and in stretch-blown bottles). In many cases the resulting product cannot be regarded as fully homogeneous.

9.4.1 Injection moulding

The injection moulding process is essentially driven by the three parameters (pressure, flow and temperature) together, resulting in products with a distribution of residual internal stresses and orientations. Some origins of orientation were made clear in section 9.3.1. We now come back on the reasoning in section 9.2.2 to understand the stress distribution of Fig. 9.7.

For this purpose we schematize (Fig. 9.21) the course of the moulding process, as observed in a cross-section of the mould, in five steps in time, during which three layers in the product cool down successively from temperature T_h (high, $> T_s$) to T_l (low, $< T_s$). As a further schematizing stage, the melt pressure is either 0 or p_h (holding pressure).

To a first approximation the materials behaviour caused by cooling can be described with

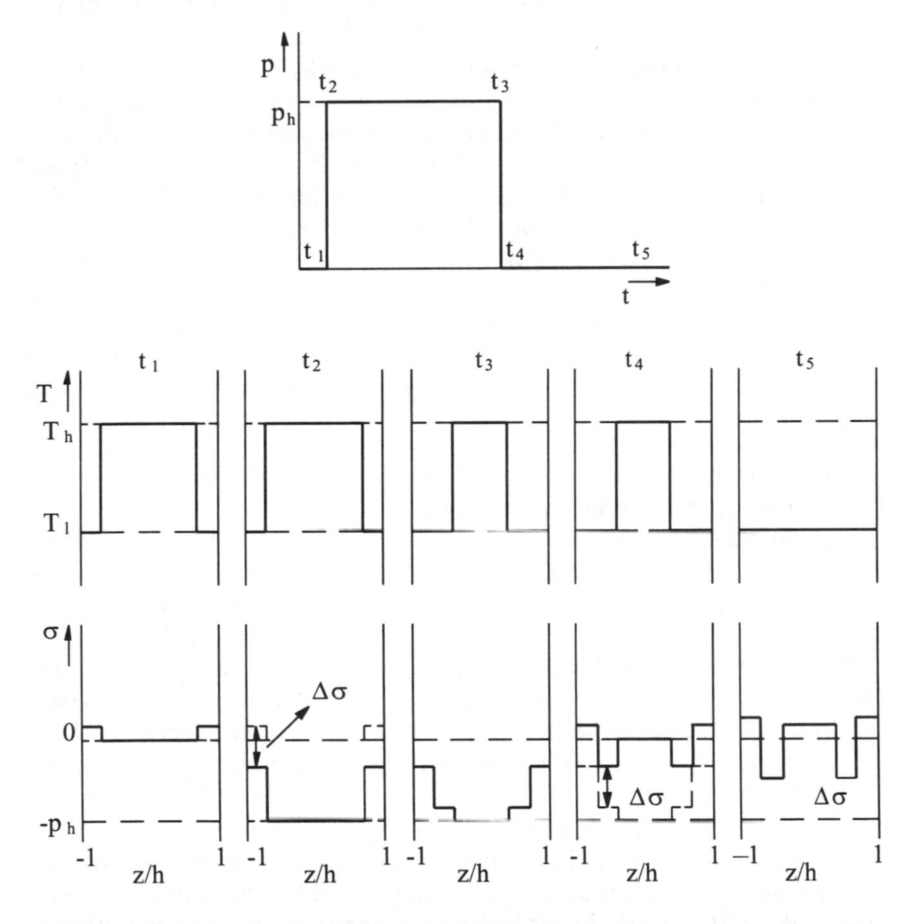

Fig. 9.21 Development of residual stress in injection-moulded products (from Zoetelief *et al.*, *Pol. Eng. Sci.*, **36**, 1887)

$$\varepsilon = \frac{\sigma}{E/(1-\nu)} + \alpha \Delta T \tag{9.1}$$

Young's modulus E, Poisson's ratio ν and the thermal linear expansion coefficient α are supposed to be constant below T_s (solidification temperature). $\Delta T = T_l - T_s$ is the temperature difference the material has to pass after solidification (note the negative sign). ε is the strain (negative shrinkage) and σ is the tensile stress in the product. Local ε and σ are both zero at solidification temperature T_s. Cooling will result either in shrinkage ($-\varepsilon$) or in stress (σ), depending on the boundary conditions.

1. The melt enters through the cross-section. Pressure at the front is zero, the hot (T_h) melt touches the wall and solidifies instantaneously. As the solidified polymer sticks to the wall it is restrained from shrinking below T_s and therefore builds up a biaxial tensile stress $\sigma = \alpha\ (T_s - T_l)\ E/(1-\nu)$ upon cooling from T_s to wall temperature T_l.
2. The mould is filled and the holding pressure p_h introduced results in a compressive stress $\sigma = -p_h$. This compresses the rigid layer at the wall. As a result the biaxial stress in this shell is decreased by $\Delta\sigma = \nu p_h/(1-\nu)$.
3. Further cooling results in a second layer, which sticks to the first one and is also restrained from shrinking during cooling from T_s to T_l, with a similar result as in step 1, but now superimposed on the p_h level.
4. The gate freezes, the holding pressure drops and the restraint at the product surface is released. This means increase of the biaxial stress in both solidified layers by $\Delta\sigma$, but in the non-solidified interior by p_h.
5. After solidification and shrinking of (or build-up of tensile stress in) the interior, the equilibrium stress distribution will show minima inside the product.

This division into three layers is clearly a very rough simplification. The origin of the minima is however illustrated.

The holding pressure causes expansion of the finished product, which counteracts thermal shrinkage. This provides a means to influence product tolerances, although influencing the internal stress distribution at the same time.

If the product is cooled asymmetrically (T_l of the two mould walls differs) the stress distribution causes a curvature. This, however, will be influenced by the magnitude of the holding pressure. It is of special importance for L-shaped products, as the holding pressure now apparently provides a means to control the deformation of the L-arc (compare problem 1 of section 9.2).

As the residual stress is biaxial by nature, and as the molecular orientation caused by flow is uniaxial, there is one stress component perpendicular to the orientation. Both the residual tensile stress and the high degree of orientation at the surface may (in particular when stress-cracking media or environments are present) give rise to surface cracking phenomena.

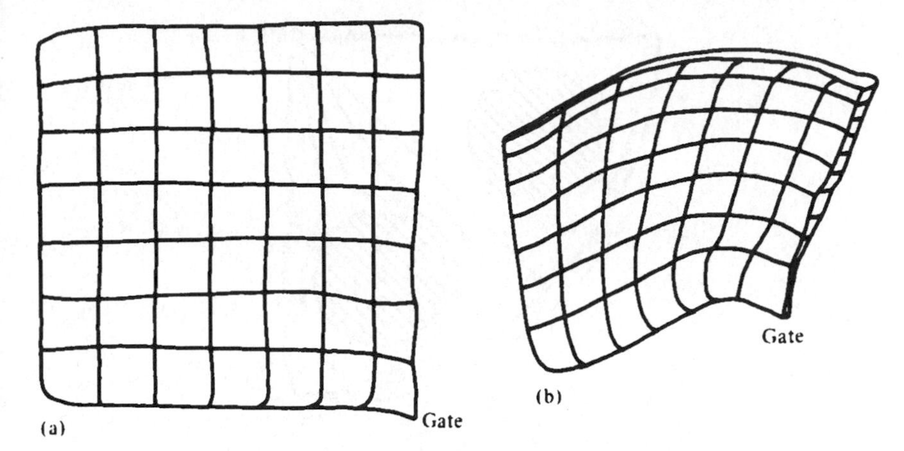

Fig. 9.22 Quarter of square plaque injection moulded in polystyrene (gate in centre of one edge) after heating to 140°C: (a) central core, (b) core and both 'skins' complete. Grid 10 mm square marked on surface before heating

As we indicated in section 9.3.1, flow from a central gate causes radial shear. However, in addition the spreading disc will show circumferential extension (combined with radial extension from the fountain effect). At constant flow rate the amounts of each vary in a different manner with distance from the gate. Material frozen-off during filling forms the surface layer of the product, and this can be highly oriented; whereas the core, which has solidified after the mould has filled, is substantially unoriented. Confirmation of this statement has been obtained by studying the retraction of identical plaques. Square plaques, originally 140 mm square and 3 mm thick, gated at the centre of one side, were moulded in general-purpose polystyrene. The plaques were cut into quarters. Two otherwise identical samples were investigated: one was a complete quarter, while the other had had removed from each surface a layer 1 mm thick. A square grid was marked on one surface and the samples heated up in talc to 140°C. Figure 9.22 shows that the core specimen has hardly retracted at all, implying that its molecules were and still are randomly oriented and curled up. However, the complete quarter has distorted a great deal because of the release of orientation frozen in during spreading disc flow.

In order to illustrate several effects, consider a flanged circular gear wheel gated at the centre. If the web is provided with moulded-in holes to reduce weight (Fig. 9.23), then weld-lines will form beyond the holes, providing a potential source of weakness. Moreover, the filling of the cavity can no longer be regarded as simple spreading disc flow, and therefore the radial shrinkage will vary round the circumference, leading to a pitch circle which is no longer circular, and hence to reduced efficiency in the transfer of power across meshing teeth.

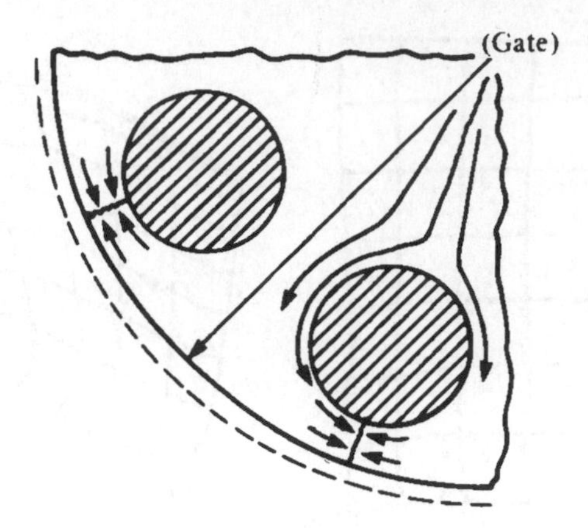

Fig. 9.23 Typical flow paths in centre-gated disc with interrupted flow

9.4.2 Biaxial orientation

Several important melt processes involve a polymer being subjected to in-plane stress while cooling occurs. It makes little sense to say that a thin, chain-like molecule becomes biaxially oriented, but the resulting polymer mass behaves as if it were preferentially aligned in the directions of both the applied stresses. It therefore shows many of the characteristics of a cross-ply laminate as discussed in section 7.4, in that properties can depend markedly on the directions in which they are measured.

In the tubular film process (section 4.3.2(b)), the extrudate is axially drawn and inflated with air pressure so that the thickness of the film is perhaps only 5–10% of that of the die land gap from which it emerged. The orientation is frozen in. In polyethylene film, it is quite possible to obtain values of tensile modulus in the axial and hoop directions ($\theta = 0°$ and $90°$) which are six times that of the unoriented polymer, while the modulus in the 45° direction may only be about one-third that of the unoriented material. By control of axial draw and inflation, either a balanced ($E_0 = E_{90}$) or an unbalanced ($E_0 \neq E_{90}$) film can be readily made to match service requirements. On heating up this material near to the melting point, the film shrinks dramatically, showing that the inflation and drawing stage of the process has an essentially elastic character: this is the basis of shrink-wrapping as used in, for example, palletized packaging, where goods are firmly held in place.

One study of anisotropy in injection-moulded products concerned the stiffness of a centre-gated flat-based seed tray moulded from a propylene–ethylene copolymer. Test samples were cut from the base of the tray with the centre of the testpiece on a 100 mm radius from the gate. Samples were cut at different angles θ to the direction of radial flow from the gate. Figure 9.24 shows that, as expected for essentially orthotropic specimens, maximum

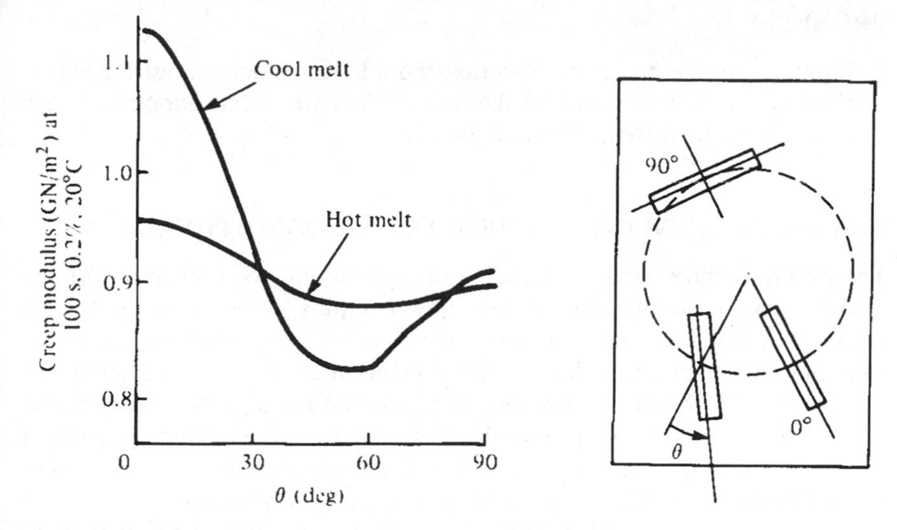

Fig. 9.24 Anisotropy of stiffness in base of centre-gated polypropylene seed tray

stiffness occurs in specimens cut in the radial direction, with a minimum at 50° to the radial direction. The 90° value is larger than the minimum, confirming that there is a small but significant degree of orientation in the circumferential direction. As expected, the anisotropy of stiffness is much less pronounced (and the product is more robust) when moulded at a high melt temperature than at a low one.

9.4.3 Inhomogeneity

From the above it should have become clear that hardly any product can be considered to be a shape in space, filled with a substance of known and homogeneous properties throughout:

- materials and their additives are not always ideally mixed; they may even tend to segregate during processing;
- all processes that involve heat transfer and chemical reactions need time; they often end far from (thermodynamic) equilibrium;
- histories of process parameters (temperature, pressure, flow) of different elements in a product are different from each other.

It is therefore clear that major product characteristics, such as chain orientation, internal stress, crystalline structure, foam density and degree of hardening are present as distributions and will determine the (local) properties of the product as being different from the bulk material. One can, however, try to make these distributions as favourable for the product as possible.

PROBLEM

1. Explain why centre-gated injection-moulded products show tangential residual tensile stress around the axis of the gate. What happens if such a product is heat treated just below T_s?

9.5 EFFECT OF CHANGE IN PROCESSING CONDITIONS

The previous three sections reveal some conflicting requirements; for example, a high output rate at low power inputs often requires a high thermal energy cost, while a high output rate at low melt temperatures runs the risk of products having undesirable qualities such as fragility or distortion. Once hardware has been designed and made to the best possible design, there are three main routes by which the resourceful technologist can overcome minor problems associated with the product made from a given type of polymer, by adjusting the melt processing conditions, changing the grade of the same type of polymer (or even the base polymer), or making changes to the hardware. The following account indicates some of the qualitative expectations associated with changes in processing conditions and does not claim to be comprehensive. Useful but incomplete insight results from changing one variable at a time (i.e. all other conditions are assumed constant).

During injection moulding of very thin-walled products from (slim and regular) crystallizable polymers there is a risk of stress- or orientation-induced crystallization at very high injection speeds or low melt temperatures. Flow stops and the product is incomplete. Cogswell (1981) puts it nicely when he says: ‘under these circumstances we understand that it is more effective to coax rather than bully a polymer melt’. People are made substantially from polymers and can easily understand his point. Polymers such as polypropylene and linear polyethylene respond better to being coaxed into the mould rather than bullied.

In extrusion, an increase in output rate introduces more die swell (which can be offset by increasing the haul-off speed) and runs the risk of sharkskin or melt fracture.

An increase in pressure with pressure flows is accompanied by an increase in output rate. An increase in hydrostatic pressure increases viscosity and may or may not be offset by viscous dissipation, depending on whether the flow is fast or slow. In free surface flows, increasing the internal air pressure gives faster shaping but the technologist will find that one upper limit is the risk of rupture of the melt.

Melt temperature is an obvious parameter in polymer processing. The lower limit (of interest in energy conservation) is set by a high viscosity and product defects; the upper limit by form-stability and thermal or oxidative degradation. A higher melt temperature leads to more shrinkage and to reduced pressure in pressure flows at constant rate. The lower viscosity reduces the likelihood of melt fracture or sharkskin occurring and leads to greater sag in parisons under self-weight loading. The more rapid relaxation of a polymer at the end of a deformation process will reduce the amount of orientation frozen in during cooling. Inflation processes will occur at lower pressures, and may lead to some instability.

In moulding processes the mould temperature is chosen to confer rapid production rates (for which it should be low for thermoplastics and high for crosslinkable systems). Too low a mould temperature may prevent filling of a cavity with a large ratio of flow path to section thickness, or may induce too coarse a crystalline texture in a crystallizable material. Too high a mould temperature for thermosets will overcure or degrade the surface of the moulding before the core is properly crosslinked. A moulding is likely to be hotter on ejection from a hot mould than a cold one and hence will shrink more. A hot mould may be needed to achieve maximum crystallinity where stiffness and strength are major factors in product performance.

As already mentioned, it is quite possible to change the grade of a given polymer type in order to achieve the best compromise in processability. One way of doing this is to reduce the molecular weight – this reduces the melt viscosity under given conditions and may be desirable in injection moulding, but not in extrusion blow moulding (beware parison sag). In rubber melt processing, it may be necessary to adjust the type and level of the crosslinking systems to achieve, say, longer safe processing times followed by more rapid cure rates without incurring the penalty of local overcuring.

More often than not, a combination of these adjustments may be necessary to achieve the desired result. For example, a dustbin injection moulded from a polyolefin will be centre-gated and the cavity has a long ratio of flow path to cavity separation. Two major features to be avoided are a short shot (the flow cross-section near the gate freezes off before the melt front has filled the cavity) and excessive orientation (which causes splitting and fragility in the side walls).

To achieve robust products, a high molecular weight polymer might seem best. A high melt temperature would permit rapid injection, thus overcoming both faults, but too long in the machine at too high a melt temperature can cause degradation which impairs robustness. An alternative is to inject faster at the optimum melt temperature, but this may not be practicable if the pump is already operating at its maximum rating. Yet another alternative is to use an easier flow, lower molecular weight grade of the same polymer – this permits faster filling at the optimum melt temperature. Practical experience suggests that the reduced orientation in large-area thin-walled mouldings swamps any differences which might otherwise be expected from molecular weight considerations.

9.6 PREFERRED SHAPES

Experience has provided a number of lessons and has resulted in shapes that avoid many problems concerning distorsion, warpage, internal stress and others. The following suggestions are a very limited choice. Look for many more in handbooks from Beck (1980), Frados (1976), Malloy (1994), Rheinfeld (1981), Rosato and Rosato (1995) and many others, and in a number of company brochures.

- Avoid differences in wall thickness; unequal cooling will result in shrinkage differences (Fig. 9.25).
- Avoid sharp corners, since they will come out sharper; round them off so that internal cooling is more effective and possible change of edge is not obvious. Rounding off corners also promotes higher strength through reduction of stress concentrations (Chapter 6; Fig. 9.26).
- Use curved surfaces to hide local deformations (especially when centre gated) and to increase stiffness (Fig. 9.27).

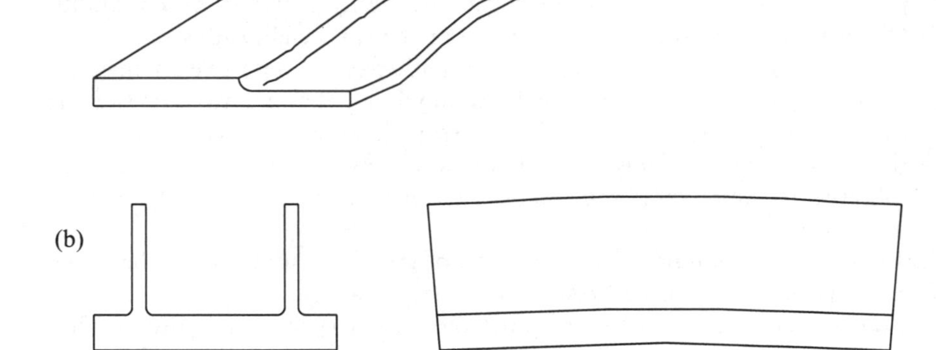

Fig. 9.25 Unequal cooling of different wall thicknesses results in distortion

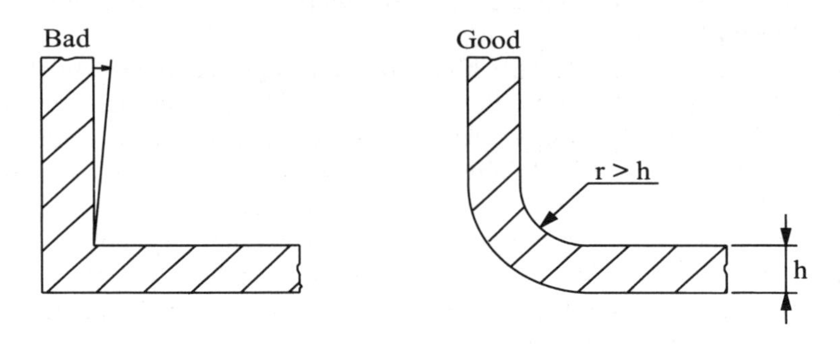

Fig. 9.26 Poor cooling at the inside of corners results in distribution of internal stress that promotes further curvature

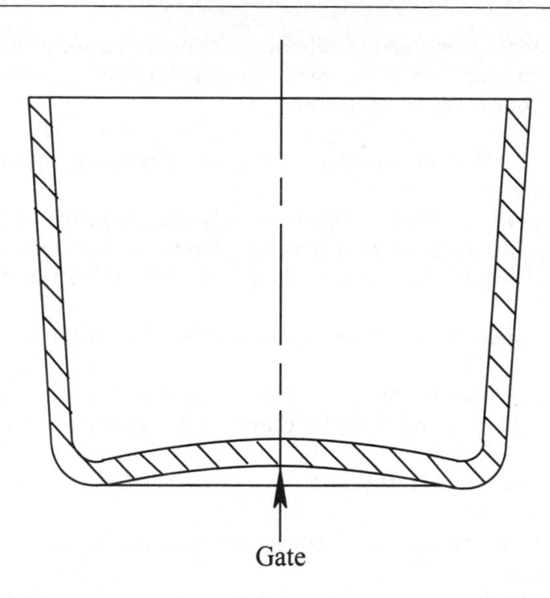

Fig. 9.27 Differences in radial and circumferential shrinkage cause internal stresses; these disappear, however, if the bottom is curved, by changing the curvature of the bottom

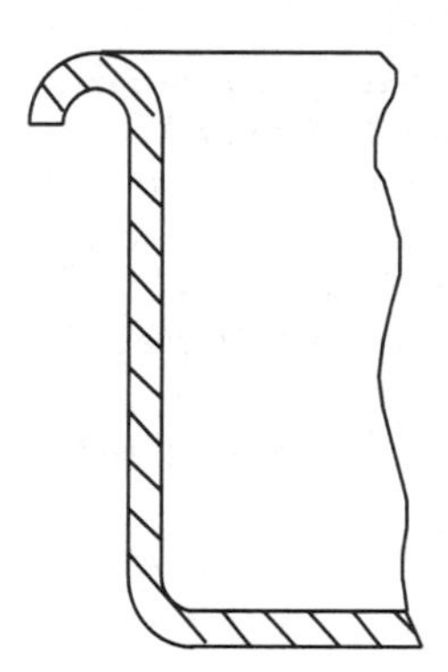

Fig. 9.28 Circumferential stress in the outer rim can result in invisible distortion

- To increase the bending stiffness of an edge of a container or crate, curve it over with equal thickness (Fig. 9.28).
- Make use of the many opportunities to combine functions in one product (Chapter 10).

FURTHER READING

The effects of processing on the various qualities of a polymer product are extensively but patchily covered in the literature. Recommendations for design for production are given in standard texts and handbooks relating to specific processes. The effect of processing on properties is documented mainly in a rich research literature, but the following books may be of interest.

Advani, S.G. (1994) *Flow and Rheology in Polymer Composites Manufacturing*, Elsevier, Amsterdam.

Beck, R.D. (1980) *Plastic Product Design*, Van Nostrand Reinhold, New York.

Cogswell, F.N. (1981) *Polymer Melt Rheology*, Godwin, Harlow.

Frados, J. (ed.) (1976), *Plastics Engineering Handbook*, Van Nostrand Reinhold, New York.

Isayev, A.I. (1987) *Injection and Compression Molding Fundamentals*, Marcel Dekker, New York.

Janson, L.-E. (1996) *Plastics Pipes for Water Supply and Sewage Disposal*, Borealis A.B., S–44401 Stenugsund Sven Axelsson A.B./Affisch & Reklamtryck A.B., Borås, Sweden.

Krevelen, D.W. Van and Hoftyzer, P.J. (1976) *Properties of Polymers*, 2nd edn, Elsevier, Amsterdam.

Malloy, R.A. (1994) *Plastic Part Design for Injection Molding, an Introduction*, Hanser, Munich.

Middleman, S. (1977) *Fundamentals of Polymer Processing*, McGraw-Hill, New York.

Osswald, T.A. and Menges, G. (1995) *Materials Science of Polymers for Engineers*, Hanser, Munich.

Rheinfeld, D. *et al.* (1981) *Injection Moulding Technology*, VDI-Verlag, Dusseldorf.

Rosato, Donald V. and Rosato, Dominick V. (1995) *Injection Molding Handbook*, Chapman & Hall, New York.

Struik, L.C.E. (1990) *Internal Stresses, Dimensional Instabilities and Molecular Orientations in Plastics*, Wiley, Chichester.

Tadmor, Z. and Gogos, C.G. (1979) *Principles of Polymer Processing*, Wiley, New York.

Ward, I.M. (1997) *Structure and Properties of Oriented Polymers*, Chapman & Hall, London.

Product design

10

10.1 INTRODUCTION

Experience over the centuries has taught carpenters their skills in finding solutions to their design problems. They know the choice of the right type of wood for an application and its protection against environmental attack; they have a general idea which beams or boards to choose; they think in terms of solutions regarding choice of direction of 'grain' (anisotropy), joints with dovetails, nails and/or glue and many others.

The same should hold for designers in plastics – if possible on a more sophisticated level. Let us consider some of their opportunities and dilemmas:

- short cycle times require thin walls (3 mm or less) but thin plastics walls have low stiffness; would ribs or changing the product shape offer the desired robustness?
- ease of manufacturing and product stability considerations usually suggest keeping the wall thickness constant;
- combination of functions in one element is often possible and offers reduced assembly costs, but needs careful consideration;
- moulded pieces have bosses for mounting with thread forming screws or ultrasonic welding which is excellent for quick assembly, but which method of assembly should be chosen?

Knowledge of a variety of these features, based on understanding of the behaviour of plastics as treated earlier in this book, forms an essential tool for every designer.

We see there is practical experience available for the designer, but theory must also be incorporated in new designs. Designing in plastics should be based on knowledge, understanding, experience and available tools. The practice of designing in plastics follows the rules of concurrent engineering (compare section 1.2):

- briefing the designer: what must the product do?
- shaping of ideas with the aid of all the information available;

- developing rough alternative solutions (is it sensible to use plastics at all?);
- working these out in terms of functionality and produceability;
- making an engineering specification of user requirements: load, temperature, tolerances, environment, life, etc.;
- choosing the most promising alternatives and working them out in terms of loading and functioning ('stiffness and strength') and manufacturing ('hot and wet'), including the choice of materials, and recognizing the interactions between the major issues: shape, polymer and process;
- evaluation of a prototype.

In one chapter of an introductory book we cannot hope to cover all aspects of product design. We have had to be selective, based on our experience. We have tried to provide a balance between the following features:

1. practical aspects of design based on commercial experience;
2. examples of principles given in earlier chapters;
3. extension of the basic principles;
4. examples which recognize and exploit unusual features;
5. examples of strategies for approaching the manufacture or design of products.

From these examples the outline design process will become clear, as well as the toolkit for predicting physical performance of the product and the toolkit for assessing process performance.

Section 10.2 addresses the practical design of features which can be readily incorporated into injection-moulded thermoplastics products.

Section 10.3 examines how to design efficient structures which have good stiffness per unit mass. Two aspects are emphasized: improving the bending stiffness by the use of stiff shapes such as ribs, and examining the behaviour of laminated composite beams.

Section 10.4 considers the role of internal stresses in products. One example involves the internal stresses in an assembled joint between two products made from different homogeneous materials, and the other example discusses the important internal stresses built up in a single product made from a composite material.

Section 10.5 discusses a range of examples where the design can tailor the process or the product to perform in the most efficient and desirable way. The main themes are how and why to place material in a process or product (or both) to make best use of it, coupling effects in laminated structures and laminated materials, and examples of design of good products.

Section 10.6 indicates a number of fields where the computer can help, and discusses the need to understand the underlying processes and theory in order to appreciate their results to their proper value.

Some of the material in this chapter is 'well known', but a substantial proportion is new and has not appeared before in the open literature in the form given here. It would be most satisfactory if readers felt encouraged to develop further their thinking and strategies in relation to their own design problems.

10.2 SPECIFIC PLASTICS ELEMENTS

Injection-moulded elements can contain integrated features, which may decrease considerably the number of parts needed and simplify the mounting procedure (Fig. 10.1) and thus reduce costs. Various features have emerged from practice; here we describe some of them.

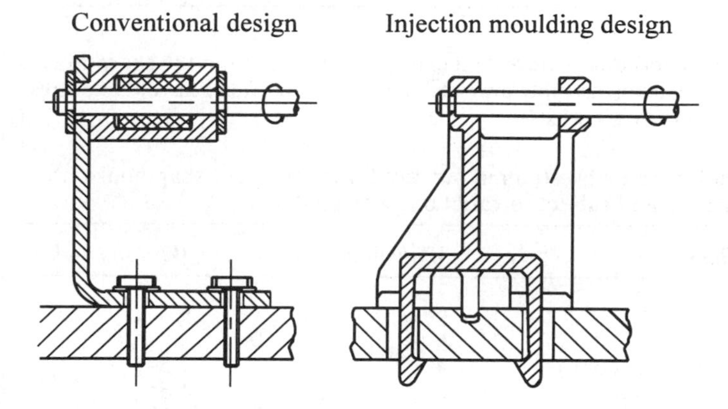

Fig. 10.1 Shaft bearing: left, conventional design; right, multifunctional injection moulding (from Ehrenstein and Erhard, 1984)

10.2.1 Snap-fittings

Assembly can be made easy and quick with snap-fits or snap-rings integrated into the moulding. Some aspects have to be considered.

(a) Maximum strain during mounting

During mounting or detaching procedures, the elastic element in a snap-fitting is deformed to a prescribed strain. If this element is a hook mounted on a rectangular cantilever beam (Fig. 10.2), the maximum strain, occurring in section A–A, is

$$\varepsilon = \frac{3Hh}{2L^2} \qquad (10.1)$$

This strain should not exceed the permissible value. Table 10.1 contains values for different thermoplastics. For plastics having a pronounced

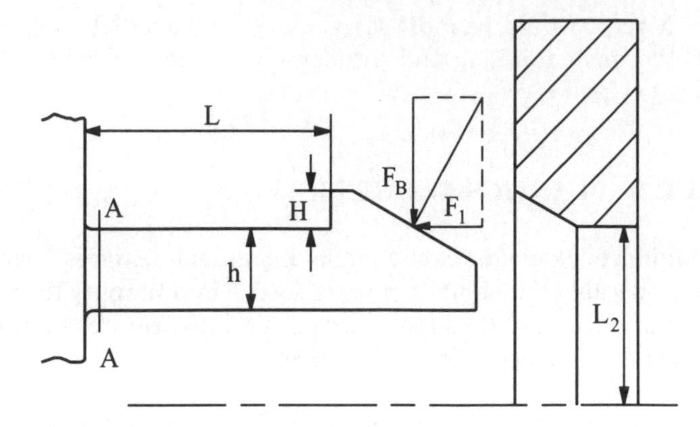

Fig. 10.2 Snap-hook with rectangular cross-section. Note the sharp internal corner which prevents disassembly and which provides an often unavoidable stress concentration (courtesy Hoechst)

Table 10.1 Permissible strain in per cent for mounting of snap-hooks (these values are indicative and subject to variations with grade)

Crystalline	*Without additives*	*With short glass fibres*
HDPE	6	6
PP	6	2
PA (eq. moisture cont.)	4	1.5–2
POM	6	1.5
PBTP	5	1.5
PS	1–1.8	
High-impact PS	2	
SAN	2	
ABS	2.5	1.2
PVC	2	
CAB	2.5	
PC	2–4	1.8
PMMA	1–4	

yielding behaviour, these values amount to about one-third of the (short-term) strain at yield, for others about one-third of the strain at fracture.

The data in Table 10.1 refer to infrequent mounting of the hook; for frequent mounting and dismounting use of 60% of these values is recommended.

(*b*) *Other designs*

Hooks can be made fit for disassembly, as shown in Fig. 10.3, by adding a return angle α_2 other than 90°.

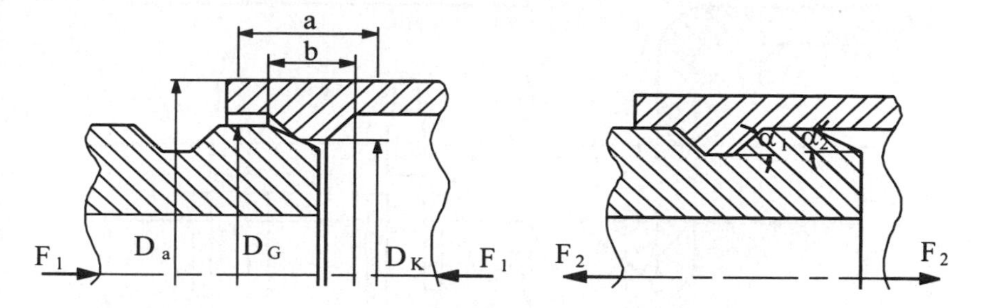

Fig. 10.3 Cylindrical snap-ring during and after assembly (courtesy Hoechst)

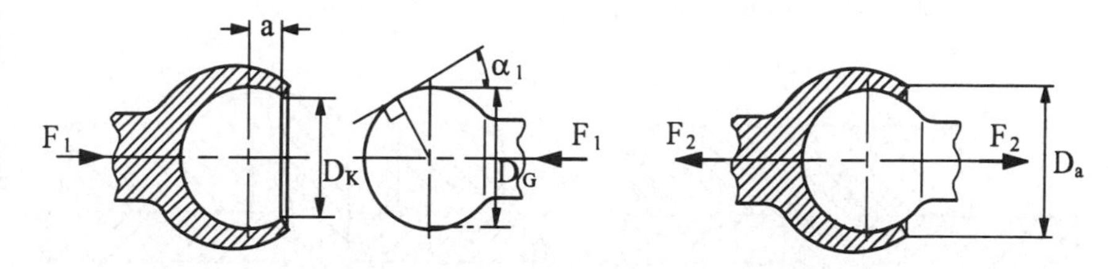

Fig. 10.4 Ball hinge. The acting force can be chosen to be in the mounting direction, but is perhaps better perpendicular to this (courtesy Hoechst)

If the snap-element is cylindrical (Fig. 10.3) the strain during mounting amounts to

$$\varepsilon = \frac{D_G - D_K}{D_K} \tag{10.2}$$

For this case only, the use of 50% of the values from Table 10.1 is permitted, as the strain will be present in all of the circumference of the element.

After being snapped into place, fingers or rings should be nearly stress-free, but in order to guarantee a tight connection and to prevent rattling the dimensions L_2 (Fig. 10.2) and D_K (Fig. 10.3) are sometimes chosen slightly smaller than those of the matching part. This induces a small permanent strain which gives rise to a relaxing stress.

Related to the cylindrical snap-element is the ball hinge (Fig. 10.4). A typical application is the connecting rod which transfers the movement of the accelerator to the throttle.

The elastic movement of a snap-element is usually derived from the bending of a beam or the straining of a ring, but other principles can be used as well, such as torsion of a rod. Another interesting design is the connection of a lid to its box with a hinge consisting of a moulded axis and a snap-fitting around it (Fig. 10.5).

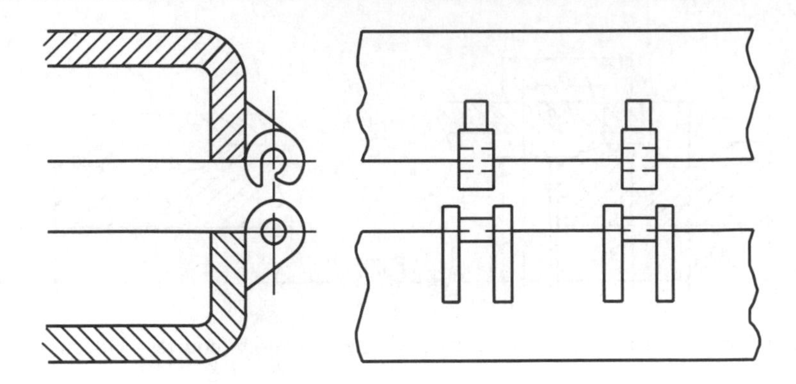

Fig. 10.5 Clicking lid around axis

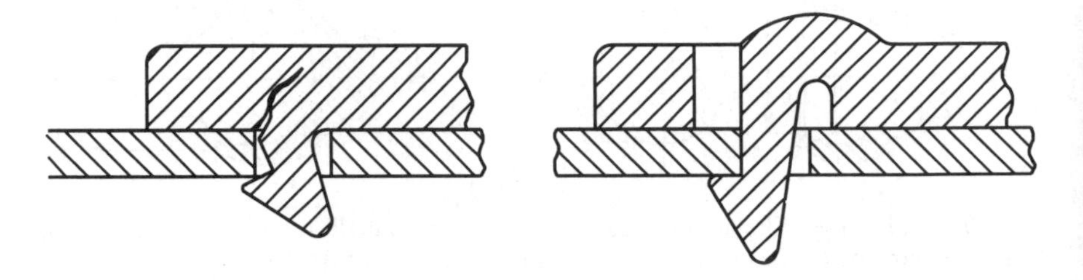

Fig. 10.6 Improved flexibility in snap-hook design (courtesy J. Spoormaker)

From equation (10.1) it becomes apparent that hooks have to have a certain length in order not to exceed the permissible strain of Table 10.1. Figure 10.6 illustrates the problem and gives a possible solution.

(c) *Insertion and disconnection forces*

Snap-connections can be made detachable if the disconnection angle on the reverse side of the hook (e.g. α_2 in Fig. 10.3) is 45° or lower.

During the insertion or disconnection movement, normal and friction forces on the inclined surface of the hook will make an equilibrium with the force to insert F_1 or to disconnect F_2 and the force F_B that is responsible for the deformation H of the hook (Fig. 10.2). From this equilibrium the force to insert or disconnect such a snap-fitting can be deduced:

$$F_{1,2} = \frac{3HE_sI}{L^3}\frac{\mu + \tan\alpha_{1,2}}{1 - \mu\tan\alpha_{1,2}} \tag{10.3}$$

α_1 is the insertion angle, α_2 the disconnection angle and μ is the coefficient of friction. As a guide, μ is 0.2 for some crystalline thermoplastics, and up to 0.6 for some amorphous thermoplastics. $3HE_sI/L^3$ is the force to bend the

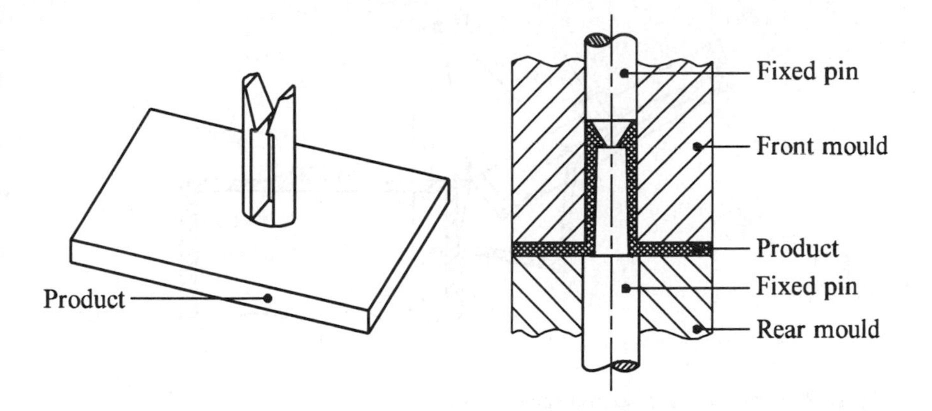

Fig. 10.7 Circular-shaped hooks, moulded in a round hole in the mould (courtesy Philips C.C.P.)

hook over a distance H, E_s is the short time secant modulus, and I represents the moment of inertia of the cross-section of the hook.

For other snap-connection designs like tapered beams, and snap rings, relationships similar to (10.3) can be derived.

(d) Pull-out strength

If the disconnection angle α_2 is 90°, the connection is tight against pull-out forces. The limit here is the shear strength of the hook section or the tensile strength of the beam of the hook, reduced by the stress concentration factor k caused by the sharp internal corner of the hook.

(e) Tooling aspects

In view of cheaper tooling (round holes are cheaper to produce than rectangular ones) non-rectangular cross-sections of the hook beam can be chosen (Fig. 10.7).

The mould in which the snap-elements are shaped has to contain the undercuts of these elements. In most cases the parting plane can be chosen skilfully (Fig. 10.7), in others sliding mould parts may be needed.

10.2.2 Integral hinges

Hinges can be integrally formed during injection moulding for various areas of application: to enable mounting for life as in electrical connectors (e.g. combined with snap-fittings), or to enable an intermittent hinge movement as in a lid on a box (Fig. 10.8) or continuous movement like in vibrating

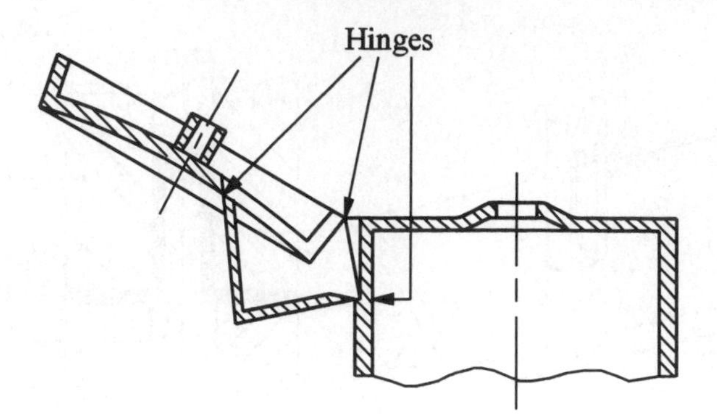

Fig. 10.8 Film hinges in soap dispenser lid.

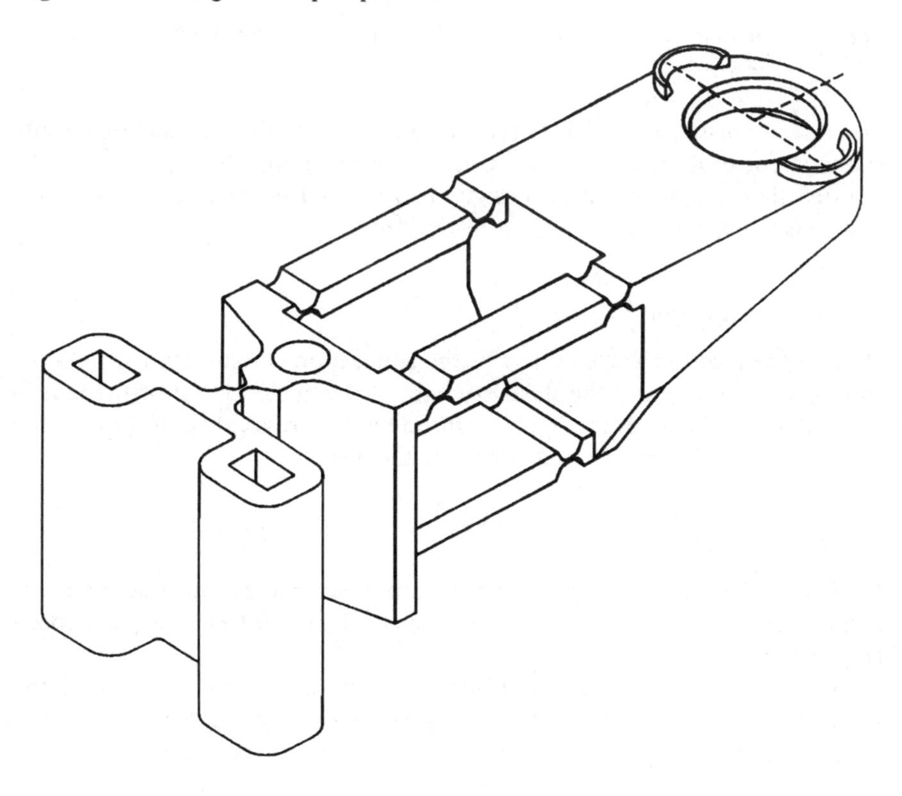

Fig. 10.9 Film hinges in actuator for compact disc player (courtesy Philips C.C.P.)

elements (e.g. a joint for crossing axes; see problem 4 at the end of this section, and the actuator of Fig. 10.9).

Integral hinges make use of improved strength of material by molecular orientation during flow through narrow passages in a mould. Additional

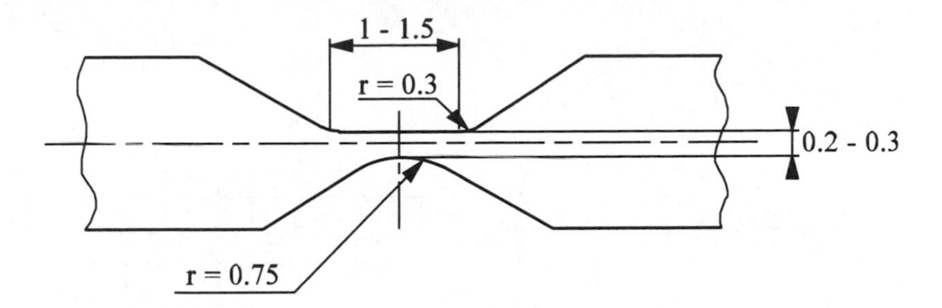

Fig. 10.10 Preferred design of polypropylene film hinge (courtesy Philips C.C.P.)

mechanically induced orientation is achieved by yielding the material during the first bending movement of the solid hinge. The preferred material for integral hinges is polypropylene (PP). Its linear molecule and its related crystalline structure provide excellent close packing of adjacent linear molecules across the line of the hinge. A closely related materials property is its striking cold-drawing behaviour (section 6.2.1(b)). PP is known for its very high fatigue life in hinges. Other linear chain polymers, such as POM and PA, are in use as well.

A preferred design is presented in Fig. 10.10. There should be a single narrowest restriction in the cross-section, in order to guide the point of first (and consecutive) bending.

Safeguards are needed to ensure that the hinge is not loaded in shear or across the direction of molecular orientation.

The cavity and the gating of the mould should be made such that preferred flow orientation will develop all over the hinge and that the formation of weld-lines in the hinge is impossible.

In some cases other shapes for the hinge are desired, such as leaf springs. If the spring hinge will be moved frequently, local deformations should stay within the strain at yield ε_s (3–15% depending on the polymer). Preferred film thickness is of the order of $t = 0.3$–0.8 mm and film length $L/t = 3$–10 or greater.

A specific example is the actuator design of Fig. 10.11. Here the S-shaped spring hinges act like membranes, permitting the inner cylinder to move axially over ± 1 mm under only a small force. Problem 5 at the end of this section deals with some of its aspects.

10.2.3 Detachable connections

Detachable connections between elements can be achieved in various ways. The most usual is the use of screws. Mounting with self-tapping screws is often attractive for connecting parts to a plastics housing or frame. The latter

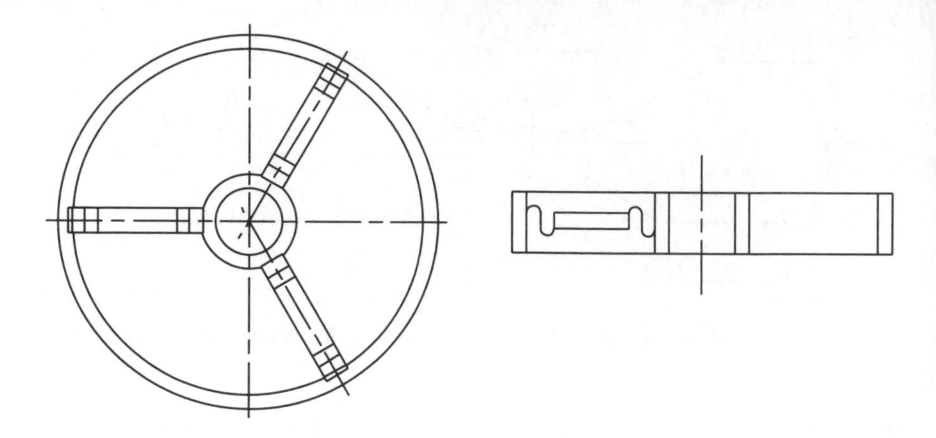

Fig. 10.11 Actuator design for axial movement under small force (courtesy Philips C.C.P.)

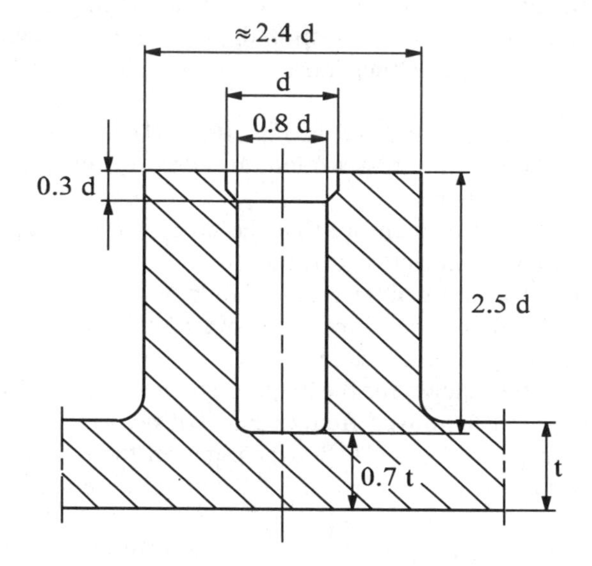

Fig. 10.12 Preferred design for the boss of a self-tapping screw (d = nominal screw diameter)

can be equipped with bosses containing a cylindrical hole. It is not always necessary to provide the hole with a thread, which simplifies the mould and the moulding. Optimum dimensions of a boss are shown in Fig. 10.12. The pull-out force for a self-tapping screw in a POM boss is dependent on screw length L and diameter d. Experience suggests that $L/d = 2.5$ is an advisable value.

10.2.4 Fixed connections

If not already made in one part, pieces may be assembled to form a unity by cementing or welding.

(*a*) *Adhesive joints*

Adhesive joints can be used for both thermoplastics and thermosets. Various types of adhesives are available: solvents, solvent-borne, water-borne, one- and two-component curing, hot melt and others. Of primary importance is that they wet the surface upon their application or otherwise are anchored into the surface; proper handling, including a thorough cleaning and preparation of all surfaces to be cemented, is essential. Second, it is important to achieve a proper solidification or curing of the adhesive or evaporation of the solvent; this will need time.

For some plastics excellent solvent cements are available, like methylene chloride and methyl ethyl ketone for PMMA, tetrahydrofuran and cyclohexanone for PVC or toluene and xylene for PS (see section 6.2.3(d) and Table 6.1 for more detail about dissolution of polymers.) Cementing of other plastics is often justified, provided suitable adhesives that will wet the surface are chosen. Polyolefins cannot be cemented without a proper pretreatment because of their apolarity and insolubility.

Of the various load-bearing adhesive joints, we mention just the single lap joint. The distribution of shear stress is not constant (Fig. 10.13) due to the elasticity of the plates. There are shear stress peaks at the ends of the joint. This has consequences for the design of the joint (see problem 6 at the end of this section). In addition a tensile force in the plates connected to it will introduce a bending moment in and near the cemented zone. A double lap joint will avoid this bending moment, but is more expensive. So the designer

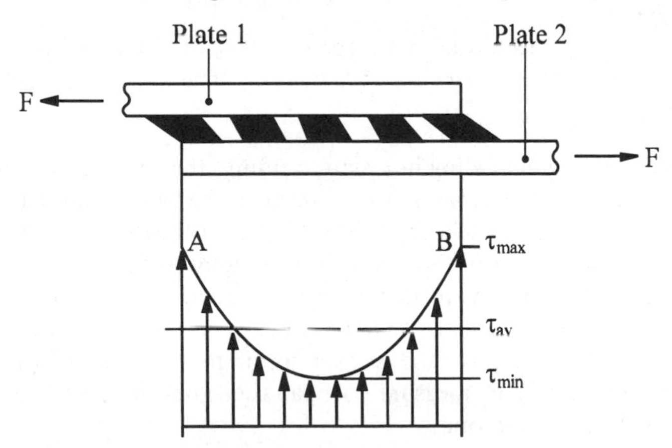

Fig. 10.13 Shear stress distribution in a lap joint

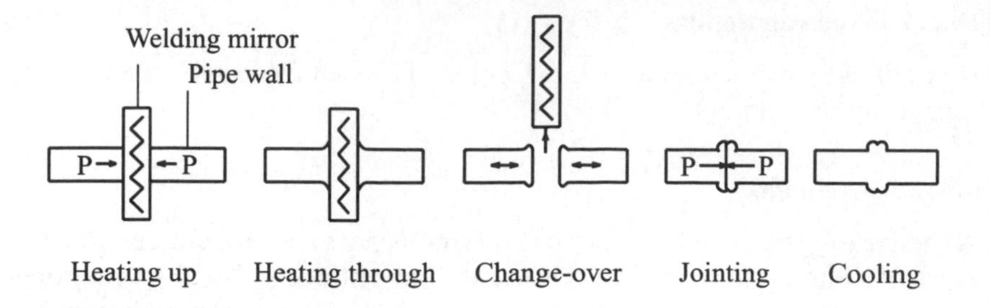

Fig. 10.14 Processing phases for making a butt-welded pipe joint (courtesy Gastec)

should carefully consider the general design of the joint. A good example of a widely used cemented connection is the PVC pipe and socket joint in the domestic installation domain. For more information on adhesive joints, see e.g. Kinloch (1987).

(*b*) *Welded joints*

For thermoplastics products a great variety of welding techniques are available. The principle of the action is (1) melting of the surfaces to be welded, (2) applying pressure and (3) cooling with the pressure still on, in order to ensure a tight and permanent connection (compare Fig. 10.14). As the surrounding material stays cool during the process, the cooling down of the heated zone causes local tensile stresses in the welded seam. Their effect may be slight but has to be considered in principle, as it adds to the design stress of the product.

Various welding techniques have been worked out, e.g. for pipe connections, based on different principles for heating. The most important are the following.

- Hot tool welding, including hot plate welding: the surface to be welded is heated by conduction from a heated surface, the heated parts are pressed together subsequently, leading to the process of healing, i.e. the interdiffusion of molecules from both surfaces, as long as these are hot enough (Fig. 10.14). A common and dependable connection is the butt-welded PE pipe joint, widely used in pressure pipeline systems for water supply. Another one is the pipe and socket joint, making use of pipe fusion heating tools made to measure to heat the outer pipe surface and the inside of the socket in one.

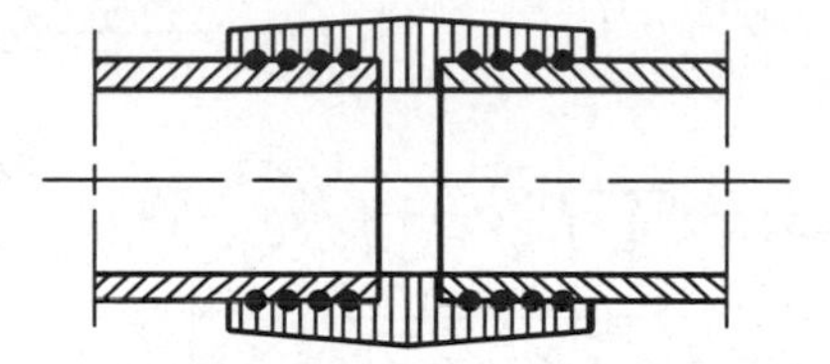

Fig. 10.15 Socket with lost electrical element

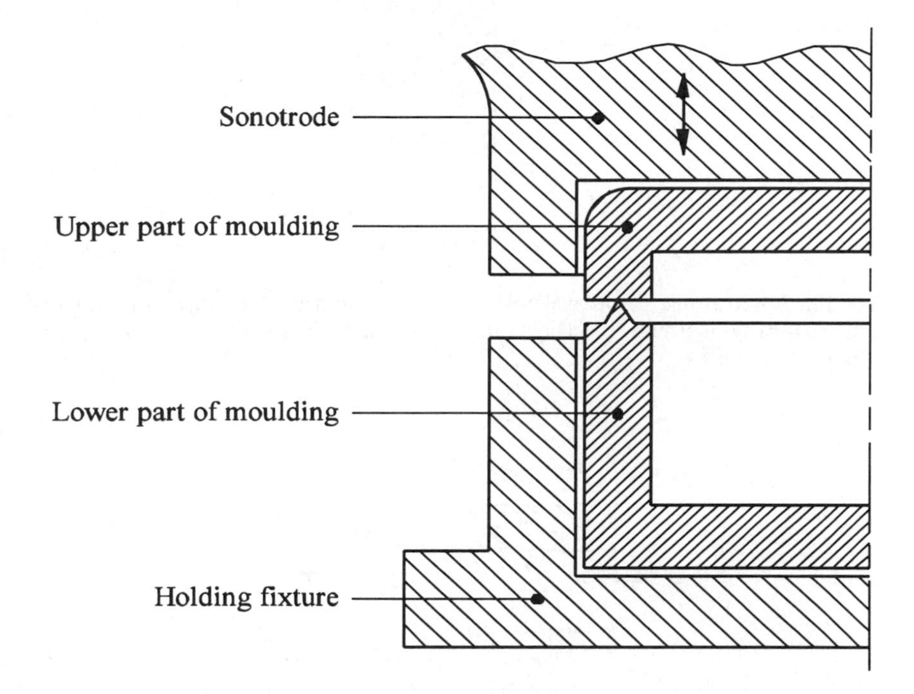

Fig. 10.16 Two parts to be ultrasonically welded, fixed between holding fixture and sonotrode. The pointed extension of the lower part is called the energy director; here the mechanical vibrations are absorbed, leading to development of friction and damping heat (courtesy German Electrical and Electronic Manufacturers' Association, ZVEI)

- Hot gas welding: V-shaped welding opening and filler rods of the same type of thermoplastic are heated by hot inert gas. They are less dependable because of the greater risk of inclusions.
- Lost electrical resistance wire: introducing this between the surfaces to be welded brings local heat to where it is needed, provided the wires are suitably located; these stay behind in the connection (Fig. 10.15).
- Other heat sources, using electricity, include inductive high frequency (needs steel wire or powder in the weld) or dielectric high

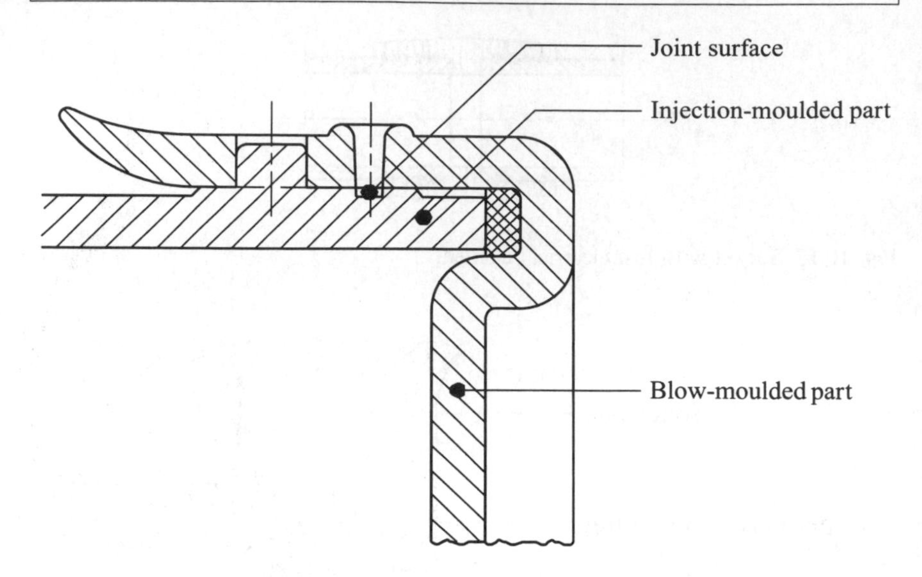

Fig. 10.17 Ultrasonic spot welding of blow-moulded and injection-moulded parts (courtesy German Electrical and Electronic Manufacturers' Association, ZVEI)

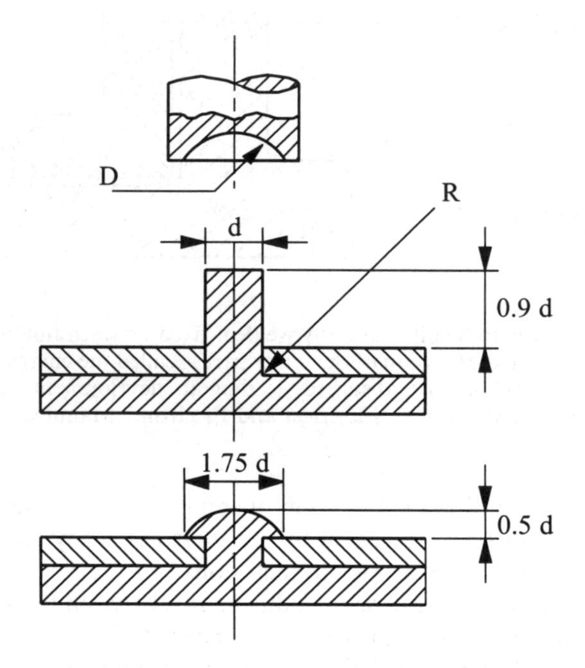

Fig. 10.18 Shape of sonotrode, component and stud for riveting (courtesy German Electrical and Electronic Manufacturers' Association, ZVEI)

frequency (only suitable for plastics with a high dielectric loss angle, such as PVC).

- Friction welding: rubbing the two surfaces together generates the heat for subsequent welding: Spinning for circular objects or vibrating for others are used.
- Ultrasonic welding: imposed movement, perpendicular to the surfaces, heats the material, because mechanical energy is transferred into heat by the internal damping ($\tan\delta$) of the material. For joining injection-moulded elements, ultrasonic welding is very suitable. The surfaces to be joined are heated by mechanical vibrations, generated electrically and transmitted to the parts by means of a sonotrode. The system is designed to concentrate the heat generation in the welding zone. For this purpose, sharp welding contact lines are preferred (Fig. 10.16). Once the material is plasticized a clamping force ensures flow and prevents movement until the joint has solidified. Welding times are of the order of several seconds.

Ultrasonic energy can also be used for spot welding (Fig. 10.17) and riveting (Fig. 10.18) of parts. With spot welding the tip of the sonotrode penetrates through the upper part into the lower part. Heat is produced on the contact surface of the parts and plasticizes the material. The expelled material flows upwards and forms a ring-shaped elevation. Riveting is often used for fixing metal parts onto injection-moulded plastics parts.

PROBLEMS

1. With reference to Fig. 10.1, indicate the primary differences between the metals and the plastics design.
2. A non-detachable snap-hook, shaped as in Fig. 10.2 and to be made of POM, has to fit in a plate with thickness $L = 10$ mm. It should have a width $b = 6$ mm, an insertion angle $\alpha_1 = 30°$ and a pull-out strength of $F_2 = 130$ N. Relevant short-time data of POM are: friction coefficient $\mu = 0.2$, secant modulus at relevant strain $E_s = 1600$ MN/m^2 (compare Fig. 5.8), tensile strength $\sigma_{max} = 62$ MN/m^2, shear strength $\tau_{max} = 0.6\sigma_{max}$, safety factor $F = 1.3$, stress concentration factor at the square ($r = 0.1$ mm) edge of the hook $k = 3$. Calculate the required thickness h, the snap depth H and the mounting force F_1.
3. (a) With reference to Fig. 10.9, how many hinges can the reader see and imagine in this design? (b) If the left-hand end of the device is to be mounted to the frame of the CD player, while the right-hand end contains the actuator head, what is the remaining freedom of movement of the head? (c) Regarding the mould design, indicate how the use of moving side cores can be avoided and suggest the position of the parting line.
4. Figure 10.19 contains the drawing of a flat piece containing a number of film hinges. This can be folded and mounted to form a universal joint of

two axles. Imagine how this can be achieved and make a carton model to demonstrate.

5. The S-shaped spring hinges of Fig. 10.11 are to be made of POM elements of thickness $t = 0.3$ mm and of width $b = 4$ mm. The radius of the S-curve is $R_1 = 1$ mm in the unloaded position. Vertical movement of the central part of the device will add extra curvature to the shape of the spring. For the sake of this problem we model the spring hinge as in Fig. 10.20. Can a displacement $v = 1$ mm be tolerated without the POM grade used reaching its strain at yield of $\varepsilon_s = 8\%$? The POM grade has a short-time modulus of $3100\ \text{MN/m}^2$.

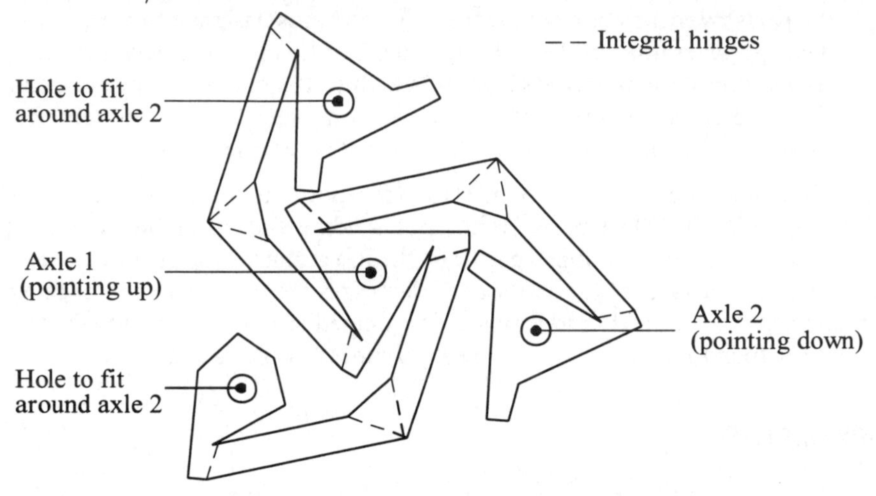

Fig. 10.19 Flat piece ready to be mounted into a universal joint (courtesy Philips C.C.P.)

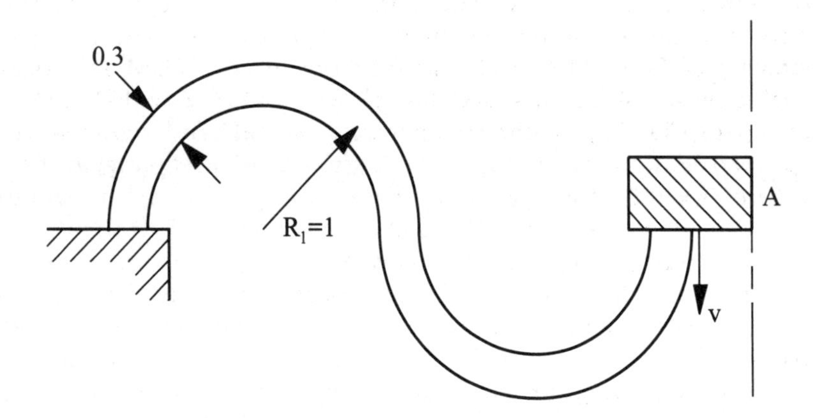

Fig. 10.20 Leaf spring of Fig. 10.11 as schematized for this problem. Note that the movement of A as part of the total construction of Fig. 10.11 is bound to be vertical.

6. A connection between two plates is to be realized as a cemented lap joint. The strength of the connection is critical. With reference to Fig. 10.13, what will be the effect of the following attempts to improve strength: (a) increase the thickness t of the plates; (b) increase the width b (perpendicular to the plane of the figure); (c) increase the length L of the overlap; (d) increase or decrease the thickness h of the cement layer; (e) choose a stiffer type of cement.
7. Indicate which factors determine the success of a welded socket and pipe joint with regard to leakage and resistance to axial forces.

10.3 DESIGNING FOR STIFFNESS

Compared with many other materials, the modulus of plastics is remarkably low. It is therefore of importance to consider the required stiffness in an early design phase and frequently thereafter. Usually the bending stiffness (EI) is involved. Therefore we have to take account of Young's modulus of the material E and the second moment of area of the cross-section I. The same however holds for the torsional stiffness GI_p, where G is the shear modulus and I_p is the polar second moment of area. It should be noted that EI and GI_p are independent properties of a design. If one is made optimal the other may well fall below the required level. We consider below $E(G)$ and I (I_p) separately.

The proper use of shape to improve bending stiffness of a given polymer is discussed in sections 10.3.2–10.3.4; this is particularly relevant to injection moulding of thermoplastics products, and a case study is given in section 10.5.5. Section 10.3.5 explains how to estimate the stiffness of beams made from symmetrical laminated composite materials, and section 10.5.3 describes one way of combining bending and twisting stiffness.

10.3.1 Increasing modulus E

Changing E or G implies changing the material. This is particularly important where higher in-plane stiffness is needed. Thermosets have a relatively high modulus. They may be considered for high stiffness and close tolerance applications, and also at higher temperatures.

The addition of short glass fibres (20–30 wt%) will cause E to increase by a factor of 2–4 (Fig. 10.21), accompanied by anisotropy of properties, more so for strength than stiffness. In some cases even carbon fibres are used. Longer fibres will increase the modulus further (see Chapter 7 for the in-plane behaviour of continuous fibre-reinforced polymers and section 10.3.5 for the behaviour of composite beams). The addition of mineral fillers can be very effective as well.

We have seen in Chapter 5 that adding mineral fillers is widely used to increase the modulus of a rubber compound. Preventing free lateral movement

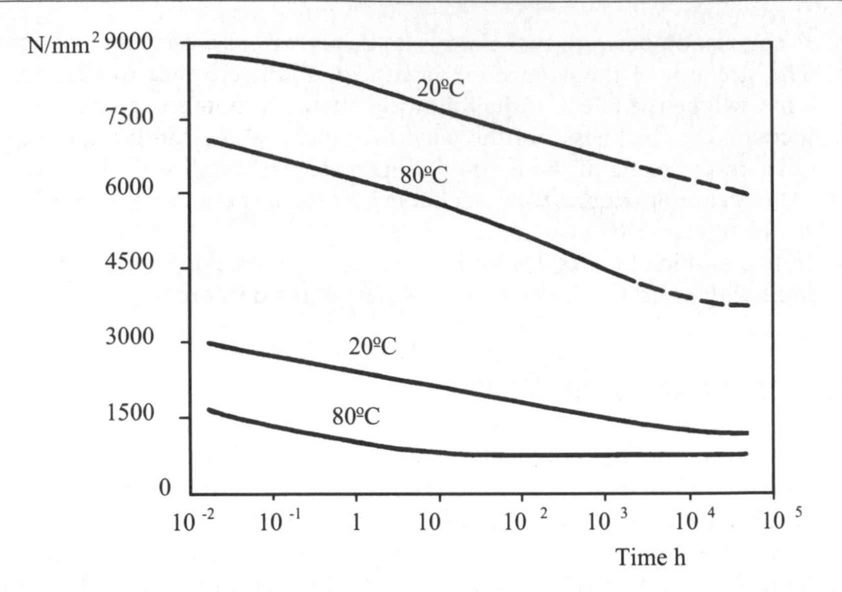

Fig. 10.21 Modulus of POM without (lower lines) or with (upper lines) 30% short glass fibres as a function of time at 20 and 80°C, respectively (courtesy Hoechst)

in rubber blocks carrying compressive loads can substantially increase the effective modulus of the rubber; this widely used effect is applied in section 10.5.2.

10.3.2 Adaptation of shape

Small loads in bending often lead to large deformations. Adaptation of shape, without changing the wall thickness, can provide considerable improvement (Figs 10.22–10.24).

If bending is involved, then second moment of area can be exploited by using corrugated (Figs 10.23 and 10.24), or I-shaped, T-shaped or U-shaped sections (Fig. 10.25), which can involve thinner walls (faster production rates). This is why many mouldings bristle with ribs. However, if the walls are too thin, webs and flanges may be prone to buckling instability. Panels can be stiffened (in bending) by the use of corrugation, ribs or curvature, or the production of a foam sandwich structure.

10.3.3 Foamed structures

Foaming can be a useful technique for large injection-moulded objects, since the closing forces of the machine can be rather small, while cycle time is not essentially longer than for massive products of the same weight.

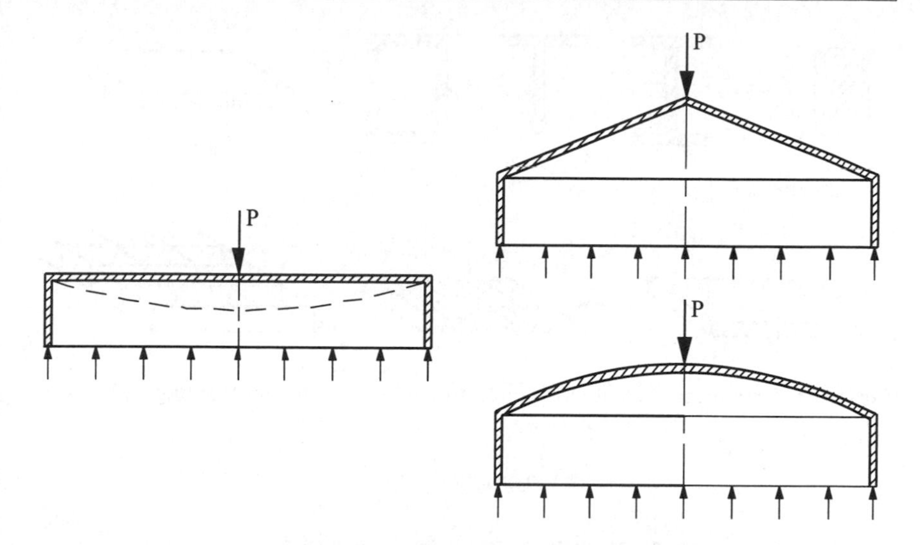

Fig. 10.22 Lid on cylindrical jar: (a) flat lid, high deformation; (b) conical lid, different distribution of stress in the object, small and not easily observable deformation; (c) spherical lid, attractive design, deformation hard to observe

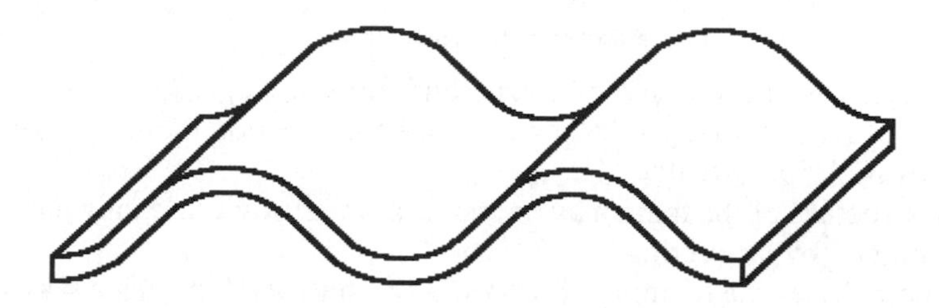

Fig. 10.23 Corrugated board.

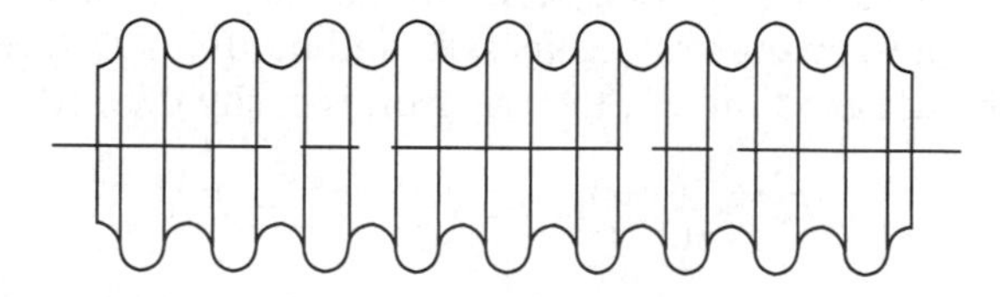

Fig. 10.24 Corrugated drain pipe

Foaming of a wall of given weight will increase its thickness *h*. The resulting foamed plastic is, however, more flexible under direct or shear stress (has lower *E* and *G*) than its unfoamed counterpart. Even if the structure of the pores is uniform all over the thickness (no skin), the resulting bending stiffness of the product,

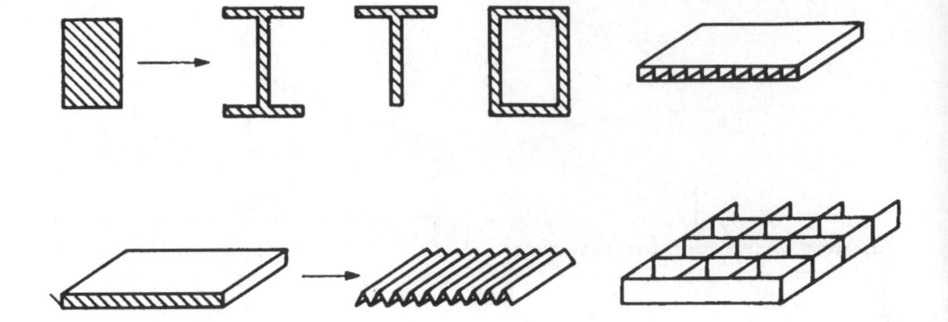

Fig. 10.25 Exploiting second moment of area to achieve higher bending stiffness per unit weight

$$E\,I = E\frac{1}{12}bh^3$$

will normally increase, in view of the high power of h.

Different approximations are given for the modulus E of foam. We provide here a simple one, found to fit open-cell data, but others may be better, dependent on the specific type of foam:

$$E_{\text{foam}} = E_{\text{solid}}(\rho_{\text{foam}}/\rho_{\text{solid}})^2 \qquad (10.4)$$

where ρ_{foam} is the density of foam, and thus $\rho_{\text{foam}}/\rho_{\text{solid}}$ is the volume fraction of plastics material. A typical value for injection moulded structural foams is $\rho_{\text{foam}}/\rho_{\text{solid}} \approx 0.5$.

The structure of the foam may be visible at the surface and may have to be camouflaged by lacquering.

In many cases the foaming process will be controlled in such a way that a sandwich with a (rather) solid skin (denoted by subscript 's') and a foamed core (subscript 'f') will arise. To replace a solid panel of modulus E and thickness h by a foam sandwich panel of the same breadth and bending stiffness having a foam core of thickness h_{f} and modulus E_{f} faced with skins of thickness h_{s} and modulus E_{s}, the criterion is easily derived:

$$\frac{1}{12}Eh^3 = \frac{1}{12}[E_{\text{s}}(h_{\text{f}} + 2h_{\text{s}})^3 - E_{\text{s}}h_{\text{f}}^3 + E_{\text{f}}h_{\text{f}}^3] \qquad (10.5)$$

The weight saving resulting from using a sandwich panel (or beam) rather than a solid panel can be readily calculated. This analysis assumes that the core has a uniform density and that there is a perfect bond between skin and core: it also assumes that the core is not so deep that failure by shear could occur. For long-term loading, the appropriate creep modulus for the foamed and unfoamed polymers must be used in calculations of stiffness.

With fibre-reinforced thermoset laminates, more sophisticated sandwich structures of higher stiffness and strength can be made, using a thin but

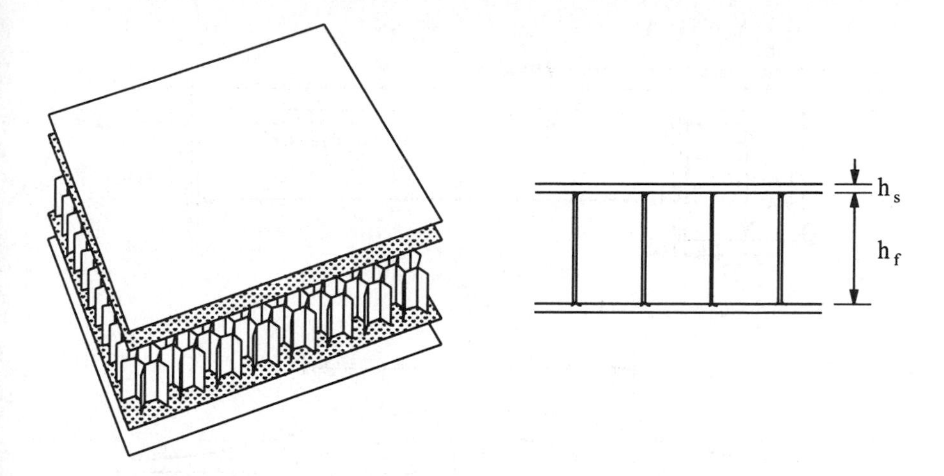

Fig. 10.26 FRP honeycomb structure.

massive (high-modulus) skin and a thick, light (low ρ_f) foam or honeycomb core (Fig. 10.26). Skin modulus and core thickness dominate bending stiffness, and other terms of equation (10.5) can be neglected:

$$\frac{1}{12}Eh^3 = \frac{1}{2}E_s h_s h_f^2 \tag{10.6}$$

In dealing with foamed structures, we have entered the field of multi-layered structures. In extrusion, injection moulding and blow moulding, a range of techniques has been developed. Layers of material with special properties, such as low diffusion and resistance to electromagnetic impulse, can be introduced, as can recycled material. The latter is of growing importance in view of recycling requirements in various countries. Moreover, it provides a low-cost means of increasing the bending stiffness.

10.3.4 Ribs

Ribs are very effective for improving bending stiffness and easy to realize in many technologies. In order to avoid visible sink marks in the flat wall opposite to the rib $t_r < 0.75t_w$ is chosen.

Ribs can be made very slender and high, up to $h_r = 20t_r$ but this may give problems with filling the mould and with buckling stability. A typical value is $h_r = 5t_r$.

(*a*) *Stiffness of ribs*

The bending stiffness of a ribbed structure is derived from the cooperation between each rib and a matching width b of the plate (Fig. 10.27a).

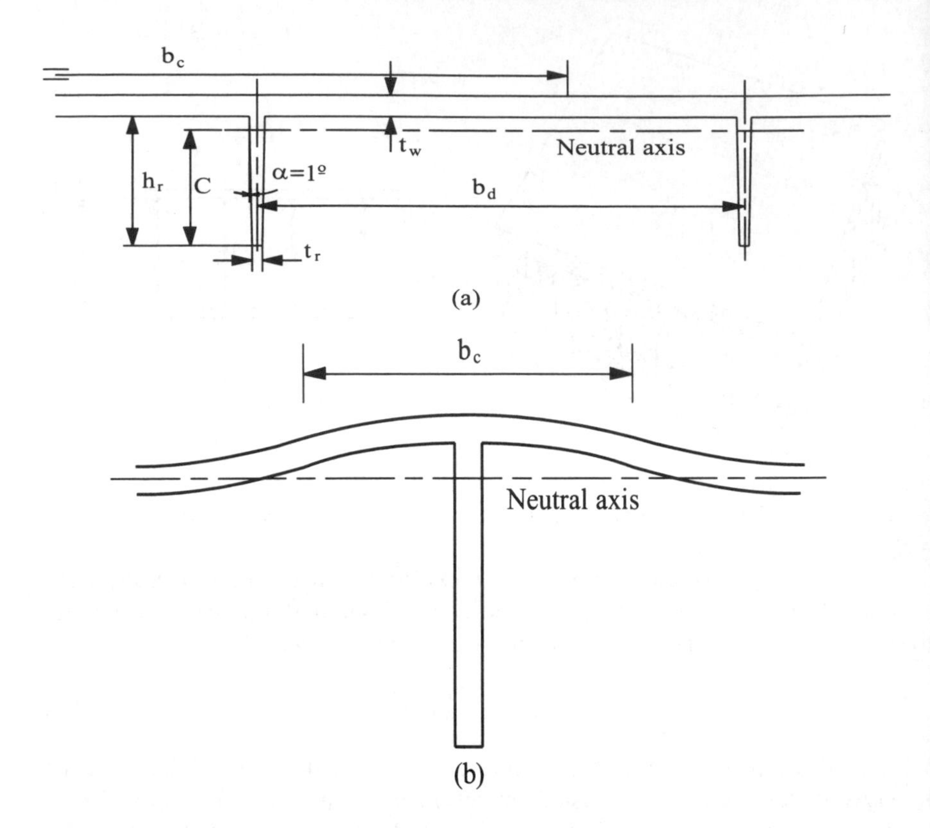

Fig. 10.27 (a) Aspects of ribs: b_c = effective web width, b_d = rib distance (plate width b/number of ribs n). (b) Effective web width b_c contributing to the joint bending stiffness (perpendicular to the plane) of plate and rib.

Classical theory can be used to calculate the neutral axis and the second moment of area of the cooperating cross-section of this apparent T-beam. We will deal with that concept below. It should, however, be borne in mind that this is just a first approximation and that in many applications of ribs the validity of this theory is surpassed. This could be the case where the structure has to be treated as thin-walled or when its flat parts might buckle or curve under load, leading to loss of the bending stiffness and strength as seen from the classic theory. Extension of the theory should deal with different effects, as follows.

- Buckling of ribs can occur if they are slender (see section 10.3.4(c)).
- If the ribs are too widely spaced (say, $b_d > 25t_w$) then, well away from the ribs, the plate tends to behave as if the ribs are not present, as it is locally very flexible under bending loads. We therefore use the concept of an effective web width b_c (Fig. 10.27b). If parallel ribs are applied, their relative distance b_d should not exceed b_c.

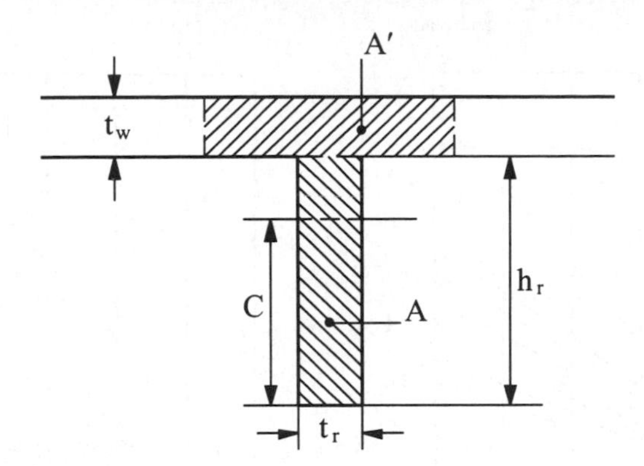

Fig. 10.28 Simple model for cooperating rib and plate.

- A high load may trigger non-linear behaviour of the structure through curving or buckling or through non-linear behaviour of the material.

A simple model for the bending stiffness states that the rib, having a cross-section A, cooperates with part of the plate, which it stiffens, having the same cross-section $A' = A$ (note that A' is generally smaller than $b_d t_w$ or $b_c t_w$) (Fig. 10.28). The neutral plane of the T-beam thus formed approximates, for $h_r \gg t_w$, a distance $C = \frac{3}{4} h_r$.

The second moment of area can be represented as

$$I = C t_r h_r^3 \tag{10.7}$$

where C for such a case ($h_r/t_w = 10$) was calculated to be 0.23; however, it will be much lower (e.g. $C = 0.15$) if a taper of, say, $\alpha = 1°$ is introduced. For lower rib heights, e.g. $h_r/t_w = 3$, higher values of C are calculated (0.31 for $\alpha = 0°$, 0.28 for $\alpha = 1°$). A straightforward calculation of second moment of area of a ribbed plate, up to the limit of the effective width b_c, is presented in Fig. 10.29a for the case of $t_r = 0.6\ t_w$.

The design of a parallel ribbed plate of suitable stiffness can be determined with Fig 10.29a:

- suggest a suitable plate wall thickness t_w to be ribbed, with regard to cycle time, etc.
- make sure a rib thickness of $t_r = 0.6\ t_w$ is required (there are data for other values of t_r in the literature);
- calculate the (imaginary) equivalent thickness t_{et} of an unribbed plate, that will match the required second moment of area $I = \frac{1}{12} b t_{et}^3$ of the ribbed plate;
- calculate t_{et}/t_w using the left-hand scale of Fig. 10.29a;
- suggest a suitable rib distance b_d and calculate b_d/t_w using the right-hand scale of Fig. 10.29b;
- read h_r/t_w from the diagram and calculate h_r;

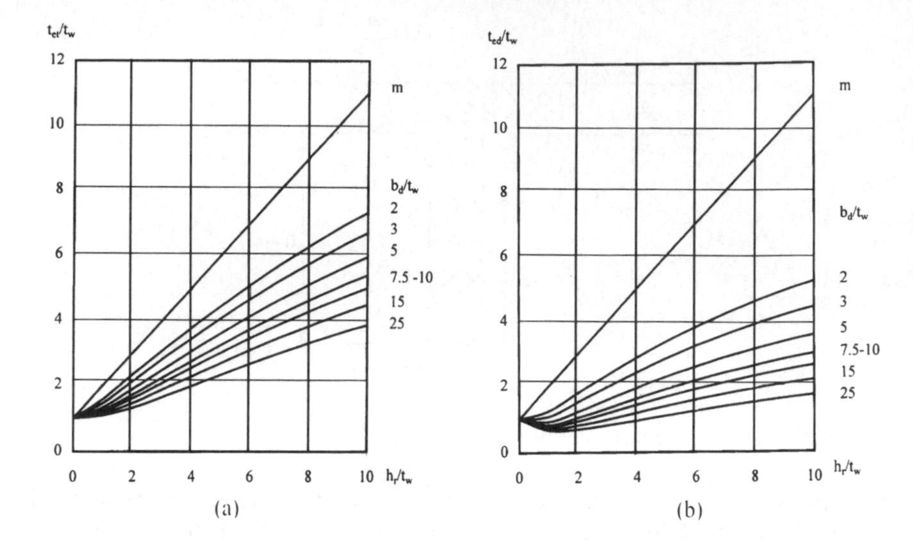

Fig. 10.29 Curves for determining wall thickness required for equivalent deflection (a) and for equivalent stress (b) for parallel ribbed sections, where rib thickness $t_r = 0.6t_w$ and taper angle $\alpha = 1°$ per side. Parameters of the lines: b_d/t_w (relative rib distance), m = solid plate of thickness h_r

- for long ribs, h_r/t_r should not exceed 6 in order to avoid buckling (Fig. 10.32a).

(b) Strength of ribs

For all elastic structures there exists a relationship between maximum deflection f and maximum strain ε. For a ribbed cantilever beam under a pure bending moment (Fig. 10.30) it can be shown to be

$$f = \frac{\varepsilon L^2}{2c} \tag{10.8}$$

Note that this is only valid within the proportional limit ε_{el}.

If ε_{el} or a lower strain (e.g. the maximum allowable strain; section 6.2.4(d)) would be accepted as allowable, an allowable deflection f is the consequence, which increases as L increases or as c decreases. The design of a ribbed plate of suitable strength can be determined with Fig. 10.29b. Since values of t_w, t_r, b_d and h_r have already been suggested, we can:

- enter values of h_r/t_w and b_d/t_w to find t_{ed}/t_w;
- realize that this value of t_{ed} stands for the equivalent thickness of a flat plate, that would have the same maximum allowable strain under the allowed deformation as the ribbed plate;
- calculate the bending stress using

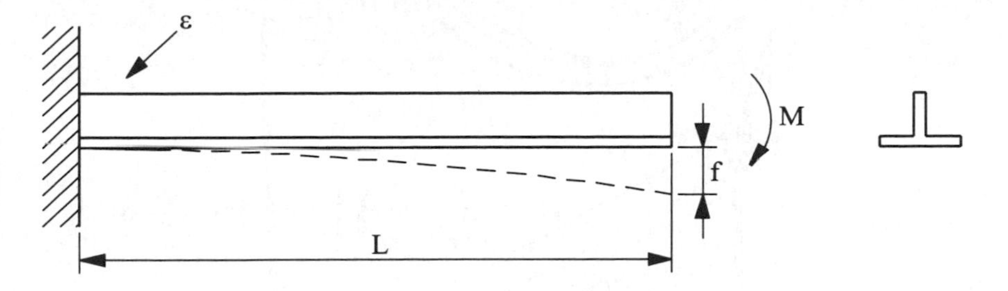

Fig. 10.30 Ribbed cantilever beam

$$\sigma = \frac{M}{\frac{1}{6}bt_{ed}^2} \tag{10.9}$$

and compare this with the allowable stress. (M is the total bending moment on the plate; note the use of the square of t_{ed} instead of the cube of t_{et} as in section 10.3.4 (a).)

Comparison of Figs 10.29a and b shows that adding ribs of any height h_r always increases t_{et} (and therefore I), but does not always increase t_{ed} (and therefore I/c). This means that the load-bearing capacity of plates can be increased or decreased by adding ribs, depending on the configuration of the ribs. This decrease will happen to a plate with low (h_r) ribs or with high base width b_d. Compared with a flat sheet, adding ribs increases the maximum distance from the neutral axis (c), and therefore the bending stress, but this may not be compensated by the increase of the second moment of area of the ribbed plate. This situation may occur with sloping ribs, as in Fig. 10.31.

(*c*) *Stability of ribs*

Ribs may buckle under compression if $h_r > 10t_r$ (i.e. an order of magnitude, depending on the load; Fig. 10.32a). Under tension a similar effect may occur, also due to the limited bending stiffness of the rib (Fig. 10.32b). Both effects give rise to a decrease of the maximum distance to the neutral axis c. This will affect the strength of the structure and diminish the stiffening effect of the ribs.

Conditions for local buckling

We have already explored Euler global buckling of struts in section 5.2.5, in order to check whether a slender structure will buckle under axial compressive loads.

There are great commercial pressures to replace bulky inefficient cross-sections by shapes having thinner walls which give better bending stiffness per unit mass. Thinner walls allow faster melt processing (faster cooling

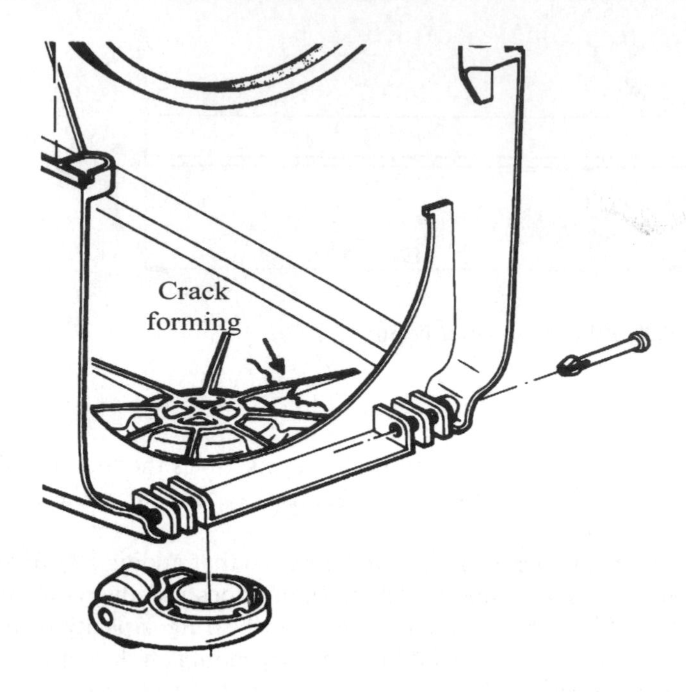

Fig. 10.31 Ribbed plate with weak points (nose wheel fixing point of a vacuum cleaner) (courtesy Philips DAP)

times) and the use of a smaller volume of material; both features reduce costs.

A key question is: how thin can walls be before failure occurs? If the wall is very thin it may be difficult to mould or extrude, or the product may warp (distort) after manufacture because the walls have low local stiffness and can no longer resist the action of internal stresses or molecular orientations (Fig. 10.33). The presence of short fibres will accentuate these problems. This is the subject of current research and definitive guidelines are awaited. Under compressive loads the structure may show local buckling, and this then is the limiting design feature, rather than the usual global bending stiffness calculations. We have changed the governing mode of failure.

What we attempt to do here is to suggest the considerations which have to be taken into account. The starting point is to assume there is no distortion from manufacture. As a simplification we choose $t_r = t_w = t$.

We can make a rough model of the compressive zone of a web or half-flange as a long thin plate under in-plane uniform axial compression (Fig. 10.34). Handbooks suggest that buckling occurs when

$$\sigma' = [KE/(1 - \nu^2)](t/b)^2 \qquad (10.10)$$

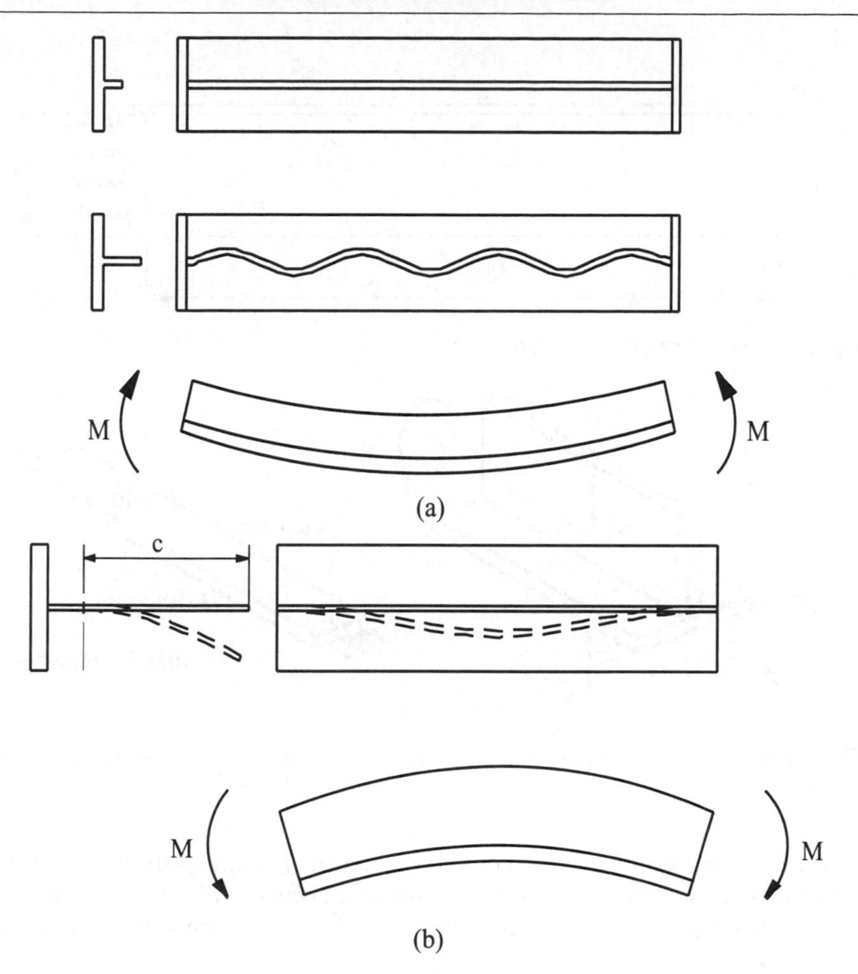

Fig. 10.32 Bucking of high ribs (a) in compression and (b) in tension ((a) courtesy J. Spoormaker)

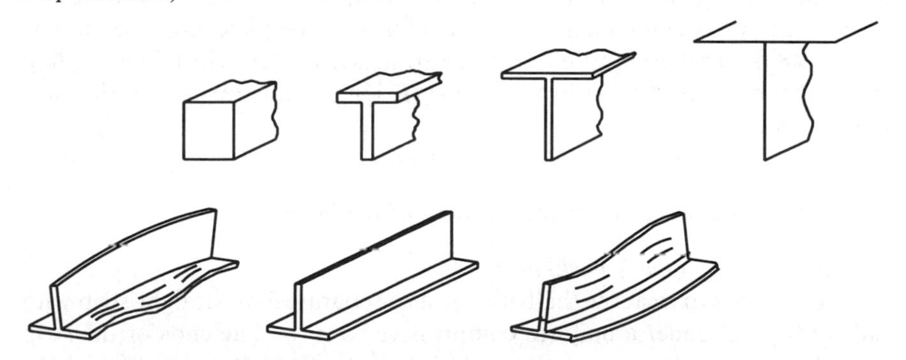

Fig. 10.33 In thinner walls there is greater risk of local deformation. The T-sections here have a horizontal flange and a vertical web or rib

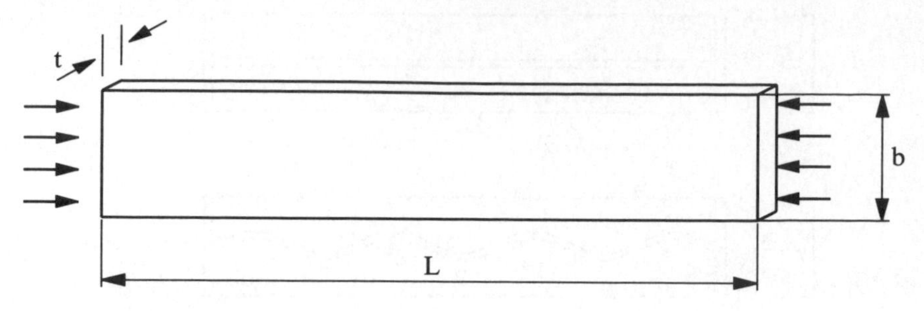

Fig. 10.34 Buckling of a thin plate

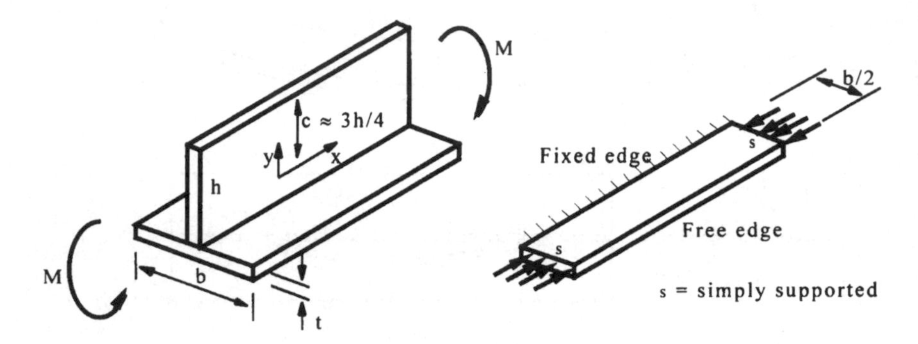

Fig. 10.35 Loading conditions for buckling calculation of the (half-) flanges of a T-beam

where K depends on L/b, edge conditions and the loading pattern. The clear principle is that as the rib becomes thinner and deeper, $(t/b)^2$ becomes small, so the buckling stress decreases dramatically. This confirms commonsense observation.

Now consider a long T-beam, simply supported at the ends and under a uniform bending moment M (Fig. 10.35). For thin walls, with $b = h$, we can show that the second moment of area of the complete cross-section is $I \simeq h^4/48[\text{m}^4]$ and the location of the centroid is at $c \simeq 3h/4$ [m]. The bending stress at the flange is given by $\sigma_f \simeq Mh/4I$. Under this stress the (midspan) maximum deflection is

$$v_{\max} = ML^2/(8EI) \tag{10.11}$$

$v_{\max}$ serves as a criterion for the stiffness of the beam.

Estimate of (half-)flange buckling

The next step is to treat the (half-)flange as a separate long straight strip with half-width $b/2$ under a uniform compressive stress σ. The ends of the strip are taken as simply supported, one long edge is fixed (to the rib) and the other edge is free.

From handbooks we can then show that for $L \gg b/2$, the buckling stress σ' is

$$\sigma' = [1.21E/(1-\nu^2)](2t/b)^2 \quad (10.12)$$

For this buckling stress σ' to occur in the flange, a bending moment on the T-beam of $M = \sigma' = I/(h-c)$ is needed, leading to

$$v_{\max} = ML^2/(8EI) = \sigma' L^2/[8E(h-c)] \quad (10.13)$$

so that buckling is predicted when

$$t^2 = 2(1-\nu^2)b^2 v_{\max}(h-c)/1.21L^2 \quad (10.14)$$

An obvious conclusion is that to avoid buckling, the wider the flange the thicker it must be.

Estimate of rib buckling

This is not so easy because the exact conditions are not given in handbooks: the dimensional edge conditions of the long strip representing the complete rib (of depth h) are the same as for the flange, but the stress varies over the depth, being partly compressive and partly tensile (Fig. 10.36).

We might argue that the free edge dominates the onset of buckling, and the bending stress at the free edge can be calculated as $\sigma = Mc/I$. Making the rather sweeping assumption (a worst case) that buckling occurs when σ' has the same value as σ taken as uniform across the web depth h, we conclude that the design of the rib is more critical than the design of the flange for this particular cross-section.

Example

A numerical example: let $t = 2\,\text{mm}$, $b = h = 30\,\text{mm}$, $L = 300\,\text{mm}$, so that $c = 0.0225\,\text{m}$ and $I = 1.69 \times 10^{-8}\,\text{m}^4$. Suppose we use a plastics material with $E = 1\,\text{GPa}$, $\nu = 0.35$.

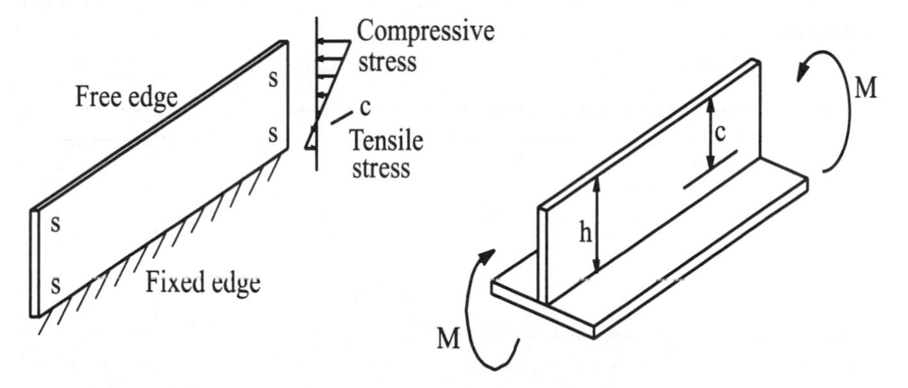

Fig. 10.36 Loading conditions for buckling calculation of the rib

The conditions of use limit the midspan deflection v_{max} to 0.004 m. If a uniform moment M is applied this would imply $M = 6$ Nm. The maximum bending stress in the flange, σ_f, is 2.67 MPa, and the maximum bending stress in the rib is 8 MPa.

If the flange is in compression, then the flange buckling stress is estimated as

$$\sigma'_f = [1.21 \times 10^9/(1 - 0.35^2)][2 \times 2/30]^2 = 24.5\,\text{MPa} \tag{10.15}$$

with the conclusion that buckling is not expected.

If the rib is in compression, and we assume that the compressive region of the rib has a uniform stress the same as the maximum bending stress, then we find

$$\sigma'_r = [1.21E/(1 - \nu^2)][t/(3h/4)]^2 = 10.9\,\text{MPa} \tag{10.16}$$

with the conclusion that buckling is not expected but that any increase in applied bending moment or any disturbances caused by manufacture would be rather risky.

(*d*) *Torsional stiffness of ribbed structures*

Many realized designs show that high bending stiffness is not a guarantee of high torsional stiffness. The torsional stiffness of, for example, an I-beam is rather low. Parallel ribs introduce a kind of anisotropy in a rectangular plate or lid. As in fibre-reinforced composites, ribs disturb the conventional balance between stiffness in bending and in torsion.

Torsional stiffness can be considerably improved by the addition of ribs at angles of +45° and −45° (compare the function of angle-ply layers in a laminate). These are more effective than diagonals with other angles. Figure 10.37 shows a torsional stiffened U-profile. Closing the U-profile with a welded-on lid will improve the torsional stiffness even more. It is clear that in such a way a balance between bending and torsional stiffness can be achieved.

A concentration of ribs in one node should be avoided.

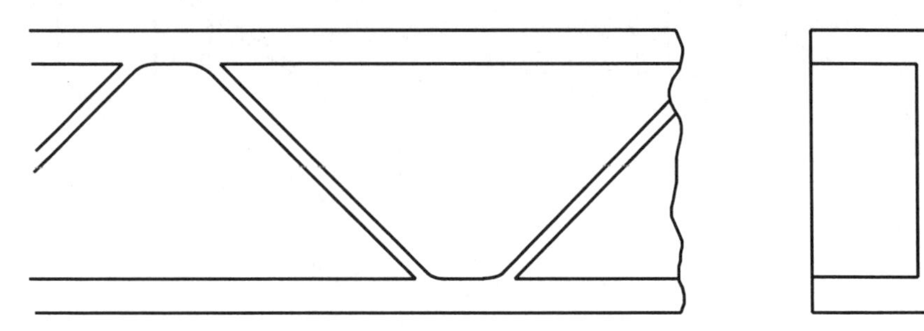

Fig. 10.37 Increasing the torsional stiffness with diagonal ribs

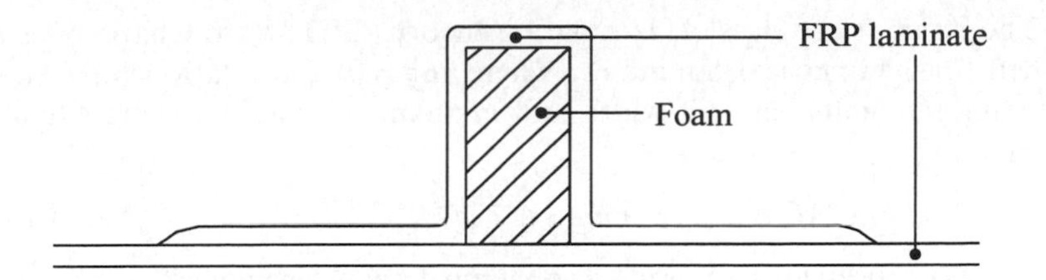

Fig. 10.38 Lost core design for rib in FRP

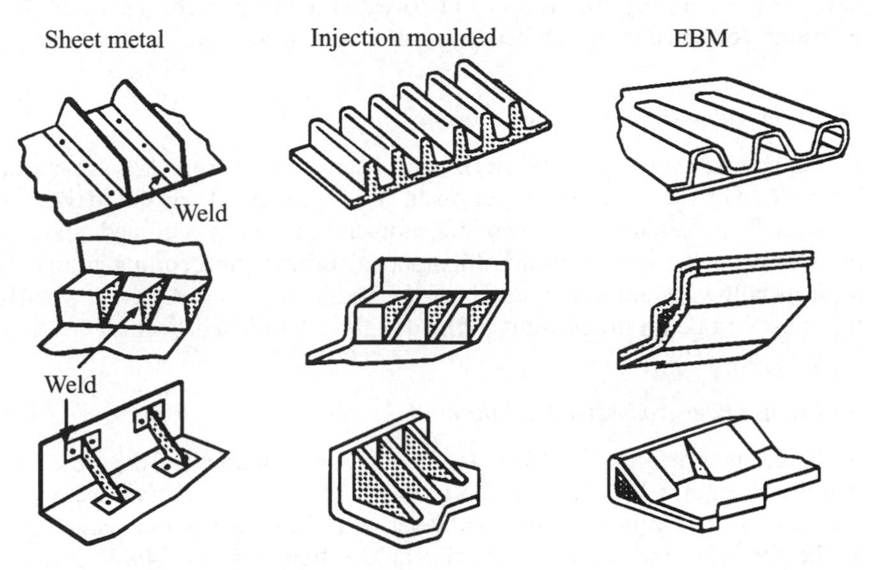

Fig. 10.39 Design for stiffness in three different technologies (courtesy General Electric Plastics)

(*e*) *Ribs in other technologies*

In reinforced laminates, mixed structures containing foamed cores have been developed that can be made within normal laminating procedures (Fig. 10.38). Ribs can be formed in extrusion-blown mouldings (EBM) in the manner shown in Fig. 10.39. Forming of box-like structures is also possible: the parison is compressed during the inflation stage and the surfaces squeezed together should weld to allow shear transfer across them, otherwise the bending stiffness is much reduced.

10.3.5 Stiffness of laminated beams

Sections 10.3.2–10.3.4 have discussed how to design efficient beams in homogeneous plastics materials. We now examine how to predict the stiffness of beams made from symmetrical laminates of unidirectional plies.

The summary of classical lamination theory (CLT) from Chapter 7 established for symmetrical laminates (where $[B] = 0$) the relationship between bending moments per unit width and midplane curvatures using equations such as

$$[M] = [D][\kappa] \quad \text{and} \quad [\kappa] = [d][M] \qquad (7.91 \text{ and } 7.93)$$

Can these expressions be used to solve most beam problems?

In general the answer is 'no' although parts of CLT can be used. So we need to explain why this is so, and to explain the procedures which are necessary for calculating the bending stiffness of simple beam situations.

(*a*) *Rectangular section laminated beams*

Let us compare two identical horizontal beams made from a symmetrical laminate with the reference direction in the midplane along the length. In Chapter 7 the discussion of bending moments in plates assumed that the bending moment was constant. In practical beams the bending moments most usually vary along the length of the beam, e.g. $M = M(x)$, so clearly this must be taken into account in the design of laminated beams.

(*b*) *Beams under transverse load through thickness*

In the beam shown in Fig. 10.40(a), the manner of loading is the closest to that described in CLT. The applied moment $M(x)$ induces curvature out of the plane of the laminate. For a cantilever with the origin of the beam $x = 0$ at the free end and loaded vertically at the free end, we know that the bending moment resultant per unit width is

$$M_x(x) = M(x)/b = -Px/b \qquad (10.17)$$

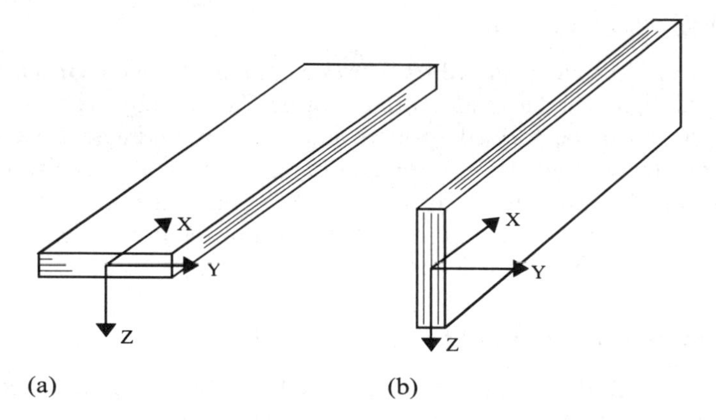

Fig. 10.40 Beam coordinate systems

Following the argument for a linear elastic isotropic beam, we ignore transverse curvature, and so for any symmetric laminate for which $d_{16} = 0$ we have an expression for the local midplane displacement from equation (7.67):

$$\partial^2 w_0(x)/\partial x^2 = -\kappa_x(x) = d_{11}Px/b \tag{10.18}$$

Double integration, using the boundary conditions that slope and deflection are zero at the fixed end, gives

$$w_0(x) = (d_{11}P/6b)(x^3 - 3Lx^2 + 2L^3) \tag{10.19}$$

and at the free end

$$w_0(0) = d_{11}PL^3/3b \tag{10.20}$$

What we have done is to use the standard beam theory for a linear elastic solid and we have substituted (b/d_{11}) for the familiar (EI). This approach can be used for other loading and support conditions provided the beam is lined up in the manner of case (a) in Fig. 10.40, i.e. the moments are applied to induce curvatures(s) out of the plane of the laminate.

Calculation of bending stresses at a given coordinate along the length of the beam then follows from CLT. Having established $\kappa_x(x)$, the bending strain at any thickness coordinate is then

$$\varepsilon_x(x, z) = z\kappa_x(x) \tag{10.21}$$

and hence for the ply 'f' having ply stiffness $[Q]_f$ we find

$$\sigma_{xf}(x, z) = Q_{11f}\varepsilon_x(x, z) \tag{10.22}$$

A similar approach can also be used for angle-ply laminates to calculate part of the bending deflection, and $[\bar{Q}]_f$ for bending stress. We must remember that the applied moment $M_x(x)$ also develops a curious twisting curvature

$$\kappa_{xy}(x) = d_{16}M_x(x) \tag{10.23}$$

which is not found in isotropic materials. If this twisting is prevented then the bending stiffness is changed somewhat.

(c) *Beam under transverse load in the plane*

The arrangement in beam (b) (Fig. 10.40) is often met in the webs of beams. Bending moments $M_x(x)$ will induce bending curvature in the plane of the laminate, which is quite a different sort of deformation from that discussed above in case (a). It is absolutely essential to distinguish the two planes of loading in (a) and (b). In beam (b) (Fig. 10.40) note that the coordinate system is changed so that the basis and coordinate system of CLT applies.

This is important, especially if you decide to use the finite element method (FEM) for more detailed analysis.

Axial properties of the laminate as a whole are independent of the depth coordinate y. We can therefore proceed with our analysis using a value of modulus based on the in-plane compliance of the symmetrical cross-ply laminate: from $\varepsilon_x = a_{11}N_x = a_{11}\sigma_x h$ we find

$$E_x = (a_{11}h)^{-1} \tag{10.24}$$

This is used with the results of the engineer's theory of bending to predict the deflections $v_0(x, y)$.

We can readily calculate an average bending stress in the beam at any depth coordinate (y) using

$$< \sigma_x(x, y) >= M(x)y/I \tag{10.25}$$

where $M(y)$ is the local bending moment [N m] and hence the axial bending strain is

$$\varepsilon_x(x, y) =< \sigma_x(x, y) > /E_x =< \sigma_x(x, y) > a_{11}h \tag{10.26}$$

The bending strain must be continuous across both the depth and the width (z) of the beam so it follows that the stress in each ply, which depends on the local ply stiffness, must vary across the width of the beam (which is the thickness of the laminate) according to

$$\sigma_{xf}(x, y, z) = Q_{11f}\varepsilon_x(x, y) \tag{10.27}$$

(d) Parallel axes theorem

For cross-sectional shapes such as I, T and □, built up from flat elements, we can in principle use approaches similar to those just described, bearing in mind the direction of loading in relation to the thickness of the laminate (Fig. 10.41). We have already seen in case (a) that the bending stiffness of a laminate about its midplane is simply

$$E_x I_y = b/d_{11} \tag{10.28}$$

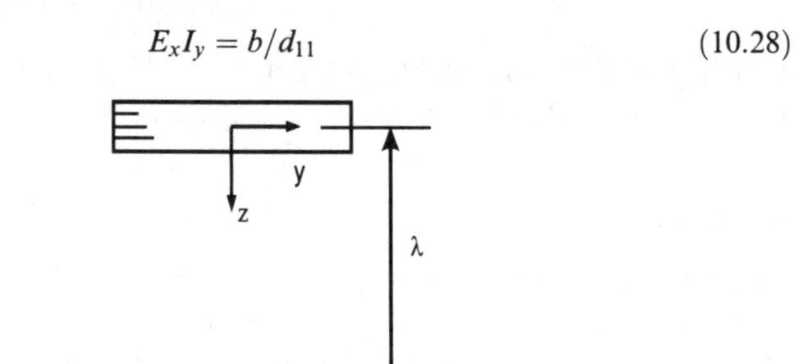

Fig. 10.41 The parallel axes theorem

The bending stiffness of the same laminate about an axis parallel to and at distance λ from its midplane follows from use of the parallel axes theorem as

$$E_x I_Y = b/d_{11} + E_x A_x \lambda^2 = b(1/d_{11} + \lambda^2/a_{11}) \tag{10.29}$$

Where λ is large compared with the laminate thickness, the conclusion follows that the term b/d_{11} can be neglected; this is of course well known in the familiar isotropic elastic analysis. It then follows that for the special case of bending stiffness of laminates about an axis remote from the midplane, the stacking sequence in the flange is also of minor importance.

Thus for bending stiffness calculations we need a_{11} (and d_{11}) and we can readily calculate the stiffness of made up sections such as I, [and □ sections.

(e) *Sections of rotational symmetry*

For closed thin-walled sections of rotational symmetry, different plies in the laminate are at such a large distance from the centroid that the laminate can, to a good approximation, be regarded as symmetrical and having the averaged properties a_{11}, a_{66}, d_{11} and d_{66}. For a thin-walled circular section, deflection under transverse loads or moments can be calculated using $E_x = (a_{11}h)^{-1}$ and axial twisting can be calculated using $G_{xy} = (a_{66}h)^{-1}$. The effect of d_{16} in thin-walled sections of rotational symmetry made from angle-ply laminates can be neglected in rough estimates of deformation behaviour.

(f) *Beam deflections by shear deformation*

The simplest calculations of bending deflections of beams assume that the beams are slender and that deflections can ignore the deformation caused by shear deformation.

The ratio of span L to depth d at which shear deflection becomes significant can be readily calculated for a linear elastic isotropic material. For example, the maximum shear and bending deflections, w_s and w_b, for a simply supported beam of rectangular cross-section under a central transverse load can be found from

$$w_s/w_b = [3PL/10Gbd]/[PL^3/4Ebd^3] = 6Ed^2/5GL^2$$

For isotropic metals $G \simeq E/2.5$ so $w_s/w_b \simeq 3d^2/L^2$, and $L/d \simeq 5.5$ for $w_s/w_b = 0.1$. Shear deflections are a significant proportion of the total deflection only for very short, deep beams of isotropic materials.

For unidirectional fibre-reinforced materials the ratio E_1/G_{12} can be much larger than for isotropic materials: for the extreme case of cfrp we obtain $E_1/G_{12} \simeq 18$, and so a beam of cfrp with quite a large L/d can still show substantial shear deformation. Thus for $w_s/w_b = 0.1$ in cfrp we have $L/d = 14.7$. Thus beams of 'normal' proportions based on laminated materials can undergo substantial shear deformation.

PROBLEMS

1. In which directions in Figs 10.23–10.25 are bending stiffness increased, not affected, or decreased?
2. A pipe of 150 mm bore with a uniform 6 mm wall extruded from solid unplasticized PVC of density 1410 kg/m^3 has been used successfully to carry drainage water through waterlogged soil. Internal pressure may be regarded as negligible. The resistance to radial buckling collapse is regarded as satisfactory, but an investigation is to be carried out to see if there would be any substantial saving in weight by using pipes of the same buckling resistance, but having a sandwich construction. The extrusion technology to be used confers a solid skin to the wall of thickness 1 mm and having the same properties as that of the PVC originally used. The density of the foamed PVC core is 950 kg/m^3. Experiments suggest that for such a system, the ratio of the moduli of foamed PVC to unfoamed PVC is approximately the same as the square of the ratio of the densities of foamed to unfoamed PVC (see equation (10.4)). Estimate the weight saving achieved by using the sandwich-walled pipe. (Hint: radial buckling depends on the second moment of area of a circumferential element of the pipe wall, not the second moment of area of the cross-section normally used in longitudinal bending calculations.)
3. A plate of total width $b = 500$ mm and thickness $t_w = 10$ mm has to be reinforced with ribs. Five parallel ribs will be used. A rib thickness of $t_r = 6$ mm and a rib height $h_r = 50$ mm are chosen. By what factor will the bending stiffness of the plate (along the rib direction) improve?

   If a load of $M = 100$ N m has to be sustained by a polyethylene ribbed plate of said dimensions for 1 year at 20°C, what would be the safety factor F (see strength data in Fig. 6.22)?
4. A beam is made from a cross-ply laminate $(0°/90°)_s$ having the following compliances:

$$[a, d] = \begin{bmatrix} 21.21 & -0.54 & 0 \\ -0.54 & 21.21 & 0 \\ 0 & 0 & 400 \end{bmatrix} \begin{bmatrix} 604 & -49.1 & 0 \\ -49.1 & 3245 & 0 \\ 0 & 0 & 19200 \end{bmatrix} [\text{mm/MN}, /\text{MN mm}]$$

   Estimate the expected modulus you would obtain from a test on a long beam in (a) axial tension, (b) three-point bending. Is there any difference in results when the 90° plies are parallel to the beam axis, compared with the 0° plies?
5. A test is to be carried out using a horizontal tube 1 m long, of 40 mm diameter, with a 2 mm thick wall. It is simply supported at each end and carries a midspan vertical load of 50 N. For the purposes of this problem you may neglect the mass of the tube. Two tubes are to be tested. One is pultruded with unidirectional fibres along the axis of the tube. The other tube is filament-wound and may be modelled as a symmetrical laminate

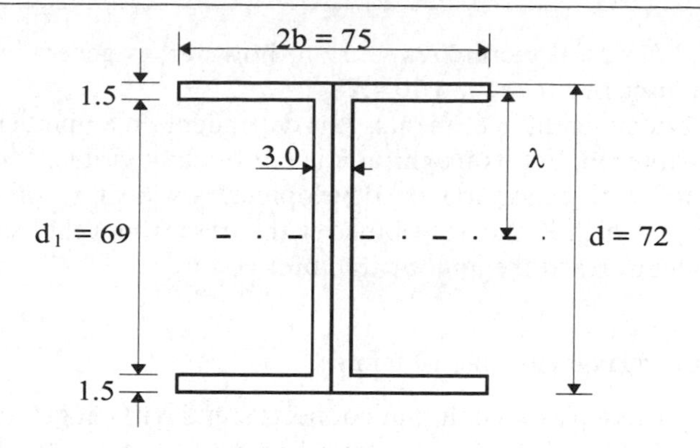

Fig. 10.42 Cross-sectional geometry of an I-beam

having the construction $(30°/-30°/30°/-30°)_s$. Both tubes may be modelled using the properties of glass fibre–epoxy material gfrp. For the purposes of this problem, assume both tubes are cured and used at room temperature. The ply transformed reduced stiffnesses and the laminate compliance coefficients are as follows.
Pultruded section:

$$[a_{11}, a_{12}, a_{22}, a_{66}] = [12.95, -3.368, 60.46, 120.8]\ \text{mm/MN}.\ a_{16} = a_{26} = 0$$
$$[Q_{11}, Q_{12}, Q_{22}, Q_{66}] = [39.17, 2.182, 8.392, 4.14]\ \text{GPa}.\ Q_{16} = Q_{26} = 0$$

Laminate $(30°/-30°/30°/-30°)_s$ with total thickness 2 mm:

$$[a_{11}, a_{12}, a_{22}, a_{66}] = [22.9, -14.81, 54.66, 54.74]\ \text{mm/MN}.\ a_{16} = a_{26} = 0.\ [\bar{Q}(+30°)] =$$
$$[\bar{Q}_{11}, \bar{Q}_{12}, \bar{Q}_{22}, \bar{Q}_{66}, \bar{Q}_{16}, \bar{Q}_{26}] = [26.48, 7.176, 11.09, 9.134, 9.546, 3.78]\ \text{GPa}.$$

What is the likely safety factor by which failure in each beam is avoided?

6. (More advanced problem) An I-beam is to be made by joining together identical][sections each of which are based on a symmetrical eight-ply laminate (Fig. 10.42). The laminate is 1.5 mm thick and can be made from one of five different arrangements based on the properties of unidirectional-ply gfrp: (a) $(0_4^\circ)_s$, (b) $(0_2^\circ/90_2^\circ)_s$, (c) $(45_2^\circ/-45_2^\circ)_s$, (d) $(0°/45°/90°/-45°)_s$, (e)$(0_2^\circ/45°/-45°)_s$. The simply supported horizontal I-beam is 2 m long and carries a uniformly distributed load of 0.5 N/mm. Calculate the midspan deflection in each of the five beams.

10.4 PRESTRESSED ELEMENTS

Stress in an element is a key aspect of mechanical engineering. It may be, however, wanted or unwanted. Stress in and between parts, in order to tighten connections or to enable exact positioning, is an example of the first and will be discussed in section 10.4.1. Internal stress in an element, caused by differential thermal expansion, however, is generally unwanted and will be discussed in section 10.4.2.

Stress is often present in elements. It is dependent on a number of aspects (see e.g. Chapter 9), but its magnitude is not directly visible. Moreover the stress is subject to relaxation or development by heating or cooling. If the stress is too high it may (combined with stress from other sources such as external load) reach the limit of fracture.

10.4.1 Stress relaxation in bolted joints

When we tighten a plastics bolt that connects metal parts, it gets a prescribed strain $\varepsilon = \Delta L/L$ (fig. 10.43). The matching stress is subject to relaxation, especially in thermoplastics. With the aid of data and models, this can be predicted. If the design brief indicates an upper limit for the strength and a lower limit for the stress to be maintained, the expected time for the strain to be increased further can be calculated.

A similar situation arises if stress is applied by metal bolts to a plastics plate or packing between flanges. Here compressive stress relaxation will occur, with the chance that, within the service life, the stress will become lower than that allowed. The treatment of this problem is analogous to that of the bolt, which we will deal with below.

(a) Undeformable flanges

The function of a bolt in a flanged pipe connection is to hold the flanges of the two pipes together, taking into account possible longitudinal forces F_P in the pipe system, and those to press the flanges tightly together, F_F, to prevent leakage along the packing (Fig. 10.46a). The force in the bolts F_B should therefore equal the force in the pipe system plus the required packing

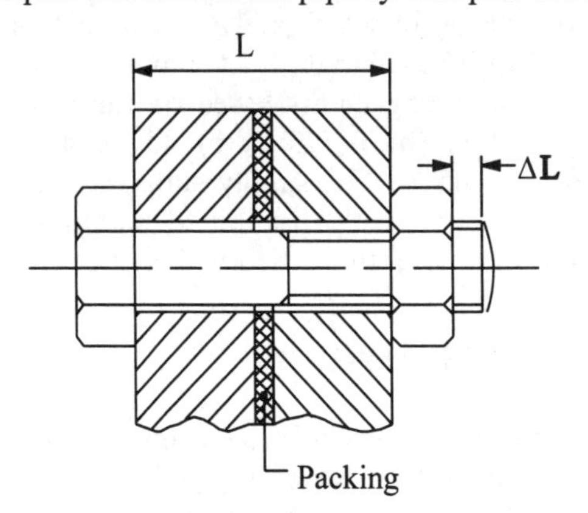

Fig. 10.43 Flange joint with tightened bolt

compression force $F_B = F_P + F_F$. This can be obtained by mounting the bolts in a non-loaded situation by hand until the packing is closed between the flanges, followed by tightening the nuts until the required force has been reached. On tightening, the active length L of the bolt is loaded to the mounting force F_0 and stretched over a fraction ΔL causing a strain of $\varepsilon = \Delta L/L$. If flanges plus packing are regarded as undeformable, the simplest analysis suggests that the stress in the bolt will be $\sigma = \varepsilon E$; if the material of the bolt is viscoelastic, this stress will relax according to $\sigma = \varepsilon E(t)$. This formula enables us to calculate the actual stress in the bolt, as long as the function $E(t)$ of the material, and the time t which has elapsed since the bolt was stretched, are known.

It may be the case that relaxation reduces the force F_B to below the required level. The longitudinal force in the pipe system F_P will remain and therefore the packing force F_F will become insufficient, leading to leakage. This is usually dealt with by tightening the bolt once more. The effect of this action can be seen in Fig. 10.44. At first the bolt is strained to ε_1, leading to σ_A at point A as the initial and upper stress level. After 100 h the stress will relax to the unacceptably low level of σ_C. By increasing the strain further to ε_2, the stress can be raised again to the upper level (D). According to the graph it will now take up to 10 000 h before the stress will reach the lower level at E again.

We now want to calculate the remaining stress in the bolt after 10 000 h of service. For this purpose we choose $E(t) = E_1 t^{-m}$ as a model of the relaxation

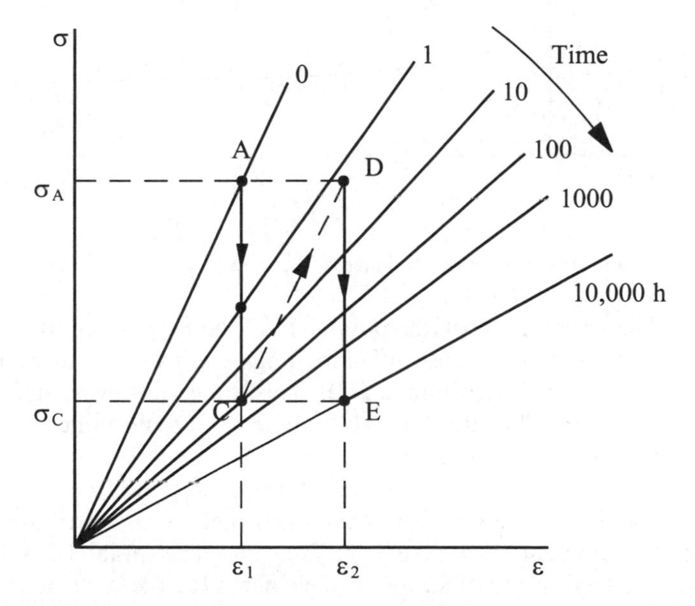

Fig. 10.44 Representation of a repeated tightening of a plastics bolt in a set of isochronous stress–strain curves

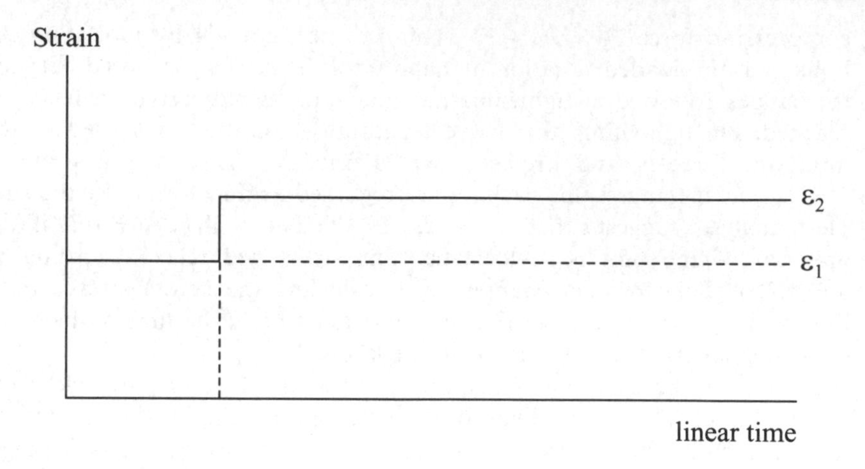

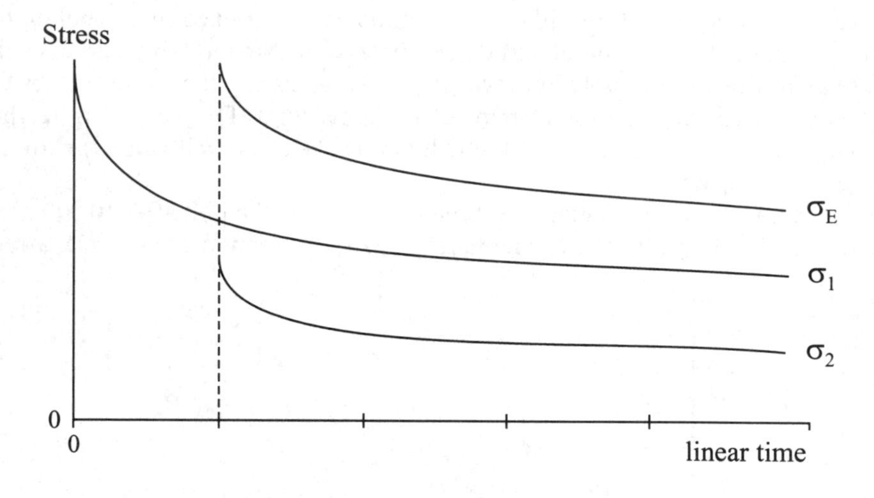

Fig. 10.45 Development of strain and stress in a bolt

modulus. The stress after 100 h (point C) will be $\sigma_C = \varepsilon_1 E_1\ 100^{-m}$. In the situation after 100 h, two consecutive loadings are involved, so we need the superposition principle (section 5.2.2). This time, however, deformation (strain) is given and the stress is to be predicted. To aid understanding we plot a graph of strain and stress versus linear time (Fig. 10.45). It is essential to consider any change in strain $\Delta\varepsilon$ and the elapsed period t during which it has been in action at the moment of the assessment as a contribution to the stress at that moment: $\sigma_i = \Delta\varepsilon E_1 t^{-m}$. That moment (point E) is 10 000 h after the first application of strain. The elements for the superposition are

$$1 : \Delta\varepsilon = \varepsilon_1 \quad t = 10\ 000 \qquad \sigma_1 = \varepsilon_1\ E_1 10\ 000^{-m}$$

$$2:\Delta\varepsilon = \varepsilon_2 - \varepsilon_1 \quad t = 10\,000 - 100 = 9900 \quad \sigma_2 = (\varepsilon_2 - \varepsilon_1)E_1\ 9900^{-m}$$
$$\Rightarrow \sigma_E = \sigma_1 + \sigma_2 = \varepsilon_1 E_1\ (10\,000^{-m} - 9900^{-m}) + \varepsilon_2 E_1\ 9900^{-m} \tag{10.30}$$

If $m = 0.1$ (an average value for many thermoplastics) we find

$$\sigma_E = \varepsilon_1 E_1 (0.3981 - 0.3985) + \varepsilon_2 E_1 \times 0.3985$$

The first term has nearly vanished; after so long a period the sequence of loading is no longer visible in the magnitude of the stress. The effect of the difference in loading times 9900 and 10 000 h is only visible in the fourth digit after the decimal point.

(b) *Elastic flanges and packing*

Normally the system of flanges plus packing is compressible, mainly due to the packing (compare section 5.3.2 (b)). Turning the nut on the bolt over a length ΔL will result in compression of the system and elongation of the bolt, each according to their own elongation stiffnesses C_F and C_B. As soon as the longitudinal forces in the pipe system, F_P, are applied, the load on the bolt increases and that on the flange decreases, according to Fig. 10.46a.

So far we have discussed the fully elastic or the initial (time $= t_0 =$ a small value, say 1 min) situation. Since the bolt material is viscoelastic, C_B is time dependent. Therefore the relevant set of isochronous stress–strain curves should be rescaled to force–elongation curves and brought into Fig. 10.46a: see Fig. 10.46b. As F_B decreases, caused by stress relaxation, and as F_P stays constant as a systems parameter, the force on the flanges F_F will diminish and eventually go below the minimum required packing force and become zero at time t_3. So this is the moment to readjust the nut (in fact preferably earlier).

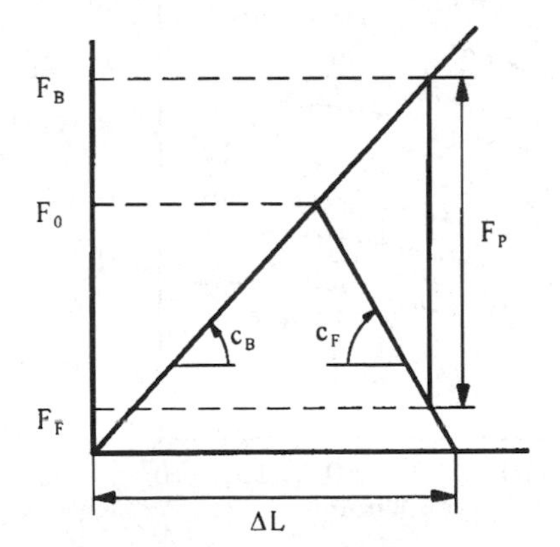

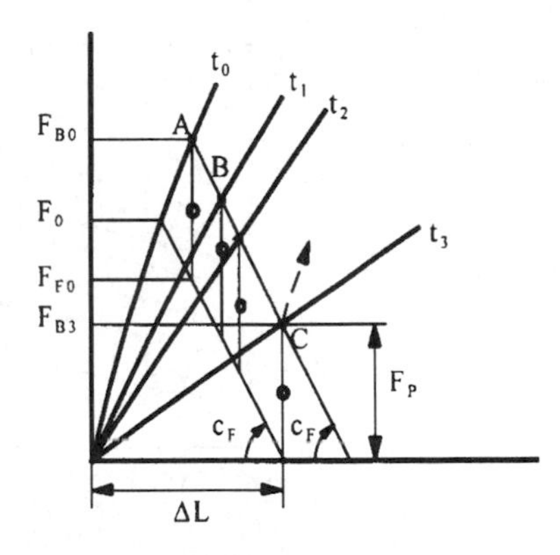

Fig. 10.46 Elasticity of bolted and loaded flange connection: (a) elastic bolt, (b) viscoelastic bolt

(c) *Allowable stress in the bolt*

Upon turning the nut, the bolt is loaded in tension, but in torsion as well, as a result of the turning of the nut and its friction against the surface of the flange. A study of this system has revealed that the magnitude of the remaining torsion moment in the bolt can be expressed as $M_0 = F_0 0.5d\ K$ where d is the diameter of the bolt and K is a friction dependent factor, usually of the order of 0.25.

Since loading of the pipe system and relaxation of the bolt cause changes in the bolt stress, and since long-term strength is almost exclusively documented in constant stress, we base the following calculation of the bolt strength on the mounting force F_0 and the related torsion moment M_0.

The tensile stress in the bolt is uniform through the diameter, but the shear stress, caused by M_0, is maximum at the surface, so we assess the stresses at the surface. Neglecting the threaded section, where stress concentration factors would need to be included, they are

$$\sigma_x = \frac{F_0 \times 4}{\pi d^2} \quad \sigma_y = 0 \quad \tau_{xy} = \frac{M_0 \times 16}{\pi d^3} = \frac{F_0 \times 8 \times K}{\pi d^2} \tag{10.31}$$

From these the principal stresses can be calculated:

$$\sigma_{1,2} = \frac{1}{2}(\sigma_x \pm \sqrt{(\sigma_x^2 + 4\tau_{xy}^2)}) = \frac{2F_0}{\pi d^2}(1 \pm \sqrt{(1 + 16K^2)}) \tag{10.32}$$

With $K = 0.25$ the values are $\sigma_1 = 1.54F_0/d^2, \sigma_2 = -0.26F_0/d^2$.

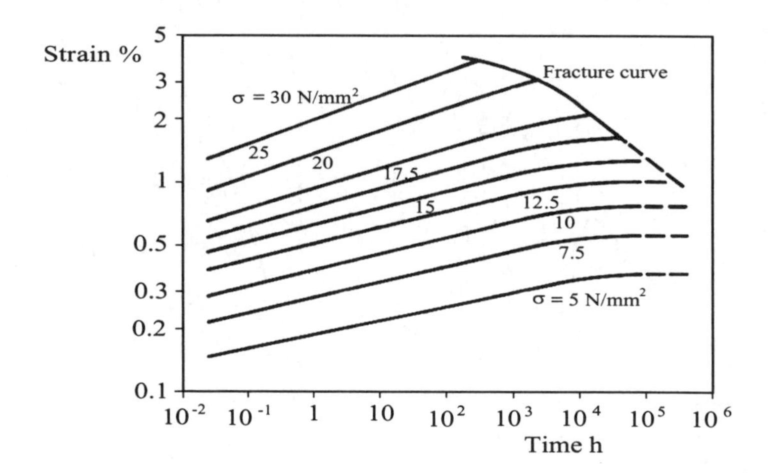

Fig. 10.47 Creep curves and fracture envelope of POM under tensile load at 20°C (courtesy Hoechst)

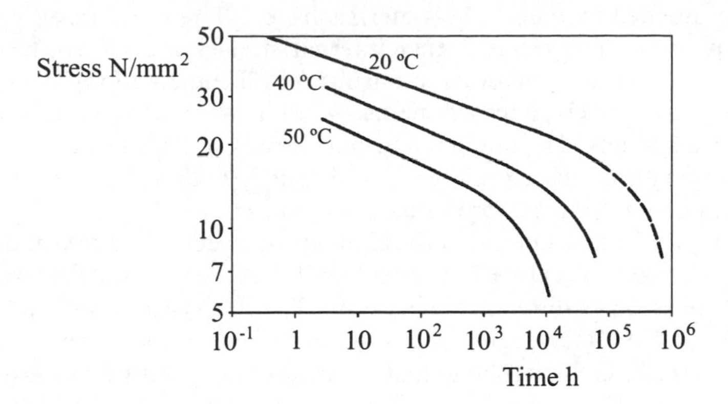

Fig. 10.48 Long-term strength (hoop stress) of POM pipes (courtesy Hoechst)

Strength calculation should be based on materials data. Since POM is a suitable material for a number of applications of bolts, and as we have the necessary data, we use them for this example. Figures 10.47 and 10.48 refer to different grades of the material. The environmental requirements are service life, 1 year; service temperature, 20°C; no aggressive media. Both figures indicate for the service requirements as allowable stress $\sigma_{max} = 22\,\mathrm{N/mm^2}$. As a factor of safety we choose $S = 1.3$.

In view of the state of stress in the bolt we need a limiting stress criterion. We use that of Tresca (section 6.2.1(d): $\sigma_1 - \sigma_3 < \sigma_{max}$. Since we calculated the stresses σ_1 and σ_2 at (and in) the (free) surface, the third element of the stress tensor is directed perpendicular to the surface with magnitude $\sigma_3 = 0$. With regard to Tresca's criterion, we have to place the stress elements in their order of magnitude: $\sigma_1 = 1.54, \sigma_2 = 0, \sigma_3 = -0.26F_0/d^2$. The allowable stress is therefore

$$(1.54 - (-0.26))F_0/d^2 = 1.80F_0/d^2 < \sigma_{max}/F = 16.9\ \mathrm{N/mm^2}$$

If F is given, the required value of d can now be calculated.

Note that the factor of 1.80 was reached by increasing the stress in two steps from the original $4/\pi = 1.27$ (left-hand part of equation (10.31), which would have been applied if the torsion moment had not been taken into account. This means that the permissible load in a given bolt is actually lower by a factor of 0.7 than if it had been calculated in tension alone! It is clear that the threaded section of the bolt needs extra attention.

10.4.2 Effect of temperature change in laminated structures

Changing the uniform temperature of a free-standing product made from an isotropic material induces predictable changes in dimensions but does not induce internal stress. But suppose the product is made from more than one material bonded together. The materials have different coefficients of thermal expansion. The result is that internal stresses are set up, as well as dimensional change. This situation arises for all combinations of materials, including laminated plates of metals or of plastics as well as for fibre-reinforced systems. The internal thermal stresses so developed use up some of the strength of the constituents, and thus it is important to know this when designing products for mechanical loading.

In Chapter 7 we assumed that the laminate had been cured cold and used at the same temperature. Often laminates have to be cured hot and then used cold, or they are used at a different temperature. The effects can be seen in Figs 7.6, 7.7, 7.9 and 7.10. Because the expansion coefficients in a ply along and across the fibres are different, it follows that a change of temperature causes different expansions in different directions if the ply is free to do so. In a bonded stack of plies, all plies must strain the same in a given direction, so some plies cannot expand as much as they would if free, and others are forced to expand more than they would do if free. The difference between the actual expansion of the laminate as a whole and the free expansion leads to thermal internal stresses; overall there is equilibrium of thermal forces in the plane of the laminate, so some plies are under tensile stress and some under compressive stress. These stresses are present whether there is any external load applied or not.

In particular we should check the expected size of these internal thermal stresses in any material which we know to be cured at elevated temperature. The higher the cure temperature (and the lower the service temperature), the bigger the effect.

We should add that after processing hot, creep and relaxation effects do occur (especially for some thermoplastics as expected, but also for cross-linkable thermosets to a noticeable extent). The following approach ignores these effects and therefore gives an upper bound to the predicted internal thermal stress levels. It is assumed that the unidirectional plies are linear elastic with thermoelastic properties which themselves do not change through the temperature range being considered.

Our focus here is on the macroscopic behaviour rather than on the micromechanics of individual fibre–polymer interactions. It is helpful to introduce two new ideas before going into details.

1. We make use of a 'notional' thermal force resultant per unit width $[N^T]$. This is an entirely fictional force used only to calculate the actual strains in global coordinates in bonded laminates which have undergone a change in (uniform) temperature.
2. We need to be able to transform the thermal expansion coefficients from principal directions to global directions. Bearing in mind that thermal strains have the form $\varepsilon_i = \alpha_i \Delta T$ it is then clear that thermal expansion coefficients must transform in the same way as strains do. Recalling that $\alpha_{12} = 0$, we can use equation (7.46) in the form

$$\begin{bmatrix} \alpha_x \\ \alpha_y \\ \alpha_{xy}/2 \end{bmatrix} = [T]^{-1} \begin{bmatrix} \alpha_1 \\ \alpha_2 \\ 0 \end{bmatrix} \tag{10.33}$$

Note carefully the factor 2 for α_{xy}.

The simplest way to estimate the internal stresses in a symmetric laminate, in which a uniform change of temperature has occurred, is based on the approach of Chapter 7 and is as follows.

1. Calculate the (notional) thermal force resultant $[N^T]$ in the regular symmetric laminate based on plies of thickness h according to

$$[N^T] = h\Delta T\Sigma[Q]_{\mathrm{f}}[\alpha]_{\mathrm{f}} \tag{10.34}$$

Sum over all plies. If the ply is off-axis, use the transformed reduced stiffness $[\bar{Q}]$ and $[\alpha_x, \alpha_y, \alpha_{xy}]$. If the ply is lined up with the principal directions use $[Q]$ and $[\alpha_1, \alpha_2, 0]$. ΔT is the temperature difference which is assumed to be uniform through the thickness.

2. Calculate laminate strains induced by $[N^T]$ in the global directions:

$$[\varepsilon^0_{x,y}] = [a][N^T] \tag{10.35}$$

If there is no bending, then the strains are uniform through the thickness.
We see that the laminate thermal expansion coefficients in global coordinates can be readily found from equation (10.35). For example $<\alpha_x>= \varepsilon^0_x/\Delta T$.

3. Calculate free expansion strains in each ply

$$[\varepsilon^0_{x,y}(\text{free})]_{\mathrm{f}} = [\alpha]_{\mathrm{f}}\Delta T \tag{10.36}$$

4. The stress-inducing strains in a given ply are therefore

$$[\varepsilon^{\text{osi}}_{x,y}]_{\mathrm{f}} = [\varepsilon^0_{x,y}] - [\varepsilon^0_{x,y}(\text{free})]_{\mathrm{f}} \tag{10.37}$$

5. The thermal stresses in global coordinates in each ply are

$$[\sigma^T_{x,y}]_{\mathrm{f}} = [\bar{Q}]_{\mathrm{f}}[\epsilon^{\text{osi}}_{x,y}]_{\mathrm{f}} \tag{10.38}$$

6. The thermal stresses in principal directions in a given ply are now obtained by stress transformation:

$$[\sigma^T_{1,2}]_{\mathrm{f}} = [T]_{\mathrm{f}}[\sigma^T_{x,y}]_{\mathrm{f}} \tag{10.39}$$

7. Now apply the Tsai–Hill criterion.

It is worth making some further and more general comments.

If the laminate is only subjected to a temperature change, then the Tsai–Hill criterion suggests the factor by which the change in temperature must be multiplied to cause failure. Remember that the properties have been assumed constant in this simple approach, and earlier chapters of this book indicate that, especially for thermoplastics, this may not always be so. Curiously, it is known that some special laminates cured at exceptionally high temperatures can in fact self-destruct on cooling down to the temperature of use, because of the very large transverse thermal stresses. This is not usual for normal laminates, although levels of transverse or shear stresses after hot cure of thermosetting resins can commonly be about 30% of the relevant ply strength.

If there is a temperature gradient through the thickness of the laminate, then the procedure given above must be adapted, to take account of the bending (Powell, 1994).

The assessment of failure caused by a combination of externally applied loads together with a change of uniform temperature is less straightforward. Adding the thermal and mechanical stresses and simply applying the Tsai–Hill criterion gives a silly result, because then the load factor applies to both load and temperature. It is more appropriate to separate out the two effects and, for example, fix the temperature in use, calculate the thermal stresses and use these to amend the actual mechanical failure stresses of individual plies; then test for failure under the mechanical loads using the Tsai–Hill criterion and the amended strengths. This approach is similar to the method described in the outline to problem 3(c) in section 7.5.

Under uniform change of temperature, a symmetrically laminated flat plate stays flat because the plate is in static equilibrium. If, however, a layer is machined away, the internal stresses are no longer in equilibrium and the plate distorts until they are. This forms the basis of the most reliable layer removal method for measuring internal stresses, which is detailed by Eijpe (1997). This confirms that it is not good practice to machine products after manufacture – and this holds not only for composites but also for extruded and injection-moulded products (Chapter 9) which contain substantial levels of internal thermal stress, particularly if thick-walled.

PROBLEMS

1. The relaxation of the bolt was represented in the isochronous stress–strain diagram of Fig. 10.44. Apparently information of the kind shown in Fig. 5.4 (bottom left) was used. Was that correct?
2. Determine the value of ε_2 in Fig. 10.44 if σ_D is required to equal σ_A.
3. Suppose there is no upper limit to σ_A in Fig. 10.44. Then in fact ε_2 could have been applied at once (Case II) instead of in two steps (case I). Derive an expression for the difference in stress $\sigma_I - \sigma_{II}$ at any time after $t_1 = 100\,\text{h}$ between the two cases, using the model $E(t) = E_1 t^{-m}$ (m is of the order of 0.1), and comment on the result.

4. What will change in Fig. 10.46 if the material of the packing is viscoelastic as well? Will this affect the time to leakage?
5. A carbon-fibre–epoxy unidirectional prepreg is 0.125 mm thick and has the following properties: $E_1 = 138$ GPa, $E_2 = 9$ GPa, $G_{12} = 7.1$ GPa, $\nu_{12} = 0.3$, $\alpha_1 = -3 \times 10^{-7}$/K, $\alpha_2 = 28.1 \times 10^{-6}$/K. A flat cross-ply laminate is to be made from this prepreg with the stacking sequence $(0°/90°/90°)_s$ and cured at 140°C. The stiffness and compliance matrices for this laminate are

$$[A](\text{kN/mm}) \quad \begin{bmatrix} 39.23 & 2.037 & 0 \\ 2.037 & 71.67 & 0 \\ 0 & 0 & 5.325 \end{bmatrix}. \qquad [a](\text{mm/MN}) \quad \begin{bmatrix} 25.53 & -0.7255 & 0 \\ -0.7255 & 13.97 & 0 \\ 0 & 0 & 187.8 \end{bmatrix}$$

Calculate the stress profiles in the laminate at 20°C.
6. The rayon–rubber laminate $(30°/-30°)_s$, based on 1 mm plies of RR, is cold cured at 20°C, and then heated to a uniform temperature of 50°C. (a) Calculate the thermal expansion coefficients for this laminate. (b) Calculate the thermal strains and curvatures, and the stresses σ_x and τ_{xy} in the 30° plies.

10.5 TAILOR-MADE ELEMENTS

We use the term 'tailor-made' to achieve a specific effect which relies on some special consideration described elsewhere in the book.

Section 10.5.1 looks at the formation of a parison needed to make blown products of defined wall thickness. Section 10.5.2 studies the effect of constraining lateral deformations on the design of rubber spring elements of desired compressive stiffness. Section 10.5.3 takes into account the combined requirements of bending and axial stiffness and strength for a golf shaft. Section 10.5.4 discusses the role of coupling effects in the design of a wind turbine blade. The last example (section 10.5.5) concerns the design of a stacking container which exploits shape to achieve high bending stiffness per unit mass.

10.5.1 Design for extrusion blow moulding

Extrusion blow moulding offers wide scope to the imaginative designer of hollow containers or items having simple or complicated shapes. The designer can within close limits realistically specify outside dimensions, because mould cavity dimensions can be specified. It is however more difficult to specify precisely the wall thickness in relation to the processing conditions.

The basic process of extrusion blow moulding has already been outlined in section 4.4.6. We begin here by adding some further practical details of processing and process design. Then we examine the likely problems during the manufacture of the parison, particularly sag during and after extrusion, and we estimate the likely effect of changes in surface temperature of the parison before inflation. We go on to look at the use of parison programming to tailor the wall thickness of the parison. The problems will enable the reader to identify some interesting design challenges relating to extrusion blow moulded structures.

Our aim is limited to highlighting some problem areas and suggesting considerations which will give useful possibilities for ball-park solutions. Detailed solutions will require the use of numerical methods and more refined modelling.

(*a*) *Some practical details*

The die and pin (mandrel) must be concentric to give a parison of uniform wall thickness round the circumference. An off-centre pin will produce a distorted parison curved along its length (sometimes called a 'hollow banana'; Fig. A7) which then does not fit over the blow pin.

The parison swells in thickness and diameter as it emerges from the die. Some estimate of this swelling can be made using the analysis given in section 8.7.4. The method there assumes there are no axial forces on the parison: this will not be a problem for short parisons, but there will be some error for very long parisons. The model for swelling does not let us predict the build-up to full swelling after emerging from the die. We might expect that the swelling is complete after a length of about one parison diameter, so for practical blow moulding, swelling can be supposed to be complete, because there is always a small amount of process scrap ('ears') at each end of the moulding, perhaps having the length of half to one parison diameter or so.

In large-diameter parisons the leading edge will fold or 'curtain' during extrusion because of the post-extrusion swelling: radial expansion introduces some circumferential orientation which relaxes, and the contraction gives some circumferential buckling.

The polymer melt must be viscous in tension to prevent excessive parison sag (especially during slow extrusion of long parisons), and have a high melt strength in tension to prevent melt rupture during inflation. It is better to use a molten polymer which is tension-stiffening rather than tension-thinning (Fig. 8.17), so that local thin spots are stabilized.

It is good to extrude the parison into still air of uniform temperature. The parison should have a temperature (profile) which is uniform round its circumference to permit even inflation during blowing.

There is a weld-line at the top and bottom of the blown product: a thick section here, say, a substantial proportion of the parison thickness, compensates for this potential weakness (see top of Fig. 4.10b).

The parison thins down as it is blown. The simplest approach is to assume that the parison is of uniform wall thickness and melt temperature, with (large) deformation at constant polymer volume. The parison has an initial diameter D_p. For a long cylindrical container, the parison inflates and its outer surface freezes on contact with the cold mould surface at a diameter D_m. The larger the blow ratio D_m/D_p, the thinner the wall becomes. In simple products the value of blow-up ratio is in the range 2–4. Sharp external corners in blow mouldings should be avoided because they are sources of local thinness as well as stress concentrations.

Inflation is mainly an elastic process, with a virtually rubber-elastic material. Extreme values of inflation pressures are best avoided. Over-fast inflation runs the risk of parison burst before the mould section is reached. If the air pressure is too low, there is a risk of unwanted sag during the early part of the blowing process, poor surface finish and even an incompletely inflated moulding.

Cooling time for the blown product (from one polymer surface only) is proportional to the square of the wall thickness, so the fastest cycle time (and the most economical product) will be achieved when the wall thickness of the product is as uniform as possible. This is usually not easy to achieve (Why not?) Thick sections of partially crystallizable thermoplastics will shrink more than thin sections. This will reduce thermal contact with the mould and hence increase the cooling part of the cycle (compared with calculations based on perfect thermal contact as described in section 8.8.4).

Ribs can be formed in extrusion blow mouldings as shown in Fig. 10.39(c). Forming box-like cross-sections is also possible: the parison is compressed during the inflation stage, and the internal surfaces thus squeezed together should weld to allow shear transfer across them, otherwise the bending stiffness is much reduced.

(*b*) *Simple parison sag*

The starting point considers that a parison has been produced 'instantaneously' having a uniform wall thickness and length L. The parison will sag while it hangs under the action of self-weight loading. Under isothermal conditions we showed (in problem 2 in section 8.6) that the axial increase in length (sag) ΔL_x after time t is given by $\Delta L_x = \rho g t L^2/2\lambda$, where λ is the tensile viscosity and ρ the density. The obvious conclusions are that to minimize sag we aim for short times and large λ (high molecular mass and low melt temperature); the sag is greatest for long parisons.

As an aside, it is sometimes desirable to co-extrude a parison consisting of two concentric tubes of different materials bonded at the interface. If the two tubes are made from materials i and o having (at the same temperature) different melt densities, tensile viscosities and cross-sectional areas, the sag $\Delta L_x{}'$ of such a tube ('instantly produced') can be shown to be

$$\Delta L'_x = (\rho_i A_i + \rho_o A_o) g L^2 t/[2(\lambda_i A_i + \lambda_o A_o] \qquad (10.40)$$

Any element of a parison may sag by an axial amount of strain ε_x. This strain will be accompanied by a decrease in diameter and thickness of the order of $\varepsilon_\theta = \varepsilon_y = -\nu\varepsilon_x$. We assume the deformations occur at constant volume, so that $\nu = 0.5$.

The parison initially at uniform melt temperature T_m extrudes into still air having a temperature T_a. The parison surface will cool relatively slowly, perhaps by 10–20°C over a period as long as 10 s. The change in surface temperature and the penetration of the change of temperature of 0.2–0.3 mm seem small, but the effect on parison sag is worth examining because the viscosity of the thin surface region is increased.

Modelling the parison as a semi-infinite solid initially at T_1 and suddenly exposed at $t = 0$ to still air at temperature T_2, we can estimate how the surface temperature T_i changes as a function of time. According to Agassant *et al.* (1991) the penetration time $t = k_1^2/\mu^2\pi\alpha_1$ where $\mu = h_2(T_i - T_2)/(T_1 - T_i)$. The symbols k_1 and α_1 are the thermal conductivity and diffusivity of the polymer, and h_2 is the heat transfer coefficient for convection in the air. It can be estimated that the amount of heat transfer by radiation from the hot polymer melt is about the same as that for convection; on this basis we can use a combined heat transfer coefficient for radiation and convection having a practical value of about $h_2 = 40\,\mathrm{W/m^2\,K}$.

For a 30°C drop in surface temperature for a LDPE melt initially at 200°C we have $\mu = 40(150/30) = 200$, so $t = (0.25/200)^2/3.14 \times 10^{-7} = 5\,\mathrm{s}$. Thus after 5 s exposure the bulk of the parison remains at 200°C, and there is a local temperature gradient at the surface. We might model the tube in a very simple manner as two concentric tubes, the inner one at constant temperature T_i and the outer one at constant temperature T_o, both temperatures assumed constant along the length. Equation (10.40) can then be used to make a rough estimate of sag under selfweight loading. If we ignore any change of density with small change of temperature, set $A_o/A_i = 0.1$, take $\lambda_o/\lambda_i = 2.28$ for LDPE (based on Fig. 8.13), and compare this sag with that of a parison of the same dimensions but at uniform temperature through the wall, we find $\Delta L'_x/\Delta L_x = 0.85$.

As expected, there is less sag with the assumed changes of surface temperature. We emphasize that this is a rough calculation. In a real parison the lower end would be cooled for the longest time, thus the penetration layer will there be thickest and coldest, whilst the just-extruded upper end of the parison will be hardly cooled at all. It is left to the interested reader to confirm analytically the commonsense observation that the reduction of sag is rather less than the 15% calculated using the constant cold shell model.

The conclusion we draw from this simple analysis is that cooling during parison formation is not itself a major problem in practice, but sag certainly remains a problem if extrusion times are long (upper boundary on times say 8–10 s) or if parisons are very long. The temperature change can also increase the pressure needed to inflate the parison, and may impose a limit on the achievable blow-up ratio.

(c) *Parison sag during extrusion*

It takes time to extrude a parison, during which the parison also sags under its own weight. We expect the sag to be least during rapid extrusion. Discontinuous extrusion at high speed is relevant for long parisons destined for making large drums or storage containers and for products having the dimensions of car bumpers.

The simplest approach to sag during extrusion at constant speed and constant temperature gives useful insight into what may be expected. We recognize that the leading end of the parison has the longest time to sag but only under a small force. The upper end of the parison has little time to sag but is under a large gravitational force. Calculating the deformations in incremental lengths, and ignoring any surface cooling effects, we find that the maximum amount of sag occurs fairly uniformly in the middle two-thirds of the length of the final parison during extrusion, and relatively little at the ends.

The starting point is that one imagines a nominal parison of length L_0, cross-sectional area A_0, which is extruded at a constant linear speed v (after post-extrusion swelling). The nominal parison has no sag. The time to produce the complete nominal parison is $t_{\text{prod}} = L_0/v$.

We designate the free (lower) end of the nominal parison as $x = 0$, with x positive upwards. At some arbitrary position between the free end and the die we consider the element of length dx. Ignoring the swelling operation, the section dx left the die for a time

$$t = [(L_0 - x)/L_0]t_{\text{prod}} = (L_0 - x)/v$$

During this time $t = x/v$, the section dx has been subjected to a load $F = xA_0\rho g$, and the element has been gradually sagging according to

$$\mathrm{d}\varepsilon/\mathrm{d}t = \sigma/\lambda = F/A\lambda = F(1+\varepsilon)/A_0\lambda = x\rho g(1+\varepsilon)/\lambda$$

If we assume that the tensile viscosity λ is independent of stress, then integration gives

$$\ln(1+\varepsilon) = (\rho g t/\lambda)x + k$$

Assuming $t = 0$ at the start of extrusion, then the section dx emerges from the die at a time $t = x/v$. Thus $\varepsilon = 0$ at $t = x/v$, and hence

$$\ln(1+\varepsilon) = (\rho g/\lambda)(t - x/v)x$$

When the parison is complete, $t = L_0/v$, so the strain in the element is

$$1+\varepsilon = \exp[(\rho g/v\lambda)(L_0 - x)x]$$

The sag in the element is

$$ds = \varepsilon dx = \{\exp[\rho g/v\lambda)(L_0 - x)x] - 1\}dx$$

and thus the total sag is $s = \int ds$ with integration over the complete length of the nominal parison. This integral is cumbersome to integrate, but if $(\rho g/v\lambda)(L_0 - x)x \ll 1$, then

$$\exp[(\rho g/v\lambda)(L_0 - x)x] \approx 1 + (\rho g/v\lambda)(L_0 - x)x + \ldots$$

so that the sag just at the end of extrusion is approximately

$$s = \rho_g L_0^3 t/6v\lambda = \rho g L_0^2 t/6\lambda \qquad (10.41)$$

Ball-park figures for the process are $\rho = 1000\,\text{kg/m}^3$, $\lambda = 10^5\,\text{N s/m}^2$, and $v = 0.1\,\text{m/s}$ for short parisons, $v = 0.5\,\text{m/s}$ for long parisons, so the approximation can be justified for most sensible conditions.

Further refinements can be made by evaluating the integral more precisely, and with more analytical complexity by taking the stress dependence of the tensile viscosity into account. With thin-walled parisons, the surface cooling during extrusion not only reduces sag somewhat during slow parison formation, but also reduces the maximum blow-up ratio.

(d) Parison programming

Sag can produce a parison which, when extruded through a constant die gap, varies in thickness along its length. Many containers have a cross-sectional shape and dimensions which vary along the length. It is desirable to make the container with a wall thickness as uniform as possible. How can this be done?

One approach commonly used involves what is called the 'parison programmer'. The parison programmer permits longitudinal movement of the die pin (die mandrel) within the die outer; there are tapers on both the pin and the outer, so as the pin moves, the die gap is adjusted and the resistance to flow changes. Thus the wall thickness changes because the die gap and the post-extrusion swelling change (assuming constant rate of extrusion). It follows that the parison diameter also changes, though this is not always mentioned in the literature. The axial movement of the pin can be preset in up to one hundred changes during the extrusion of one parison, so fine adjustment of parison cross-sectional dimensions is possible. The extrusion conditions are now not entirely steady state, so the swelling calculations according to section 8.7.4 are not quite accurate.

The parison programmer can also be used to compensate for parison sag, thus producing a parison which when inflated will produce a product of more uniform wall thickness – for an axisymmetric long product, and ignoring the ends which are much more difficult to analyse.

(e) Wall thickness design

We conclude this section with an outline for solving a problem which brings together many melt processing topics discussed in this book.

A product to be extrusion blow moulded from a propylene–ethylene copolymer is 1 m long and of 150 mm outside diameter inclusive of hemispherical ends. For the purposes of discussion only, the extrusion die exit dimensions are set at $D_{di} = 50$ mm and $D_{do} = 55$ mm; in the die exit region the flow channel is 40 mm long and untapered. How may we estimate the wall thickness of such a product away from the ends?

We suggest a series of steps in the strategy without going into fine detail. Some but not all possible refinements are indicated.

1. First a suitable polymer (grade) and melt temperature must be chosen. For the purposes of this problem the simplest is to choose the propylene–ethylene copolymer for which data are given in Chapter 8. It is not obvious that 230°C is the best melt temperature; among the many compromises are the balance between avoiding too much sag in the long parison which is necessary and the avoidance of defects in the parison caused by flow just before the die exit region described in the problem. We could well discuss which grade of polymer to use with the raw material supplier, but this will not affect the strategy.
2. We assume the parison has to be 15% longer than the product, with ends to be trimmed away and recycled. Thus the length of the parison including sag is $L_s = 1150$ mm. We shall ignore shrinkage during cooling to room temperature; this can be easily taken into account.
3. We assume for the purposes of this discussion that the blowing starts immediately the leading end of the parison is in position over the blow pin. This is not entirely realistic, but readers can assess the effect of some sag after extrusion stops, if they wish to pursue this.
4. The parison sag ΔL_0 is based on the original unsagged length L_0 produced during extrusion at assumed constant volumetric rate Q. Thus $L_0 + \Delta L_0 = L_s$. We know from equation (10.41) that $\Delta L_0 = \rho g t_0 L_0^2 / 6\lambda = \beta t_0 L_0^2$ in which the tensile viscosity λ is assumed constant over the range of tensile stresses in the problem. This leads to a quadratic equation in $L_0 : \beta t_0 L_0^2 + L_0 - L_s = 0$ which can be solved for L_0 using a sensible choice of extrusion time, t_0. Typical extrusion times are bounded: too short will give melt fracture problems and too long will allow excessive sag (which we allow for in the quadratic equation) and will allow the parison to cool to the point where it is too stiff to blow. An extrusion time in the range 2–4 s is sensible.
5. The parison cross-section dimensions H_p and D_p must now be calculated. The volume of the required parison is $V_p = \pi D_p H_p L_0$, and the volumetric flow rate is $Q = V_p / t_0$. Thus the required value of Q is chosen such that flow through the die having the given dimensions D_d and H_d will

give a parison of dimensions D_p and H_p; the iterative method is given in problem 4 at the end of section 8.7, and convergence is rapid. The result is $D_pH_p = D_dH_dB_w^3$ where B_w is the swelling ratio for the flow rate Q.

6. We can now estimate the wall thickness of the blown product away from the ends. The simplest estimate is to assume that the sag is more or less uniform along the length. Thus the sag ΔL_0 will reduce both parison cross-section dimensions by a factor $(1 - 0.5\Delta L_0/L_0)$. Assuming instantaneous inflation of the sagged parison as soon as the extrusion from the ram accumulation stops, and ignoring parison bounce at the free end, and assuming inflation occurs at constant parison volume, we have the wall thickness of the moulding

$$h_m = D_pH_p/D_m = D_dH_dB_{wo}^3(1 - 0.5\Delta L_0/L_0)^2/D_m$$

7. The method can be much further developed, and lends itself to numerical modelling.

10.5.2 Rubber block springs and seals

Rubber block springs are often used to support machine elements, e.g. the motor in a car or the wheel under a wagon, in a flexible but defined manner relative to a frame. This function includes the position and the mobility in space, i.e. in the three principal directions (Fig. 5.39). Each direction may be required to have its own stiffness K.

(*a*) *Design of bonded block springs loaded in the principal directions*

The principal directions in a block spring are in two mutually perpendicular directions, in the plane and normal to the plane. The two orthogonal in-plane shear stiffnesses K_s are identical. The compressive stiffnesses K_c can be varied over a wide range by correct choice of the effective thickness, h, of the rubber. To achieve high stiffness, metal plates can be interleaved with the rubber to increase the restraint. This does not affect the shear stiffnesses K_s if these are calculated using the total thickness of rubber.

The design procedure consists of the following steps.

1. Choose the plan dimensions of the block from information given.
2. For a given hardness of rubber (normally in the range 40–60 IRHD), read G from Fig. 5.35 $\Rightarrow E_0 = 3G$.
3. Calculate the thickness of rubber needed to satisfy the prescribed shear stiffnesses K_s using equation (5.64).
4. Choose a suitable number of interleavings to satisfy the overall compressive stiffness K_c using equations (5.90) and (5.92). This may require

adjustment to the hardness of the rubber (and revision of K_s), or it may require the use of layers of rubber having different thicknesses, so that both K_c and K_s can be achieved.

(b) Stiffness of blocks loaded in non-principal directions

There is often a requirement for the stiffnesses in three different directions to be different (rather than have two the same). This can be achieved by inclining the bonded rubber blocks at an angle to one, or more, of the loading directions. The straightforward example discussed here (Fig. 10.49) involves the use of two identical rubber blocks which are inclined at 0° to the horizontal direction and therefore at 90–0° to the vertical z-direction. The stiffness in the y-direction (through the plane of the paper) is evidently a principal stiffness, and can easily be calculated as before. The stiffnesses K_x and K_z must be related to the stiffnesses K_s and K_c by resolving applied loads and deformations into the principal directions.

Consider one rubber block (Fig. 10.50a) which carries a vertical load P_z, and requires a horizontal load H to be applied to keep the half-unit in equilibrium. P_z is related to the equivalent direct and shear loads on the one block, P_c and P_s, by

$$P_z = P_c \cos\theta + P_s \sin\theta \tag{10.42}$$

and the components of the vertical deformation δ_z are

$$\delta_c = \delta_z \cos\theta \text{ and } \delta_s = \delta_z \sin\theta \tag{10.43}$$

Using the definitions for stiffness, $P_z = K_z\delta_z, P_c = K_c\delta_c$ and $P_s = K_s\delta_s$, and combining equations (10.42) and (10.43) gives

$$K_z = K_c \cos^2\theta + K_s \sin^2\theta \tag{10.44}$$

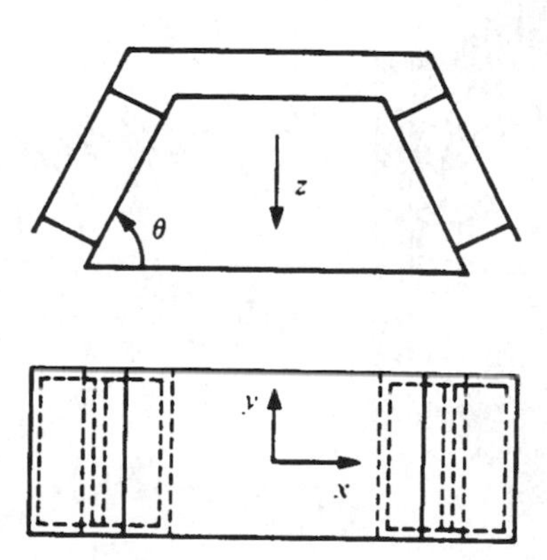

Fig. 10.49 Inclined shear mounting (schematic)

If the same rubber block now carries only the horizontal load, P_x, with the components of compressive and shear forces acting as before (Fig. 10.50b), then

$$P_x = P_c \sin\theta - P_s \cos\theta$$

and

$$\delta_c' = \delta_x \sin\theta \text{ and } \delta_s' = -\delta_x \cos\theta$$

i.e.

$$K_x = K_c' \sin^2\theta + K_s' \cos^2\theta$$

However, the spring stiffnesses in the principal directions 'c' and 's' are for the same block as before, so $K_c = K_c'$ and $K_s = K_s'$, i.e.

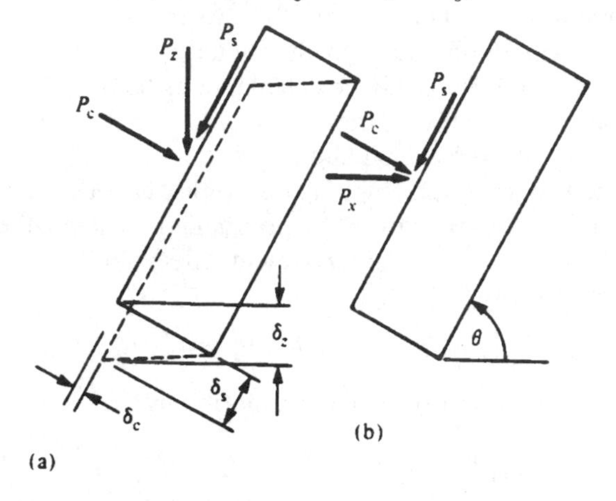

Fig. 10.50 Resolution of force on one inclined block

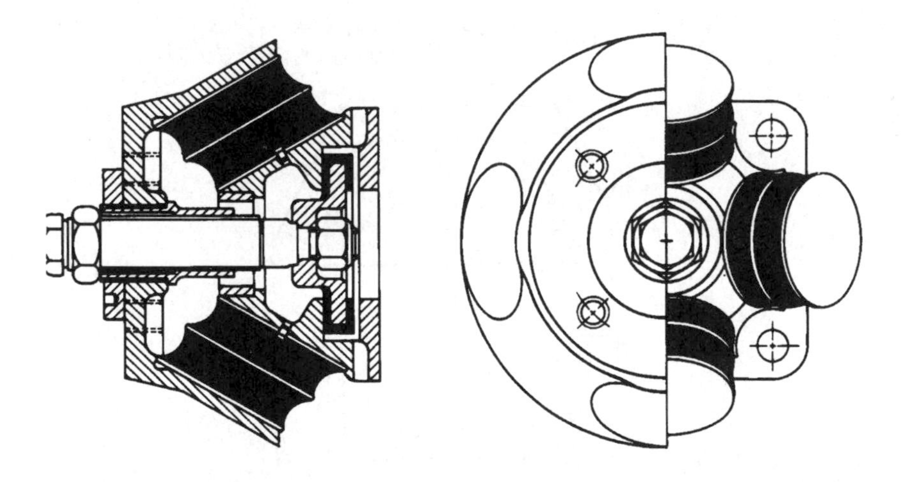

Fig. 10.51 A complicated inclined shear assembly (courtesy Dunlop)

$$K_x = K_c \sin^2 \theta + K_s \cos^2 \theta \qquad (10.45)$$

The shear stiffness in the y-direction under a force P_y is a principal (shear) stiffness given by equation (5.64):

$$K_y = \frac{GA}{H}$$

where H is the total thickness of the rubber block.

Thus, given a design of the kind specified in Fig. 10.49, the stiffnesses K_x, K_y and K_z can be satisfied for given value of A by suitable combinations of H, hardness and θ. Other solutions are also possible. The principle of inclined units can be extended, resulting in complicated assemblies to meet difficult requirements, one example of which is shown in Fig. 10.51.

(c) *Other types of rubber springs and coupling units*

Many shapes of spring are of practical interest. They include discs, cones, plates, tubes and solid rollers, and spheres. These may be subjected to compression or shear in any principal direction and also to bending. Analysis of these forms is not attempted in this introductory text, but the results of the analysis are given in Göbel (1974) and in Freakley and Payne (1978).

(d) *Rubber seals*

In sealing applications, the ability of the seal to recover can be important if there is movement in the surrounding metalwork. The seal should ensure the minimum required compression force between the flanges if they move within a certain time over a certain distance. Clearly the permanent set, as indicated in section 5.3.1(c), may give an indication of the fitness of the sealing rubber for the purpose, but the speed of the movement of the flanges (for example, caused by temperature changes in the pipe system) and the speed of recovery of the rubber may or may not match. So more dedicated experiments as described in section 5.3.1(c) may be necessary.

In order to understand the requirements for a flange sealant (say an O-ring) it may be useful to consider its function; namely to prevent leakage. The force between rubber and flange will result in a certain fit between their surfaces, but not an ideal one. Leakage Q will be a function of the clamping force P and this will be greater as the pressure p of the fluid or gas in the pipe is greater, but smaller as P is higher (Fig. 10.52).

The value of permanent set, as measured for the relevant conditions, will enable the designer of, for example, tightening strips for doors or windows, to design suitable profiles.

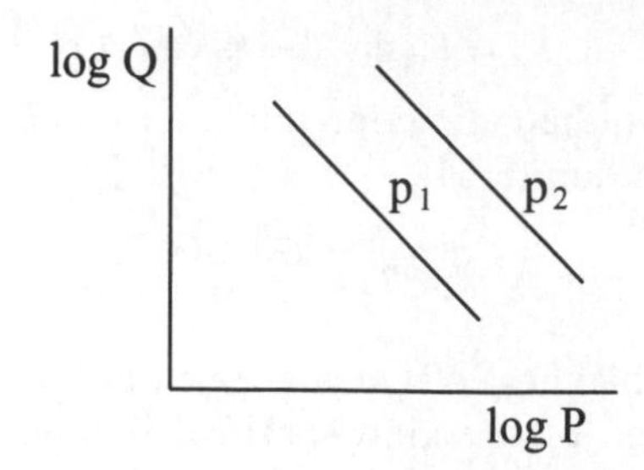

Fig. 10.52 Schematic indication of leakage Q through the seal of a flanged pipe connection

10.5.3 Golf club shaft

A golf club shaft has to cope with several requirements, including minimum mass, bending stiffness and strength, and torsional stiffness and strength. Composite club shafts with carbon fibres have excellent properties. For these shafts, different kinds of laminate construction can be chosen. Which procedure should we follow in order to reach the optimum, in a minimal number of calculating steps?

A first step is to define the requirements for the shaft. These can be derived from an analysis of the functions of the club, but here, as is often done, we choose to derive them from the properties of the steel shaft it will replace. So, if we notice that a golf club shaft is subjected to bending and torsion, we do not do an analysis of the forces of the ball (Fig. 10.53), the grip and possible misuse. If the composite carbon fibre-reinforced epoxy shaft provides the same bending stiffness and strength, and torsional stiffness and strength, as the steel shaft with a lower mass, it certainly would represent an improvement. It will later be seen whether all of these together may mean an over-requirement.

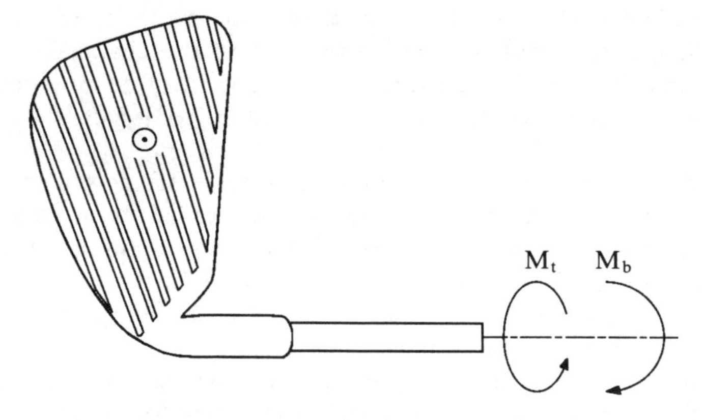

Fig. 10.53 Load on the lower part of a golf club

In technical terms the above means that the carbon shaft should have at least the same bending stiffness EI, the same torsional stiffness GI_p, the same bending strength $M_{max} = \hat{\sigma} W_b$ (where $W_b = I/e$) and the same torsional strength $M_{tmax} = \hat{\tau} W_t$ (with $W_p = I_p/r$) as the steel shaft. These requirements should be applied to the shaft at its connection to the clubhead, where the shaft outside diameter $d_s = 8$ mm and the local wall thickness is of the order of $h_s = 1.25$ mm. The properties of a typical steel for the shaft are $E_s = 210\,kN/mm^2$, $G_s = 80\,kN/mm^2$, $\nu_s = 0.31$, $\hat{\sigma}_s = 410\,N/mm^2$, $\hat{\tau}_s = 205\,N/mm^2$, $\rho_s = 7830\,kg/m^3$.

It is proposed to use carbon fibre-reinforced epoxy (cfrp) plies of 0.125 mm thickness with the properties given in the data for materials section 7.2.8. We will use the index 's' to indicate the steel and the index 'c' to indicate the cfrp.

The (simple) idea is that the carbon shaft should be built up from 0° plies in order to deal with the bending stresses and from ± 45° plies in order to deal with the torsional stresses. As the 0° plies will not contribute to the torsional stiffness and strength, but in contrast the ± 45° plies do contribute to bending stiffness, we start with the 45° plies for torsion. Further, as stiffness requirements are usually more difficult to meet than strength requirements, we start with stiffness.

(*a*) *Torsional stiffness*

We start by choosing the right number of ± 45° plies in order to satisfy GI_p. A first approximation of I_p is $\pi d^3 h/4$. With given shaft diameter ($d = 8$ mm) for both materials we conclude that we have to match a value of $G_s h_s = 80 \times 1.25 = 100\,kN/mm$ in the carbon fibre laminate of the same value. A $(\pm 45°)_s$ laminate of cfrp will have $E_x = E_y = 25\,kN/mm^2$, $G_{xy} = 46.54\,kN/mm^2$ and $\nu_{xy} = 0.75$. Using G_{xy} we find that a laminate thickness of $t_c = 2.15$ mm or 18 plies (2.25 mm) would bring the required torsional stiffness. For reasons of symmetry, we reduce the number of plies to 16 ($h_c = 2.0$ mm), hoping that the future addition of 0° plies will restore the required torsional stiffness.

(*b*) *Torsional strength*

The requirement for torsional strength $M_t = \hat{\tau} W_t \simeq \hat{\tau} \pi d^2 h/2$ reduces to matching $\hat{\tau}\, h$ for the cfrp shaft with that of steel. Since $h_c = 2$ mm, $\hat{\tau}_c$ should equal at least $205 \times 1.25/2 = 128\,N/mm^2$. Calculations with laminates are performed using force per unit width, so we have to relate this stress to the wall thickness $t (= h_L$ in equation (7.81)). We consider the shaft as thin walled, so the shear stress τ_{xy} is supposed to be constant through the wall, if the material were homogeneous. Using equation (7.73) we find $N_{Lxy} = \tau_{xy} h_c = 128 \times 2 = 256\,N/mm$.

If we now perform a laminate calculation with this load (in equation (7.92) all left-hand forces and moments per unit width are zero except N_{Lxy}), we can find not only laminate deformations, but also local stresses in the plies. For the +45° plies we find $[\sigma_1 = 246, \sigma_2 = 10, \tau_{12} = 0\,\mathrm{N/mm^2}]$, and for the −45° plies $[\hat{\sigma}_1 = -246, \hat{\sigma}_2 = 10, \tau_{12} = 0\,\mathrm{N/mm^2}]$. In order to compare these with the strength data of the cfrp ply we can use the Tsai–Hill failure criterion (equation (7.19)). It predicts failure first in the −45° plies, when the load is increased by a factor 2.79, assuming the tube is made using a cold-curing resin. This means a clear safety margin against fracture by torsion.

(*c*) *Bending stiffness*

The shaft as it stands now will have a certain bending stiffness EI, but perhaps not enough to match $E_s I_s$. We now add 0° plies to the ±45° plies designed under section 10.5.3(a). In analogy with equation (7.87), their stiffnesses can be added: $E_{45}I_{45} + E_0 I_0 = E_s I_s$. The second moment of inertia I can be approximated with $\pi d^2 h/8$. The requirement is therefore

$$\begin{aligned} &E_{45}h_{45} + E_0 h_0 = E_s h_s \\ &25 \times 2 + 181 \times h_0 = 210 \times 1.25 = 262.5\,\mathrm{kN/mm} \\ &h_0 = 1.17\,\mathrm{mm} \end{aligned} \tag{10.46}$$

This means that we have to use 10 plies ($h_0 = 1.25$ mm) in the axial direction.

(*d*) *Bending strength*

We now have to make sure that this laminate $(0^\circ_5/+45^\circ_4/-45^\circ_4)_s$ with total thickness 3.25 mm can cope with the load. Just to see whether these 10 plies would hold the bending load, we state the requirement:

$$\begin{aligned} &\hat{\sigma}_c h_0 \geqslant \hat{\sigma}_s h_s \\ &1000 \times 1.25 \geqslant 410 \times 1.25 \end{aligned} \tag{10.47}$$

Note that we use the compression strength of the ply in its principal (0°) direction; as this is lower than the tensile strength, the shaft will fail in bending at the side which is in compression. The result is more than sufficient for our requirements. Because of their axial stiffness compared with the ±45° plies, the 0° plies will preferentially bear the bending load.

This calculation by itself is not, however, sufficient. We have just guessed that the bending load will not affect the ±45° plies, nor the torsional load the 0° plies, but in view of the complexity of the system we have to make sure!

(*e*) *Incorporating the requirements into the design*

The laminate will be built up of 26 plies: 10 at 0°, 8 at + 45° and 8 at −45° with a total wall thickness of $26 \times 0.125\,\mathrm{mm} = 3.25\,\mathrm{mm}$. In order to reduce

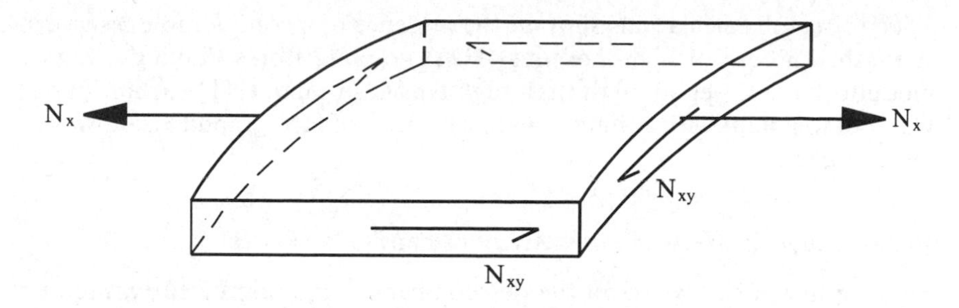

Fig. 10.54 Stresses on a wall element of the shaft of a golf club

interlaminar stresses at the ends of the shaft the ply directions will be interchanged as much as possible. This leads to the preferred lay-up: $[0, +45, -45, 0, +45, -45, 0, +45, -45, 0, +45, -45, 0]_s$. The loading scheme of the laminate of the shaft wall can be represented by Fig. 10.54.

The required loading is $[N_x = 410 \times 1.25 = 512\,\text{N/mm}, N_y = 0\,\text{N/mm}, N_{xy} = 205 \times 1.25 = 256\,\text{N/mm}]$. Under this combined loading, the worst case value of the Tsai–Hill criterion predicts that failure will occur first in the $-45°$ plies at 2.01 times this loading; therefore enough safety is present.

In order to compare the overall stiffness, we need to know the effective moduli. Using $E_x = 1/(a_{11}t)$, etc., where $[a_{ij}] = [A_{ij}]^{-1}$, average values for the modulus of the wall can be calculated: $E_x = 85.8$, $E_y = 31.8$, $G_{xy} = 31.4\,\text{kN/mm}^2$, $\nu_{xy} = 0.70$. Now we can verify the stiffness requirements:

$$E_x h_c = 85.5 \times 3.25 \geqslant E_s h_s = 256 \Rightarrow \text{correct}$$
$$G_{xy} h_c = 31.4 \times 3.25 \geqslant G_s h_s = 99 \Rightarrow \text{correct}$$

In order to verify whether we have in fact designed a lighter shaft, we have to compare the weights. With the outside diameter of the shaft given, we calculate

$$\rho_c h_c / \rho_s h_s = 1600 \times 3.25/7830 \times 1.25 = 0.53$$

so this shaft really will be much lighter.

An 8 mm diameter shaft is no longer thin-walled with a 3.25 mm thick wall. This makes the last formula and in fact some earlier ones incorrect. This, together with the remarks below, would require a second refined assessment and calculation of the shaft. This, however, would fall outside the scope of this introductory book.

(*f*) *Concluding remarks*

In practice the number of $\pm 45°$ plies is much lower. Apparently the torsion requirement (which the steel meets) is only addressed on a much lower level.

The strength calculations showed the presence of a considerable reserve in strength. The use of high-modulus (HM) graphite fibres (having a higher modulus but a lower strength than high tensile strength (HT) carbon fibres) will lead to a shaft with a thinner wall. The price of HM graphite is, however, much higher.

10.5.4 Coupling effects in symmetrical composite laminates

By 'coupling effect' we mean the development of a strain or curvature in a direction or sense different from that of the applied force or moment, and vice versa. The Poisson effect is thus a familiar form of coupling, and is almost always present in both isotropic and composite materials.

In laminated composites based on unidirectional plies of different materials or the same materials arranged at different angles, a wide range of additional coupling effects is available (section 7.1.2). We focus here on symmetrical angle-ply laminates.

We first consider the in-plane compliance matrix $[a]$ for a symmetrical laminate, in which the coupling effects are given by the non-diagonal terms. The coefficient a_{12} indicates the Poisson effect, and only the magnitude is sometimes a surprise to the designer familiar with isotropic materials – values greater than $\nu_{xy} = 0.5$ are quite common. We can conclude from section 7.4 (for example, problem 7) that we can tailor thin-walled tubes based on angle-ply laminates which under internal pressure will not change either their diameter or their length.

The more interesting coupling effects are represented by a_{16} and a_{26}, which relate direct strains to applied shear forces or shear strain to applied direct forces. For a single ply under off-axis load $a_{16} \neq a_{26}$ unless the ply is at $+45°$ or $-45°$ to the load N_x. For symmetrical angle-ply laminates we need to distinguish between balanced and unbalanced: balanced means the same number of plies in each direction $+\theta$ and $-\theta$. Only for unbalanced laminates are a_{16} and a_{26} non-zero.

In problem 8 at the end of section 7.3 we examined the behaviour of a composite tube made from fibres wound under a single helical angle. We can now introduce a more interesting example. A windmill blade for generating electricity turns faster, the harder the wind blows. In extremely high winds there is a risk of structural damage, and the conventional approach is to twist the blade along its length partially out of the line of action with the wind, using a separate mechanism. But why not design a blade which thinks for itself and which avoids the need for a separate mechanism?

We consider a blade which is fixed at one end to the generator axis; the blade thus rotates in the vertical plane. Suppose we start simply, by modelling the blade as a circular tube – we must refine this later, of course. Our design could exploit the coupling effect of the angle of twist (and thus the shear strain) induced by the centrifugal force from blade rotation. What we seek is a high value of a_{16}. A tube made from a single helical winding would certainly give a useful change of angle of attack of the blade which would vary along the length of the blade and which would vary with the wind speed. We should, however, bear in mind that such a unidirectional construction would have low transverse and shear strengths. More sensible would be to use a laminate construction; it is attractive to consider an angle-ply laminate construction with plies at, say, +20° and −70° to the length of the blade. Such a laminate would give a good balance of stiffness and strength properties, and can be manufactured from balanced or unbalanced woven cloth which, with care, is reasonably easy to handle.

We can now consider another type of coupling effect, that based on out-of-plane coupling in symmetrical angle-ply laminates, using the bending compliance $[d]$. The coefficient d_{12}, which always has a negative value, is responsible for the familiar anticlastic curvature when a single moment is applied. More interestingly, we have already remarked in section 10.3.5 that the coefficient d_{16} is responsible for the development of twisting curvature when a single moment is applied: this curious effect occurs even in balanced laminates.

Returning to the windmill blade, we might now idealize the structure as a flat laminate under the bending moment applied by the wind (case (a) in section 10.3.5). The blade bends out of plane (using d_{11}) and twists about its axis (using d_{16}). We should also consider the effect of self-weight loading of the idealized beam during a complete revolution, using case (b) of section 10.3.5: for a symmetric laminate this will give bending but not twisting.

Of course we now need to combine these ideas with a blade with a decent aerodynamic profile: substantial refinement is necessary, but this is beyond the scope of our introduction. The application of the coupling effects in the design of 'smart' wind turbine blades has reached the prototype stage with blades of aerodynamic profile several metres wide and 30 m long. The aerodynamics and detailed mechanics are quite involved, but the basic principles of using coupling effects are essentially those described above.

There is a much broader spectrum of special coupling effects in non-symmetrical laminates where $[B] \neq 0$; most of these effects are unfamiliar to designers used to isotropic materials and their exploitation rests with the imaginative designer.

10.5.5 Case study of a potato tray

(a) Introduction

For many years, wooden trays have been used to store seed potatoes so that root systems can sprout before the potatoes are sown in the ground (Fig. 10.55). Wood rots, and the absorbent surface can retain disease. It is not easy to make wooden trays cheaply by hand to the precise reproducible dimensions which permit interlocking for safe high stacking (for best use of

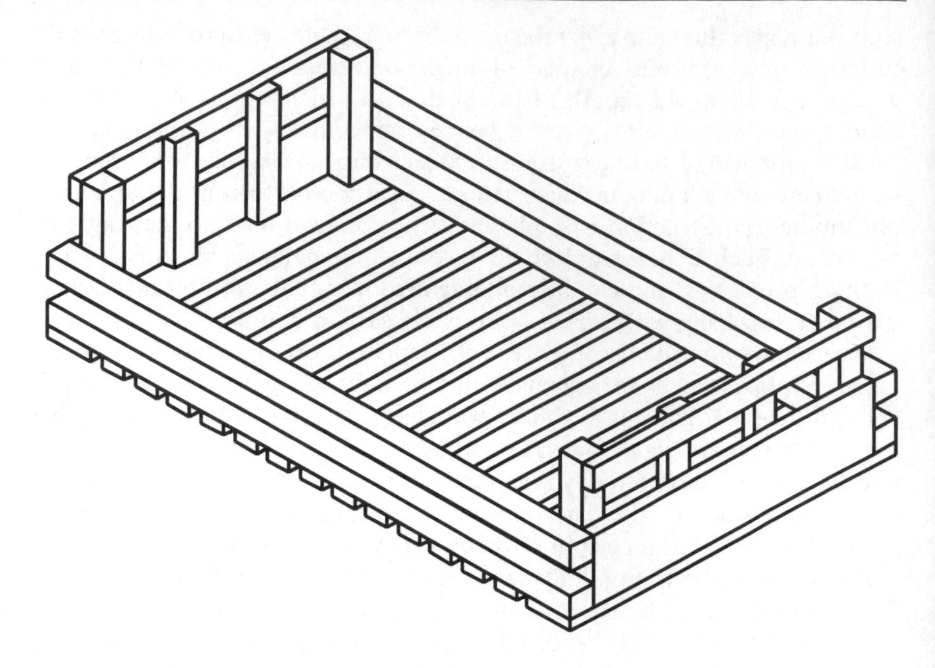

Fig. 10.55 Wooden potato tray

storage space). Many years ago the price of wood and labour rose dramatically; the heavy wooden trays were therefore attracting not only high initial costs, but also excessive maintenance costs. The question then arose: 'Can you design a low-cost plastics tray to replace the wooden one?'

This case study involves the use of many of the principles described in this book, although only an outline will be given here. Clearly key issues include identifying what the tray has to do, concept design, material choice, process and machinery choice, and economics.

(*b*) *What must the tray do?*

Market research suggested the following. The maximum outside dimensions should be 725 × 450 × 150 mm. It should stack safely 20 high and carry 20 kg of potatoes of minimum size 12 mm without excessive sag in the base. The tray should be robust in near-frost conditions. Several million trays are required, each of which should last at least 10 years in service, including storage outdoors when not in use. A target maximum selling price was identified.

(*c*) *Design approaches*

There are many starting points when trying to satisfy the design brief in section 10.5.5(b). Design is certainly not a linear process: much iteration is needed to sort out the interactions between the key issues. Where to start – the product design, the material choice or the processing issues – depends on your own background; the iterations should still lead to the same conclusion if the knowledge base is sufficient.

(*d*) *Choice of materials*

Our starting point is that the quantities required strongly suggest that injection moulding would be a suitable process, if a suitable polymer could be found which permitted an acceptable design. The low cost of the tray suggested that a commodity polymer offering a high volume/cost ratio would be a likely choice.

Of the commodity polymers, styrene-based polymers are likely to be too brittle, PVC-based polymers are too difficult to process, and low-density polyethylene is too flexible. This leaves high-density polyethylene and polypropylene as worth considering further. Non-commodity polymers offer too small a volume of material to be useful.

Although high-density polyethylene is more robust, it creeps more than polypropylene, an important factor in avoiding failure by excessive sag in the base of the tray or by buckling of the lowest tray in a complete stack of full trays. For the best compromise between robustness and creep resistance, a propylene–ethylene copolymer was chosen, and a small amount of (white) titanium dioxide was added to confer good resistance to outdoor exposure.

Experience of injection moulding suggested that it would be realistic to aim for a 45 s cycle, which suggested a wall thickness of 2.5 mm. The target selling price suggested that the upper limit on the mass of polypropylene would be about 1.35 kg.

(*e*) *Concepts for structural design*

An exact copy of the wooden tray (Fig. 10.56) would be unacceptable because it would be far too heavy and would be too expensive to mould in

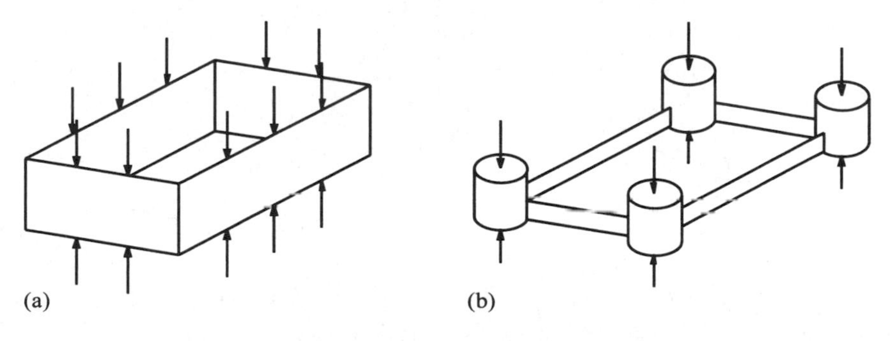

Fig. 10.56 Loaded side walls (a) versus loaded pillars (b)

such thick sections. It would be far better to exploit the advantages of plastics, and go for a one-piece moulding of uniform wall thickness, with a major design criterion being maximum stiffness per unit mass.

Two basic concepts were proposed. The first (Fig. 10.56a) carries the vertical loading uniformly round the side walls, which therefore completely enclose the potatoes; the potatoes are supported on a base. The second (Fig. 10.56b) uses corner pillars to carry the stacking loads. The shallower side walls locate the corner pillars and support the base. Both concepts can be worked out in detail; experience shows that both will lead to a satisfactory product. In this account we look only at the second concept.

(*f*) *Stiffness considerations*

We look at the three major elements separately: the corners, the side walls and the base.

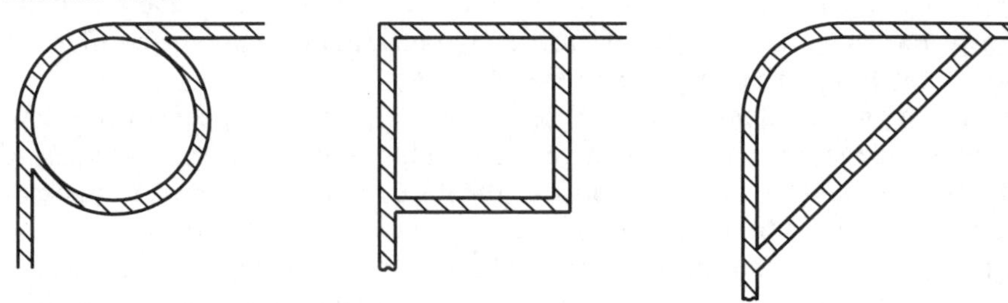

Fig. 10.57 Concepts in corner pillar shape

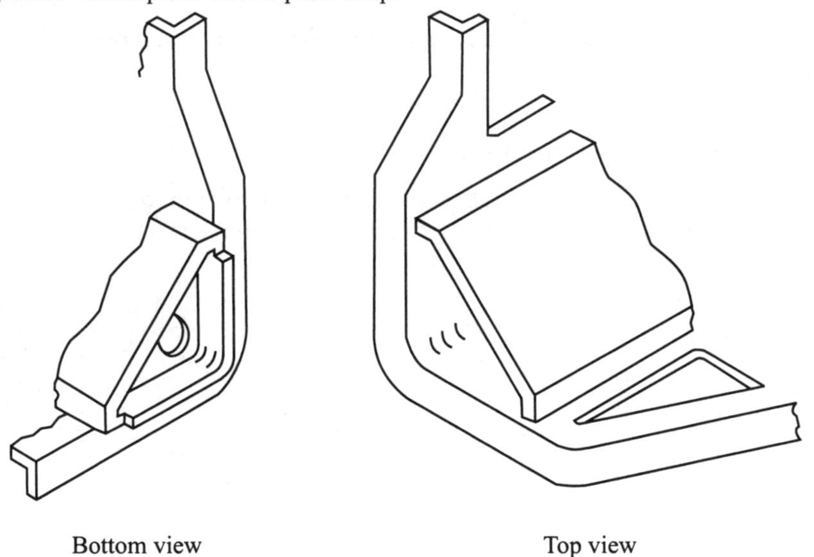

Fig. 10.58 Interlocking detail on corners: (a) bottom view, (b) top view

The corners must support the load of 19 full trays without buckling. Three concepts are shown in Fig. 10.57: a cylindrical pillar would be structurally efficient, the sharp edge of a square pillar would damage potatoes, and a triangular pillar offers a useful compromise. To ease extraction from the mould core, the corner design is slightly tapered in the line of movement of the tool. An interlocking feature was also incorporated at top and bottom of the corner, to ensure safe stacking (Fig. 10.58), and a drainage hole was provided. The flanges used to transmit vertical loads between adjacent corners dictate the use of side cores: a necessary and expensive complication in the tool construction.

The purpose of the side walls is to tie the corner pillars together and carry the bending moments and twisting forces exerted by the base. The volume of potatoes suggested that the side wall need be only half the depth of the tray, and this would have the added advantage of letting light in (which stabilizes root growth). A single-plane vertical web is easy to mould, but prone to buckling at the upper edge as the result of compression from the bending. Flanges were introduced to improve the buckling stability: the channel section can be demoulded using the side cores already needed for the corner details (Fig. 10.59).

The base must support the corners without sagging too much. A limit of 10 mm maximum sag seemed reasonable. A uniform 2.5 mm thick base is far too flexible, and provides no drainage. To stiffen the base, the principle is to

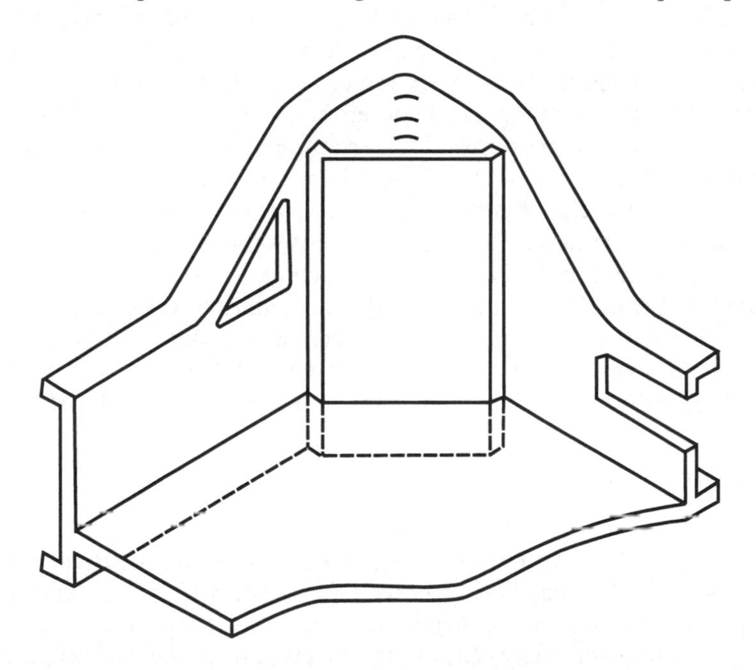

Fig. 10.59 Channel sections on side walls; inadequate flat base

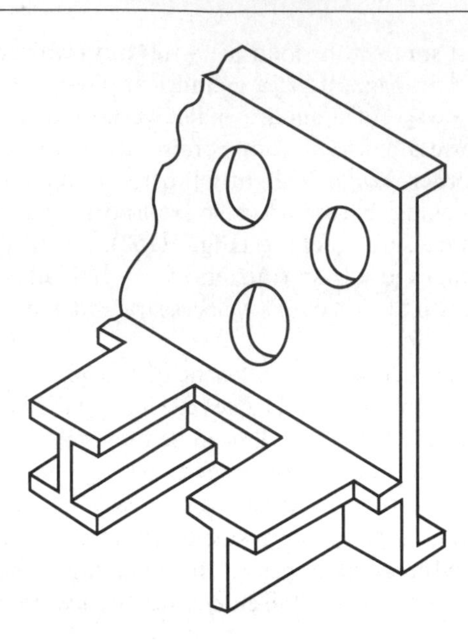

Fig. 10.60 I- and T-beams in base; holes in side walls above and below neutral axis

increase the effective thickness, using beams between the short sides (Fig. 10.60). I-beams are the best structural choice but cannot be demoulded at reasonable cost. T-beams can be readily demoulded and are still reasonably efficient. To dimension the cross-section of the 2.5 mm thick T-beams, it was assumed that the minimum gap between beams was 12.5 mm, and that all beams were simply-supported and under a uniformly distributed load.

Estimates of the mass of the proposed design were disappointing: it was still too heavy. Material was saved in three ways. Holes were provided in the side walls near the neutral axis. Extra windows were provided in the end walls on either side of the handle. The end conditions for the T-beams were reviewed: at the middle of the side walls the condition was certainly simply-supported, but near the ends the rigid corners would make for (nearly) built-in ends. The depths of T-beams at each end of the tray were therefore reduced. Some of these changes are shown in Figs 10.60 and 10.61.

(*g*) *Prototype*

At this stage a prototype was made by welding strips cut from 2.5 mm natural propylene sheet. The purpose of making the prototype was to provide a model for assessing market reactions and check the stiffness of the base. The prototype would not be tested to destruction to assess its robustness under impact loading, because for many reasons the behaviour of the prototype would probably be quite different from that of the moulded tray.

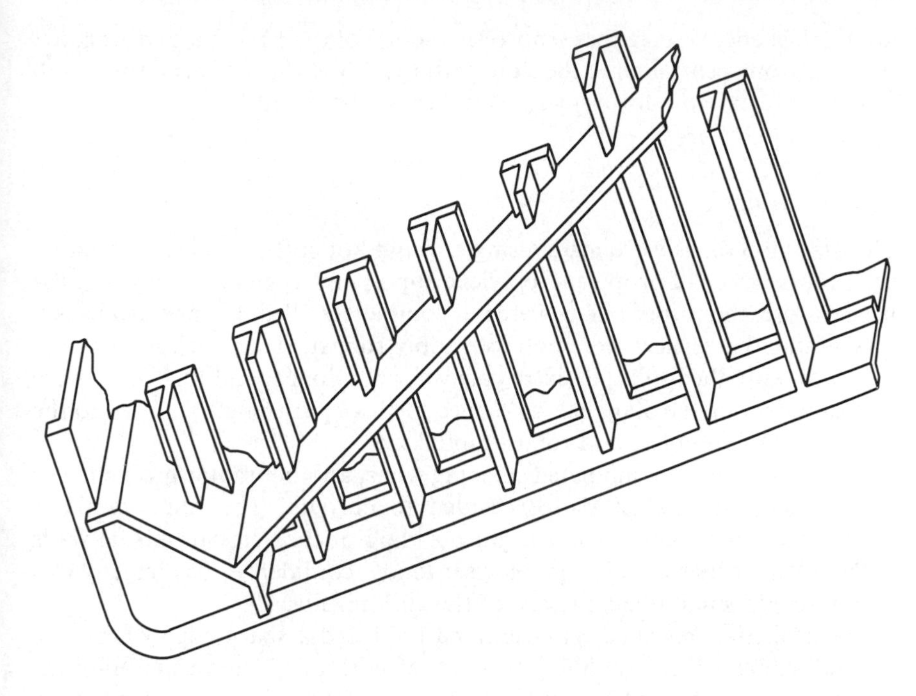

Fig. 10.61 The diagonal brace and beams of different depth

The bending stiffness of the base of the prototype was satisfactory, and market reaction was extremely favourable. The main adverse comment was that the tray distorted excessively under (improbable) shear loads in the plane of the base. This shear loading had not been expected, and was taken into account by adding diagonal braces – this added extra mass, but improved feed of melt to the corners of the moulded tray; this called for further creativity in removing mass from side walls and end walls, and adjusting the depth of the T-beams.

(*h*) *Injection machine and mould*

The company already had available an injection-moulding machine with almost ideal characteristics: just sufficient clamping force (350 tonnes), plasticizing capacity, shot volume and platen area. There was the added advantage that the accountant had already written off the cost of the machine, so its capital cost was advantageous for this exercise. At the time, software for melt flow simulation was not available, and the experienced technologist was aware that there was little room for error.

The mould consisted of six parts defining the cavity shape and dimensions, with conventional arrangements for cooling channels and for ejector pins. Ejector pins would act on the lower end of the web of the T-beams, which were locally thickened to take the compressive action of the pin without distortion. Assuming one central gate, the flow path ratio (the ratio of maximum path from gate to extreme corner) was estimated at about 230:1.

Trial mouldings

The first mouldings were made using a tough but stiff-flow (i.e. high molecular mass) grade of propylene–ethylene copolymer; even at the highest melt temperature, the corner pillars were not completely filled. Even with an easy-flow grade, the corners were not acceptably robust. It was therefore necessary to modify the tool: the centre gate was abandoned, and four gates were introduced along the diagonals to reduce the flow path length; this led to the production of complete and robust mouldings.

Towards the end of simulated stacking trials, another minor problem was noted: the lower corner flanges tended to curl up. The result was not dangerous, but it caused some slight buckling of the vertical outside walls of the pillars, which could cause the user to lose confidence. The remedy was simply to add small fillets to prevent the curling (Fig. 10.62).

Practical trials revealed that the hand-hold had a sharp edge which was uncomfortable when handling a tray full of potatoes. This was an inherent feature of the tooling which could not be put right by simple adjustments to or redesign of the tooling. The practical remedy was for the operator to trim away the sharp edge while removing the four sprue gates.

A sketch of the complete tray is given in Fig. 10.63. This end result is a successful product which has met its commercial targets. More significantly, it opened up new markets for a range of (now well-established) lightweight containers.

The following extracts from the advertising brochures for this product provide a marketing perspective to this account:

- One-piece moulding: no repairs, no nails (rusting), no joints or cracks.

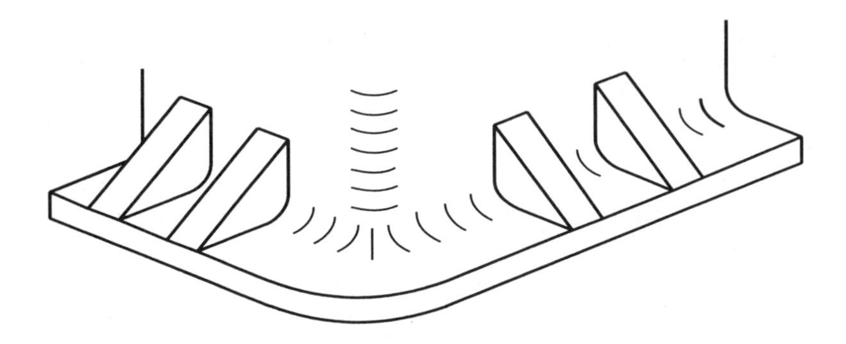

Fig. 10.62 Fillets at base of corner

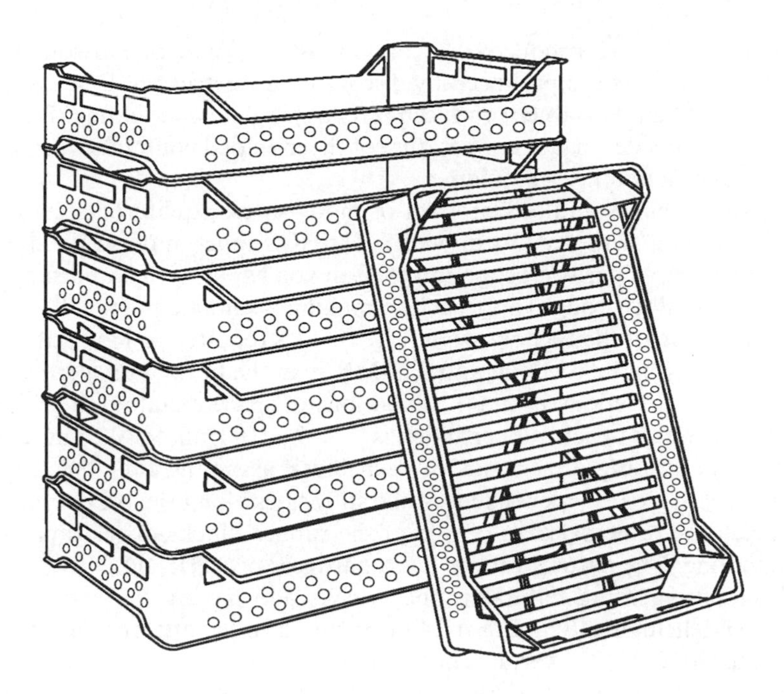

Fig. 10.63 The proposed tray

- Made from polypropylene: long-life, sterilizable – no disease carry-over, virtually unbreakable, rot-proof, non-absorbent.
- Self-coloured white: maximum reflectivity, no painting, white all through.
- Custom design: for greater process control, improved air circulation, better light penetration, interlocking corner feature, thin-walled for space saving.
- Easier and safer handling: lightweight even when wet, smooth surfaces – easy to clean, no splinters.
- Load-bearing: 20 kg load, stackable 20 high when loaded.

If the product were to be redesigned today, the uncertainties of moulding would have been anticipated by the routine use of melt flow simulation software, perhaps even suggesting the use of thinner walls, and some saving in mass would also result from the use of numerical methods such as the finite element method to support the structural calculations.

PROBLEMS

1. Most bottles are produced by continuous extrusion of parison(s), but large containers are produced by discontinuous extrusion. Discuss why.
2. A 200 l container is to be produced by extrusion blow moulding. Describe how the outside surface temperature of the moulded container varies with time during and after moulding.
3. Making a blow-moulded product of more-or-less square or rectangular cross-section with a uniform wall thickness is much more difficult than using a simple parison programmer. Can you explain why, and suggest a practical (albeit rough) approach to making a suitable parison?
4. A mounting of the type shown in Fig. 10.49 is required to have stiffnesses of 3.66 MN/m, 3.66 MN/m and 0.24 MN/m in the horizontal, vertical and transverse directions (x, z, y). Limitations of space dictate that the load-bearing area of each rubber block is $L \times B = 100\,\text{mm} \times 100\,\text{mm}$. Determine the leading dimensions of each block assuming that the stiffness along each axis is independent of the stiffnesses along the other two axes. Sketch the principal axes of one of the rubber blocks in relation to the mounting. The natural rubber has a hardness of 38 IRHD.
5. A mounting using natural rubber blocks and having the arrangement shown schematically in Fig. 10.49 is required to have stiffnesses of 2.0, 6 and 0.4 MN/m in the vertical, horizontal and transverse directions at 20°C. The load-bearing face of each rubber block is to be 100 mm in diameter. You may assume that the stiffness along each axis is independent of the stiffnesses along the other two axes. Recommend a design for the mounting units, and the hardness of the rubber compound to be used, and indicate the angle that the rubber units make to the horizontal in your design.
6. In section 10.5.3(a) explain why the elastic properties are based on $(\pm 45°)_s$ and not on $(+45°)_T$ or $(+45°/-45°)_T$.
7. Show why in section 10.5.3(b) the Tsai–Hill failure criterion indicates a safety factor of 5.80 for the +45° plies, which is different from that (2.79) for the −45° plies while the numbers of the relative stress values are apparently equal.

10.6 COMPUTER-AIDED ENGINEERING WITH POLYMERS

The design process involves the use of computer programs more and more frequently. Often, finite element simulations are used before a prototype is made, in order to make more reliable estimates of product behaviour and production performance. The goal is to make the product 'right the first time', thus reducing development time and costs.

For polymers, major applications of the finite element method are found in simulations of structural mechanics and the injection moulding process. Both will be discussed briefly after the basics have been elucidated.

10.6.1 Finite element method

Many excellent textbooks can be found on the finite element (FE) theory (e.g. Zienkiewicz and Taylor, 1991). Here only quasi-static formulations will be briefly reviewed. The method is based on subdividing a continuum of interest (i.e. the product being designed) into a number of discrete elements. In each element the displacements of the nodes (e.g. the corner points) are related to internal strains and stresses.

In a two-dimensional linear triangular element the six nodal displacement components lead to a uniform strain in the element (Fig. 10.64). By considering the constitutive behaviour of the material under consideration, the stress in this element can be calculated. Usually this is done for discrete points in the element, for the triangle considered here at its centre of gravity. The stresses must be in equilibrium with the forces applied at the nodal points. These reaction forces are found by integrating the stresses over the volume of the element. Each nodal point has force contributions from all elements in which it is contained. A node is in equilibrium when the sum of the reaction forces is equal to the applied boundary condition. Hence, for a 'free' node the total reaction force must be zero. A loading situation has been solved when all equilibrium equations are satisfied. In the most simple case, where stresses are related linearly to the nodal displacements (i.e. a linear stress–strain relationship and a linear strain–displacement relationship) the total system of equilibrium equations can be solved directly, while in the case of non-linearities an iterative procedure must be followed. Plastics are often in the second category.

10.6.2 Structural mechanics

Most products are designed to carry a load during some time span. The initial stiffness upon applying the load can in general be predicted by a linear FE simulation, using a linear elastic material model with appropriate stiffness constants. The resulting stresses can be compared to a failure criterion such as yield stress or fracture stress.

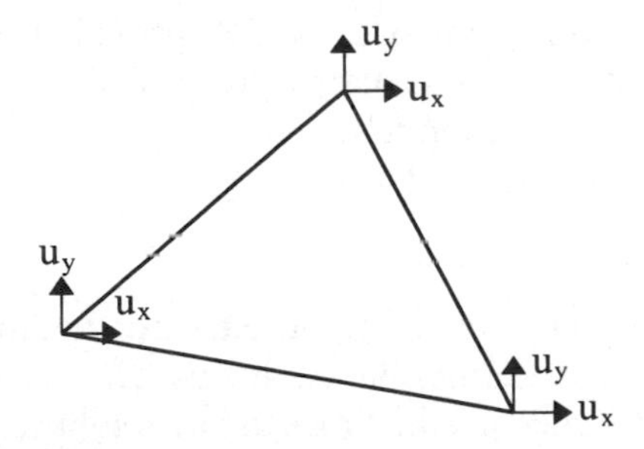

Fig. 10.64 Linear triangular element with two degrees of freedom in each node

However, plastics very often show time-dependent behaviour such as stress relaxation and creep. These processes accelerate when the temperature increases. The materials behaviour must be described properly to give a sensible prediction of the product behaviour for its total estimated lifetime. This implies that the available FE package must incorporate appropriate (non-linear) constitutive models and that sufficient experimental data must be available to fit the material property data in the simulation. Both prerequisites are by no means trivial. The central issue here is that no generally applicable constitutive equation exists for different plastics under various load conditions, for which the materials property data can be obtained by a few simple tests. Of course foams, rubbers, glassy polymers and fibre-reinforced plastics behave differently. However, for a particular material the non-linear behaviour under a strain rate of 0.01 s may be described reasonably by an elastoplastic model, whereas the creep behaviour under static loading needs a different description.

Elastoplastic models, designed for metal plasticity, are available in many FE packages. They may incorporate strain-rate dependency, which is characteristic for polymer behaviour. Whether the behaviour of a plastic is described adequately by an available constitutive model can only be answered by performing experiments on the specific material under different strain rates and observing how well the model and experiment fit together. To date, the material suppliers have not usually been able to provide this information.

The same story holds, more or less, for viscoelastic models for solids. The difference here is that, if available, every FE package uses a different model and each of these needs different constants. Again, tests must be conducted under different strain rates to provide the information to determine the required material property data. Linear dynamic measurements may or may not be sufficient, depending on the constitutive model and the amount of strain: linear viscoelastic theory cannot be expected to give accurate predictions above 1% strain.

For rubbers the constitutive models are reasonably uniform (classical rubber elasticity or Mooney–Rivlin theory) and materials data can be determined by tensile tests. Finally, fibre-reinforced materials can be described accurately by an orthotropic elastic model, provided all elastic constants have been determined. Note that matrix creep, which may be significant, is usually not included in these models.

10.6.3 Injection moulding simulations

The second major application of finite element methods in plastics is in injection moulding simulations. The goal is usually to shorten the development time of a new product or sometimes to find a solution for a problem with a new mould. A number of injection moulding simulation programs are available. Basically they can simulate the filling stage, the holding and cooling stage and the cycle-averaged temperature in the mould, but other additions may be available. The input parameters required are material property data, such as the viscosity as a function of shear rate and temperature, the elastic properties of the solidified melt, and a *pvT* diagram. The process parameters, such as injection and holding pressure, the temperature of the mould or the cooling medium and holding and cooling times, are the boundary conditions of the simulation. The solution of this non-linear problem can vary significantly with only a moderate change in input parameters.

The analysis starts by subdividing the geometry defined into (usually) triangular plate elements, using the fact that injection-moulded products are in general thin-walled structures (Fig. 10.65).

A thin-film approximation can be used, in which the conservations laws of mass and momentum are combined with a shear-thinning viscous model for the polymer, resulting in a formulation with only the pressure as a nodal degree of freedom. The temperature shows a large variation in the thickness direction. This can be accounted for by taking multiple points in the thickness direction in which the temperature is evaluated by means of a finite difference discretization of the heat balance. Both pressures and temperatures are approximated iteratively in a discrete number of time steps.

The position of the flow front can be predicted with a reasonable accuracy, provided that the material property data are correct. The absolute values of the pressures must be interpreted with some care. Viscoelastic phenomena in the polymer melt are not taken into account in this type of simulation. First, elongational flow occurring in the gates may well lead to a higher actual pressure drop. Second, the steady-state viscosities may not be reached during filling due to the transient nature of the process. Thus pressures may be overpredicted as well as underpredicted.

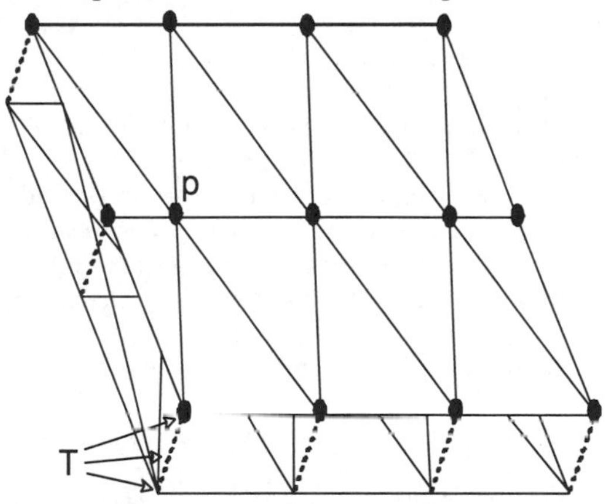

Fig. 10.65 Plate elements for injection moulding simulations, with pressure degrees of freedom in the nodal points and several temperature degrees of freedom in thickness direction

If the pressure is predicted well, the solidification during the packing and cooling stage can also be described with reasonable accuracy. Note that cooling is very poor when the polymer material is no longer pressed on to the mould, i.e. when the pressure has dropped to the ambient value. The calculated shrinkage, residual stresses and warpage of the product depend on the solidification process and hence depend heavily on the accuracy of the pressure prediction.

Commercial FE programs for injection moulding simulations are under constant development. Extensions predicting fibre distributions (and resulting residual stresses, shrinkage and warpage) due to injection moulding are already available, as well as models for multicomponent injection moulding, gas-assisted injection moulding and injection of thermosets. The design engineer can expect a steadily growing number of tools for product optimization before the first prototype has been made. It remains to the designer to judge the validity of the predictions of these tools.

10.6.4 Conclusion

Although many difficulties have been sketched in the foregoing section, finite element simulations can give a useful insight in product behaviour, provided the user knows their possibilities and restrictions. Moreover, this insight can be attained in a relatively fast and cheap way. Certainly when the design is pushed to the limits of performance with minimal costs, simulation can be a very effective technique.

FURTHER READING

Adams, R.D., Comyn, J. and Wake, W.C. (1997) *Structural Adhesive Joints in Engineering*, Chapman & Hall, London.

Agassant, J.-F., Avenas, P., Sergent, J.-Ph. and Carreau, P.J. (1991) *Polymer Processing: Principles and Modelling*, Hanser, Munich.

Beck, R.D. (1980) *Plastic Product Design*, Van Nostrand Reinhold, New York.

Cracknell, P.S. and Dyson, R.W. (1993) *Handbook of Thermoplastics Injection Mould Design*, Blackie Academic & Professional / Chapman & Hall, London.

DuBois, J.H. and Pribble, W.J. (1987) *Plastics Mold Engineering Handbook*, Van Nostrand Reinhold, New York.

Eckold, G. (1994) *Design and Manufacture of Composite Structures*, Woodhead, Cambridge.

Ehrenstein, G.A. and Erhard, G. (1984) *Designing with Plastics*, Carl Hanser Verlag, Munich.

Eijpe, M.P.I.M. (1997) *A Modified Layer Removal Method for Determination of Residual Stresses in Polymeric Composites*, University of Twente, Enschede.

Frados, J. (ed.) (1976) *Plastics Engineering Handbook*, Van Nostrand Reinhold, New York.

Freakly, P.K. and Payne, A.R. (1978) *Theory and Practice of Engineering with Rubber*, Applied Science, London.

Gibson, L.J. and Ashby, M.F. (1988) *Cellular Solids, Structure & Properties*, Pergamon Press, Oxford.

Göbel, E.F. (1974) *Rubber Springs Design*, Newnes Butterworth, London.

Kinloch, A.J. (1987) *Adhesion and Adhesives: Science and Technology*, Chapman & Hall, London.

Levy, S. and DuBois, J.H. (1977), *Plastics Product Design Engineering Handbook*, Van Nostrand Reinhold, New York.

Malloy, R.A. (1994) *Plastic Part Design for Injection Molding, An Introduction*, Hanser, Munich.

McCrum, N.G., Buckley, C.P. and Bucknall, C.B. (1988) *Principles of Polymer Engineering*, Oxford University Press, Oxford.

Menges, G. and Mohren, P. (1986) *How to Make Injection Molds*, Hanser, Munich.

Muccio, E.A. (1993) *Plastic Part Technology for Assembly*, ASM International, Materials Park, Ohio.

Powell, P.C. (1994) *Engineering with Fibre/Polymer Laminates*, Chapman & Hall, London.

Tres, P.A. (1994) *Designing Plastic Parts*, Hanser, Munich.

Zienkiewicz, O.C. and Taylor, R.L. (1991) *The Finite Element Method*, McGraw-Hill, New York.

A number of suppliers of plastics and rubber raw materials have excellent brochures on engineering and design available, providing basic information and greatly illustrated with many examples. For instance:

BASF
Bayer
Du Pont
GE Plastics
Hoechst
ICI
Malaysian Rubber Producers' Research Association

Outline Answers

CHAPTER 2

1. If the backbone chain is stiff (e.g. polycarbonate), then the amorphous polymer is inherently stiff, especially if below T_g. A flexible chain in a partially crystalline polymer (e.g. polyethylene) above T_g is likely to confer flexibility even when reinforced by crystallites. Even high-density polyethylene (up to 90% crystallinity) is less stiff than polycarbonate at room temperature.
2. (a) Degree of polymerization is $10^6/68 = 14\,700$ or higher.
   (b) Degrees of polymerization are (9000 to 15 000)/113 = 80 to 133.
   (c) The degree of polymerization is 160 000/28 = 5700. Taking the distance between carbon atoms as 0.15 nm, the axial distance x in Fig. A1 is $x = 0.15 \sin 54° = 0.12$ nm, so one mer is 0.24 nm long, and the maximum length of a fully extended molecule is $5700 \times 0.24 = 1380$ nm. However, molecules are entangled and coiled up; crystallites are also present.
3. The lower the temperature, the less mobile the molecules (i.e. a more viscous fluid) and hence the less readily they will move towards a crystalline site.
4. Crosslinks restrict mobility, make the polymer more glassy, and inhibit the formation of crystallites. A high crosslink density gives a totally glassy polymer with no tendency to achieve a rubbery phase before

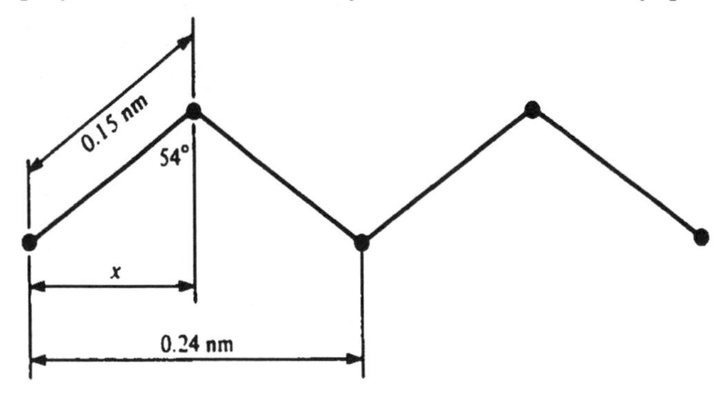

Fig. A1 Model of zig-zag backbone for a 'straight' chain

decomposition sets in. A low crosslink density inhibits mobility to a lesser extent, and some slim polymers (particularly natural rubber) will actually crystallize under high tensile stress, i.e. molecules become aligned after straightening up; on removing the stress the molecules coil up again and the crystallinity disappears.

5. Higher molecular weight promotes greater entanglement which confers higher strength in an isotropic polymer mass. However, higher entanglement gives a higher melt viscosity, so the melt is more difficult to process and it is likely that the shaped polymer will be oriented in the directions of principal stress arising during shaping – it will be stronger along the direction of orientation but weaker across it because the Van der Waals' forces between chains are much smaller than the covalent forces holding the backbone atoms of the chain together. In commercial practice a balance has to be struck between strength and processability.
6. Polymer scientists measure T_g under (near) equilibrium conditions. Factors which can change the temperature at which a polymer undergoes a ductile–brittle transition include rate of testing (raises transition temperature), orientation (increases brittleness), stiff additives (impair molecular mobility) and plasticizing additives (promote mobility).
7. Enthalpies are higher for partially crystalline materials because of the latent heat of fusion. Other information would include melting points and solidification temperatures, the temperature at which the solid polymer became form-stable, and the thermal diffusivity in order to calculate cooling times.

CHAPTER 3

1. (a) Gear wheels: low mass provides low inertia, low friction even when unlubricated, 'good fatigue', readily moulded in one piece with repeatable precision, quiet in operation, tough enough to resist high start-up shock torques.
   (b) Telephone handsets: light weight, readily moulded (but look at the handheld microphone/earpiece case and try to envisage the nature of the tooling in which it is made), resists abrasion from atmospheric dust, range of fast colours, good surface finish, toughness.
   (c) Rear lamp housings: resistance to oil, petrol, water, salt and UV, fast transparent colours, intricate design moulded to very close tolerances (especially the Fresnel lenses) with three colours (natural, red and orange) in one finished moulding.
   (d) Distributors: moulded-in inserts, dimensional stability over a wide temperature range, good electric arc tracking resistance.
2. (a) O-ring seals: chemical resistance, high recoverable elastic compression, acceptable stiffness, low stress relaxation, readily moulded in large quantities.

(b) Sealing strip: resistance to screen wash, antifreeze, UV, water, oil and grease, flexibility, low stress relaxation, readily extrudable.
(c) and (d) Fenders and buffers: resistance to large shock loads, absorption of large amounts of energy per unit volume, resistance to sea water and oils.

3. The main problem would be to conduct the heat away from the rubbing surface through the plastic. Use as thin a plastic layer as possible.
4. Treat the pipe as a strut: Euler buckling theory works if the slenderness ratio L/k is 100 or larger. Distortion occurs when

$$\varepsilon = \alpha \Delta T = \sigma_c / E = 4\pi^2/(L/k)^2$$

i.e. $\Delta T \simeq 40°C$. (A more accurate analysis would take into account the dependence of modulus on temperature, e.g. of the form $E = E_0(1 - k\Delta T)$.)

CHAPTER 4

1. The main attraction is high stiffness per unit weight and ease of manufacture with a good surface finish, achieved with glass fibre–crosslinkable plastic resin systems.

   Idealizing the aerofoil section, you will recognize that the wing has to resist bending loads (weight of wing before take-off, and uplift in flight) and twisting loads. The bending loads dictate fibres aligned along the length of the wing for high stiffness – but be careful to avoid buckling failure from the compression duties. Good twisting resistance requires the fibres to be cross-plied at 45° to the length of the wing.

2.

	Compression moulding	Extrusion blow moulding
Feed	Solid, takes time to heat up	Already molten
Shaping	Slow spreading, inhomogeneous; all dimensions precisely fixed	Rapid deformation, homogeneous melt; only the mould surface defined and wall thickness depends on blow-up ratio
Stabilizing	Crosslinking by conduction from mould surfaces	Cooling from one surface only
Moulds	Ejector pin marks and flash	Pinch-off pad witness marks

3. The compression moulding would show some line where the flash had been trimmed away – seen as a narrow line of lower gloss perhaps. The injection moulding would show a gate scar where the runner had been severed. Both mouldings would show mould parting lines and ejector pin marks.
4. The blow-moulded dustbin would be heavier because of its uneven wall thickness – you would have to design for the thinnest section, and elsewhere

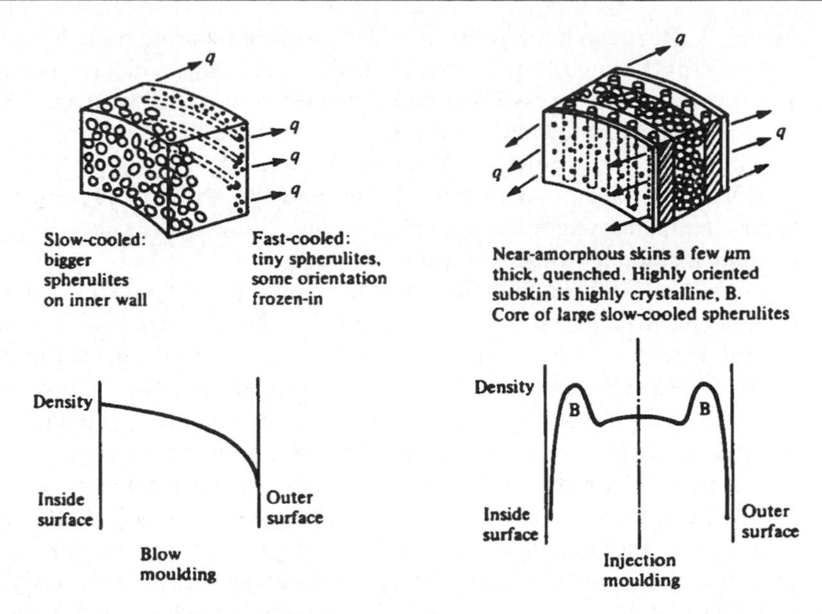

Fig. A2 Orientation and morphology in side walls of blown and injection-moulded dustbins

the wall would be thicker in the blown product but not in the injected dustbin. Neglecting taper angles, the orientation in the blown bin will be circumferential caused by inflation (except near the base), but axial in the injected bin due to pressure flow and high shear rate at the mould wall. Further comments are made in Fig. A2.

In the injected bin it would be best to aim for an easy-flow melt which could be rapidly pumped into the cavity, i.e. a low molecular weight. The parison for blow moulding must not sag too much under its own weight, so aim for a higher molecular weight polymer – it is also extruded at lower shear rates; hence lower pumping power is needed compared with that for the injected bin.

5. Moulds are heavy and have a high heat content, so heating and cooling moulds for a thermoplastic would be expensive and slow, and would be an inefficient use of energy.
6. There are some common modifications abbreviated as T = add axial taper angle; R = round corners. (a) Extrude, no modifications needed. (b) Injection or compression mould; T, R. (c) Extrusion blow mould; R, make wall thickness as uniform as process allows (assuming bore is critical). (d) Same as (b). (e) Injection mould; T from each end; if length of block is unnecessary, close with web of same thickness as cylinder; if length is important, core out with longitudinal ribs from each end, R. (f) Extrusion blow mould; R, make wall thickness as uniform as process allows, if outside diameter is critical, mould it with rounded corners. (g) Injection

mould; T, R, radial holes to be moulded-in using separate movable pins (thus a 4-part mould). (h) Extrusion blow mould; R, no taper needed.

7. The polymer would crosslink in the barrel if overheating occurred. This would stop production and the machine would have to be stripped down to remove the scrap polymer. The remedy would be to provide cooling channels and coolant to cool the barrel and the screw to prevent too great a temperature caused by shear heating.
8. Extrude a suitable rubber compound.
9. The outer rim is thicker than the rest of the moulding: it would take longer to cool and would crystallize more and hence shrink more. (a) The thin central panel would be compressed and would distort or buckle out-of-plane. (b) Polyethylene is above its T_g at room temperature, so it would continue to crystallize, albeit slowly, hence a little more shrinkage and distortion. (c) The design would be improved by using a uniform wall thickness, and by radiusing internal corners to improve robustness generally. (Later we shall see that a different gating system would further reduce the tendency to buckle even in a lid having a uniform wall thickness.)
10. The thick web would also cause the thin outside wall to distort – remedy as in 9. During filling, flow will occur preferentially in thick sections leading to a weld-line X and an area of weakness, as shown in Fig. A3, through which no fibres pass. This may also lead to venting problems at Y in Fig. A4, where air cannot escape as melt is pumped into the cavity.

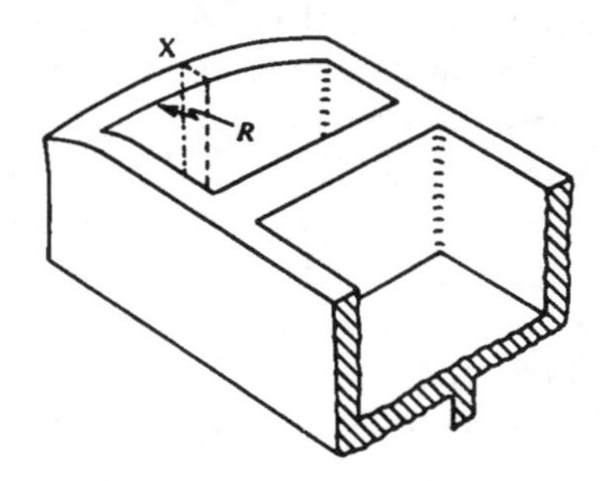

Fig. A3 Thick web causes thinner web to distort

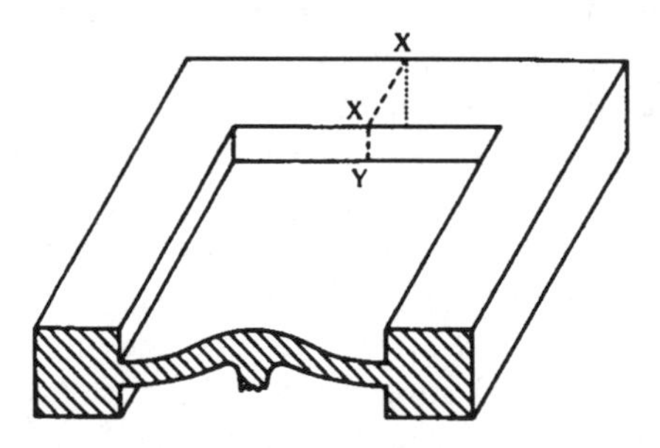

Fig. A4 Distortion in web caused by greater shrinkage in thick rim

Fig. A5 Revised bottle design with rounded corners and conical shoulder (a small degree of curvature in the plan view would also be advantageous)

11. Radius all internal corners to improve robustness and prevent excessively thin corners and edges; the bigger the radius, the better. A concave base will promote improved stability when the bottle stands upright. See comments in Fig. A5.
12. Unidirectionally aligned fibres can obviously pack closer together than cross-plied or random mat or woven cloth fabrics. In hand lay-up there is little pressure to squeeze out excess resin. See Fig. A6.
13. The tube wall emerges fastest from the wide gap in the die. The tube therefore curls up like a banana, or even like a coil spring in extreme cases. If the bottle could be blown at all, it would have an uneven wall thickness round the circumference. See Fig. A7.
14. Avoid deadspots where the polymer melt is stagnant for long periods – the melt is hot and will degrade. Eventually blobs of degraded polymer become dislodged and act as hard lumps of stress-concentrating matter, and the resultant product will be more brittle.
15. One approach is to use counter-rotating dies with slots, as shown in Fig. A8. A helical fibre pattern emerges, and where the slot apertures momentarily coincide a big blob is extruded to hold the helices of strands together.

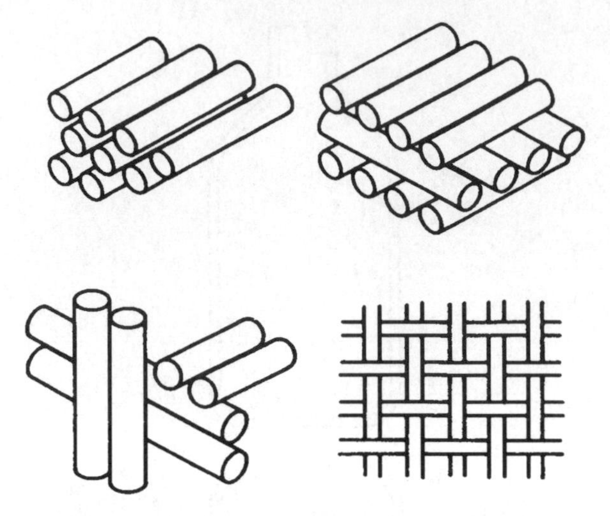

Fig. A6 Packing arrays of fibres

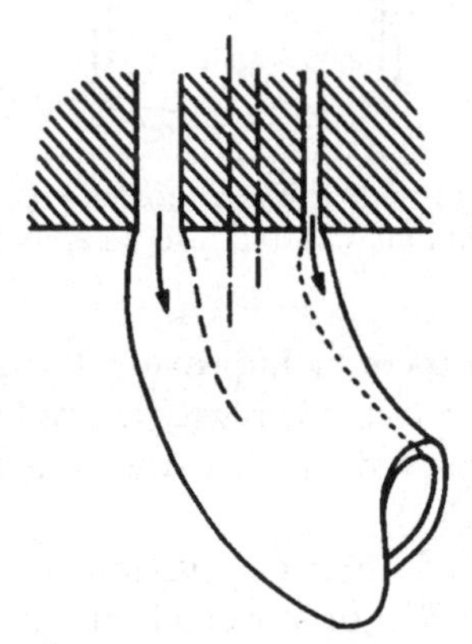

Fig. A7 Curved extrudate caused by off-centre mandrel in die

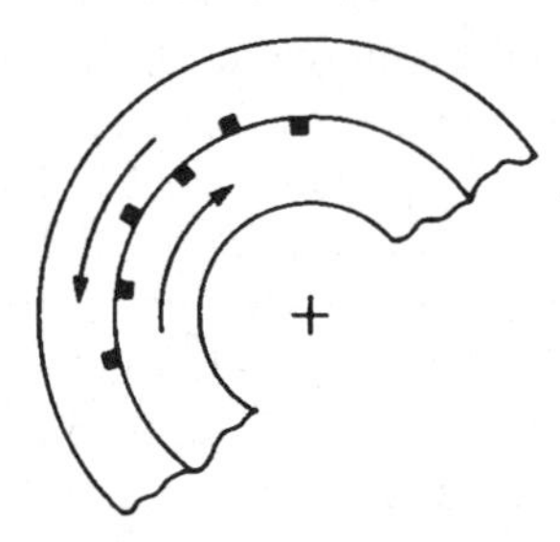

Fig. A8 Slots in counter-rotating mandrels can be used to extrude a net

CHAPTER 5

Section 5.1

1. (a) $\varepsilon = 0.1\%, E = 2.45/10^{-3} = 2.45\,\text{GN/m}^2$.
   (b) $\sigma = 23.5\,\text{MN/m}^2, E = 2.35\,\text{GN/m}^2$.
   (c) $\sigma = 13.75\,\text{MN/m}^2$ gives $E = 1.375\,\text{GN/m}^2$.
   (d) 20°C, 1 year, 0.5%: $\sigma = 4.5\,\text{MN/m}^2$ gives $E = 0.9\,\text{GN/m}^2$.
   60°C, 1 year, 0.5%: $\sigma = 3.65\,\text{MN/m}^2$ gives $E = 0.73\,\text{GN/m}^2$.
   Linear interpolation, at 40°C, $E = 0.815\,\text{GN/m}^2$ (Fig. 5.9 supports linear interpolation over this temperature range).
   (e) 2 weeks = 1.2 Ms, 20°C. Construct an isochronous stress–strain curve for 2 weeks, 20°C: 4.85, 1%; 8.2, 2%; 9.35, 3%. So $4\,\text{MN/m}^2$ corresponds to 0.8%, i.e. $E = 0.5\,\text{GN/m}^2$.
   (f) From graph for (e), 20°C, 1.2 Ms, $3\,\text{MN/m}^2 \rightarrow 0.6\%$, $E = 0.5\,\text{GN/m}^2$. 60°C, 1.2 Ms data give (σ, ε): 2.45, 1%; 4.3, 2%; 5.85, 3%. $3\,\text{MN/m}^2$ gives 1.25%, i.e. $E = 0.24\,\text{GN/m}^2$.

   Here the temperature dependence is not quite linear (Fig. 5.11). With no other information to hand, assume the temperature dependence of modulus at 1.2 Ms is the same as that at 100 s in Fig. 5.11. For the 100 s data, $E_{20} \doteq 1.34, E_{60} = 0.7, E_{50} = 0.82$; hence

$$E_{50} - E_{20} - \left(\frac{E_{20} - E_{50}}{E_{20} - E_{60}}\right)(E_{20} - E_{60}) = E_{20} - 0.8125(E_{20} - E_{60})$$

   Use the same scaling factor 0.8125 for the 1.2 Ms data:

$$E'_{50} = E'_{20} - k(E'_{20} - E'_{60}) = 0.5 - 0.8125(0.5 - 0.24) = 0.29\,\text{GN/m}^2$$

   where the prime denotes the 1.2 Ms values of modulus.

2. Treat as a thin-walled cylinder. Mean diameter $D_m = 53$ mm. Hoop stress is

$$\sigma_H = pD/2h = 1 \times 0.053/2 \times 0.003 = 8.83\,\text{MN/m}^2$$

   Longitudinal stress is

$$\sigma_L = 4.41\,\text{MN/m}^2$$

   Creep data for uniaxial strains after 1 year, 20°C from Fig. 5.12. $8.83\,\text{MN/m}^2 \rightarrow 0.58\% = \varepsilon_H$; $\varepsilon_L = 0.29\%$. No data given for ν but, from Table 2.1, cast acrylic is well below its T_g, hence $\nu \simeq 0.33$.

   The actual hoop strain ε'_H under biaxial stress from Hooke's law:

$$\varepsilon'_H = \varepsilon_H - \nu\varepsilon_L = 0.58 - 0.33 \times 0.29 = 0.48\%$$

   After 1 year, 20°C, new outside diameter is $56 \times 1.0048 = 56.27$ mm.

3. Assume the beam is weightless.

   Formal approach: $\hat{M} = wL^2/8 \rightarrow \sigma_B = 3wL^2/4bh^2$, hence from Fig. 5.12,

$$\varepsilon(20°\text{C}, 3 \text{ years}, \hat{\sigma}_B) \rightarrow E_1(20°\text{C}, 3 \text{ years}, \hat{\sigma}_B).$$

For the same thickness calculate $E_2 = 5wL^4/384I\delta$. The thickness is satisfactory if $E_1 = E_2$:

$$h = 0.02\,\text{m}: \quad \sigma = 3 \times 11 \times 9.81 \times 0.6^2/4 \times 0.12 \times 0.02^2 = 0.6\,\text{MN/m}^2$$

The problem now is to estimate ε from the data: the lowest curve for which data are available is $\sigma = 3.5\,\text{MN/m}^2$. You would have to assume that the acrylic shows linear σ versus ε at smaller stresses (you can see this by drawing a 3 year isochronous stress–strain curve). At $\sigma = 3.5\,\text{MN/m}^2$, $\varepsilon \simeq 0.225\%$, so $E_1 = 1.55\,\text{GN/m}^2$.

$$E_2 = 11 \times 9.81 \times 0.6^4/32 \times 0.12 \times 0.02^3 \times 0.003 = 1.52\,\text{GN/m}^2$$

Quick approach: Assume linear stress–strain curve at 20°C, 3 years and work with a convenient stress or strain. $14\,\text{MN/m}^2$ gives 1.1%, i.e. $E_1 = 1.27\,\text{GN/m}^2$.

This is not very accurate for this polymer as the isochronous curve shows.

Moral: it is best to work as formally as possible with the given data.

4. The maximum bending stress is $\sigma = 6PL/bh^2 = 9\,\text{MN/m}^2$.
   (a) 1%, 10 s, 20°C, Fig. 5.7: $\sigma = 25.5\,\text{MN/m}^2$, i.e. $E_1 = 2.55\,\text{GN/m}^2$. 10 s, 20°C, $9\,\text{MN/m}^2$, Fig. 5.7: $\varepsilon = 0.33\%$. Therefore $E_2 = 9 \times 10^6/3.3 \times 10^{-3} = 2.73\,\text{GN/m}^2$.

   $$\text{Discrepancy} = (E_1 - E_2)/E_2 = -6.6\%$$

   (b) 1%, 3 years, 20°C, $\sigma = 8.2\,\text{MN/m}^2$, $E_3 = 0.82\,\text{GN/m}^2$. 3 years, 20°C, $9\,\text{MN/m}^2$, $\varepsilon = 1.12\%$. Therefore $E_4 = 0.8\,\text{GN/m}^2$.

   $$\text{Discrepancy} = (E_3 - E_4)/E_4 = +2.5\%$$

5. The given I-beam is inadequate because (1) no radius on internal corners; (2) local thick sections will crystallize more, causing bowing of the extrudate; (3) the beam is not stiff enough. The second moment of area is $4.68 \times 10^{-9}\,\text{m}^4$.

   From Fig. 5.10 at 1% strain, $E = 0.35\,\text{GN/m}^2$ so $P = 3EI\delta/L^3 = 3.07\,\text{N}$ after 1 year. To achieve a minimum thrust of 4.0 N, $I = PL^3/3E\delta = 6.09 \times 10^{-9}\,\text{m}^4$.

   For a uniform thickness of x, overall breadth B, depth H (and neglecting the essential radii of 2.5 mm on internal corners), x is found from

   $$12I = BH^3 - (B - x)(H - 2x)^3$$

   i.e. $x \simeq 2.5\,\text{mm}$.

   For short-term loading at 100 s, 2.0% strain, $E \simeq 0.76\,\text{GN/m}^2$, so the thrust would be $P = 3EI\delta/L^3 = 8.7\,\text{N}$, i.e. well within specification.

6. The second moment of area for the original beam is

$$I = bh_1^3/12 = 3 \times 10^{-3} \times (1.2 \times 10^{-2})^3/12 = 4.32 \times 10^{-10}\,\mathrm{m}^4$$

Deformation is prescribed, so the problem is one of relaxation. The creep and stress relaxation moduli are assumed to be identical. The creep modulus at 1% strain at 20°C has dropped after 1 year to about $0.9\,\mathrm{GN/m^2}$ (from Fig. 5.7). Hence the thrust drops to the minimum value:

$$P = 3EI\hat{y}_1/L^3 = 3 \times 0.9 \times 10^9 \times 4.32 \times 10^{-10} \times 2 \times 10^{-3}/0.125^3$$
$$= 1.2\,\mathrm{N}$$

Section 5.2

1. (a) Stress relaxation is the time-dependent stress resulting from an applied deformation. Recovery is the change in deformation after removing load at end of (say) creep test.
   (b) Residual strain is the strain remaining at any time after removing load at end of creep test. Recovered strain is the maximum strain at end of creep test minus residual strain.
2. The mechanics:

$$\sigma_{H1} = pD/4h = 1.03 \times 0.41/4 \times 0.01 = 10.56\,\mathrm{MN/m^2}$$
$$\sigma_{H2} = 0.7 \times 0.41/0.04 = 7.17\,\mathrm{MN/m^2}$$

Creep data for uniaxial load (Fig. 5.12) then give the required strains:

$$\varepsilon(\sigma_{H1}, 15\ \text{days}) = 0.51\%$$
$$\varepsilon(\sigma_{H1}, 5\ \text{days}) = 0.47\%$$
$$\varepsilon(\sigma_{H2}, 4\ \text{days}) = 0.29\%$$

Uniaxial strain after 15 days is $\varepsilon_H = 0.51 - 0.47 + 0.29$ (equation (5.6)). Actual hoop strain by Hooke's law is

$$\varepsilon'_H = \varepsilon_H(1 - \nu) = 0.33 \times 0.6 = 0.198\%$$

After 15 days new internal diameter is

$$400 \times (1 + \varepsilon'_H) = 400.79\,\mathrm{mm}$$

Volume of gas is

$$V = \frac{1}{8} \times \frac{4}{3}\pi(400.79)^3 \times 10^{-9}\,\mathrm{m}^3 = 0.0337\,\mathrm{m}^3$$

3. Highest strain occurs in last case. This is well below 1%, so linear curve (strain 1%) applies.
   (a) Equation (5.6) applies: $\varepsilon_1 + \varepsilon_2 = 1 \times C(1.1 \times 10^5) + 1 \times C(10^4) = (1/600) + (1/750) = 0.0017 + 0.0013 = 0.0030 = 0.3\%$

(b) $\varepsilon_1 + \varepsilon_2 = 1 \times C(11 \times 10^5) + 1 \times C(10^6) = (1/475) + (1/485) = 0.00210 + 0.00206 = 0.00416 = 0.416\%$

(c) $\varepsilon = 2 \times C(1.1 \times 10^6) = 2/475 = 0.00421 = 0.421\%$, a minimal difference from the result under (b). Compare the comment under equation (5.6).

4. First calculate the magnitude of

$$\varepsilon_1(t_1) = \sigma_1 C(t_1) = \sigma_1 C_1 t_1^n$$

Then calculate with equation (5.7):

$$\varepsilon_R = \sigma_1 C(t_1 + t_R) - \sigma_1 C(t_R)$$

(note the times and compare them with Fig. 5.15), and then with equation (5.39):

$$\varepsilon_R = \sigma_1 C_1[(t_1 + t_R)^n - (t_R)^n]$$

It was given that

$$t_1 = t_R \Rightarrow \varepsilon_R/\varepsilon_1(t_1) = [(2t_1)^n - t_1^n]/t_1^n = 2^n - 1$$
$$\text{For } n = 0.1 \Rightarrow 2^n - 1 = 0.07$$
$$\text{For } n = 0.2 \Rightarrow 2^n - 1 = 0.15$$

Much more strain remains (much less recovery) in the second case.

5. No! Compared with a halfway point, the end of the tensile curve is reached at a later time. So the end of the curve is related to an isochronous curve for a longer time, which will show a higher strain for the full stress, than follows from proportionality with the halfway point. (Note that an exact description of this answer requires the use of the principle of superposition, reaching the same conclusion.)

6. Use Fig. 5.7,

$$E_c = E_0 t^n \quad \text{and} \quad \log E_c = \log E_0 + n \log t$$

With t in seconds,

$$E_c = E_0 \quad \text{at} \quad t = 1\,\text{s}$$

The 10 MN/m^2 curve is flat, so $\varepsilon = 0.365$, $E_0 = 2.74\,\text{GN/m}^2$. Hence

Time (s)	*log t*	ε(%)	E(GN/m^2)	*log E*
10	1	0.365	2.74	9.438
10^3	3	0.44	2.27	9.356
10^5	5	0.645	1.55	9.19
10^8	8	1.23	0.813	8.91

For two intervals t_1, t_2,

$$n = \frac{\log E_1 - \log E_2}{\log t_1 - \log t_2}$$

$10 \rightarrow 10^8$ s :

$$n = (9.438 - 8.91)/(1 - 8) = -0.0754$$

$10 \rightarrow 10^3$ s :

$$n = (9.438 - 9.356)/-2 = -0.044$$

$10^5 \rightarrow 10^8$ s :

$$n = (9.19 - 8.91)/-3 = -0.0933$$

Conclusion: the power law is not a very good representation in this polymer.

7. The approach is to estimate the relaxation time at which the modulus is halfway between E_1 and $E_1E_2/(E_1 + E_2)$ – see section 5.2.3(a). E_1 and E_2 will be for the glassy and rubbery values, i.e. $E_1 \sim 10^9\,\mathrm{N/m^2}$, $E_2 \sim 10^6\,\mathrm{N/m^2}$. The creep curve at $20\,\mathrm{MN/m^2}$ for this uPVC is swinging up sharply in the timescale 10^4 to 10^5 h, so use this range as an estimate for the relaxation time

$$\eta = E_2 t = 10^6 \times (3.6 \times 10^7 \text{ to } 10^8) = 3.6 \times 10^{13} \text{ to } 10^{14}, \text{ say } 10^{14}\,\mathrm{N\,s/m^2}$$

(For $30\,\mathrm{MN/m^2}$, $t \simeq 1000$ h, so $\eta \simeq 3.6 \times 10^{12}\,\mathrm{N\,s/m^2}$.)

8. The same as that of the 1 h isochronous stress–strain curve, provided the scale of the distance to the neutral axis is taken as being the same as the strain (these being proportional). Applying the load means applying the stress in every part of the cross-section in the same time.

9. All data for the equation for f are given, except E. One should be able to find this through $E(\sum_i t_i/a_{T_i}, T_{\mathrm{ref}})$ and equation (5.55). If we choose $T_{\mathrm{ref}} = 60°\mathrm{C}$, then $a_{60} = 1$, and from Fig. 5.10 (at the level $E_c = 400\,\mathrm{N/mm^2}$)

$$a_{20} = t_{60}/t_{20} = 10^7/0.8 \times 10^3$$
$$\Rightarrow a_{20} = 1.25 \times 10^4$$

Working out $\sum_i (t_i/a_{T_i})$ gives

$$t = 20 \times 365(23/1.25 \times 10^4 + 1/1)$$
$$\Rightarrow t = 7300(1.84 \times 10^{-3} + 1)\,\mathrm{h} = 26 \times 10^6\,\mathrm{s}$$
$$\Rightarrow E_c = 210\,\mathrm{N/mm^2}$$

With this value $f = 5.8$ mm, only a little more than was allowed.

The maximum strain can be calculated to be 0.23%, much less than the parameter $\varepsilon = 1\%$ for the E_c curve and apparently in the linear σ–ε domain.

Section 5.3

1. (a) We have $\lambda = 1 + \varepsilon$ so

$$\lambda^{-2} = \frac{1}{(1+\varepsilon)^2} \simeq 1 - 2\varepsilon + 3\varepsilon^2$$

and so

$$\lambda - 1/\lambda^2 = (1+\varepsilon) - (1-2\varepsilon) = 3\varepsilon$$

so equation (5.58) becomes

$$\sigma = \frac{1}{3}E_0(\lambda - \lambda^{-2}) = \frac{1}{3}E_0 \times 3\varepsilon = E_0\varepsilon$$

(b) $\varepsilon_1 = \varepsilon_1; \varepsilon_2 = \varepsilon_3 = -\nu\varepsilon_1$ for uniaxial deformation in one direction. New sides of deformed unit cube are $L_1(1+\varepsilon_1)$, $L_1(1-\nu\varepsilon_1)$, $L_1(1-\nu\varepsilon_1)$.
New volume is

$$V_2 = L_1^3(1+\varepsilon_1)(1-\nu\varepsilon_1)^2$$

Old volume is

$$V_1 = L_1^3$$

Constant volume requires $V_1 = V_2$, i.e.

$$1 = (1+\varepsilon_1)(1-\nu\varepsilon_1)^2$$

$$1 = 1 - 2\nu\varepsilon_1 + \nu^2\varepsilon_1^2 + \varepsilon_1 - 2\nu\varepsilon_1^2 + \nu^2\varepsilon_1^3$$

i.e.

$$\varepsilon_1(1-2\nu) + \nu\varepsilon_1^2(\nu - 2) + \nu^2\varepsilon_1^3 = 0$$

But $\varepsilon_1^2 \simeq 0 \approx \varepsilon_1^3$ if ε_1 is small, hence $\nu = \frac{1}{2}$.
(c) From Chapter 2 (section 2.8): $G \propto T$, i.e.

$$G_1/G_2 = T_1/T_2 = 273/323 = 0.8452$$

i.e. 8% change about the mean at 25°C.
(d) Block $L \times L \times H$. Displacement x, therefore

$$\tan\gamma = x/H = \tau/G$$

Strain energy/volume $= \frac{1}{2}\tau\tan\gamma = Px/2L^2H = \frac{1}{2}G(\tan\gamma)^2$
Therefore

$$u = \frac{1}{8}G$$

Take $G = 1$ MN/m^2, $u = 0.125$ MN m/m^3 = 125 kJ/m^3
(e) Creep in plastics is the (total) strain which is time dependent resulting from the applied stress (section 5.1.3(a)). Creep in rubber is the percentage change in the original elastic strain under the applied stress (section 5.3.1(c)).
(f) 50 years $\simeq 2.63 \times 10^7$ min, i.e. 7.4 decades (in minutes), i.e. 14.8% creep meaning 14.8% of the strain observed when the bearing pad was

originally loaded to the same stress. If for the sake of argument the pad were elastically compressed to 12% strain, then after 50 years the creep strain would be 14.8% of 12%, i.e. a strain of 1.78% so the total strain after 50 years would be $12 + 1.78 = 13.78\%$.

(g) For a wheel 0.5 m in diameter moving at 110 km/h,

$$\text{frequency} = 110 \times 1000/\pi \times 0.5 \times 3600 = 19.5\ \text{Hz}$$

2. (a) $S = A_L/A_F = 100 \times 100/(4 \times 100 \times 25) = 1.$
   (b) $S = 100 \times 100/(4 \times 100 \times 10) = 2.5.$
   (c) $S = 300 \times 10/(2 \times 310 \times 6) = 0.8.$
   (d) $S = \pi R^2/2\pi RH = R/2H = 50/20 = 2.5.$
   (e) The thickness of one block of rubber is $18/3 = 6$ mm. Therefore $S = 50/12 = 4.2$ for one 6 mm thick block of rubber. Shape factor is not defined for the overall assembly.
3. It is good practice to work on one individual layer of rubber, i.e. $H = 10$ mm. Each block compressed by 1.5 mm under 100 kN compressive load. For one block $S = 200 \times 100/600 \times 10 = 3.333$, $S^2 = 11.1$. For one block

$$E' = HP_z/Lb\delta_z = 10^{-2} \times 10^5/0.2 \times 0.1 \times 0.01 = 33.4\ \text{MN/m}^2$$

There is no need to introduce E'_c (equation (5.89)): compression was 1.5 mm per 10 mm, so exactly 15%, and the level of the modulus is rather low.

From equation (5.90):

$$E_0 = E'/(1 + 2S^2) = 33.4/(1 + 22.2) = 1.44\ \text{MN/m}^2$$

$$\Rightarrow G = E_0/3 = 0.48\ \text{MN/m}^2$$

Fig 5.35 gives 35 IRHD.

4. From Fig. 5.35, $G = 1.37\ \text{MN/m}^2$; using equation (5.71)

$$x = P \ln(R_2/R_1)/2\pi L_1 G = 2.44\ \text{mm} \quad \text{with } \hat{\tau} = P/2\pi R_1 L_1 = 0.4\ \text{MN/m}^2$$

The volume of the bush is

$$V_1 = \pi L(R_2^2 - R_1^2)$$

For the new bush you have to apply the argument about constant stress forms (pp. 129–30) to the analysis on pp. 127–28. If the inner dimensions of the rubber are R_3, L_3, and outer dimensions are R_4, L_4, then $\tau = P/2\pi R_3 L_3 = P/2\pi R_4 L_4 = P/2\pi RL$ at any intermediate radius.

$$dx/dR = \tan\gamma = P/2\pi R_3 L_3 G$$

So

$$K'_a = 2\pi R_3 L_3 G/(R_4 - R_3)$$

The volume of such a bush is

$$V_2 = \pi(R_3R_4(L_3 - L_4) + L_4R_4^2 - L_3R_3^2)$$

For the two units to have the same axial stiffness,

$$2\pi L_1G/\ln(R_2/R_1) = 2\pi R_3L_3G/(R_4 - R_3) = 2\pi R_1L_1/(R_4 - R_1)$$

(The last expression comes from $R_3 = R_1$ and $L_3 = L_1$ in the problem.) Thus

$$R_4 = R_1[1 + \ln(R_2/R_1)] = 33.4 \text{ mm}$$
$$L_4 = L_1/[1 + \ln(R_2/R_1)] = 30 \text{ mm}$$

so the unit is slimmer and has a shorter outside length: it therefore uses less rubber and in manufacture will cure more quickly.

The saving in weight (WS) achieved using the constant-stress unit is

$$WS = 1 - \frac{V_2}{V_1} = 1 - \frac{R_3R_4(L_3 - L_4) + L_4R_4^2 - L_3R_3^2}{L_1(R_2^2 - R_1^2)}$$

$$WS = 1 - 2\ln(R_2/R_1)/[(R_2/R_1)^2 - 1] = 0.3, \text{ i.e. } 30\%$$

5. The outside radius R_2 of a unit of constant cross-sectional area must satisfy the ratio of the stiffnesses:

$$\frac{K_\theta}{K_a} = \frac{4\pi LGR_1^2R_2^2/(R_2^2 - R_1^2)}{2\pi LG/\ln(R_2/R_1)} = 2R_2^2\ln(R_2/R_1)/(R_2^2/R_1^2 - 1)$$

This can be solved iteratively to give $R_2 = 27.5$ mm; the required shear modulus derives from one of the stiffness equations, giving $G = 0.97$ MN/m^2, i.e. a compound with a hardness of about 53° will be suitable, and will use 5.16×10^{-5} m^3.

An alternative approach is to explore the possibilities of using a constant-stress form. Basing the design on a constant shear stress form for torsional stiffness gives

$$K'_\theta = 2\pi R_3^2L_3G/\ln(R_4/R_3)$$

The axial displacement of an annular element is

$$\tan\gamma = dx/dR = P/2\pi RLG = PR/2\pi GR_3^2L_3$$

and on integration between R_3 and R_4 this gives the axial stiffness K'_a of the unit designed for constant stress in torsion as

$$K'_a = 4\pi GR_3^2L_3/(R_4^2 - R_3^2)$$

Following the previous method

$$K'_\theta/K'_a = (R_4^2 - R_3^2)/2\ln(R_4/R_3)$$

By iteration $R_4 = 21$ mm with $L_4 = 5.7$ mm (which is rather short), and $G = 1.59$ MN/m^2 corresponding to about 69° for the hardness of the compound. Only 1.12×10^{-5} m^3 of rubber will be needed.

Basing the design on a constant shear stress form for axial stiffness gives

$$K_a'' = 2\pi R_3 L_3 G/(R_4 - R_3) \quad R_3 L_3 = R_4 L_4 = RL$$

For this form the torsional stiffness can be calculated from

$$\tan \gamma = T/2\pi R^2 LG = T/2\pi R R_3 L_3 G = R\mathrm{d}\theta/\mathrm{d}R$$

Integrating gives

$$K_\theta'' = 2\pi R_3^2 R_4 L_3 G/(R_4 - R_3)$$

from which $R_4 = K_\theta''/K_a'' R_3 = 23.33$ mm, with $L_4 = 10.7$ mm, and from $G = 1.27$ MN/m^2 the hardness has to be 60 IRHD. Thus 2.09×10^{-5} m^3 of compound is required.

CHAPTER 6

Section 6.2

1. The sketch is consistent with
   (a) slightly off-axis loading;
   (b) a smooth region of slow crack growth;
   (c) a defect growing from within because the inside core of material is under some tensile stress from constrained cooling;
   (d) a rough region from fast crack growth, orientation along the bar together with the compressed surface inhibiting slow crack growth.
2. The waisted part of the bar has plastically deformed beyond yield and has voided, thus decreasing its density. The drop in density makes it more flexible than an untested bar, and the plastic deformation makes it non-elastic – it stays bent when flexed and feels rather soggy under load.
3. The highly oriented polypropylene is much weaker in shear because of the low strength (poor bonding between molecular chains) in the transverse direction. The untested bar will be more nearly isotropic and will have a higher shear strength.
4. There will be a high local stress in the transverse direction just ahead of the crack when it is loaded in the opening mode. This will try to orient the rubber perpendicular to the direction of crack growth. The natural rubber will crystallize and hence impede crack growth more, whereas the SBR will not offer so much resistance because it is not so closely ordered ahead of the crack tip.

5. Perishing is an oxidation phenomenon which occurs on the surface of the rubber. A thin band has a larger ratio of surface to volume than a thick one.
6. The main problem is the weld-line formed along the length of the pipe as melt recombines after passing the spider: there is some opportunity for the polymer to form a good weld, but there is no real flow mechanism for encouraging the diffusion of glass fibres across the weld-line. With no fibres bridging the weld in the hoop direction, the weld-line becomes a continuous source of weakness in a pipe carrying fluid under internal pressure.
7. After gelling occurs, the outside of the block may be able to cool by conduction to the mould, but the inside is effectively insulated. Thus the core can heat up and the expansion may cause the outside skin to crack after it has become rigid by crosslinking. Alternatively, the outside may not crosslink to a rigid state until the core has become quite hot. The outside then constrains the slow shrinking of the core and the resulting core tension may induce cracking within the block.
8. From Table 6.1 the solubility parameters in $(\mathrm{cal/cm^3})^{1/2}$ are: styrene–butadiene rubber 8.7, polysulphone 10.8, water 25, ethylene glycol 17.1, isobutanol 10.5, carbon tetrachloride 8.6 and *n*-pentane 7.05. Small differences in solubility parameter between polymer and liquid suggest that the liquid will be a good solvent. Thus (a) the SBR is likely to be dissolved by carbon tetrachloride, *n*-pentane and isobutanol, and (b) the polysulphone by isobutanol and carbon tetrachloride.
9. Gear durability is enhanced by
   (a) constrained cooling leaving the surface of the tooth form in compression;
   (b) some molecular orientation along the surface of the tooth which is in line with the tooth bending stress.
10. When stressed, any oxidation causes development of a crack-like defect which will grow if stressed in tension in the opening mode.
11. Answers should not be necessary.
12. (a) From Fig. 6.22, the failure stress after 50 years under continuous pressure is $6.5\,\mathrm{MN/m^2}$ at 20°C. The minimum wall thickness can then be calculated from

$$h = pD/2\sigma_\mathrm{d}$$

    where $\sigma_\mathrm{d} = 6.5/1.3 = 5\,\mathrm{MN/m^2}$. This gives a wall thickness of 52 mm. The pipe weighs about 220 kg/m and could be extruded at about 5 m/h using a 150 mm screw extruder.

    (b) Extrapolating the early (ductile failure) part of the failure curve out to 50 years suggests (incorrectly, of course) a failure stress of about $7.8\,\mathrm{MN/m^2}$. Applying the safety factor of 1.3 would then give a design stress of about $6.5\,\mathrm{MN/m^2}$, whereas the long-term brittle failure data suggest that this is just the stress at which failure would occur.

(c) The failure stress after 1 year at 20°C is about 9.7 MN/m^2, and that at 50 years is 6.5 MN/m^2: the uprating factor is $9.7/6.5 = 1.49$.
(d) The 50 year failure stress at 40°C is about 2.4 MN/m^2 and hence the pressure rating for a given pipe must be multiplied by $2.4/6.5 = 0.37$. In fact the Code of Practice (CP312:Part 3:1973) errs on the safe side, with a conservative factor of 0.4.

Section 6.3

1. (a) Using equation (6.39) with $a = 11/4 : r = 1, k = 4.3; r = 1/4, k = 7.6$.
   (b) In LEFM C = extension/load; in stiffness, C = reciprocal of modulus.
   (c) (i) $C = L_0/AE$; (ii) $C = L^3/3EI$.
   (d) From equation (6.32), if n is very large, and the final crack size is much larger than the initial crack size, then the initial crack size dominates the failure time. Small cracks grow slowly and hence dominate the failure time.
   (e) Substituting equation (6.19) gives

$$\hat{\tau} = \sqrt{\{[(\sigma_x - \sigma_y)/2]^2 + \tau_{xy}^2\}} = K \sin\theta/[2\sqrt{(2\pi r)}]$$

   (f) The crossover occurs at $7 \times 10^{-6} K^{24} = 1.3 \times 10^{-4} K^{12.5}$, giving $K = 1.29\ MN/m^{3/2}$ and $a = 3.1$ mm/s.
   (g) The short-term fracture stress σ_F of 67 MN/m^2 relates to a material having no deliberate flaw in it. The short-term fracture toughness of the same polymer with a deliberate crack in it is $K_{Ic} = 1.6\ MN/m^{3/2}$. So the inherent flaw size a_{IFS} can be calculated from equation (6.23)

$$K_{Ic} = \sigma_F \sqrt{(\pi a_{IFS})}$$

   and for a centre defect, $2a_{IFS} = 0.22$ mm. The polymer behaves as if it contained defects of this size, even though they cannot be detected under a microscope.
   (h) The longitudinal crack has to cope with the hoop stress which is higher than the longitudinal stress. A flaw on the inside of the pipe is worse because the pressure in the pipe acts on the crack faces, as well as the stress in the walls acting on the tip; on the outside only the stress is acting.
   Adapting the argument given in section 6.2.3(c), cooling the pipe from the outside would confer a residual thermal stress distribution in the pipe wall: the outside surface would be in compression, which is helpful, and the bore would be in tension, which is detrimental.
2. Assume the stress–strain curve for the polymer is linear up to the yield stress σ_Y. From equation (6.38),

$$r_Y = (2/60)^2/2\pi = 1.77 \times 10^{-4}\ \text{m}$$

Combining equations (6.23) and (6.38):

$$\sigma^2/\sigma_Y^2 = (K_I^2/\pi a)/(K_I^2/2\pi r_Y)$$

and so

$$(\sigma/\sigma_Y^2) = 2r_Y/a = 2 \times 1.77 \times 10^{-4}/(5 \times 10^{-4})$$

Hence $p/p_Y = \sigma/\sigma_Y = 0.84$.

3. A plot of energy absorbed versus $HD\phi$ has a slope $G_{Ic} = 3.5\ \mathrm{kJ/m^2}$.
4. First, note the unusual choice of symbols for length L, thickness H and depth D. They are however consistent with the choice made in Fig. 6.24, which is important for proper use of the formulae! The displacement of one of the beams under load P is:

$$\frac{1}{2}x = \frac{Pa^3}{3E\frac{1}{12}H(\frac{1}{2}L)^3} \quad (6.11) \Rightarrow C(a) = \frac{x}{P} = \frac{8a^3}{EH(\frac{1}{2}L)^3}$$

$$\frac{\mathrm{d}C}{\mathrm{d}a} = \frac{24a^2}{EH(\frac{1}{2}L)^3} \quad (6.17) \Rightarrow \phi(a) = \frac{a}{3D}$$

Note that $C = 0$ if the crack $a = 0$, which is not the case for most other specimen shapes: compare C_0 for Figs 6.24 and 6.43.

Note that this is an approximate solution: for a short crack length the cantilever behaves more like a block under shear loading.

5. Plotting the data gives $\dot{a} = 2 \times 10^{-9} K_I^{8.6}$. The bending stress remote from the crack is $15\ \mathrm{MN/m^2}$. Substituting $K_I = 2\sigma\sqrt{a}$ in $\dot{a} = CK^n$ and integrating gives

$$t = 2(a_i^{1-n/2} - a_f^{1-n/2}/(n-2)C(2\sigma)^n = 3050\ \mathrm{s}$$

6. $\sigma_{bend} = 6M/bh^2 = 10.32\ \mathrm{MN/m^2}$. Cracking will be dominated by slow growth, as noted in problem 1(d). Using values of n and C from problem 1(f),

$$t \approx 2a_i^{1-n/2}/(n-2)C(Y\sigma\sqrt{\pi})^n = 2.58\ \mathrm{Ms}$$

With a failure time of about 1 month, recommend weekly inspection and take the mechanism out of service either now or next week.

7. Gas engineers design pipes to avoid the pessimistic situation where a longitudinal crack has grown right through the wall and then unzips the pipes for a few kilometres. The strain energy δU to extend such a crack along the pipe by an incremental amount δa is given by

$$\delta U = (\sigma^2/2E) \times \pi D_m H \delta a$$

where σ is the hoop stress in a pipe of mean diameter D_m and wall thickness H. Catastrophic fracture occurs when (equation (6.13))

$$\delta U = G_{Ic} H \delta a$$

With equation (6.25) we then find

$$D_m = 2K_{Ic}^2/\pi\sigma^2$$

With the prescribed D/H we find $\sigma = pD/(2H) = 10p$, so for the pressure test $\sigma = 6\,\text{MN/m}^2$. Thus the maximum allowable diameter is

$$D_m = 0.16\text{ m}$$

Concerning long-term growth of surface defects in these pipes, the current view is that cracks will not grow if the stress intensity factor is less than 0.51 $\text{MN/m}^{3/2}$, leading to a design value of 0.34 $\text{MN/m}^{3/2}$, i.e. a safety factor of 1.5.

CHAPTER 7

Section 7.1

1. No. Just consider that an in-plane load under 45° to the warp direction will give a much lower modulus that E_1 (partly indicated in Fig. 7.3) and a much higher Poisson's ratio than ν_{12} as the cloth will show a high shear deformation.

Section 7.2

1. (a) Hexagonal packing: $V_f = \pi\sqrt{3}/6 = 90.7\%$.
   (b) Square packing: $V_f = \pi/4 = 78.5\%$.
   (c) Fig. A9 shows the element: $V_f = 3\pi/16 = 58.9\%$.
2. From equation (7.27), $L_c/R = \hat{\sigma}_f/\tau_i$ so the average stress is

$$\bar{\sigma} = \hat{\sigma}_f(1 - L_c/2L), \quad \text{i.e. } L = 10L_c.$$

3. $V_f = (w_f/\rho_f)/(w_f/\rho_f + w_p/\rho_p + w_c/\rho_c) = 0.124$
4. For a beam of length L, breadth b, depth h, simply supported, loaded at midspan, simple bending theory gives $\hat{\sigma}_B = 3WL/2bh^2$ and $\hat{\tau} = 3W/4bh$, from which $L/h = 2\hat{\sigma}_B/\hat{\tau}$. For isotropic materials, $\hat{\sigma}/\hat{\tau}$ is about 1.5 so shear stresses are only significant when the beam is very deep. For orthotropic materials with fibres along the beam, $\hat{\sigma}_1/\hat{\tau}_{12} \simeq 10$ to 20 typically, so shear failure is much more likely even in slender beams, especially if of sandwich construction with a foam core. $\hat{\tau}_{12}$ is not greatly affected by the volume fraction of fibres but $\hat{\sigma}_1$ is, so the higher the volume fraction of fibres the more likely is failure by shear unless allowed for in design.
5. (a) $F_1 = 80 \times 10^3$ N. Using L_2^0 for original length in 2-direction

$$\sigma_1 = F_1/hL_2^0 = 80 \times 10^3/(2 \times 10^{-3} \times 0.1) = 400\,\text{MN/m}^2$$
$$\varepsilon_1 = \sigma_1/E_1 = 400 \times 10^6/208 \times 10^9 = 0.00192$$
$$\varepsilon_2 = -\nu_{12}\varepsilon_1 = -0.3 \times 0.00192$$
$$L_2 = (1 + \varepsilon_2)L_2^0 = 99.94\,\text{mm}$$

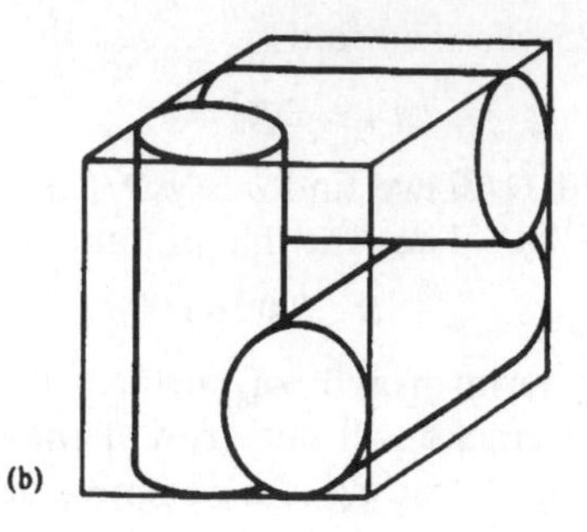

Fig. A9 Three-dimensional reinforcement

(b) $F_1 = 80 \times 10^3$ N
$F_2 = 5\,\text{kN}$

$$\sigma_2 = F_2/hL_1^0 = 12.5\,\text{MN/m}^2$$
$$\nu_{21} = \nu_{12}E_2/E_1 = 0.3 \times 7.6/208 = 0.011$$
$$\varepsilon_1 = \sigma_1/E_1 - \nu_{21}\sigma_2/E_2 = 1.9019 \times 10^{-3}$$

so

$$L_1 = (1 + \varepsilon_1)L_1^0 = 200.38\,\text{mm}$$

(c) $F_{12} = 10\,\text{kN}$

$$\tau_{12} = F_{12}/h_2L_1^0 = 25\,\text{MN/m}^2$$

Applying the Tsai–Hill criterion with $\sigma_1 = 0$,

$$\sigma_2 = 41\sqrt{[1 - (25/55)^2]} = 36.5\,\text{MN/m}^2$$

(d) It is necessary to relate weight fraction to volume fraction. If W is the total weight, and V the total volume, then $W = W_f + W_p$, and the weight fraction of fibres is

$$w_f = W_f/W = \rho_f A_f L/(\rho_f A_f L + \rho_p A_p L)$$

and hence

$$W_f = V_f\rho_f/(\rho_f V_f + \rho_p V_p)$$

Using $E_f \simeq 390\,\text{GN/m}^2$, $V_f \simeq 208/390 = 0.533$. With $\rho_f = 1950\,\text{kg/m}^3$ and $\rho_p = 1200\,\text{kg/m}^3$ (say), $w_f = 0.65$.

6. Let the thickness of the composite beam be h_c. Its stiffness is proportional to $E_1h_c^3$ and $E_1 = V_fE_f + V_pE_p$. The mass of beam per unit cross-sectional area is

$$m = \rho_c h_c = \rho_f h_f + \rho_p h_p$$

i.e. $\rho_c = \alpha V_f + \rho_p$ where $\alpha = \rho_f - \rho_p$. Hence

$$E_1 h_c^3 = [V_f E_f + (1 - V_f)E_p]m^3/(\alpha V_f + \rho_p)^3$$

and for a given mass per unit area, the maximum stiffness is given by

$$\frac{d(E_1 h_c^3)}{dV_f} = 0 = -\frac{3\alpha[V_f E_f + (1 - V_f)E_p]}{(\alpha V_f + \rho_p)^4} + \frac{E_f - E_p}{(\alpha V_f + \rho_p)^3}$$

If $E_p \ll E_f$,

$$E_f(2 \propto V_f - \rho_p) = 0$$

so $\rho_p = 2\alpha V_f$. Taking $\rho_f \simeq 2500\,\text{kg/m}^3$ and $\rho_p \simeq 1000\,\text{kg/m}^3$, $V_f \sim 0.33$.

7. Using m for mass, D for diameter, subscripts g and c for gfrp and cfrp, we find:

	Criterion	D_g	m_g/m_s	D_c	m_c/m_s
(a)	$\Delta = PL/AE$	23.2	1.24	10.72	0.24
(b)	$P = \pi^2 EI/L^2$	15.24	0.54	10.35	0.22
(c)	$T/\theta = \pi GR^4/2L$	21	1.02	18.33	0.69
(d)	$w = kPL^3/E\pi R^4$	15.24	0.54	10.35	0.22
(e)	$P = \sigma_{max} A$	4.34	0.043	3.65	0.027
(f)	$T = \tau_{max}\pi R^3/2$	11.2	0.287	11.37	0.265
(g)	$\sigma_{max} = kM/R^3$	6.9	0.11	5.84	0.07

In (b) we assume failure occurs by buckling before failure in compression. Be aware that the longitudinal modulus in compression may be less that that in tension. In (g) we assume failure occurs in compression before failure in tension.

8. The guiding principle is that axial strains in the composite and in the individual plies is the same. Let the thickness of each ply be h_g or h_c. Using subscript 1 for the laminate, we find

$$\varepsilon_1 = \sigma_1/E_1 = \sigma_g/E_g = \sigma_c/E_c$$

Expressing stresses in terms of forces we find

$$P_1/bh_1E_1 = (P_g + P_c)/[b \times 2(h_g + h_c)E_1] = P_g/b \times 2h_gE_g = P_c/b \times 2h_cE_c$$

Simplifying leads to

$$E_1 = (h_gE_g + h_cE_c)/(h_g + h_c) = 145\,\text{GPa}$$

9. If the closed coil spring is loaded in torsion, coils deform mainly by bending. Strengths in the fibre direction are high but the shear strength is low and we would expect shear failure. Failure from torsion loading of the spring would occur with a loud bang and the spring would divide along its helical length with one piece neatly inside the other. By twisting the parallel bundles of fibres as they are fed into the mould, the shear strength can be improved substantially.

Section 7.3

1. (a) For steel, $E = 208\,\text{GN/m}^2$, $\nu = 0.29$, so $G = 81\,\text{GN/m}^2$:

$$S_{\text{steel}} = \begin{bmatrix} 4.81 & -1.39 & 0 \\ -1.39 & 4.81 & 0 \\ 0 & 0 & 12.3 \end{bmatrix} 10^{-12}\,\text{m}^2/\text{N}$$

$$Q_{\text{steel}} = \begin{bmatrix} 227 & 66 & 0 \\ 66 & 227 & 0 \\ 0 & 0 & 81 \end{bmatrix} 10^{9}\,\text{N/m}^2$$

For aluminium, $E = 70\,\text{GN/m}^2, \nu = 0.33$, so $G = 26\,\text{GN/m}^2$:

$$S_{\text{Al}} = \begin{bmatrix} 14.3 & -4.7 & 0 \\ -4.7 & 14.3 & 0 \\ 0 & 0 & 38.5 \end{bmatrix} 10^{-12}\,\text{m}^2/\text{N}$$

$$Q_{\text{Al}} = \begin{bmatrix} 78.6 & 25.9 & 0 \\ 25.9 & 78.6 & 0 \\ 0 & 0 & 26 \end{bmatrix} 10^{9}\,\text{N/m}^2$$

These calculations are based on equations (7.30) and (7.31).

(b) No timescale is given, so choose to work to a short-term figure such as 20°C, 0.1 h, 0.1% strain; from Fig. 5.12 $E \simeq 3.2\,\text{GN/m}^2$ and hence $G \simeq 1.19\,\text{GN/m}^2$ for PMMA:

$$S_{\text{PMMA}} = \begin{bmatrix} 0.312 & -0.109 & 0 \\ -0.109 & 0.312 & 0 \\ 0 & 0 & 0.84 \end{bmatrix} 10^{-9}\,\text{m}^2/\text{N}$$

The compliance would increase with time under load: after 1 year E would drop to about 1.5 GN/m^2 so that $S_{11} = 0.67 \times 10^{-9}\,\text{m}^2/\text{N}$.

(c) From Fig. 5.35 and $E_0 = 3G$, $E_0 = 1.62\ \text{MN/m}^2$, $G = 0.54\ \text{MN/m}^2$, $\nu = 0.5$:

$$Q_{\text{rubber}} = \begin{bmatrix} 2.16 & 1.08 & 0 \\ 1.08 & 2.16 & 0 \\ 0 & 0 & 0.54 \end{bmatrix} 10^{6}\,\text{N/m}^2$$

(d) $\nu_{21} = E_2\nu_{12}/E_1 = 0.0594$; $\lambda = 1 - \nu_{12}\nu_{21} = 0.985$. Inverting equation (7.42):

$$Q = \begin{bmatrix} E_1/\lambda & \nu_{12}E_2/\lambda & 0 \\ \nu_{12}E_2/\lambda & E_2/\lambda & 0 \\ 0 & 0 & G_{12} \end{bmatrix}$$

$$= \begin{bmatrix} 35.53 & 2.11 & 0 \\ 2.11 & 8.12 & 0 \\ 0 & 0 & 3.5 \end{bmatrix} 10^{9}\,\text{N/m}^2$$

From equation (7.42)

$$S = \begin{bmatrix} 28.57 & -7.43 & 0 \\ -7.43 & 125 & 0 \\ 0 & 0 & 286 \end{bmatrix} 10^{-12}\,\mathrm{m^2/N}$$

$E_f = 75\,\mathrm{GN/m^2}$ from Table 7.1 so $V_f \simeq E_1/E_f = 0.47$ approximately.
More precisely $V_f = (E_1 - E_p)/(E_f - E_p) = 0.44$ with $E_p = 3\,\mathrm{GN/m^2}$.

2. From equation (7.60) we have

$$g = 1/G_{xy} = A\cos^2\theta\sin^2\theta + B(\cos^4\theta - 2\cos^2\theta\sin^2\theta + \sin^4\theta)$$

Substituting $\cos^2\theta = 1 - \sin^2\theta$ leads to a quadratic in $\sin^2\theta$:

$$g = -C\sin^4\theta + C\sin^2\theta + B$$

where $C = (A - 4B)$.

$$dg/d\theta = 2C\cos\theta\sin\theta(1 - 2\sin^2\theta) = 0$$

for maximum or minimum. One root is $\sin^2\theta = \frac{1}{2}$, i.e. $\theta = \pm 45°$ (other roots are 0°, 90°).

3. $E_1 = 4.4\,\mathrm{GN/m^2}, E_2 = 2.75\,\mathrm{GN/m^2}$ and $E_x(45) = 3.4\,\mathrm{GN/m^2}$. From equation (7.58):

$$4/E(45) = 1/E_1 + 1/E_2 + (1/G_{12} - 2\nu_{12}/E_1)$$

so in general

$$1/E(\theta) = \cos^4\theta/E_1 + \sin^4\theta/E_2 + [4/E(45) - 1/E_1 - 1/E_2]\sin^2\theta\cos^2\theta$$

This gives $E(30) = 3.84\,\mathrm{GN/m^2}$ and $E(60) = 3.04\,\mathrm{GN/m^2}$.

4. This problem is analogous to that developed in equation (7.52). The procedure is to start with what you are looking for and transform it to what you are given: $\bar{Q}$ relates given strains to desired stresses

$$\begin{bmatrix} \sigma_x \\ \sigma_y \\ \tau_{xy} \end{bmatrix} = T^{-1}\begin{bmatrix} \sigma_1 \\ \sigma_2 \\ \tau_{12} \end{bmatrix} = T^{-1}Q\begin{bmatrix} \varepsilon_1 \\ \varepsilon_2 \\ \gamma_{12} \end{bmatrix} = T^{-1}QR\begin{bmatrix} \varepsilon_1 \\ \varepsilon_2 \\ \frac{1}{2}\gamma_{12} \end{bmatrix}$$

$$= T^{-1}QRT\begin{bmatrix} \varepsilon_x \\ \varepsilon_y \\ \frac{1}{2}\gamma_{xy} \end{bmatrix} = T^{-1}QRTR^{-1}\begin{bmatrix} \varepsilon_x \\ \varepsilon_y \\ \gamma_{xy} \end{bmatrix} = \bar{Q}\begin{bmatrix} \varepsilon_x \\ \varepsilon_y \\ \gamma_{xy} \end{bmatrix}$$

5. Following the approach in problem 4 and noting $\sin^2 45° = \cos^2 45° = \frac{1}{2}$,

$$\begin{bmatrix} \sigma_1 \\ \sigma_2 \\ \tau_{12} \end{bmatrix} = QRTR^{-1}\begin{bmatrix} \varepsilon_x \\ \varepsilon_y \\ \gamma_{xy} \end{bmatrix}$$

$$= \begin{bmatrix} \frac{1}{2}(Q_{11} + Q_{12}) & \frac{1}{2}(Q_{11} + Q_{12}) & \frac{1}{2}(Q_{11} - Q_{12}) \\ \frac{1}{2}(Q_{12} + Q_{22}) & \frac{1}{2}(Q_{12} + Q_{22}) & \frac{1}{2}(Q_{12} - Q_{22}) \\ -Q_{66} & +Q_{66} & 0 \end{bmatrix} \begin{bmatrix} \varepsilon_x \\ \varepsilon_y \\ \gamma_{xy} \end{bmatrix}$$

$$\nu_{21} = \nu_{12}E_2/E_1 = 0.011, \qquad \lambda = 1 - \nu_{12}\nu_{21} = 0.9967,$$
$$Q_{11} = E_1/\lambda = 208.69\,\text{GN/m}^2, \qquad Q_{22} = 7.625\,\text{GN/m}^2,$$
$$Q_{66} = G_{12} = 4.8\,\text{GN/m}^2, \qquad Q_{12} = \nu_{21}Q_{11} = 2.296\,\text{GN/m}^2;$$

hence

$$\begin{bmatrix}\sigma_1\\ \sigma_2\\ \tau_{12}\end{bmatrix} = 10^9\begin{bmatrix}105.49 & 105.49 & 103.19\\ 4.961 & 4.961 & -2.665\\ -4.8 & +4.8 & 0\end{bmatrix}\begin{bmatrix}100\times10^{-6}\\ 150\times10^{-6}\\ 200\times10^{-6}\end{bmatrix}$$

Hence $\sigma_1 = 47.03\,\text{MN/m}^2$, $\sigma_2 = 0.707\,\text{MN/m}^2$ and $\tau_{12} = 0.24\,\text{MN/m}^2$. The Tsai–Hill criterion can now be tested:

$$(47.03/828)^2 - (47.03\times0.707/828)^2 + (0.707/41)^2 + (0.24/55)^2$$
$$=1/\text{F}^2 = 3.4936\times10^{-3}$$

Failure occurs when $1/F^2 = 1$, so the safety factor is

$$(1/3.4936\times10^{-3})^{\frac{1}{2}} = 16.9.$$

6. For a panel of thickness h, width w under a load P, the tensile stress σ and force resultant N are given by $\sigma = P/wh = N/h$. Hence $N = P/w = 9\times10^3/0.3 = 30\,\text{kN/m}$.
7. To sketch the deformed sheet we need to calculate the midplane strains and curvatures. We can use expressions such as $\varepsilon_x = z\kappa_x$ and thus (using t and b to denote top and bottom) we find $\varepsilon_x^0 = (\varepsilon_x^{\text{t}} + \varepsilon_x^{\text{b}})/2$:

$$\varepsilon_x^0 = (0.3121 + 0.3046)/2 = 0.030835\%$$
$$\varepsilon_y^0 = -0.25415\%; \qquad \gamma_{xy}^0 = -0.22525\%$$

From $\varepsilon_x^{\text{t}} = (-h/2)\kappa_x$ and $\varepsilon_x^{\text{b}} = (+h/2)\kappa_x$ we have

$$\kappa_x = -(\varepsilon_x^{\text{t}} - \varepsilon_x^{\text{b}})/h$$
$$\kappa_x = 0.01875\,/\text{m}; \quad \kappa_y = 0.10325\,/\text{m}; \quad \kappa_{xy} = 0.5068\,/\text{m}$$

We can transform the strains from global to principal directions, to find $(\varepsilon_1, \varepsilon_2, \gamma_{12})$. The stresses then follow from

$$\begin{bmatrix}\sigma_1\\ \sigma_2\\ \tau_{12}\end{bmatrix} = \begin{bmatrix}E_1/J & \nu_{12}E_2/J & 0\\ \nu_{12}E_2/J & E_2/J & 0\\ 0 & 0 & G\end{bmatrix}\begin{bmatrix}\varepsilon_1^0\\ \varepsilon_2^0\\ \gamma_{12}^0\end{bmatrix}$$

where $J = (1 - \nu_{12}\nu_{21})$, from which we see that

$$\sigma_1/\sigma_2 = (E_1\varepsilon_1^0 + \nu_{12}E_2\varepsilon_2^0)/(\nu_{12}E_2\varepsilon_1^0 + E_2\varepsilon_2^0)$$

so that ν_{12} and the anisotropy ratio E_1/E_2 are needed to calculate the stress ratios. (For an isotropic material we see that only ν is needed.)

8. For off-axis loading of a lamina we can use either

$$\begin{bmatrix} \varepsilon_x \\ \varepsilon_y \\ \gamma_{xy} \end{bmatrix} = \begin{bmatrix} S^*_{11} & S^*_{12} & S^*_{16} \\ S^*_{12} & S^*_{22} & S^*_{26} \\ S^*_{16} & S^*_{26} & S^*_{66} \end{bmatrix} \begin{bmatrix} \sigma_x \\ \sigma_y \\ \tau_{xy} \end{bmatrix}$$

or the form to be developed for laminate analysis, which we use here:

$$\begin{bmatrix} \varepsilon^0_x \\ \varepsilon^0_y \\ \gamma^0_{xy} \end{bmatrix} = \begin{bmatrix} a_{11} & a_{12} & a_{16} \\ a_{12} & a_{22} & a_{26} \\ a_{16} & a_{26} & a_{66} \end{bmatrix} \begin{bmatrix} N_x \\ N_y \\ N_{xy} \end{bmatrix}$$

(a) Let x be the hoop direction and y the axial direction in the tube. Under axial tension N_y we have

$\varepsilon^0_x = -3.92N_y$ hoop contraction

$\varepsilon^0_y = +9.09N_y$ axial extension

$\gamma_{xy} = 1.53N_y$ small positive shear (positive twist)

(b) Under internal pressure p, we know that $N_x = 2N_y$, and hence

$$\varepsilon^0_x = (2a_{11} + a_{12})N_y = 9.4N_y$$
$$\varepsilon^0_y = (2a_{12} + a_{22})N_y = 1.25N_y$$
$$\gamma^0_{xy} = (2a_{16} + a_{26})N_y = -13.41N_y$$

We see that the tube increases in length and diameter, and develops large negative shear.

(c) Under opposing torques N_{xy} at each end:

$$\varepsilon^0_x = a_{16}N_{xy} = -7.47N_{xy}$$
$$\varepsilon^0_y = a_{26}N_{xy} = 1.53N_{xy}$$
$$\gamma^0_{xy} = a_{66}N_{xy} = 12.2N_{xy}$$

If N_{xy} is positive, the tube contracts in diameter, extends slightly along its axis and twists in a positive sense. If N_{xy} is negative, the tube expands in diameter, decreases in length and twists in a negative sense.

9. (a) Applying $M_x = 1\,\text{N}$, so that $M_y = M_{xy} = 0$ gives immediately

$$\kappa_x = d_{11}M_x = 7.144/\text{m}, \kappa_y = d_{12}M_x = -2.195/\text{m}, \kappa_{xy} = d_{16}M_x = -3.421\ /\text{m}$$

(ii) Setting $\kappa_y = 1\ /\text{m}$ can be done in one of two ways, and it is important to decide which you intend.

(b) If we assume that $M_x = M_{xy} = 0$, which is the normal assumption, then $\kappa_y = 1 = d_{22}M_y$ so that $M_y = \kappa_y/d_{22} = 0.14\,\text{N}$.

(b) If we decide to suppress other curvatures so that $\kappa_x = \kappa_{xy} = 0$, then we must solve the equations

$$\begin{aligned} 0 &= d_{11}M_x + d_{12}M_y + d_{16}M_{xy} \\ 1 &= d_{12}M_x + d_{22}M_y + d_{26}M_{xy} \\ 0 &= d_{16}M_x + d_{26}M_y + d_{66}M_{xy} \end{aligned}$$

The easiest way is to invert $[d]$, preferably using software, so that we find

$$M_x = D_{12}\kappa_y = 0.1114\,\text{N}; \quad M_y = D_{22}\kappa_y = 0.2185\,\text{N}; \quad M_{xy} = D_{26}\kappa_y = 0.09244\,\text{N}$$

9. Bearing in mind that thermal strains are directly related to expansion coefficients if the temperature change is uniform thoughout the ply, we need to transform expansion coefficients from principal to global directions using equation (7.46)

$$\begin{bmatrix} \alpha_x \\ \alpha_y \\ \alpha_{xy}/2 \end{bmatrix} = [T]^{-1} \begin{bmatrix} \alpha_1 \\ \alpha_2 \\ 0 \end{bmatrix}$$

Thus, filling in for the transformation matrix $[T]^{-1}$:

$$\alpha_x = \alpha_1 \cos^2 20 + \alpha_2 \sin^2 20 = 2.65 \times 10^{-6}\ /\text{K}, \text{ and similarly}$$
$$\alpha_y = 19.87 \times 10^{-6}\ /\text{K}, \text{ and } \alpha_{xy} = -14.48 \times 10^{-6}\ /\text{K}$$

Section 7.4

1. (All $\bar{Q}$ are for $+\theta$ unless otherwise indicated.) Use equations (7.63), (7.87), (7.88) and (7.90).
   (a) Non-regular symmetric angle-ply laminate.
   $A_{11} = h\bar{Q}_{11}$ because $Q_{11}(+\theta) = +Q_{11}(-\theta)$; similarly A_{12}, A_{22} and A_{66}.
   $A_{16} = A_{26} = 0$ and $B_{ij} = 0$ by symmetry.
   $D_{11} = h^3\bar{Q}_{11}/12$ and $D_{16} = h^3\bar{Q}_{16}/16$.
   (b) Regular symmetric cross-ply laminate.

$$A_{11} = (h/3)(2Q_{11} + Q_{22}) \quad A_{12} = hQ_{12} \quad A_{22} = (h/3)(2Q_{22} + Q_{11})$$
$$A_{66} = hQ_{66} \quad A_{16} = A_{26} = 0 \quad B_{ij} = 0$$
$$D_{11} = h^3(52Q_{11} + 2Q_{22})/(3 \times 6^3) \quad D_{12} = h^3(54Q_{12})/(3 \times 6^3)$$
$$D_{16} = D_{26} = 0$$
$$D_{22} = h^3(2Q_{11} + 52Q_{22})/(3 \times 6^3) \quad D_{66} = h^3(54Q_{66})/(3 \times 6^3)$$

   (c) Regular antisymmetric angle-ply laminate.

$$A_{11} = h\bar{Q}_{11} \quad A_{12} = h\bar{Q}_{12} \quad A_{22} = h\bar{Q}_{22}$$
$$A_{66} = h\bar{Q}_{66} \quad A_{16} = A_{26} = 0$$
$$B_{11} = B_{12} = B_{22} = B_{66} = 0 \quad \text{but} \quad B_{16} = -h^2\bar{Q}_{16}(+\theta)/4$$
$$D_{11} = h^3\bar{Q}_{11}/12 \quad D_{16} = D_{26} = 0$$

(d) Regular antisymmetric cross-ply laminate.

$$A_{11} = A_{22} = \tfrac{1}{2}h(Q_{11} + Q_{22}) \quad A_{12} = hQ_{12} \quad A_{66} = hQ_{66} \quad A_{16} = A_{26}$$
$$B_{11} = h^2(Q_{22} - Q_{11})/8 \quad B_{22} = -B_{11} \quad B_{12} = B_{66} = B_{16} = B_{26} = 0$$
$$D_{11} = D_{22} = h^3(Q_{11} + Q_{12})/(3 \times 8) \quad D_{12} = h^3(2Q_{12})/(3 \times 8)$$
$$D_{16} = D_{26} = 0$$

2. The laminate is regular, so the general coordinate of the fth lamina lower boundary will be $h_f = -(h/2) + fh/n$, where n is the total (even) number of laminae, and $h_{f-1} = -h/2 + (f-1)h/n$. From equation (7.88)

$$B = \frac{1}{2}\sum_{f=1}^{n} \bar{Q}(h_f^2 - h_{f-1}^2) = \frac{1}{2}\sum \bar{Q}(h_f + h_{f-1})(h_f - h_{f-1})$$

and

$$(h_f + h_{f-1})(h_f - h_{f-1}) = -h^2/n[1 - (2f - 1)/n]$$

For B_{11}, B_{12}, B_{22} and B_{66}, $\bar{Q}(+\alpha) = \bar{Q}(-\alpha)$, and these $B_{ij} = 0$. For B_{16} and B_{26} $\bar{Q}(-\alpha) = -\bar{Q}(+\alpha)$, so it is necessary to write

$$B_{16} = -(h^2\bar{Q}_{16}/2n)\sum\{(-1)^f[1 - (2f - 1)/n]\} = h^2\bar{Q}_{16}/2n$$

i.e. B_{16} becomes small as n becomes large so that the B matrix tends to zero and the laminate becomes sensibly symmetric.

3. From the data

$$Q = \begin{bmatrix} 209.9 & 7.98 & 0 \\ 7.98 & 209.9 & 0 \\ 0 & 0 & 7.6 \end{bmatrix} \text{GN/m}^2$$

and hence for 30° (in GN/m^2)

$$\bar{Q}(+30) = \begin{bmatrix} 126.21 & 42.67 & 60.92 \\ 42.67 & 33.58 & 20.87 \\ 60.92 & 20.87 & 44.81 \end{bmatrix}$$

and

$$\bar{Q}(-30) = \begin{bmatrix} 126.21 & 42.67 & -60.92 \\ 42.67 & 33.58 & -20.87 \\ -60.92 & -20.87 & 44.81 \end{bmatrix}$$

$$A_{11} = (h/3)(2\bar{Q}_{11} + Q_{11}) = 154.1 \times 10^9 h\,\text{N/m}$$
$$A_{12} = 31.1 \times 10^9 h\,\text{N/m} \quad A_{22} = 29.38 \times 10^9 h\,\text{N/m}$$
$$A_{66} = 32.4 \times 10^9 h\,\text{N/m}$$
$$A_{16} = A_{26} = 0$$

Hence

$$a = \begin{bmatrix} 8.17 & -8.65 & 0 \\ -8.65 & 43.2 & 0 \\ 0 & 0 & 30.86 \end{bmatrix} \times 10^{-12} h^{-1}\,\text{m/N}$$

Using equation (7.95)

$$E_{Lx} = 1/h_L a_{11} = 1/8.17 \times 10^{-12} = 122.4\,\text{GN/m}^2$$
$$E_{Ly} = 1/h_L a_{22} = 23.15\,\text{GN/m}^2 \quad G_{Lxy} = 1/h_L a_{66} = 32.4\,\text{GN/m}^2$$

To find the major Poisson's ratio ν_{xy}, apply only N_x to the laminate: equation (7.94) then gives $\varepsilon^0_{Lx} = a_{11}N_x$ and $\varepsilon^0_{Ly} = a_{12}N_x$ so by definition

$$\nu_{xy} = -\varepsilon^0_{Ly}/\varepsilon^0_{Lx} = a_{12}/a_{11} = +8.65/8.17 = 1.059$$

To find the in-plane strain response to the general applied stresses, equation (7.94) is used in the form

$$[\varepsilon^0]_L = [a][N]_L = [a][\sigma]h$$
$$\varepsilon^0_{Lx} = (8.17 - 8.65) \times 10^{-12} \times 60 \times 10^6 = -28.8 \times 10^{-6}$$
$$\varepsilon^0_{Ly} = (-8.65 + 43.2) \times 10^{-12} \times 60 \times 10^6 = 2.07 \times 10^{-3}$$
$$\gamma^0_{Lxy} = 30.9 \times 10^{-12} \times 35 \times 10^6 = 1.08 \times 10^{-3}$$

The very large value of ε^0_{Ly} is expected because the laminate is not very stiff in the y-direction. In view of the high strains, resins would need to be chosen with care.

4. The balanced symmetrical laminate of thickness h is wound as $h/4$ at $+\theta$, $h/2$ at $-\theta$, and $h/4$ at $+\theta$, where θ is the angle to the axial x-direction. All laminae have the same stiffness properties. The hoop stress is $\sigma_y = pD_m/2h$, and the axial stress in the casing is $\sigma_x = pD_m/4h$. For a symmetric laminate $[\varepsilon^0] = [A]^{-1}[N]$, where $A_{11} = h\bar{Q}_{11}, A_{12} = h\bar{Q}_{12}$, $A_{22} = h\bar{Q}_{22}, A_{66} = h\bar{Q}_{66}$ and $A_{16} = A_{26} = 0$. Denoting $J = (\bar{Q}_{11}\bar{Q}_{22} - \bar{Q}^2_{12})$ for convenience,

$$\begin{bmatrix} \varepsilon^0_x \\ \varepsilon^0_y \\ \gamma^0_{xy} \end{bmatrix} = \begin{bmatrix} \bar{Q}_{22}/J & -\bar{Q}_{12}/J & 0 \\ -\bar{Q}_{12}/J & \bar{Q}_{11}/J & 0 \\ 0 & 0 & \bar{Q}_{66} \end{bmatrix} \begin{bmatrix} \sigma_x \\ \sigma_y \\ \tau_{xy} \end{bmatrix}$$

Hence the required condition is $2\bar{Q}_{11} = \bar{Q}_{12}$ provided $J \neq 0$. Substituting for $\bar{Q}_{ij}$ in terms of Q_{ij} gives a quadratic of the form $a\sin^4\theta + b\sin^2\theta + c = 0$. The condition is satisfied if $b^2 > 4ac$. Substituting numerical values, suitable winding angles would be 69° or 65.27°. This is not a complete solution to the problem of course; for example, the strength also has to be checked, especially in the axial direction.

5. Consider a unidirectional lamina stressed uniaxially in the x-direction at θ to its principal direction: $\sigma_x = \bar{Q}_{11}(\theta)\varepsilon_x$. For three identical laminae a, b, c inclined at θ_a, θ_b, θ_c, the total stress will be

$$\sigma_x = (1/3)[\bar{Q}_{11}(\theta_a) + \bar{Q}_{11}(\theta_b) + \bar{Q}_{11}(\theta_c)]\varepsilon_x$$

For a very large number of laminae randomly oriented in the plane, the total stress is given by

$$\sigma_x = (1/\pi)\int_{-\pi/2}^{+\pi/2} \bar{Q}_{11}(\theta)\mathrm{d}\theta\varepsilon_x$$

Substituting

$$\bar{Q}_{11} = Q_{11}\cos^4\theta + 2(Q_{12} + 2Q_{66})\sin^2\theta\cos^2\theta + Q_{22}\sin^4\theta$$

gives

$$\begin{aligned} E_r = \sigma_x/\varepsilon_x &= [3Q_{11} + 2(Q_{12} + 2Q_{66}) + 3Q_{22}]/8 \\ &= \frac{3E_1}{8(1-\nu_{12}\nu_{21})} + \frac{3E_2}{8(1-\nu_{12}\nu_{21})} + \frac{\nu_{12}E_2}{4(1-\nu_{12}\nu_{21})} + \frac{G_{12}}{2} \end{aligned}$$

Using $\nu_{12} \simeq 0.3$ and assuming $E_f \gg E_p$, it follows that $1 - \nu_{12}\nu_{21} \simeq 1$; from equations (7.4) and (7.11), $G_{12} \simeq E_2/2.6$ and $\nu_{12}E_2/4 = 0.075E_2$. Thus these terms in E_2 amount to $0.267E_2$, i.e. about $E_2/4$.

6. (a) We know that under N_x alone, $\varepsilon_x = a_{11}N_x$ and $\varepsilon_y = a_{12}N_x$, so

$$\nu_{xy} \equiv -\varepsilon_2/\varepsilon_1 = -a_{12}/a_{11} = -(-0.452/21.23) = +0.02129$$

(b) $[\varepsilon_x^0, \varepsilon_y^0] = [0.2123, -0.004516]\%$

$$[\sigma_x(0^\circ), \sigma_x(90^\circ)] = [383.1, 16.94] \text{ MPa}$$
$$[\sigma_y(0^\circ), \sigma_y(90^\circ)] = [3.895, -3.895] \text{MPa}$$

(c) $\kappa_y = d_{26}M_{xy} = 0$; $\kappa_{xy} = d_{66}M_{xy} = 1.92/\text{m}$

7. (a) $\nu_{xy} = -a_{12}/a_{11} = +8817/3225 = 2.734$. This is an unusually large value of the major Poisson's ratio: the material behaves like a sort of pin-jointed latticework of fibres.

(b) $[\varepsilon_x^0, \varepsilon_y^0, \gamma_{xy}^0] = [0.3225, -0.8817, 0]\%$
$[\sigma_x, \sigma_y, \tau_{xy}(+30^\circ), \tau_{xy}(-30^\circ)] = [0.25, 0, 0.1431, -0.1431]\text{MPa}$. The stress profile $\tau_{xy}(z)$ arises because the shear strains in each ply are suppressed. We know that $Q_{16}^*(-30^\circ) = -Q_{16}^*(+30^\circ)$, and hence we find (Fig. A10)

$$\tau_{xy}(+30^\circ) = Q_{16}^*(+30^\circ)\varepsilon_x^0 + Q_{26}^*(+30^\circ)\varepsilon_y^0 = +0.1644 \text{ MPa}$$
$$\tau_{xy}(-30^\circ) = Q_{16}^*(-30^\circ)\varepsilon_x^0 + Q_{26}^*(-30^\circ)\varepsilon_y^0 = -0.1644 \text{ MPa}$$

(c) $[\kappa_x, \kappa_y, \kappa_{xy}] = [0.2729, -0.6736, -0.06294]/\text{m}$. The bending strain profiles are linear through the thickness of the laminate. This means

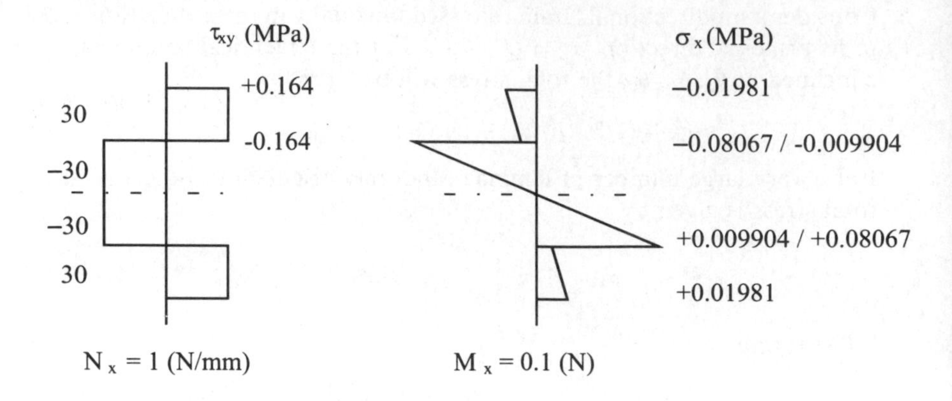

Fig. A10 Stress profiles (ii) and (iii) (refer to problem 7)

that we must have the same strain at the interface between the $+30°$ and $-30°$ plies. The discontinuity at the interface $z = i$ is caused only by the change in sign of Q^*_{16}:

$$\sigma_x(i, +30°) = Q^*_{11}(+30°)i\kappa_x + Q^*_{12}(+30°)i\kappa_y + Q^*_{16}(+30°)i\kappa_{xy}$$
$$\sigma_x(i, -30°) = Q^*_{11}(-30°)i\kappa_x + Q^*_{12}(-30°)i\kappa_y + Q^*_{16}(-30°)i\kappa_{xy}$$

Section 7.5

1. The suggested approach is valid for isotropic materials, but for an anisotropic lamina the angle θ to the principal direction is essential as well.
2. As this is a balanced symmetrical laminate, only in-plane stresses result from the applied strains.
   (a) For the 0° laminae which are stressed in the principal directions, equation (7.71) gives

$$\begin{bmatrix}\sigma_x\\ \sigma_y\\ \tau_{xy}\end{bmatrix} = \begin{bmatrix}\sigma_1\\ \sigma_2\\ \tau_{12}\end{bmatrix} = \begin{bmatrix}56 & 4.6 & 0\\ 4.6 & 18.7 & 0\\ 0 & 0 & 8.9\end{bmatrix}\begin{bmatrix}1 & \times & 10^{-3}\\ 1 & \times & 10^{-3}\\ 1 & \times & 10^{-3}\end{bmatrix} \times 10^9$$

Hence

$$\sigma_1 = 56 \times 10^6 + 4.6 \times 10^6 = 60.6\ \text{MN/m}^2$$
$$\sigma_2 = 4.6 \times 10^6 + 18.7 \times 10^6 = 23.3\ \text{MN/m}^2$$
$$\tau_{12} = 8.9\ \text{MN/m}^2$$

Applying the Tsai–Hill criterion

$$\frac{1}{F^2} = \left(\frac{60.6}{1000}\right)^2 - \frac{60.6 \times 23.3}{(1000)^2} + \left(\frac{23.3}{30}\right)^2 + \left(\frac{8.9}{45}\right)^2 = 0.64$$

and hence $F = 1.25$, a slender margin occasioned by the low transverse strength. The 90° laminae have the same value of F by virtue of the symmetry of the laminae and the strain.

(b) $\tau_{12} = 8.9 \times 10^9 \times 3 \times 10^{-3} = 26.7\ \mathrm{MN/m^2}$
$1/F = \tau_{12}/\hat{\tau}_{12} = 26.7/45 = 0.593$ and so $F = 1.69$.

(c) The applied strains are not symmetrical so laminae in both directions will have to be examined.
0° laminae:

$$\begin{bmatrix} \sigma_x \\ \sigma_y \\ \tau_{xy} \end{bmatrix} = \begin{bmatrix} \sigma_1 \\ \sigma_2 \\ \tau_{12} \end{bmatrix} = \begin{bmatrix} 56 & 4.6 & 0 \\ 4.6 & 18.7 & 0 \\ 0 & 0 & 8.9 \end{bmatrix} \begin{bmatrix} 1 \times 10^{-3} \\ 5 \times 10^{-4} \\ 1 \times 10^{-3} \end{bmatrix} \times 10^9$$

hence $\sigma_1 = 58.3\ \mathrm{MN/m^2}, \sigma_2 = 13.95\ \mathrm{MN/m^2}$ and $\tau_{12} = 8.9\ \mathrm{MN/m^2}$ leading to $F = 1.97$.

90° laminae: the 90° laminae are stressed at 90° to their principal directions, so what is required is $\bar{Q}(90)$. This can be obtained from Q by applying equation (7.63) or simply by inspection, and gives

$$\begin{bmatrix} \sigma_1 \\ \sigma_2 \\ \tau_{12} \end{bmatrix} = \begin{bmatrix} 18.7 & 4.6 & 0 \\ 4.6 & 56 & 0 \\ 0 & 0 & 8.9 \end{bmatrix} \begin{bmatrix} 1 \times 10^{-3} \\ 5 \times 10^{-4} \\ 1 \times 10^{-3} \end{bmatrix} \times 10^9$$

hence $\sigma_1 = 21\ \mathrm{MN/m^2}$, $\sigma_2 = 32.6\ \mathrm{MN/m^2}$ and $\tau_{12} = 8.9\ \mathrm{MN/m^2}$.
Evidently the 90° laminae fail, because of the high stiffness.

3. Let x be the hoop direction, y the axial direction (an unusual convention, but clearly defined).
   (a) Axial load $P_y = 4\ \mathrm{kN}$. We add the subscript A to denote axial load.

$$N_{y\mathrm{A}} = 4000/2\pi \times 50 = 12.74\ \mathrm{N/mm}$$
$$\varepsilon_{x\mathrm{A}} = a_{12} N_{y\mathrm{A}} = -65.7 \times 10^{-6} \times 12.74 = -0.837 \times 10^{-3}$$
$$\varepsilon_{y\mathrm{A}} = a_{22} N_{y\mathrm{A}} = 128.9 \times 10^{-6} \times 12.74 = +1.642 \times 10^{-3}$$
$$\gamma_{xy\mathrm{A}} = 0 \text{ because } a_{16} = 0$$

$$\sigma_{x\mathrm{A}} = \bar{Q}_{11}\varepsilon_x + \bar{Q}_{12}\varepsilon_y = 20.97 \times -0.837 + 10.69 \times 1.642 = 0$$
$$\sigma_{y\mathrm{A}} = \bar{Q}_{12}\varepsilon_x + \bar{Q}_{22}\varepsilon_y = 10.69 \times -0.837 + 20.97 \times 1.642 = 25.486\ \mathrm{MPa}$$
$$\tau_{xy\mathrm{A}} = \bar{Q}_{16}\varepsilon_x + \bar{Q}_{26}\varepsilon_y = 8.874\,(-0.837 + 1.642) = 7.144\ \mathrm{MPa}$$

$$\begin{bmatrix} \sigma_1 \\ \sigma_2 \\ \tau_{12} \end{bmatrix} = \begin{bmatrix} 0.5 & 0.5 & +1 \\ 0.5 & 0.5 & -1 \\ -0.5 & 0.5 & 0 \end{bmatrix} \begin{bmatrix} 0 \\ 25.486 \\ 7.144 \end{bmatrix} \text{(MPa)}$$

and then find the stresses in principal directions:

$$(\sigma_{1A}, \sigma_{2A}, \tau_{12A}) = (19.887,\ 5.599,\ 12.743)\text{MPa}$$

Applying the Tsai-Hill criterion:

$$\begin{aligned} 1/F^2 &= (19.887/1000)^2 - 19.887 \times 5.599/10^6 + (5.599/100)^2 + (12.743/80)^2 \\ &= 0.02879 \Rightarrow F = 5.89 \end{aligned}$$

The reader can confirm that the strength of the $-45°$ ply is the same as that of the $+45°$ ply.

(b) This part of the problem can be solved using unit pressure $p = 1$ N/mm^2. Then multiply unit pressure by the Tsai–Hill load factor to obtain the predicted failure pressure. We use subscript ‘p’ to denote pressure load.

Using $R =$ mean radius, and assuming thin walls, the pipe wall stresses are $\sigma_x = pR/h = 100$ N/mm^2, $\sigma_y = 50$ N/mm^2, thus $N_x = \sigma_x h =$ 50 N/mm, $N_y = 25$ N/mm, and by inspection $N_{xy} = 0$.

Global strains follow:

$$\begin{aligned} \varepsilon_x &= a_{11}N_x + a_{12}N_y = 4.8 \times 10^{-3} \\ \varepsilon_y &= a_{12}N_x + a_{22}N_y = -6.25 \times 10^{-3} \\ \gamma_{xy} &= 0 \end{aligned}$$

The stresses in the $+45°$ ply in global coordinates are

$$\begin{aligned} \sigma_x &= \bar{Q}_{11}\varepsilon_x + \bar{Q}_{12}\varepsilon_y = 100 \text{ N/mm}^2 \\ \sigma_y &= \bar{Q}_{12}\varepsilon_x + \bar{Q}_{22}\varepsilon_y = 50 \text{ N/mm}^2 \\ \tau_{xy} &= \bar{Q}_{16}\varepsilon_x + \bar{Q}_{26}\varepsilon_y = 42.04 \text{ N/mm}^2 \end{aligned}$$

Some students miss the ply shear stress which arises because shear strain is prevented in a bonded balanced laminate.

Transforming ply stresses to principal directions gives

$$(\sigma_{1p}, \sigma_{2p}, \tau_{12p}) = (117.04, 32.96, -25)\text{N/mm}^2$$

Applying Tsai–Hill gives:

$$\begin{aligned} 1/F^2 &= (117.04/1000)^2 - 117.04 \times 32.96/10^6 + (32.96/100)^2 + (-25/80)^2 \\ &= 0.21614 \end{aligned}$$

hence $F = 2.15$.

The expected failure pressure is $Fp = 2.15\ \mathrm{N/mm^2}$.

From the stress/strength ratios ($\sigma_1/\sigma_{1\max} = 0.117$, $\sigma_2/\sigma_{2\max} = 0.329$, $\tau_{xy}/\tau_{xy\max} = 0.313$) we see that failure is likely to be dominated by a mixture of transverse tension with shear. We cannot be more definite about this because two ratios have similar values.

(c) The load is constant, but the pressure changes. Note that applying the failure criterion to the combination is not correct because then the factor applies to the load as well as the pressure. Using subscripts A and p for the axial load and the pressure we have already found that

$$(\sigma_{1\mathrm{A}}, \sigma_{2\mathrm{A}}, \tau_{12\mathrm{A}}) = (19.887, 5.599, 12.743)\ \mathrm{N/mm^2}$$
$$(\sigma_{1\mathrm{p}}, \sigma_{2\mathrm{p}}, \tau_{12\mathrm{p}}) = (117.04, 32.96, -25)\ \mathrm{N/mm^2}$$

Failure is expected when

$$\sigma_1' = \sigma_{1\mathrm{A}} + k\sigma_{1\mathrm{p}};\ \sigma_2' = \sigma_{2\mathrm{A}} + k\sigma_{2\mathrm{p}};\ \tau_{12}' = \tau_{12\mathrm{A}} + k\tau_{12\mathrm{p}}$$

where k is the pressure at which failure occurs. The value of k must satisfy the Tsai–Hill criterion.

$$(\sigma_1'/\sigma_{1\max})^2 - \sigma_1'\sigma_2'/\sigma_{1\max}^2 + (\sigma_2'/\sigma_{2\max})^2 + (\tau_{12}'/\tau_{12\max})^2 = 1$$

and for this material we know

$$\sigma_{1\mathrm{T}\max} = 1000\ \mathrm{MPa}, \sigma_{2\mathrm{T}\max} = 100\ \mathrm{MPa}, \tau_{12\max} = 80\ \mathrm{MPa}$$

Thus we need to solve

$$(\sigma_{1\mathrm{A}} + k\sigma_{1\mathrm{p}})^2 - (\sigma_{1\mathrm{A}} + k\sigma_{1\mathrm{p}})(\sigma_{2\mathrm{A}} + k\sigma_{2\mathrm{p}}) + 100(\sigma_{2\mathrm{A}} + k\sigma_{2\mathrm{p}})^2 + 156.25(\tau_{12\mathrm{A}} + k\tau_{12\mathrm{p}})^2 - 10^6 = 0$$

and we find $k = -1.987$ or $+2.26$. Only the second term is sensible, so failure is expected at 2.26 MPa. The stress strength ratios are ($\sigma_1/\sigma_{1\max} = 0.28, \sigma_2/\sigma_{2\max} = 0.8, \tau_{xy}/\tau_{xy\max} = 0.55$), so failure is likely to be dominated by transverse tension.

Note that the quadratic expression in k can also be solved for problems (a) and (b) to confirm what has already been found by the simpler approach used there.

4. The bending strain profiles under $M_x = 10\ \mathrm{N}$ are linear and continuous through the thickness of this symmetrical cross-ply laminate. The bending stress σ_x reflects the high modulus of the top and bottom plies and the smaller modulus of the middle plies. If we consider the interface between the two bottom plies, both have the same strain at the interface. If at the

same strain but not bonded together, the bottom 0° ply would want to contract more than the 90° ply at the interface. To achieve the same strain at the bonded interface, the 0° ply is stretched more in the y-direction, and the 90° ply is compressed a little in the y-direction. At the midplane the bending strains ε_x and ε_y are zero so $\sigma_y = 0$. The net transverse force is zero as there is no external force in the y-direction.

Under $M_x = 10\,\mathrm{N}$ the first failure is dominated by the large transverse compressive stress at the lower surface of ply 3. It is interesting to note that, in comparison with the behaviour of an isotropic plate, first ply failure in this laminate does *not* occur at the lower surface.

CHAPTER 8

Section 8.2

1. See Fig. A11.

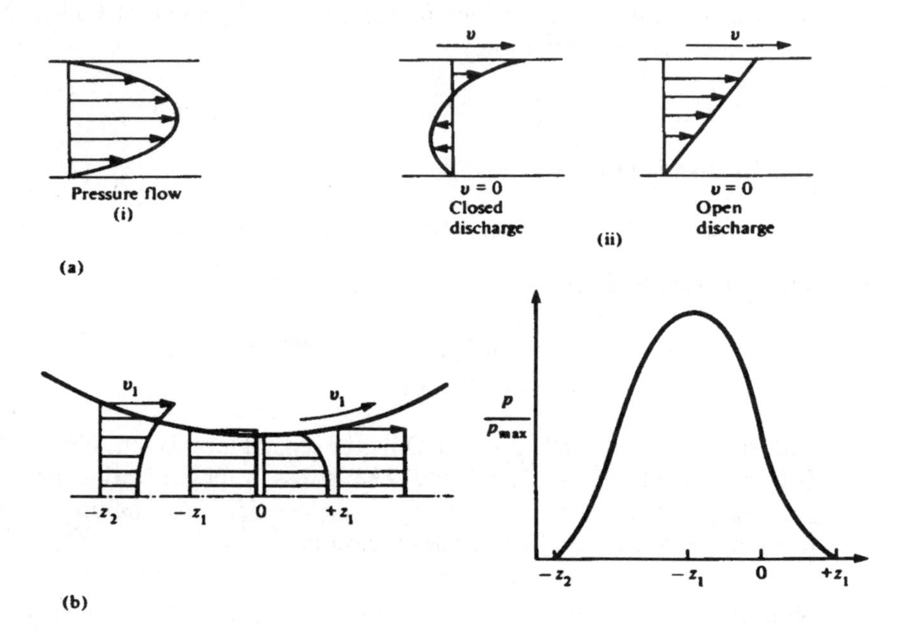

Fig. A11 (a) Velocity profiles between parallel plates. (b) Velocity profiles and pressure distribution in the two-roll mill

2. (a) See Fig. A12.
   (b) Considering radial flow only, the maximum shear rate at the gate is

$$\dot{\gamma}_0 = 6Q/WH^2 = 6Q/2\pi R_0 H^2$$

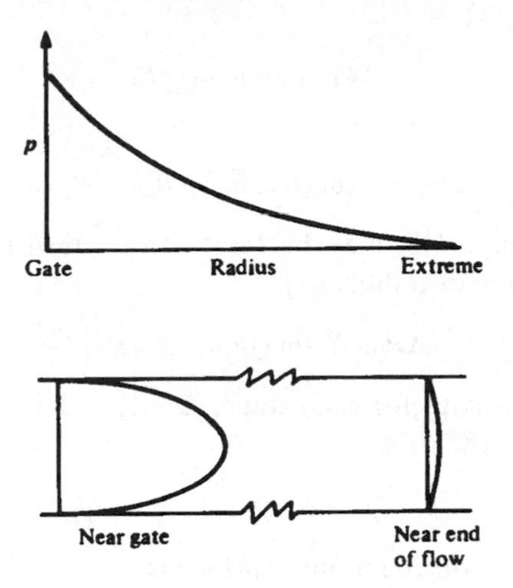

Fig. A12 Pressure profile along a radial flow path and velocity profiles for spreading disc flow

and the minimum rate at the end of the flow is

$$\dot{\gamma}_1 = 6Q/2\pi R_1 H^2$$

so $\dot{\gamma}_0/\dot{\gamma}_1 = R_1/R_0$.

(c) Cavity volume $V = \pi R_1^2 H$, and $t_{\text{fill}} = V/Q = \pi R_1^2 H/Q$ approximately.

(d) Locking force is pressure × projected area of cavity. This is not the same as the force needed to keep the mould closed during injection because then the pressure decreases along the flow path (problem 2(a)).

3. In a single-start extruder $W = \pi D \sin\theta$ ignoring e. Equation (8.5) gives

$$Q = \frac{1}{2}\pi^2 D^2 HN \sin\theta\cos\theta - \pi DH^3 \Delta P \sin^2\theta/(12L)$$

For a capillary die, Table 8.1, equation (O) gives $\Delta P = \eta Q/k$ with $k = \pi R_d^4/8L_d$. Substituting for $\Delta P/\eta$:

$$Q = (6kL\pi^2 D^2 HN \sin\theta\cos\theta)/(12kL + \pi DH^3 \sin^2\theta)$$

If only θ and H can be varied, the maximum output can be found from $\partial Q/\partial H = 0$ and from $\partial Q/\partial\theta = 0$; i.e.

$$H^3 = 6kL/\pi D\sin^2\theta \quad \text{and} \quad \sin^2\theta = 12kL/(24kL + \pi DH^3)$$

Solving simultaneously, $\sin^2\theta = 0.25$, i.e. $\theta = 30°$ with $H^3 = 24kL/\pi D$.

4. Runner (stage I)

$$\Delta P_{\mathrm{I}} = -8\eta Q L_{\mathrm{I}}/\pi R_{\mathrm{I}}^4$$

Disc (stage II)

$$\Delta P_{\mathrm{II}} = -(6\eta Q/\pi H_{\mathrm{II}}^3)\ln(R_{2\mathrm{II}} - R_{1\mathrm{II}})$$

The cylinder (stage III) may be treated as a thin slit with $W = 2\pi R^*$ where R^* is the mean radius:

$$\Delta P_{\mathrm{III}} = -6\eta Q L_{\mathrm{III}}/\pi R_{\mathrm{III}}^* H_{\mathrm{III}}^3$$

Adding pressure drops for each stage, $\Delta P_{\mathrm{total}} = 85\,\mathrm{MN/m^2}$. The general form of equation (8.26) is

$$p_1 - p_r = (6\eta Q/\pi H^3)\ln(r/R_1)$$

so the force acting on the mould surface is

$$F_x = \int_{R_1}^{R_2} 2\pi r p_r \mathrm{d}r$$

i.e.

$$F_x = \pi p_1(R_2^2 - R_1^2) - (6\eta Q/\pi H^3)[R_2^2\ln(R_2/R_1) - \tfrac{1}{2}(R_2^2 - R_1^2)] = 1.464\ \mathrm{MN}$$

5. For the die: $\Delta P = \eta Q/k$ with $k = \pi R^4/8L = 1.23 \times 10^{-9}\,\mathrm{m^3}$. For the screw: $\tan\theta = 1/\pi$, $\sin\theta = 0.303$, $\cos\theta = 0.953$. From problem 3,

$$6kL\pi^2 D^2 HN \sin\theta\cos\theta = 5.22 \times 10^{-15}$$
$$12kL = 2.952 \times 10^{-9}$$

and

$$\pi D H^3 \sin^2\theta = 5.766 \times 10^{-11}$$

Hence

$$Q = 5.22 \times 10^{-15}/(2.952 \times 10^{-9} + 5.766 \times 10^{-11}) = 1.7344 \times 10^{-6}\,\mathrm{m^3/s}$$

i.e. 5 kg/h.

$$W \simeq \pi D \sin\theta = 23.8\,\mathrm{mm}$$

so

$$q_{\mathrm{d}} = \tfrac{1}{2} W v_1 h = 1.87 \times 10^{-6}\,\mathrm{m^3/s}$$

This gives $q_{\mathrm{p}} = Q - q_{\mathrm{d}} = -0.136 \times 10^{-6}\,\mathrm{m^3/s}$, and $q_{\mathrm{p}}/q_{\mathrm{d}} = -0.072$.

6. $R = 0.05\,\mathrm{m}$, $L = 0.4\,\mathrm{m}$, $H_0 = 0.001\,\mathrm{m}$, $H_1 = 0.00116\,\mathrm{m}$, $N = 1/\mathrm{s}$: hence $v_1 = \pi DN = 0.314\,\mathrm{m/s}$ and from $z_1^2 = R(H_1 - H_0) = 8 \times 10^{-6}\,\mathrm{m^2}$, $z_1 = 2.83\,\mathrm{mm}$.

$$p_{\max} = 16 \times 10^4 \times 0.134 \times 0.00283^3/(0.005 \times 10^{-9}) = 22.75\,\mathrm{MN/m^2}$$

This maximum pressure occurs 2.83 mm upstream from the nip. To find where the rolls engage the feed, put $p = 0$ in equation (8.18) with $B = 16\alpha_1^3/3$: this gives an equation in α_2, in which $\alpha_1 = 0.4$. From

$$\alpha_2(0.52\alpha_2^2 - 1.8)/(1 + \alpha_2^2)^2 = -0.52\tan^{-1}\alpha_2 - 0.341$$

$\alpha_2 = -1.35$ or $z_2 = -9.54$ mm, upstream from the nip.

At the point where the rolls engage the feed, the inclination of the roll is at $\tan^{-1}$ (9.54/50), i.e. 10.8°, so the lubrication approximation is still just about valid.

Ignoring roll curvature,

$$F_x = L\int_{z_0}^{z_1} p\mathrm{d}z = L\surd(RH_0)\int_{\alpha_0}^{\alpha_1} p(\alpha)\mathrm{d}\alpha$$

Substituting for $p(\alpha)$, integrating and tidying up gives

$$\begin{aligned}\mathrm{F}_x = (3\eta v_1 LR/2H_0)[(\alpha_2^2 - \alpha_1^2)/(1 + \alpha_2^2) + 16\alpha_1^3(\alpha_1 - \alpha_2)/3 \\ + (1 - 3\alpha_1^2)(\alpha_1\tan^{-1}\alpha_1 - \alpha_2\tan^{-1}\alpha_2)\end{aligned}$$

from which $F_x = 49.8$ kN, about 5 tonnes.

7. This rough model is of interest where a (double-) tapered annular cross-section is to be used in a pipe die. We assume that $W_\mathrm{i} \gg H_\mathrm{i}$. At any downstream coordinate z we have

$$\begin{aligned}W_z &= W_1(1 - az) \quad \text{with} \quad a = (\tan\phi)/L = (W_1 - W_2)/W_1 L \\ H_z &= H_1(1 - bz) \quad \text{with} \quad b = (\tan\theta)/L = (H_1 - H_2)/H_1 L\end{aligned}$$

and thus for small taper angles we find

$$\begin{aligned}\Delta P &= \int[12\eta Q/W_z H_z^3]\mathrm{d}z = [12\eta Q/W_1 H_1^3]\int[(1 - az)^{-1}(1 - bz)^{-3}]\mathrm{d}z \\ &= [12\eta Q/W_1 H_1^3]I\end{aligned}$$

The integral over the die length 0 to L can be evaluated by standard forms or by partial fractions. After some tedious detail we find

$$\begin{aligned}I =& [-1/2(a - b)][(1 - bL)^{-2} - 1] - [a/(a - b)^2][(1 - bL)^{-1} - 1] \\ &+ [a^2/(a - b)^3][\ln(1 - bL/(1 - aL)]\end{aligned}$$

Noting that $1 - bL = +H_2/H_1$, and $1 - aL = +W_2/W_1$, we find

$$\begin{aligned}I =& [1/2(a - b)][1 - (H_1/H_2)^2] + [a/(a - b)^2][1 - H_1/H_2] \\ &+ [a^2/(a - b)^3][\ln(H_2/H_1)(W_2/W_1)]\end{aligned}$$

As a rough check we may suppose that one taper is suppressed.

(a) if $W_1 \approx W_2$, $a \approx 0$, $I \approx (1/2b)[(H_1/H_2)^2 - 1]$

(b) if $H_1 \approx H_2$, $b \approx 0$, $I \approx (1/a)\ln(W_1/W_2)$

Both are sensible reductions to known special cases.

(Anticipating the remarks in section 8.4.2, a similar approach is not possible for a power-law fluid, and the result must be calculated numerically.)

Section 8.3

1. LDPE at 170°C below $10^3\,\mathrm{N/m^2}$, i.e. about 0.01/s. LDPE at 210°C below $10^3\,\mathrm{N/m^2}$, i.e. about 0.03/s. Polyacetal at 200°C below $4 \times 10^4\,\mathrm{N/m^2}$, i.e. about 50/s.
2. Nylon: at 285°C: $n - 1 = \ln(165/145)/\ln(10/100)$ gives $n = 0.944$. Estimating $4\,\mathrm{N\,s/m^2}$ at 10^5/s gives $n - 1 = \ln(30/4)/\ln(10^4/10^5); n = 0.125$.
3. For the extruder $Q = C_1 - C_2\Delta P/\eta$, and for the die $Q = C_3\Delta P/\eta$. The conclusion is that Q is independent of viscosity (see problem 3 in section 8.2).
4. $\dot{\gamma}_{\mathrm{T}} = [(2n + 1)/3n]\dot{\gamma}_{\mathrm{a}} = C_{\mathrm{R}}\dot{\gamma}_{\mathrm{a}}$
   At low shear rates less than 5/s, $n = 1$ hence $C_{\mathrm{R}} = 1$. At high shear rates n is approximately constant at about 0.255: $C_{\mathrm{R}} = 1.97$. At an apparent rate of 100/s, shear stress is $1.4 \times 10^5\,\mathrm{N/m^2}$, but at 197/s the stress is $1.85 \times 10^5\,\mathrm{N/m^2}$.
5. For convenience work with data for $0.1\,\mathrm{MN/m^2}$ and 1/s. $T^* = 150$°C, $\eta_{\mathrm{a}}^* = 2.8 \times 10^4\,\mathrm{N\,s/m^2}$; $T = 210$°C, $\eta_{\mathrm{a}} = 1.3 \times 10^4\,\mathrm{N\,s/m^2}$: $b_{\mathrm{T}} = 8.26 \times 10^{-2}$/°C.

Section 8.4

1. (b) Equation (8.28) defines apparent viscosity based on an assumed Newtonian fluid with an apparent shear rate. Equation (8.29) defines power-law behaviour.
   (c) The ratio of the shear rates is $(3n + 1)/4n$, the Rabinowitsch correction.
2.

$$\dot{\gamma}_{\mathrm{a1}} = 1180/\mathrm{s};\ \eta_{\mathrm{a1}} = 280\,\mathrm{N\,s/m^2},\ \tau_1 = 3.2 \times 10^5\,\mathrm{N/m^2}$$

$$\dot{\gamma}_{\mathrm{a2}} = 9430/\mathrm{s};\ \eta_{\mathrm{a2}} = 56\,\mathrm{N\,s/m^2}$$

$$n - 1 = \ln(280/56)/\ln(1180/9430) = -0.774$$

$$n = 0.226$$

   Therefore

$$\Delta P = -(2 \times 3.2 \times 10^5)(2^{0.678} - 1)/(3 \times 0.226 \times 0.03) = -18.9\,\mathrm{MN/m^2}$$

   It is important to note that increasing the flow rate by 10 does not increase the pressure drop by 10.
3. For an untapered capillary, equation (8.51) can be rearranged and integrated; after some manipulation, $\Delta P = (2L/R)\tau_{\mathrm{w}}$, where τ_{w} is the shear

stress at the wall corresponding to a shear rate of $4Q/\pi R^3$. Using $Q = 10^{-5}\,\text{m}^3/\text{s}$, $R = 2.5 \times 10^{-3}\,\text{m}$, the shear rate is 815/s which corresponds to a shear stress of $1.6 \times 10^5\,\text{N/m}^2$, and $\Delta P = 38.4\,\text{MN/m}^2$.

Figure 8.13 shows that viscosity rises by rather more than a factor of 2 at constant flow rate when the pressure rises from atmospheric to $200\,\text{MN/m}^2$. The simplest approach to take pressure into account would be to split the runner into short lengths and use the viscosity for the average pressure in each length; an alternative approach is to derive a formula for pressure drop using a pressure-dependent viscosity in a linear form. This approach would not be significant at low flow rates.

4. The molten acrylic is evidently non-Newtonian and so to solve this problem it is necessary to derive the power-law equivalent of equation (8.26). The early part of the derivation has been done in equations (8.38)–(8.42) subject to replacement of $\partial p/\partial z$ by $\partial p/\partial r$ as explained in Fig. 8.7 and equation (8.23). The 'width' W of an element of channel is $W = 2\pi r$, so the modified (8.42) can be integrated:

$$p_1 - p_2 = \frac{2}{H}\left(\frac{2n+1}{3n}\frac{6Q}{2\pi H^2}\right)^n \frac{\eta_{T_0}}{\dot{\gamma}_{T_0}^{n-1}} \int_{R_1}^{R_2} \frac{dr}{r^n}$$

$$= \frac{2}{H}\left(\frac{2n+1}{3n}\frac{6Q}{2\pi R_1 H^2}\right)^n \frac{\eta_{T_0}}{\dot{\gamma}_{T_0}^{n-1}} \frac{R_1}{1-n}\left[\left(\frac{R_2}{R_1}\right)^{1-n} - 1\right]$$

but the first term in the parentheses is the true shear rate at the wall at the inlet radius R_1, so equation (8.38) can be used to give the shear stress at the wall at the inlet, τ_{xR_1}; giving

$$p_1 - p_2 = \frac{2}{H}\frac{R_1}{1-n}\tau_{xR_1}\left[\left(\frac{R_2}{R_1}\right)^{1-n} - 1\right]$$

Using subscripts 1 and 2 to refer to inlet and exit conditions at the flow boundary,

$$\dot{\gamma}_{a1} = 6Q/2\pi R_1 H^2 = 6 \times 2 \times 10^{-4}/(2\pi \times 3 \times 4 \times 10^{-9}) = 1.59 \times 10^4/\text{s}$$

so (from Fig. 8.11)

$$\tau_1 = 6 \times 10^5\,\text{N/m}^2 \text{ and } \eta_1 = 35\,\text{N s/m}^2;\ \text{and } \dot{\gamma}_{a2} = 136/\text{s}$$

so

$$\eta_2 = 1100\,\text{N s/m}^2$$

Hence $n = 0.275$ and numerical substitution gives

$$p_1 - p_2 = 75.5\,\text{MN/m}^2$$

(This does not include the pressure drop needed to pump material from the end of the barrel to the gate into the cavity.)

Progressive freezing off of polymer onto the cold mould surfaces would restrict the flow channel and therefore increase the pressure needed to fill the cavity.

5. For constant taper angle, pressure gradients are proportional to $R_1^{-3n}[(R_1/R_2)^{3n} - 1]$. The pressure gradient is greatest at the narrow end of the complete flow channel, as expected. Compared with the Newtonian fluid, the pressure gradient is more smoothed out along the length of the flow channel for a power-law index of $\frac{1}{3}$.
6. (i) Newtonian fluid. For a given constant volumetric flow rate Q, the average velocity is $\langle v \rangle = Q/\pi R^2$, so the average residence time $\langle t \rangle = L/\langle v \rangle$. The velocity profile is $v_z = 2\langle v \rangle(1 - r^2/R^2)$. So at $r = 0, v_{\max} = 2\langle v \rangle$, and $t_{\min} = \langle t \rangle/2$. At $r = 0.99R, v \approx 0.04\langle v \rangle$, so $t_{0.99R} = 25\langle t \rangle$.
   (ii) For a power-law fluid $\langle v \rangle$ is the same as for a Newtonian fluid, so the average residence time $\langle t \rangle$ is unchanged. We can find

$$v_z = [(3n + 1)(n + 1)]\langle v \rangle[1 - (r/R)^{1+1/n}]$$

   Thus at the centre line the residence time $t_{\min} = \langle t \rangle/1.4$, which is greater than for a Newtonian fluid. At $r = 0.99R, t_{0.99R} = 14.57\langle t \rangle$, thus a lower residence time nearer the wall for a power-law fluid, a useful result for thermally sensitive melts.

Section 8.5

1. Two phenomena can be observed: a difference in residence time between the core and outer layer due to the velocity distribution (Table 8.1, B) and also (if the pipe is relatively short) a difference in temperature due to the low heat conduction. Both contribute to a low degree of reaction in the core, while at the pipe wall over-reaction is likely. Both can be improved either by introducing static mixing elements which, along the pipe, exchange wall and core layers, or by introducing in the pipe a rotating extruder screw or a rotating distributive mixing device, which cleans the pipe surface regularly and exchanges wall and core material.

Section 8.6

1. See answer to problem 4, Chapter 4.
2. Let the parison length be L_0 and assume a constant cross-section A along the complete length. Consider an element of the parison of length $\mathrm{d}x$ a distance x up from the free end. The stress σ_x on the lower face of the element is $\sigma_x = \rho g x$, where ρ is the density of the molten parison. The strain rate $\dot{\varepsilon}_x$ in the element is $\dot{\varepsilon}_x = \sigma_x/\lambda$. If λ is constant, $\varepsilon_x = \sigma_x t/\lambda$ and hence the change in length of the element is $\mathrm{d}L_x = (\sigma_x t/\lambda)\mathrm{d}x$. The change in length of the parison is therefore

$$\Delta L_x = (\rho g t/\lambda)\int_0^L x\mathrm{d}x = \rho g t L^2/2\lambda$$

From Fig. 8.22 the specific volume of low-density polyethylene at 170°C is about $1.28\,\mathrm{cm^3/g}$, i.e. $\rho = 780\,\mathrm{kg/m^3}$, so $\hat{\sigma}_x = \rho g L = 7650\,\mathrm{N/m^2}$. At this stress the elongational viscosity is $\lambda = 2.8\times10^5\,\mathrm{N\,s/m^2}$ from Fig. 8.17. The change in parison length after hanging for 3 s is therefore

$$\Delta L_x = 780\times9.81\times3\times1^2/(2\times2.8\times10^5) = 41\,\mathrm{mm}$$

This corresponds to an average strain of 4.1%; the actual strain at the die end would be much larger than this, although the effect would be to some extent compensated by die swell.

3. If the radius R at some distance z downstream from R_1 is $R = R_1(1-kz)$ where $k = (R_1 - R_2)/R_1 L$, then $k = (\tan\theta)/R_1$. At constant volume flow rate $Q = A\langle v\rangle$, then $\mathrm{d}Q = 0 = A\mathrm{d}\langle v\rangle + \langle v\rangle\mathrm{d}A$, and hence

$$\dot{\varepsilon} = (1/L)(\mathrm{d}L/\mathrm{d}t) = \mathrm{d}\langle v\rangle/\langle v\rangle = -\mathrm{d}A/A\mathrm{d}t$$

Now $A = \pi R_1^2(1-kz)^2$ so

$$\mathrm{d}A/\mathrm{d}t = -2kR_1^2(1-kz)\mathrm{d}z/\mathrm{d}t$$

and

$$\mathrm{d}z/\mathrm{d}t = Q/A$$

and hence

$$\dot{\varepsilon} = -\mathrm{d}A/A\mathrm{d}t = 2kQ/[\pi R_1^2(1-kz)^3]$$

Putting $1 - kz = R/R_1$,

$$\dot{\varepsilon} = 2Q\tan\theta/\pi R^3 = \tfrac{1}{2}\dot{\gamma}_w\tan\theta$$

Section 8.7

1. (a) If $W \gg H$ then the shear rate across the midplane will be negligible compared with that through the thickness H, in the ratio H^2/W^2. There would be some local effect near the edge of the width. (b), (c) and (d) – see Fig. A13.
2. $\dot{\gamma} = 6Q/WH^2 = 6Q/2\pi R_m H^2 = 6\times5\times10^{-6}/2\times3.14\times0.06\times4\times10^{-6}$ $= 20/\mathrm{s}$

$\tau_w = 6\times10^4\,\mathrm{N/m^2}$ gives $\gamma_R = 1.7$, hence $B_{SH} = 1.24$ and $B_{SW} = 1.11$. The pipe dimensions after swelling are $H_e = 1.24\times2 = 2.48\,\mathrm{mm}$, and $R_e = 66.6\,\mathrm{mm}$ (mean), both measured hot. The cross-sectional area of the hot pipe is $2\pi R_e H_e = 1.04\times10^{-3}\,\mathrm{m^2}$. The volumetric flow rate is $Q = Av$, so the haul-off speed is $v = 5\times10^{-5}/1.04\times10^{-3} = 48\,\mathrm{mm/s}$. The mass flow rate from the given data is $Q\rho = 135\,\mathrm{kg/h}$.

Shape of extrudate from wide slot die

Shape of extrudate from square-section die

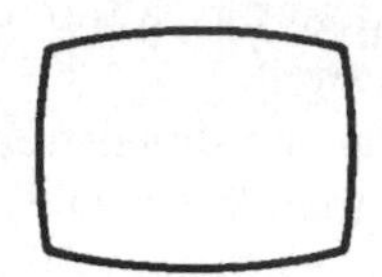

Shape of die needed to make extrudate of rectangular cross section

Fig. A13 The effect of die swell

3. Combining equations (B) and (C) in Table 8.1,

$$v_z = -(2Q/\pi)(1/R^2 - r^2/R^4)$$

hence

$$dv_z/dr = -(4Q/\pi R^3)(r/R)$$

i.e. a linear strain rate profile. At any radius r the time taken to pass through the die of length L is

$$t(r) = L/v_z(r)$$

The amount of shear strain is

$$\gamma = \dot{\gamma}t = (L/2R)(r/R)[1/(1 - r^2/R^2)]$$

Thus the amount of shear is negligible for $r/R < 0.5$, i.e. the centre of the channel, but near the wall ($r/R > 0.9$) the amounts of shear strain can be enormous.

4. If R_r = molten rod radius after swelling and $\bar{v}_r$ is the average rod velocity then

$$Q = \pi R_r^2 \bar{v}_r = 7.5 \times 10^{-7} \text{m}^3/\text{s}$$

The die radius R_d must be chosen so that the shear rate at the wall $\dot{\gamma}$ gives the required die swell B. This requires a small amount of iteration.

To start the exercise assume (incorrectly of course) that there is no die swell ($B = 1$). Then $R_{d1} = 4 \times 10^{-3}$ m, $\dot{\gamma}_1 = 15$/s, so from Fig. 8.11

$\tau_1 = 3.2 \times 10^4\,\mathrm{N/m^2}$, and from Fig. 8.18 $\gamma_{R1} = 1.55$ and so from Fig. 8.20 $B_1 = 1.21$. From our incorrect assumption the resulting rod would be too big ($R_{r1} = R_{d1}B_1$).

For the second attempt, reduce the die radius by B_1 and multiply the average die velocity by B_1^2 to maintain constant Q. Hence $R_{d2} = 4 \times 10^{-3}/1.21 = 3.3 \times 10^{-3}\,\mathrm{m}$; $\bar{v}_{d2} = 15 \times 10^{-3} \times 1.21^2 = 22 \times 10^{-3}\,\mathrm{m/s}$, $\dot{\gamma}_2 = 4\bar{v}_{d2}/R_{d2} = 26.7/\mathrm{s}$, $\tau_2 = 4.2 \times 10^4\,\mathrm{N/m^2}$, $\gamma_{R2} = 1.72$, $B_2 = 1.24$. So the rod would still be too big (4.092 mm).

To reduce the die further, try $B_3 = 1.25$. $R_{d3} = 4 \times 10^{-3}/1.25 = 3.2 \times 10^{-3}\,\mathrm{m}$. $v_{d3} = 23.4 \times 10^{-3}\,\mathrm{m/s}$, $\dot{\gamma}_3 = 29.3\,/\mathrm{s}$, $\tau_3 = 4.3 \times 10^4\,\mathrm{N/m^2}$, $\gamma_{R3} = 1.75$, $B = 1.25$. Thus the die land radius should be $R_{d3} = 3.2 \times 10^{-3}\,\mathrm{m}$.

Section 8.8

1. The midplane corresponds with $\xi = 0$, and the plate thickness is $2H$. The first step is to calculate the dimensionless temperature which is the same for all three conditions: using equation (8.70) we find $Y = (80 - 20)/(120 - 20) = 0.6$.
   (a) From Fig. 8.25 for $Y_1 = 0.6$ we estimate $Fo = 0.32$, and hence from equation (8.68) we have $t = H^2 Fo/\alpha = (5 \times 10^{-3})^2 \times 0.32/10^{-7} = 80\,\mathrm{s}$.
   (b) From equation (8.64) we have $Bi = hH/k$. Assuming $h \approx 1500\,\mathrm{W/m^2\,K}$ from Table 8.2, $Bi = 1500 \times 5 \times 10^{-3}/0.2 = 37.5$. Using Fig. 8.27 with $Bi \approx \infty$, for $Y_2 = 0.6$ we find $Fo = 0.3$, so $t = (5 \times 10^{-3})^2 \times 0.3/10^{-7} = 75\,\mathrm{s}$.
   (c) Ignoring radiation and convection, and thus using $h \approx 20\,\mathrm{W/m^2K}$ we find $Bi = 20 \times (5 \times 10^{-3})^2/0.2$ giving $Fo \approx 1.35$ so that $t = 337\,\mathrm{s}$.
   These results confirm several common sense observations: cooling in notionally still air is much slower than cooling in stirred cold water or between cold metal plates; cooling in stirred water is, under these circumstances, efficient and almost as efficient as cooling between cold metal plates. Cooling of polymer melts by immersion in cold water is commonly used after extrusion, and cooling between metal plates is relevant to the injection moulding process.
2. (a) Assume perfect thermal contact between the polymer and the mould on both polymer surfaces: the midplane cools most slowly. $Y_1 = (60 - 20)/(220 - 20) = 0.2$, leading to $Fo = 0.8$. Cooling from both sides means $H = 1.5\,\mathrm{mm}$, so that $t = H^2 Fo/\alpha = (1.5 \times 10^{-3})^2 \times 0.8/10^{-7} = 18\,\mathrm{s}$.
   (b) We assume the buckets are made in pairs and cut in half, or that the bucket is blown with a lid which is then cut away. There is effectively no cooling from the free inside surface of the blown product: the heat capacity of the air is very small and we have already seen (problem 1) that exposure to air provides inefficient cooling. Thus cooling can

only take place from the outer surface. This is equivalent to having a notional product which is twice as thick as the real product with the same kind of cooling on both surfaces. We assume perfect thermal contact. Thus $Y_1 = 0.2$ as before, but now $H = 3\,\text{mm}$. Thus $t = (3 \times 10^{-3})^2 \times 0.8/10^{-7} = 72\,\text{s}$.

The main conclusion is that extrusion blow moulding is thermally less efficient than injection moulding for the same wall thickness and temperatures used. However, bear in mind that in practice injection moulding usually takes place at higher temperatures with polymers having different molecular masses and mass distributions, that the shape possibilities for both processes are quite different, and that the capital investment for blow moulding is much less than for injection moulding, a feature attractive if smaller quantities are required.

3. For typical extrusion pressures of 20 MPa, 8.73 suggests $\Delta T = \Delta P/\rho c_p = 20 \times 10^6/2 \times 10^6 = 10\,\text{K}$. For moulding typically $\Delta P \approx$ 100 MPa, sometimes even 200 MPa, thus $\Delta T \approx$ 50–100 K. These are average temperature rises and locally the temperature rise can be larger: if relatively high melt temperatures are used, then there are real concerns about the possibility of local degradation in thermoplastics or premature crosslinking in thermosetting polymers or rubbers.
4. For LDPE $T_m = 115°\text{C}$, so we could usefully process at 170°C for which data are available in this book. From Fig 8.22 $\rho_{20} = 917\,\text{kg/m}^3$ and $\rho_{170} = 775\,\text{kg/m}^3$. The volume flow rate is $Q_{20} = 23.5/917 \times 3600 = 7.12 \times 10^{-6}\,\text{m}^3\,\text{s}$. The cross-sectional area is $A_{20} = \pi R_{20}^2 = 1.77 \times 10^{-4}\,\text{m}^2$, thus the product velocity after cooling is $v_{20} = Q_{20}/A_{20} = 0.040\,\text{m/s}$.

We see a volume expansion of 15% and therefore an average linear thermal expansion coefficient from 20°C to 170°C of 5%. Thus the polymer enters the water cooling bath with a radius of $R_{170} = 1.05 \times R_{20} = 7.88 \times 10^{-3}\text{m}$. Using well-stirred water in the cooling bath at 20°C and assuming the rod will be form stable when the axial temperature (at $r = 0$) has reached 100°C, $Y_2 = (100 - 20)/(170 - 20) = 0.53$. For a heat transfer coefficient $h = 1500\,\text{W/m}^2\,\text{K}$ and an average value $k = 0.28$ from Fig 8.24, we find

$$Bi = hR/k = 1500 \times 7.88 \times 10^{-3}/0.28 = 42.2$$

Hence $Fo \approx 0.2$ and a cooling time $t = R^2 Fo/\alpha = (7.88 \times 10^{-3})^2 \times 0.2/10^{-7} = 124\,\text{s}$. This suggests a bath length of 5 m, but it might be wiser to use a rather longer one to allow some process flexibility.

The volume flow rate is $Q_{170} = 1.15 Q_{20} = 8.12 \times 10^{-6}\,\text{m}^3/\text{s}$. The fully swollen hot radius of the rod is $R_{170} = 7.88 \times 10^{-3}\,\text{m}$. We are now in a position to use the iterative method given in section 8.7.4 (used in problem 4 in section 8.7) to estimate the die exit radius R_{do}. We expect R_{do} to be substantially less than R_{170}, and we might therefore look to a die land length of some 180 mm. We can choose a taper (half-) angle to connect the extruder barrel diameter to the die land diameter by ensuring that the maximum average extensional rate (calculated in the manner of problem 3 at the end of section 8.6) does not exceed the allowable rate given in Fig. 8.17. From Fig. 8.15 we see that $d\varepsilon/dt_{max} = \sigma/\lambda = 1.66/\,s$, and we might choose a design value of 1 /s to allow some margin of error. We cannot choose the length of the taper until we have chosen the size of extruder, but we would be open to the idea that we could reduce the taper length considerably by using a series of tapers rather than just one.

It is easier to ask what the size of extruder should be than to calculate it. Our approach is the simple one of assuming that the flow in the single-screw extruder is governed by the capacity of the metering zone. We shall choose a standard screw, as recommended by Rauwendaal, in which $L/D = 24$, with a single start and square pitch so that $\theta = 17.66°$, having an axial length of metering zone of $L_m = 10D$ and a depth of metering channel $H_m = 0.05D$. Making all of the assumptions in section 8.2.2, equation (8.7) becomes

$$Q = (0.0713N - 3 \times 10^{-7}\Delta P/\eta)D^3$$

where ΔP is in N/m^2. To begin the choice of screw size D, we start by suggesting the reasonable values of screw speed $N = 1$ /s and die pressure drop $\Delta P = 20$ MPa.

Choice of the value of viscosity is less straightforward, as the polymer melt is clearly non-Newtonian, and in this book we have not analysed drag flows using power-law fluids. We shall base our rough estimate on the relevant apparent viscosity. There are two approaches. The first estimate of viscosity assumes the flow in the metering zone is dominated by pressure-driven flow. Suppose we choose $D = 50$ mm to start with. Then $W_m = \pi D \sin\theta = 50$ mm, and $H_m = 0.05D = 2.5$ mm. We know $Q_{170} = 8.12 \times 10^{-6}\,m^3/s$, so the average velocity down the developed channel is $\langle v \rangle = Q_{170}/A = 0.065$ m/s. This suggests an average shear rate under pressure flow of the order $d\gamma_w/dt = 6Q/WH^2 = 6\langle v \rangle/H_m = 156$ /s. From Fig. 8.11 we see that $\eta_a = 800\,N\,s/m^2$. This leads to a screw diameter close to 50 mm, which confirms that our original guess was good. The second estimate of apparent viscosity assumes that the flow in the channel is governed by drag flow. The screw root velocity in the down-channel direction is $v_z = \pi ND\cos\theta = 0.15$ m/s, giving a shear rate uniform between root and barrel of $v_z/H_m = 60$ /s. This value is not so very different from the value for pressure flow, and thus the small change in melt viscosity will still lead to the conclusion that $D = 50$ mm is reasonable.

We are now in a position to calculate the taper length in the die, and more than one taper angle will be necessary to achieve a reasonable overall length of die.

CHAPTER 9

Section 9.2

1. The 90° angle of the L in the product tends to diminish: the inside surface of the product corner will solidify later than its outside and therefore will develop more tensile stress during cooling in the mould than the outside. After release from the mould the stress will cause an inward bending curvature of both limbs of the L-section and hence a decrease of the 90° angle (compare Fig. 9.26).
2. The extension will, because of its thickness, solidify and cool earlier in the mould than the plate. It has therefore fully shrunk at a moment that the core of the plate still has to cool down after solidification. After complete cooling the thin extension will be subject to compression forces and may eventually buckle (compare Fig. 9.25a).
3. Holding pressure during solidification implies a local compression stress in the product. This works out as expansion of the product after release from the mould, which counteracts shrinkage due to cooling from solidification to room temperature.
4. Release frees the length L from its restraint in the mould, which had ended up as the constant tensile load σ_1. This σ_1 can be supposed to have brought the length from the free shrink magnitude to the L prescribed by the mould. Because time is involved the superposition principle (section 5.2.3) should be used. This considers every change and its consequences at the time of consideration. The release action should be considered as putting a second (compression) load of $-\sigma_1$ on top of the existing σ_1 on the specimen. From both actions σ_1 will not change L any more, $-\sigma_1$ however will do so, with magnitude $\varepsilon = \Delta L/L = \sigma_1 C(t)$. The bar will therefore show shrinkage according to the creep compliance.

Section 9.3

1. (a) The crosslinks prevent free flow of molecules; (b) disorientation is prevented if the temperature is kept as low as possible; (c) cooling speed in the thin film is too high for disorientation; (d) the blow-moulded bottle may be rather thick and cools slowly; for blowing of preforms for bottles see (b).
2. At first it will increase in length according to the (low) anisotropic thermal expansion coefficient, until it approaches T_g. Next it will shrink, approximately according to the degree of orientation it had.
3. High melt temperature, high injection velocity (heat developed by high shear rate, with little time to cool during flow), high (local) mould temperature, local wall insulation to retard the cooling process, low molecular weight, a certain not-too-high holding pressure. Most of

these will result in a higher cycle time. Just a longer cooling time without any of the above measures is not helpful.

4. It is supposed that the second treatment will not change the orientation of the fibres. The orientation of the fibres at the walls will be lengthwise, which leads to a greater thermal expansion coefficient in the transverse direction, compared with the longitudinal direction. Because of the heat treatment the internal stresses at the radiated surface will be in tension. The permanently cooled surface will, however, still contain a possible tensile stress peak at the outside. The resulting biaxial bending moment on a section through the thickness will cause a curvature concave at the radiated side, which by nature should have one preferred axis. This axis will be in the direction of polymer flow, as the thermal expansion is greatest perpendicular to the fibre direction.

Section 9.4

1. Thermal expansion coefficient α in the direction of orientation of chain molecules is much lower than perpendicular to this direction. In the cavity two characteristic flow patterns can be identified: radial (away from the gate) a shear flow as described in section 9.3.1 and Fig. 9.8, resulting in a radial orientation which is not evenly distributed across the thickness (Fig. 9.9) and circumferential as the melt during the filling stage has to spread out over an ever-growing circumference. The product can only be stress-free with equal shrinkage ε_{rad} and ε_{circ}. Practice with quarter-disks showed $\varepsilon_{circ} > \varepsilon_{rad}$ which makes it clear that the radial orientation was (as an average) greater than the circumferential one. For a full disk this would mean a residual circumferential tensile stress σ_{rad}. Cases with a radial slit through the gate have been observed. They were apparently the consequence of (environmental) stress cracking based on this residual stress.

   The heat treatment may affect ε_{rad} more than ε_{circ} as ε_{rad} is based on a distribution with peaks in molecular orientation. This may balance or even reverse the sign of σ_{rad}.

CHAPTER 10

Section 10.2

1. Number of parts, multifunctionality of the axle holder part, and ease of assembly.
2. In view of the required pull-out strength we calculate the thickness h based on the tensile strength:

$$F_2 = \sigma_{max} hb/kF \Rightarrow h = 1.36\,\text{mm}$$

This is rather thin; if the hook is also subject to other forces a choice of $h = 2\,\text{mm}$ might make it more robust.

The snap depth H follows from the permissible strain ε (see equation (10.1)) for which Table 10.1 indicates, for POM, as being 6%. Therefore $H \leqslant 0.06 \times 2 \times 10^2/3 \times 2 = 2\,\text{mm}$.

Now we can verify the pull-out force based on shear strength: the relevant surface is $6 \times 2/\tan 30° = 20.78\,\text{mm}^2$. Allowable shear force is therefore $0.6 \times 62 \times 20.78/(3 \times 1.3) = 198\,\text{N}$. This exceeds the required F_2.

Finally we calculate the mounting force F_1 with equation (10.3): $F_1 = 33.7\,\text{N}$. It should be noted that the stress concentration factor k is strongly dependent on the radius r at the edges of the hook. The order of its magnitude for the data used in this problem should be correct for the calculation based on tensile stress; it is, however, less well documented but probably higher for the calculation based on shear stress.

3. (a) Six horizontal and one vertical hinge.
   (b) With small displacements along a vertical axis which lies on a cylinder around the vertical hinge.
   (c) The horizontal hinges are parallel but in different planes. The parting line should follow these planes in order to allow vertical movement of the mould halves. It thus necessitates moving cores to contain the vertical hinge.
4. See caption to cover figure.
5. If a flat strip with thickness t is bent to a curve with radius R_2 the surface will have a strain of $\varepsilon = 0.5t/R_2$. In our case the strip is not flat but curved with a radius R_1, which was changed by bending to R_2. This change will induce the following strain at the surface of the element:

$$\varepsilon = 0.5t\left(\frac{1}{R_2} - \frac{1}{R_1}\right)$$

The deformation of one half-bow of Fig. 10.20 into the curve with length πR_2 is related to the displacement of its free end over a distance $0.5v$. We therefore suggest as a first approximation

$$R_2 = R_1 - \frac{0.5v}{\pi}$$

As $R_1 = 1\,\text{mm}$ and $v = 1\,\text{mm}$ we find $\varepsilon = 0.0284 = 2.84\%$: there is plenty of safety.

A closer investigation of this system, including symmetry and boundary conditions shows the following relationship:

$$v = 4.33FR_1^3/EI \quad \text{and} \quad M_{\max} = 0.619FR_1$$

The latter induces a bending stress in the spring; the related strain in the spring is

$$\varepsilon = M \times 0.5t/EI = 0.31FR_1t/EI$$

Combining these two gives

$$\varepsilon = 0.07136tv/R_1^2$$

which becomes, with the data given, $\varepsilon = 2.14\%$, somewhat smaller than the result obtained above.

Given the formula $v = 4.33FR_1^3/EI$ and the modulus of the POM grade, we can calculate the force $3F$ for the whole device to reach the displacement $v = 1$ mm. The result is 19.3 N, to be verified by the reader.

6. Increase of L and b will diminish the average shear stress τ_{av}. Of high importance, however, is the influence of the proposed changes on the stress peaks at the ends of the joint. A theoretical study of the magnitude of the local shear stress in a lap joint shows that a major determining factor is $(G_aL_a^2)/(E_st_sh_a)$ where a is adhesive, s is substrate (i.e. the lap), L is the lap length, h is the thickness of the adhesive layer and t is the thickness of each plate. Stress peaks are higher if this factor is higher.

   The proposed changes (a), (c), (d) and (e) all affect the stress peaks at the beginning and the end of the joint. Relative to the average shear stress τ_{av} in the seam, increasing t and h and decreasing L and choosing a weaker cement all decrease the stress peaks τ/τ_{av} and vice versa. Just increasing b does not affect the stress peaks. With regard to (a), this may give an improvement with respect to τ_{av}, but from a certain length L upwards increasing L will not affect the absolute value τ of the stress peaks any more.
7. The following factors are important:
   - correct clearance between pipe and socket, including the matching dimensions of the tool;
   - correct tool temperature, including evenness of temperature over the full heating surface;
   - correct heating time;
   - quick, pure axial and nearly forceless mounting of heated pipe and socket, regarding the right depth of the mounting in order to catch the polymer scraped off from the surface;
   - allowance of enough cooling time before applying an external load.

Section 10.3

1. The directions in which second moment of inertia I is affected by changing shape from a flat plate.
2. Apply equation (10.5) to a foam sandwich pipe of unit length, i.e. unit breadth of circumferential element:

$$Eh^3 = E_s(h_f + 2h_s)^3 - E_sh_f^3 + E_fh_f^3$$

The problem states that $E = E_s$ and $E_f/E_s = (\rho_f/\rho_s)^2 = (950/1450)^2 = 0.43$. Therefore

$$h^3 = (h_f + 2h_s)^3 - (1 - 0.43)h_f^3$$
$$6^3 = (h_f + 2)^3 - 0.57h_f^3$$
$$h_f = 4.4\text{ mm} \quad h_f + 2h_s = 6.4\text{ mm}$$

Weight saving per unit length based on same bore:
Weight of old wall,

$$W_1 = (\pi/4)(162^2 - 150^2) \times 10^{-6} \times 1410 = 5.279\,(\pi/4)$$

Weight of sandwich,

$$W_2 = (\pi/4) \times 10^{-6}[(162.8^2 - 160.8^2) \times 1410 + (160.8^2 - 152^2) \times 950 + (152^2 - 150^2) \times 1410] = 4.378(\pi/4)$$

Weight saving = (4.378 – 5.279)/5.279 = 17%

3. The problem states $t_r = 0.6t_w$. If a taper of $\alpha = 1°$ could be advised, then Fig. 10.29 can be used: $b_d = b/n = 100$ mm. From the entries in Fig. 10.29a: $h_r/t_w = 5$ and $b_d/t_w = 10$, the diagram reads $t_{et}/t_w = 3 \Rightarrow t_{et} = 30$ mm. By adding these ribs the second moment of area of the plate was increased from $500 \times 10^3/12$ to $500 \times 30^3/12$, i.e. by a factor of 27.

   Regarding the strength diagram Fig. 10.29b indicates $t_{ed}/t_w = 1.67 \Rightarrow t_{ed} = 16.7$ mm. According to Fig. 6.22 creep rupture strength of PE is 8 MN/m^2. The bending stress is

$$\sigma = \frac{100 \times 10^3}{\frac{1}{6} \times 500 \times 16.7^2} = 4.3\text{ N/mm}^2$$

   leading to a safety factor $F = 8/4.3 = 1.86$

4. (a) Axial tension:

$$E_x = (a_{11}h)^{-1} = [21.21 \times 10^{-9}\text{ m/N} \times 5 \times 10^{-4}\text{ m}]^{-1} = 94.3\text{ GPa} = E_y$$

   (b) Bending test:

$$E_{x\text{bend}} = 12/d_{11}h^3 = 12/[604 \times 10^{-3}\text{ N m} \times 125 \times 10^{-12}\text{ m}^3] = 159\text{ GPa}$$
$$E_{y\text{bend}} = 12/d_{22}h^3 = 12/[3245 \times 10^{-3} \times 125 \times 10^{-12}] = 29.6\text{ GPa}$$

   With the 0° plies on the outside, the bending stiffness is large. With 90° plies outside, the stiffness is low, as expected.

5. The first step is to calculate the stresses of interest, using standard simple stress analysis. The maximum bending moment is at midspan, with $M_{max} = PL/4 = 12.5$ N m.

The maximum bending stress at top and bottom of the tube is therefore

$$\sigma_{x\max} = \pm M_{\max}/2\pi R^2 h = \pm 2.4\,\text{N/mm}^2,\ \text{hence } N_x = \pm 4.2\,\text{N/mm}$$

The shear force in the tube is $V = P/2$. Based on the equation $\tau_{\max} = VQ/Ib'$ where Q is the first moment of the area of tube above the neutral axis, the breadth of the tube at the neutral plane is $b' = 2h$, and using $I \approx \pi R^3 h$ for a thin-walled tube, we may calculate

$$\tau_{\max} \approx P/2\pi R_m h = 0.2\,\text{N/mm}^2,\ \text{so } N_{xy} = 0.4\,\text{N/mm}$$

The second step is to examine the composite laminates. For the system $(30°/-30°/30°/-30°)_s$,

$$a_{11} = 22.9, a_{12} = -14.8, a_{22} = 54.66, a_{66} = 54.74\,\text{mm/MN, hence}$$

	$N_x = +4.2\,\text{N/mm}$	$N_x = -4.2\,\text{N/mm}$
$(\varepsilon_x^0, \varepsilon_y^0, \gamma_{xy}^0) =$	$(0.011, -0.0071, 0)\%$	$(-0.011, +0.0071, 0)\%$
$(\sigma_x, \sigma_y, \tau_{xy}) =$	$(2.4, 0, \pm 0.781)$ MPa	$(-2.4, 0, \pm 0.781)$ MPa

Using $[T(+30°)]$ and $T(-30°)]$ we find

$(\sigma_1, \sigma_2, \tau_{12}) = $	$(2.476, -0.0759, \pm 0.649)$	$(-2.476, +0.0759, \pm 0.649)$
$(\sigma_i/\sigma_{i\max}) =$	$(2.3\times10^{-3}, 6.4\times10^{-4}, 9\times10^{-3})$	$(4.1\times10^{-3}, 2.5\times10^{-3}, 9\times10^{-3})$
Tsai–Hill factor	107	98

For $N_{xy} = 0.4\,\text{N/mm}$ we have,

$(\sigma_x, \sigma_y, \tau_{xy}) =$	$(0.209, 0.0828, 0.2)$ MPa
$(\sigma_1, \sigma_2, \tau_{12}) =$	$(\pm 0.351, \pm 0.059, 0.04532)$ MPa
Tsai–Hill factor	477

The conclusion is that this pipe fails at the lowest of the three load factors: in compression at the top surface of the pipe. We cannot say much about the likely mode of failure, because all the stress/strength ratios are of similar magnitude.
For the system $(0_8^\circ)_T$ we find:

	N_x (N/mm)	N_y (N/mm)	N_{xy} (N/mm)
Force/width	+4.2	−4.2	0.4
F	442.5	254.2	360

The conclusion is that the unidirectional pipe fails in compression at the upper surface.

The overall conclusion is that the filament-wound pipe is closest to failure by a factor of 98 according to these calculations.

6. We examine first the conventional mechanics analysis, then identify the principles needed from composites.

For bending of a horizontal simply-supported beam under a uniformly distributed load w_1 per unit length, the midspan bending and shear deflections are

$$v_b = 5w_1L^4/384EI \quad \text{and} \quad v_s = w_1L^2/8k_sGb'd$$

where $b' = 3\,\text{mm}$ is the total thickness of the web of the beam.

The essence of the composite analysis is to work out the in-plane compliance coefficients on which the modulus values $E_x = (a_{11}h)^{-1}$ and $G_{xy} = (a_{66}h)^{-1}$ can be found for the relevant stiffness calculations. We shall see that in this example the laminate bending stiffness b/d_{11} can be neglected.

For the 1.5 mm laminate based on 0.1875 mm thick gfrp plies, the laminate compliance coefficients (in units of 10^{-6} mm/N and 10^{-6}/N mm) are as follows:

Laminate	a_{11}	a_{12}	a_{22}	a_{66}	d_{11}
(a) $(0^\circ_4)_s$	17.27	−4.491	80.61	161	92.11
(b) $(0^\circ_2/90^\circ_2)_s$	28.27	−2.594	28.27	161	101.8
(c) $(45^\circ_2/-45^\circ_2)_s$	53.1	−27.42	53.1	61.74	304.5
(d) $(0^\circ/45^\circ/90^\circ/-45^\circ)_s$	35.15	−9.474	35.15	89.25	130.4
(e) $(0^\circ_2/45^\circ/-45^\circ)_s$	25.88	−11.18	57.09	89.25	100.4

We now present the design method for the 0° laminate, and summarize the results for all five laminates.

Under in-plane loading N_x we know that

$$E_x = (a_{11}h)^{-1} = (17.27 \times 10^{-6} \times 1.5)^{-1} = 38.6 \times 10^3\ \text{N/mm}^2$$

For the web of thickness 1.5 mm we therefore find

$$(EI)_w = E_x bh^3/12 = 38.6 \times 10^3 \times 1.5 \times 69^3/12 = 1.586 \times 10^9\ \text{N/mm}^2$$

For one flange of the [section we can use the parallel axes theorem to obtain the stiffness of the flange element about the complete beam neutral axis:

$$(EI)_{fl} = b(1/d_{11} + \lambda^2/a_{11})$$

Using $b = 37.5$ mm and $\lambda = 35.25$ mm we find

$$(EI)_{fl} = 37.5(10.86 \times 10^3 + 71.97 \times 10^6) = 2.699 \times 10^9\ \text{N mm}^2$$

The stiffness of the complete I section is thus

$$EI = 2(EI)_w + 4(EI)_{fl} = 13.97 \times 10^9\ \text{N mm}^2$$

The bending deflection at midspan is

$$v_b = 5w_1L^4/384EI = 5 \times 0.5 \times 16 \times 10^{12}/384 \times 13.97 \times 10^9 = 7.46\,\text{mm}$$

The shear modulus G_{xy} of the laminate loaded in the plane under N_{xy} is

$$G_{xy} = (a_{66}h)^{-1} = (161 \times 10^{-6} \times 1.5)^{-1} = 4141\,\text{N/mm}^2$$

and hence, using the analysis for isotropic materials and $k_s = 1$, the midspan deflection of the complete cross-section caused by shear deformation in the web of dimension 3×72 is

$$v_s = w_1L^2/8k_sGb'd = 0.5 \times 4 \times 10^6/8 \times 1 \times 4141 \times 3 \times 72 = 0.27\,\text{mm}$$

The summary of the bending calculations for all five laminate constructions is:

Laminate	E_x (N/mm²)	G_{xy} (N/mm²)	v_b (mm)	v_s (mm)	EI (N/mm²)
(a) $(0^\circ_4)_s$	38 600	4 141	7.46	0.26	13.97×10^9
(b) $(0^\circ_2/90^\circ_2)_s$	23 580	4 141	12.19	0.26	8.53×10^9
(c) $(45^\circ_2/-45^\circ_2)_s$	12 550	10 798	22.91	0.1	4.54×10^9
(d) $(0^\circ/45^\circ/90^\circ/-45^\circ)_s$	18 970	7 470	15.16	0.143	6.86×10^9
(e) $(0^\circ_2/45^\circ/-45^\circ)_s$	25 760	7 470	11.16	0.143	9.32×10^9

We confirm that bending caused by shear deformation in this particular example is small. Can you explain why G_{xy} has the same value for beams (a) and (b), and a different same value for beams (d) and (e)?

Section 10.4

1. No. The isochronous stress–strain diagrams are usually obtained from creep data. There is a distinct difference between creep modulus and relaxation modulus (see section 5.2.3, compare equations (5.34) and (5.37)). This difference is, however, small if the material is used remote (in time/frequency and temperature) from the relaxation time of its nearest transition region (section 5.2.6). The difference may, however, grow to say 10% if conditions come nearer.
2. $\sigma = \varepsilon E(t) \Rightarrow \sigma_A = \varepsilon_1 E(0)$ and $\sigma_C = \varepsilon_1 E(100)$. According to the superposition principle the additional strain $\varepsilon_2 - \varepsilon_1$ has to cause an immediate increase in stress of $\sigma_A - \sigma_C$, thus $(\sigma_A - \sigma_C) = (\varepsilon_2 - \varepsilon_1)E(0)$. The only unknown here is ε_2.
3. $\sigma_I - \sigma_{II} = \varepsilon_1 E_1(t^{-m} - (t - t_1)^{-m}) + \varepsilon_2 E_1(t - t_1)^{-m} - \varepsilon_2 E_1 t^{-m}$
   The first term as well as the difference between the last two terms vanish as $t \gg t_1$ and therefore also the difference $\sigma_I - \sigma_{II}$ between the two loading paths.

4. The compression stiffness C_F now will be time dependent as well so, similarly to the family of isochronous curves for the bolt, a family of isochronous curves for the flange has to be drawn. The line A–B–C will now be a curve but, as F_p does not change and as $F_F = 0$ on the moment of leakage, the time to leakage will not be affected.
5. Identify the plies by angle to the reference x-direction, to avoid confusion. $\nu_{21} = \nu_{12}E_2/E_1 = 0.01957$, so $(1 - \nu_{12}\nu_{21}) = 0.994$. Hence

$$[Q(0°)] = \begin{bmatrix} 138 & 2.716 & 0 \\ 2.716 & 9.05 & 0 \\ 0 & 0 & 7.1 \end{bmatrix} \quad (\text{kN/mm}^2)$$

$$[Q(90°)] = \begin{bmatrix} 9.05 & 2.716 & 0 \\ 2.716 & 138 & 0 \\ 0 & 0 & 7.1 \end{bmatrix} \quad (\text{kN/mm}^2)$$

$$\alpha_1(0) = -0.3 \times 10^{-6}/\text{K} \quad \alpha_1(90) = 28.1 \times 10^{-6}/\text{K}$$

$$\alpha_2(0) = 28.1 \times 10^{-6}/\text{K} \quad \alpha_2(90) = -0.3 \times 10^{-6}/\text{K}$$

$$\Delta T = 20 - 140 = -120\,\text{K}.$$

Now calculate the (notional) thermal force resultants:

$$N_x^{\text{T}} = \Delta T \sum [Q]_{\text{f}}[\alpha]_{\text{f}}(h_{\text{f}} - h_{\text{f}-1})$$

$$\begin{aligned} \sum [Q]_{\text{f}}[\alpha]_{\text{f}}(h_{\text{f}} - h_{\text{f}-1}) &= (2 \times 0.125)\{Q_{11}(0)\alpha_1(0) + Q_{12}(0)\alpha_2(0)\} \\ &\quad + (4 \times 0.125)\{Q_{11}(90)\alpha_1(90) + Q_{12}(90)\alpha_2(90)\} \\ &= 0.00867 + 0.1267 = 0.1354\,\text{N/mm K} \end{aligned}$$

$N_x^T = -120 \sum = -16.25\,\text{N/mm}$. Similarly $N_y^{\text{T}} = -9.685\,\text{N/mm}$. The laminate cure strains are therefore

$$\varepsilon_x^{\text{oc}} = a_{11}N_x^{\text{T}} + a_{12}N_y^{\text{T}} = -4.079 \times 10^{-4}$$

$$\varepsilon_y^{\text{oc}} = a_{12}N_x^{\text{T}} + a_{22}N_y^{\text{T}} = -1.2351 \times 10^{-4}$$

The stress inducing strains are found as

$$\varepsilon_x^{\text{si}}(0) = \varepsilon_x^{\text{oc}} - \alpha_1(0)\Delta T = -4.439 \times 10^{-4}$$

$$\varepsilon_y^{\text{si}}(0) = \varepsilon_y^{\text{oc}} - \alpha_2(0)\Delta T = 3.249 \times 10^{-3}$$

$$\varepsilon_x^{\text{si}}(90) = 2.9641 \times 10^{-3}$$

$$\varepsilon_y^{\text{si}}(90) = -1.595 \times 10^{-4}$$

Hence the thermal stresses are found as

$$[\sigma_x(0), \sigma_y(0)] = [-52.79, 28.2]\,\text{MPa}$$

$$[\sigma_x(90), \sigma_y(90)] = [26.39, -14.09]\,\text{MPa}$$

The (thermal) shear stresses are zero in this cross-ply laminate. The cure temperature used is reasonably representative of normal practice. At 20°C we see that the transverse stress $\sigma_2 = \sigma_y(0°)$ is very large in comparison with a transverse strength of the order 40 MPa (Table 7.2). Such a laminate is not therefore a good design for hot curing. (How would you go about improving the laminate?)

6. We need to calculate first the values of thermal expansion coefficients of a lamina in global coordinates. Using equations (7.45) and (10.33):

$$[T(+30°)]^{-1} = \begin{bmatrix} 0.75 & 0.25 & -0.866 \\ 0.25 & 0.75 & +0.866 \\ 0.433 & -0.433 & 0.5 \end{bmatrix}$$

$$\alpha_x(+30°) = 0.75 \times 1 \times 10^{-5} + 0.25 \times 20 \times 10^{-5} = 5.75 \times 10^{-5}$$

$$\alpha_y(+30°) = 0.25 \times 1 \times 10^{-5} + 0.75 \times 20 \times 10^{-5} = 15.25 \times 10^{-5}$$

$$\alpha_{xy}(+30°)/2 = 0.433 \times 1 \times 10^{-5} - 0.433 \times 20 \times 10^{-5} = -8.227 \times 10^{-5}$$

$$\alpha_x(-30°) = \alpha_x(+30°), \alpha_y(-30°) = \alpha_y(+30°), \alpha_{xy}(-30°) = +16.454 \times 10^{-5}$$

Now we calculate the notional thermal force resultants.

$$\begin{aligned} N_x^T &= \Delta T\{2[Q_{11}(+30°)\alpha_x(+30°) + Q_{12}(+30°)\alpha_y(+30°) + Q_{16}(+30°)\alpha_{xy}(+30°)] \\ &\quad + 2[Q_{11}(-30°)\alpha_x(-30°) + Q_{12}(-30°)\alpha_y(-30°) + Q_{16}(-30°)\alpha_{xy}(-30°)]\} \\ &= 4\Delta T[Q_{11}(+30°)\alpha_x(+30°) + Q_{12}(+30°)\alpha_y(+30°) + Q_{16}(+30°)\alpha_{xy}(+30°)] \\ &= 4 \times 30[0.9867 \times 5.75 + 0.3326 \times 15.25 - 0.5622 \times 16.454] \times 10^{-2} = 1.795\ \text{N/mm} \end{aligned}$$

$N_y^T = 0.8299$ N/mm. $N_{xy}^T = 0$ because terms in Q_{ij} for +30° and −30° cancel out.

$$\varepsilon_x^T = a_{11}N_x^T + a_{12}N_y^T + a_{16}N_{xy}^T = -1.529 \times 10^{-3}$$

$$\varepsilon_y^T = a_{12}N_x^T + a_{22}N_y^T + a_{26}N_{xy}^T = +5.885 \times 10^{-3}$$

$$\gamma_{xy}^T = a_{16}N_x^T + a_{26}N_y^T + a_{66}N_{xy}^T = 0$$

The stress-inducing strains are worked out for +30° plies:

$$\varepsilon_x^e = \varepsilon_x^T - \alpha_x(+30°)\Delta T = -1.529 \times 10^{-3} - 5.75 \times 10^{-5} \times 30 = -3.254 \times 10^{-3}$$

$$\varepsilon_y^e = \varepsilon_y^T - \alpha_y(+30°)\Delta T = +1.31 \times 10^{-3}$$

$$\gamma_{xy}^e = \gamma_{xy}^T - \alpha_{xy}(+30°)\Delta T = +4.936 \times 10^{-3}$$

$$\sigma_x(+30°) = Q_{11}\varepsilon_x^e + Q_{12}\varepsilon_y^e + Q_{16}\gamma_{xy}^e = 1.92 \times 10^{-5}\ \text{MPa} \simeq 0$$

$$\tau_{xy}(+30°) = Q_{16}\varepsilon_x^e + Q_{26}\varepsilon_y^e + Q_{66}\gamma_{xy}^e = 0.032\ \text{N/mm}^2$$

Section 10.5

1. It is common practice in bottle production to continuously extrude the parison. Once the mould has been closed and the parison has been

(automatically) cut off, the mould(s) are moved down and to one side so that the blowing process and cooling can proceed; meanwhile the next parisons are being produced. There are usually two to eight sets of moulds in use at the same time. Parisons for bottles are usually short, say up to 350 mm long, and over extrusion times of a few seconds there should not be too much sag, especially if a parison programmer is used.

For large containers with parison lengths of 1 m or more, continuous extrusion would lead to excessive sag under gravity forces and excessive cooling before blowing, leading to unacceptable parison quality, even where parison programming is used. It is then usual to extrude polymer melt directly into a melt accumulator. Sufficient quantity of melt is stored until it is needed: when the mould is ready, a piston in the accumulator forces the melt out at high speed (5 kg/s is common for big containers) to produce the long thick-walled parison in just 2–4 s. The mould is closed and the parison inflated in the usual way. Taper angles in the die exit region are essential to avoid melt fracture at these high flow rates.

2. During extrusion from a ram accumulator, the surface temperature will remain at about the melt temperature. On inflation there will be a slight drop in temperature. When the parison reaches the mould surface, there will be an instant drop to about the temperature of the mould, typically at 20°C. The solid surface will advance slowly into the wall of the container until (say) the inside wall is just frozen solid. The moulding is then ejected from the mould. The inside is now still much hotter than the outside: thus there will be a flow of heat radially outwards and hence the temperature on the outer surface rises perhaps by 40 K, and after some time it drops again until the container reaches thermal equilibrium at ambient temperature.
3. Parisons are almost always extruded with an annular cross-section of constant wall thickness. Away from the ends, inflation into a mould having a square cross-section must give a product with thinner corner-edges because of the larger inflation. Rounding the corners of the product helps a little (but not much). It is possible and practical to use an annular die where the exit gap varies round the circumference to match the inflation (and swelling) requirements. Thus there are 4 thick gaps and 4 thin gaps coinciding with the corner-edges and the middle of the sides of the blown product.

   It is an interesting and substantial challenge to determine the variation in wall thickness round the circumference of the die and the fully swollen parison.

   Now look at flow of melt inside the die. If the flow channels feeding the non-uniform die exit are of uniform resistance up to the land, then polymer melt will be preferentially extruded through the thick channel (for the corners) and insufficiently through the narrow gap: this produces an unwanted mess. To achieve uniform parison velocity round the circumference (after extrusion swelling), there must be provision in the body of the die for choking-off the flow to the wider gap, thus locally reducing the flow rate. One reasonable approach – commercially proven – is to combine a locally longer die land combined with narrower upstream taper angles for the thick-gap-flow, and shorter land with wider taper angles for the thin-gap-flow.

The approach described above works for products of rectangular section except at the ends, where the inflation is biaxial and is distorted by the pinch-off of the parison. For long rectangular containers the parison sag problem remains unresolved. Using parison programming in its simplest form interferes with the choking-off principle used to make a parison of non-uniform wall thickness. We should also recall that the simplest choking-off procedure assumes that melt in the die only flows downstream. In practice there will be some flow round the circumference because of local changes in pressure drop between the two flows (to the thin and thick gap). Interesting problems still remain, which are outside the scope of this book.

4. It is helpful to work with one assembly:

$$K_x = K_z = 1.83\,\mathrm{MN/m}; K_y = K_s = 0.12\,\mathrm{MN/m}$$

From Fig. 5.35 for 38 IRHD, $G = 0.54\,\mathrm{MN/m^2}$, $E_0 = 3G = 1.62\,\mathrm{MN/m^2}$. From equation (5.64):

$$H = GA/K_s = 0.54 \times 10^6 \times 10^{-2}/0.12 \times 10^6 = 45\,\mathrm{mm}$$

Combining the expression for K_x with that for K_z (equation (10.44)) gives

$$K_x + K_z = K_c + K_s$$

But, noting equations (5.64), (5.90) and (5.91) and assuming that the compression will be less than 15%:

$$K_c = E_0 A(1 + 2S^2)/H \quad \text{and} \quad K_s = GA/H$$

Combine these:

$$K_x + K_z = 2K_x = (A/H)[E_0(1 + 2S^2) + G]$$

Putting in numerical values, $S^2 = 4.416$, $S = 2.10$, giving a single layer thickness $h = 11.89\,\mathrm{mm}$. This h does not correspond with $H = 45\,\mathrm{mm}$ found earlier, but it is necessary to tune E' and thus K_c to the required level (compare equation (5.92)). In order to reach the required H we need four layers of rubber with three interleaf plates.

Combining equations (10.44) and (10.45) gives

$$\sin^2\theta = (K_x - K_s)/(K_c - K_s), \text{ i.e. } \theta = 45^\circ \text{ to the horizontal}$$

5. For one block, $K_x = 3\,\text{MN/m}$, $K_y = K_s = 0.2\,\text{MN/m}$, $K_z = 1\,\text{MN/m}$. Using equations (10.44) and (10.45), $K_z + K_x = K_c + K_s$. Therefore $K_c = 3.8\,\text{MN/m}$, and hence

$$\sin^2\theta = (K_x - K_s)/(K_c - K_s) = 0.77 \Rightarrow \theta = 62^\circ$$

The choice of compound is not specified, but as $K_c \gg K_s$, some interleaving may be required. The shear stiffness is satisfied by equation (5.64):

$$K_s = GA/H = GA/nh$$

where n is the number of layers of rubber each of thickness h. The shape factor for one layer of rubber is $S = \pi R^2/2\pi Rh = R/2h$.
The compressive stiffness K_c is given by combining equations (5.64), (5.90) and (5.91):

$$K_c/K_s = (E_0/G)/(1 + R^2/2h^2)$$

The required value of thickness of one rubber layer h is therefore 15.3 mm.
Taking $n = 1$ (one single slab, which would offer the most economical solution to this problem), G is $0.2 \times 0.0153 \times 4(\pi \times 0.1^2) = 0.48\,\text{N/m}^2$. Thus the required hardness is 34 IRHD (Fig. 5.35) and $E_0 = 3G = 1.44\,\text{N/m}^2$.

For $n=2$, G must be $0.2\times2\times0.0153 \times 4/(\pi \times 0.1^2) = 0.96\,\text{N/m}^2 \Rightarrow$ 53 IRHD.

6. For a symmetrical laminate $(\pm45^\circ)_s$ the B matrix is zero by definition. Therefore the relation between σ_x and ε_x is unique for E_x (no interaction of moment or curvature) and similar for E_y, G_{xy} and ν_{xy}. Experiment and definition of these values are clear.

Applying N_x to one single ply at $[+45^\circ]$ will induce shear deformation γ_{xy} as well as ε_x and ε_y. This is not desirable. Using two bonded plies $(+45/-45^\circ)_T$ eliminates the γ_{xy} response but the laminate is not symmetric, thus $[B] \neq 0$ and N_x induces an unwanted twisting curvature κ_{xy}. Using four plies $(+45/-45^\circ)_s$ provides a symmetric laminate giving the desired ε_x and ε_y on which E_x is based, whilst eliminating both γ_{xy} and κ_{xy}.

7. Tensile and compression strengths $\hat{\sigma}_{1,t}$ and $\hat{\sigma}_{1,c}$ have different values, as have $\hat{\sigma}_{2,t}$ and $\hat{\sigma}_{2,c}$.

Index

See table on p 18 for names and abbreviations of polymers.